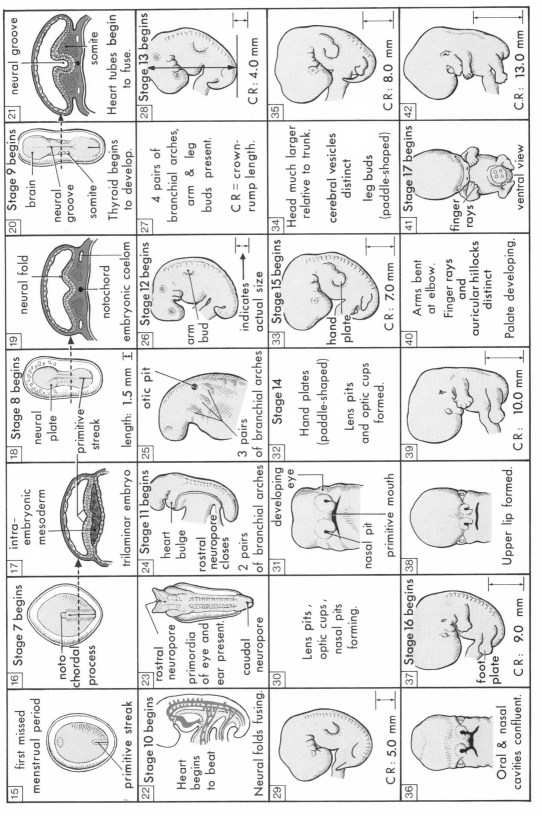

The Developing Human

Third Edition

CLINICALLY ORIENTED EMBRYOLOGY

KEITH L. MOORE, M.Sc., Ph.D., F.I.A.C., F.R.S.M.

Professor and Chairman, Department of Anatomy,
University of Toronto, Faculty of Medicine,
Toronto, Ontario, Canada

Illustrated primarily by
GLEN REID B.Sc., A.A.M.

Medical Illustrator, Faculty of Medicine,
University of Manitoba
Winnipeg, Manitoba, Canada

W. B. SAUNDERS COMPANY

Philadelphia/London/Toronto/Mexico City/Rio de Janeiro/Sydney/Tokyo

W. B. Saunders Company: West Washington Square
Philadelphia, PA   19105

1 St. Anne's Road
Eastbourne, East Sussex BN21 3UN, England

1 Goldthorne Avenue
Toronto, Ontario M8Z 5T9, Canada

Apartado 26370 — Cedro 512
Mexico 4, D.F., Mexico

Rua Coronel Cabrita, 8
Sao Cristovao Caixa Postal 21176
Rio de Janeiro, Brazil

9 Waltham Street
Artarmon, N.S.W. 2064, Australia

Ichibancho, Central Bldg., 22-1 Ichibancho
Chiyoda-Ku, Tokyo 102, Japan

Library of Congress Cataloging in Publication Data

Moore, Keith L.

The developing human.

Includes bibliographies and index.

1. Embryology, Human.   2. Abnormalities, Human.
   I. Title.

QM601.M76 1982   612'.64   81–40900

ISBN 0-7216-6472-5   AACR2

Listed here is the latest translated edition of this
book together with the language of the translation
and the publisher.

French — Edisem, Inc., St. Hyacinthe, Quebec,
        Canada
Spanish — Nueva Editorial Interamericana,
        S.A. México, Mexico
Portuguese — Editora Interamericana do Brasil
        Ltda., Rio de Janeiro, Brazil
German — F. K. Schattauer Verlag, Stuttgart,
        Germany
Italian — Nicola Zanichelli Editore, Bologna,
        Italy
Japanese — Ishiyaku Publishers Inc., Tokyo, Japan

Front cover: Photograph of a 13-week-old human fetus.

The Developing Human — Clinically Oriented Embryology   ISBN   0-7216-6472-5

Last digit is the print number: 9   8   7   6   5   4   3   2

To our first grandchild

MELISSA CATHERINE MOORE

daughter of Warren and Cathy

PREFACE TO THE THIRD EDITION

The continued worldwide acceptance of *The Developing Human,* as indicated by its many reprintings and foreign translations, is most gratifying. This acceptance has provided the encouragement to prepare a better textbook by restructuring it and adding new material and illustrations in this third edition.

Knowledge of embryology has expanded considerably since the second edition was published. The extreme vulnerability of the embryo to certain drugs and other agents and the practice of in vitro fertilization provide strong stimuli to embryologists studying early stages of human development. I have attempted to bring the book abreast of current literature, but I have adhered to my original aim of writing a book for undergraduate students and not a reference work for specialists.

To incorporate new information, all sections have been reviewed. In some chapters there have been extensive modifications, but in others relatively few changes were necessary. All this has entailed little increase in the number of pages because judicious pruning has been done. To improve readability and obviate possible ambiguity, many sentences and paragraphs have been rewritten. Throughout the book, the importance of embryology to the clinician is emphasized; this orientation is very likely to appeal to any person concerned with human development.

To reinforce the information given in the text, several *clinically oriented problems* have been added at the end of each chapter. The problems and the answers to them, given at the back of the book, illustrate the kinds of practical considerations every physician may be required to address. Care has been taken to keep the problems simple and the comments at a level that is suitable to beginning medical students without much clinical experience. To emphasize the clinical significance of embryology further, description and discussion of congenital malformations follow discussion of the normal development of each organ instead of appearing at the end of the chapter as in previous editions.

New illustrations have been added in this edition, and a number of figures have been redrawn or modified in light of teaching experience. *Color has also been added to several more drawings* to facilitate understanding. I owe thanks to Dorothy Irwin for this work, but I should again like to express my appreciation to Glen Reid, who is primarily responsible for the illustrations in this book.

Italics have been used more freely than in the past editions to indicate important terms; officially recognized synonyms or alternatives appear in parentheses, e.g., *syncytiotrophoblast* (syntrophoblast). *Italics* have also been used to emphasize important terms, concepts, and statements.

v

The *Nomina Embryologica,* approved by the Tenth International Congress of Anatomists in Tokyo, 1975, has been followed, and, in accordance with international agreement, the terminology is anglicized, departing from strict Latin in most cases. There is also some use of eponyms (e.g., Meckel's diverticulum and Down syndrome); students will need to know such terms because they are used in specialty texts and by clinical teachers.

While working on this edition, I have had the benefit of receiving helpful criticisms from students in many parts of North America and suggestions from a number of embryologists who have kindly written me or sent reprints of their publications. To all these people I express my most sincere thanks. Several colleagues have been very helpful with this edition: Dr. J. W. A. Duckworth, Dr. D. L. McRae, and Dr. I. M. Taylor. Dr. T. V. N. Persaud, Professor and Head of Anatomy at the University of Manitoba in Winnipeg, Dr. Douglas E. Kelly, Professor and Head of Anatomy, University of Southern California, and Dr. Kunwar Bhatnagar, Associate Professor of Anatomy at the University of Louisville, have also made good suggestions for improving the book. Mrs. Jill Weinheimer and my wife, Marion, have carefully and cheerfully typed corrections and additions to the text. Roberta Kangilaski, Albert Meier, and Walter Bailey of the W. B. Saunders Company have given me much help with this edition. To all these people, I express my sincere appreciation.

KEITH L. MOORE

FROM THE PREFACE TO THE FIRST EDITION

This book began as a set of illustrated notes for a *core course in medical embryology* that was intended to give students a base from which to develop. The purpose of this book is to present a synopsis of human development and related information. Although each chapter gives a relatively concise account of developmental processes, it is followed by numerous references for students wishing further information. Each chapter, except the first, is followed by a summary of the main events. An attempt has been made to "bridge the gap" between embryology and adult anatomy, histology, pathology, obstetrics, pediatrics, and surgery. Congenital malformations of each system are described, with emphasis on the common ones, and an entire chapter has been devoted to a discussion of the *causes of congenital malformations*.

The book is freely illustrated because much of the difficulty encountered by students beginning to study embryology results from their inability to visualize developmental processes and time sequences. Most illustrations are diagrammatic, some in color, and show *progressive stages of development*, conveying ideas and processes as blackboard sketches do during lectures. Numerous photographs are also included, similar to those used in case presentations at clinical seminars.

Text material has been used mainly for the following purposes: (a) to emphasize important points, (b) to discuss opposing views, and (c) to summarize concepts and processes. *Basic, or core, material is set in regular type*, whereas less important information is shown in small type or added as footnotes.

While writing this book, I have attempted to keep in mind what the naturalist John Ray said in the seventeenth century: "*He that useth many words for the explaining of any subject, doth like the cuttle-fish hide himself . . . in his own ink*"; and the often-quoted Chinese proverb, "*A little picture is worth a million words.*"

CONTENTS

INTRODUCTION
Terms and Concepts

Interest in developing humans is wide-spread largely because of curiosity about the subject and a desire to improve the quality of human life. The processes by which a child develops from a single cell are miraculous, and there is no more exciting event than that of a human birth. The adaptation of the newborn infant to its new life is also exhilarating to witness. We can readily observe and fully enjoy the process of human development during this period of life.

Human development is a continuous process that begins when an ovum from a female is fertilized by a sperm from a male. Growth and differentiation transform the *zygote*, a single cell formed by the union of the ovum and the sperm, into a multicellular adult human being. Most developmental changes occur during the embryonic and the fetal periods, but important changes also occur during the other periods of development: childhood, adolescence, and adulthood.

DEVELOPMENTAL PERIODS

Although it is customary to divide development into *prenatal* and *postnatal* periods, it is important to realize that birth is merely a dramatic event during development resulting in a distinct change in environment. Development does not stop at birth; important developmental changes, in addition to growth, occur after birth, e.g., development of the teeth and the female breasts. The brain triples in weight between birth and 16 years. Most developmental changes are completed by the age of 25.

PRENATAL PERIOD

The important changes occurring before birth are illustrated in the *Timetables of Human Prenatal Development* (Figs. 1–1 and 1–2), which are mainly based on studies by Streeter (1942), O'Rahilly (1973), and Gasser (1975). Note that the most striking advances in development occur during the first eight weeks, in which the embryonic period is included. The following terms are commonly used in discussions of the prenatal period.

Abortion (L. *aborto,* to miscarry). This term refers to the birth of an embryo or a fetus before it is viable (mature enough to survive outside the uterus). All terminations of pregnancy that occur before 20 weeks are called *abortions.* About 15 per cent of all recognized pregnancies end in *spontaneous abortions* (ones that occur naturally), usually during the first 12 weeks. Legal *induced abortions* are brought on purposefully, usually by *suction curettage* (evacuation of the embryo and its membranes from the uterus). *Therapeutic abortions* are induced because of the mother's poor health, or to prevent the birth of a severely malformed child.

Abortus. This term describes any product or all products of an abortion. An embryo or a *nonviable fetus* and its membranes weighing less than 500 gm is called an *abortus.*

Oocyte. This term refers to the immature ovum, or female germ cell. When mature, the female germ cell is called an *ovum* (L., egg). A secondary oocyte finishes its second meiotic division and becomes a mature ovum, immediately after entry of the sperm during fertilization (see Fig. 2–2).

Zygote. This cell results from fertilization of an oocyte, or ovum, by a sperm, or spermatozoon, and is *the beginning of a human being.* The expression "fertilized ovum" refers to the zygote.

Cleavage. This term refers to mitotic division of the zygote, which results in the formation of daughter cells called *blastomeres.* At each succeeding division, the blastomeres become smaller and smaller.

Text continued on page 6

TIMETABLE OF HUMAN PRENATAL DEVELOPMENT
1 to 6 weeks

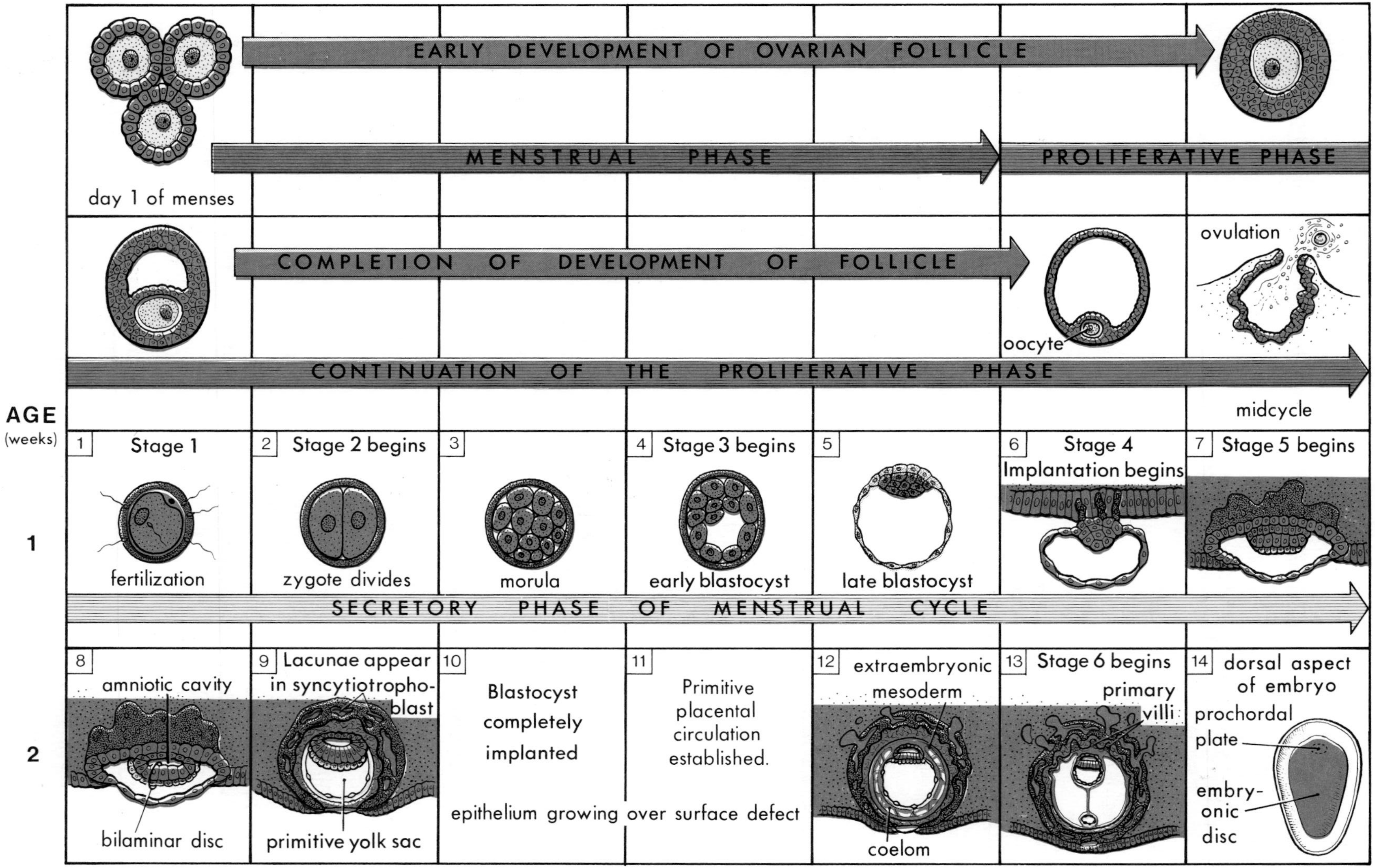

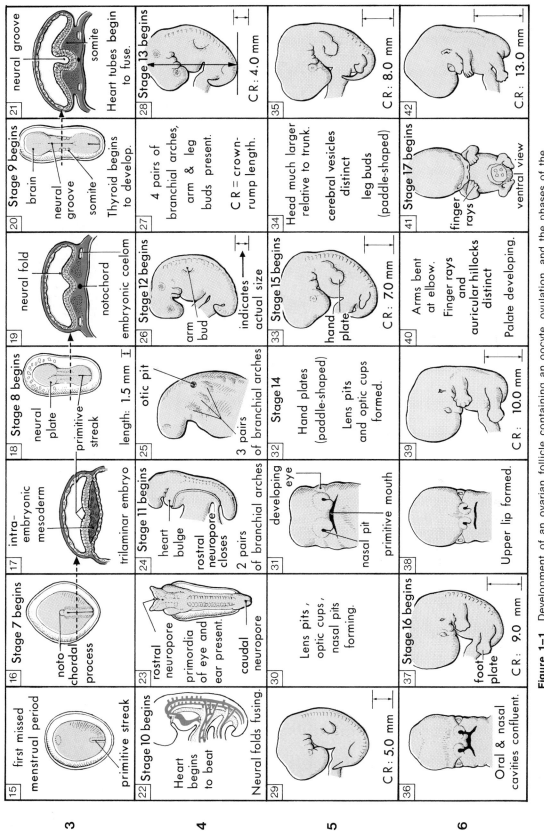

Figure 1-1 Development of an ovarian follicle containing an oocyte, ovulation, and the phases of the menstrual cycle are illustrated. *Development begins at fertilization*, about 14 days after the onset of the last menstruation. Cleavage of the zygote in the uterine tube, implantation of the blastocyst, and early development of the embryo are also shown. The main features of developmental stages in human embryos are illustrated. For a full discussion of embryonic development, see Chapter 5.

Panel contents:

15 first missed menstrual period — primitive streak

16 Stage 7 begins — notochordal process

17 intra-embryonic mesoderm — trilaminar embryo

18 Stage 8 begins — neural plate, primitive streak — length: 1.5 mm

19 neural fold — notochord, embryonic coelom

20 Stage 9 begins — brain, neural groove, somite. Thyroid begins to develop.

21 neural groove — somite. Heart tubes begin to fuse.

22 Stage 10 begins — Heart begins to beat. Neural folds fusing.

23 rostral neuropore — primordia of eye and ear present. caudal neuropore

24 Stage 11 begins — heart bulge, rostral neuropore closes. 2 pairs of branchial arches

25 otic pit — 3 pairs of branchial arches

26 Stage 12 begins — arm bud. ↔ indicates actual size

27 4 pairs of branchial arches, arm & leg buds present. CR = crown-rump length.

28 Stage 13 begins — CR: 4.0 mm

29 CR: 5.0 mm

30 Lens pits, optic cups, nasal pits forming.

31 developing eye — nasal pit, primitive mouth

32 Stage 14 — Hand plates (paddle-shaped). Lens pits and optic cups formed.

33 Stage 15 begins — hand plate. CR: 7.0 mm

34 Head much larger relative to trunk. cerebral vesicles distinct, leg buds (paddle-shaped)

35 CR: 8.0 mm

36 Oral & nasal cavities confluent.

37 Stage 16 begins — foot plate. CR: 9.0 mm

38 Upper lip formed.

39 CR: 10.0 mm

40 Arms bent at elbow. Finger rays and auricular hillocks distinct. Palate developing.

41 Stage 17 begins — finger rays. ventral view

42 CR: 13.0 mm

3

4

5

6

3

TIMETABLE OF HUMAN PRENATAL DEVELOPMENT
7 to 38 weeks

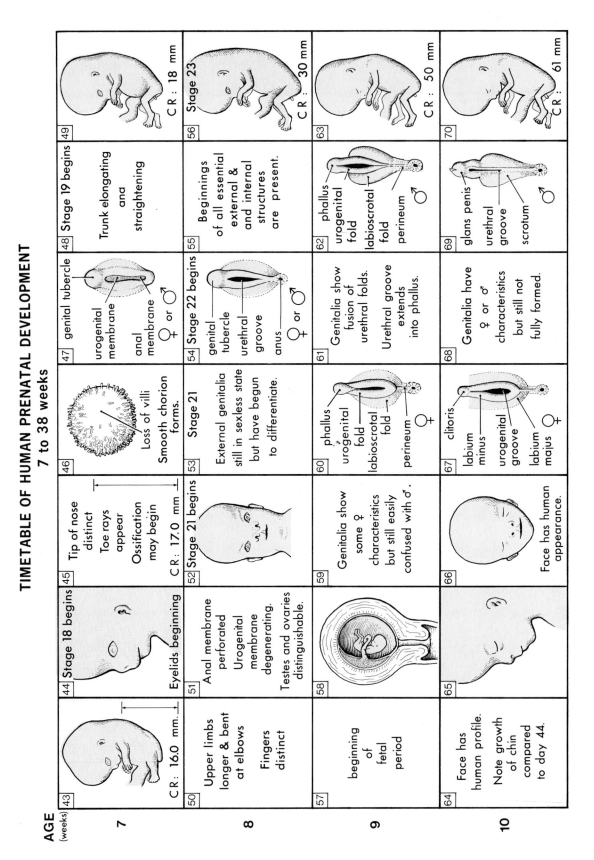

AGE (weeks)						

7

43 — CR: 16.0 mm.

44 — **Stage 18 begins** — Eyelids beginning

45 — Tip of nose distinct / Toe rays appear / Ossification may begin — CR: 17.0 mm.

46 — Loss of villi / Smooth chorion forms.

47 — genital tubercle / urogenital membrane / anal membrane — ♀ or ♂

48 — **Stage 19 begins** / Trunk elongating and straightening

49 — Stage 23 — CR : 18 mm

8

50 — Upper limbs longer & bent at elbows / Fingers distinct

51 — Anal membrane perforated / Urogenital membrane degenerating. / Testes and ovaries distinguishable.

52 — **Stage 21 begins**

53 — **Stage 21** / External genitalia still in sexless state but have begun to differentiate.

54 — **Stage 22 begins** / genital tubercle / urethral groove / anus — ♀ or ♂

55 — Beginnings of all essential external and internal structures are present.

56 — CR : 30 mm

9

57 — beginning of fetal period

58 —

59 — Genitalia show some ♀ characteristics but still easily confused with ♂.

60 — phallus / urogenital fold / labioscrotal fold / perineum — ♀

61 — Genitalia show fusion of urethral folds. / Urethral groove extends into phallus.

62 — phallus / urogenital fold / labioscrotal fold / perineum — ♂

63 — CR : 50 mm

10

64 — Face has human profile. / Note growth of chin compared to day 44.

65 —

66 — Face has human appearance.

67 — clitoris / labium minus / urogenital groove / labium majus — ♀

68 — Genitalia have ♀ or ♂ characteristics but still not fully formed.

69 — glans penis / urethral groove / scrotum — ♂

70 — CR : 61 mm

The Fetal Period

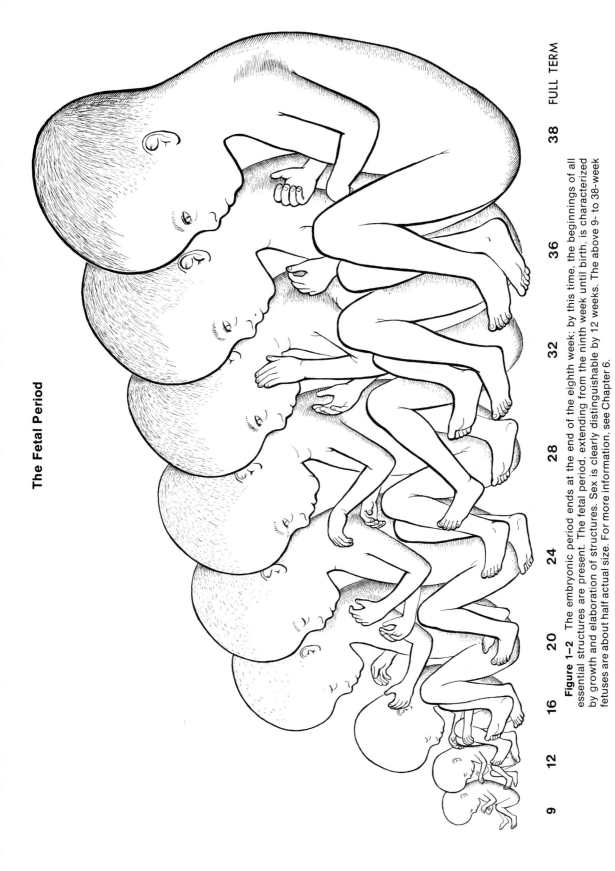

9   12   16   20   24   28   32   36   38   FULL TERM

Figure 1-2 The embryonic period ends at the end of the eighth week; by this time, the beginnings of all essential structures are present. The fetal period, extending from the ninth week until birth, is characterized by growth and elaboration of structures. Sex is clearly distinguishable by 12 weeks. The above 9- to 38-week fetuses are about half actual size. For more information, see Chapter 6.

Morula. When 12 to 16 blastomeres have formed by division of the zygote, the *solid ball of cells* is called a morula. It was so named because it resembles the fruit of a mulberry tree (the Latin word *morus* means mulberry). The morula stage occurs about three days after fertilization. Its centrally located cells, called the *inner cell mass,* will form the embryo.

Blastocyst. After the morula enters the uterus, a cavity develops inside it and fills with fluid; this converts the morula into a blastocyst.

Gastrula. During gastrulation, the period during which the trilaminar embryonic disc forms (see Fig. 4–2), the embryo is sometimes called a gastrula.

Neurula. During neurulation, the period during which the neural plate forms and closes to form the neural tube, the embryo is sometimes called a neurula.

Embryo. This term refers to the developing human during the early stages of development. The term is usually not used until the second week, after the *embryonic disc* forms (see Fig. 3–1). The *embryonic period* extends until the end of the eighth week, by which time the beginnings of all major structures are present.

Fetus. After the embryonic period, the developing human is called a fetus. During the *fetal period* (ninth week to birth), many systems develop further. Though developmental changes are not so dramatic as those occurring during the embryonic period, they are very important. The rate of body growth is remarkable, especially during the third and the fourth months, and weight gain is phenomenal during the terminal months.

Conceptus. This term refers to the embryo (or fetus) and its membranes, i.e., the *products of conception.* It includes all structures that develop from the zygote, both embryonic and extraembryonic. Hence it includes not only the embryo or the fetus, but also the placenta and the embryonic, or fetal, membranes.

Primordium (L. *primus,* first + *ordior,* to begin). This term refers to the first trace or indication of an organ or structure, i.e., its earliest stage of development. The term *anlage* has a similar meaning.

Miscarriage. This word is used colloquially to refer to any interruption of pregnancy that occurs before term. In medical description, it is most accurate to use the term *spontaneous abortion* for the birth of an embryo or a fetus prior to about 20 weeks; thereafter the event is called a *premature birth.*

Trimester. Obstetricians commonly divide the nine calendar months, or period of gestation (stages of intrauterine development), into three 3-month periods called *trimesters.*

POSTNATAL PERIOD

Changes occurring after birth are more or less familiar to most people. Explanations of frequently used terms follow.

Infancy. This refers to the first year or so after birth, the first two weeks of which are designated as the *newborn* (or *neonatal*) *period.* Transition from intrauterine to extrauterine existence requires gradual changes, especially in the cardiovascular and respiratory systems. If an infant survives the first crucial moments after birth, his or her chances of living are usually good (90 per cent of *neonates* do well).

The body as a whole grows particularly rapidly during infancy; total length increases by about one half, and weight is usually trebled.

Childhood. This is the period from about 15 months until 12 to 13 years. The primary (deciduous) teeth appear and are later replaced by the secondary (permanent) teeth. During early childhood there is active ossification, but as the child becomes older, the rate of growth slows down. Just before puberty it accelerates; this is known as the *prepubertal growth spurt.*

Puberty. This is the period between the ages of 12 and 15 years in girls and between 13 and 16 years in boys during which the *secondary sexual characteristics develop.* The legal ages of presumptive puberty are 12 years in girls and 14 years in boys.

Adolescence. This is the period of three to four years after puberty, extending from the earliest signs of sexual maturity until the attainment of physical, mental, and emotional maturity. The general growth rate decelerates, but growth of some structures accelerates (e.g., the female breasts).

Adulthood. Ossification and growth are virtually completed during *early adulthood,* 18 to 25 years. Thereafter, developmental changes occur very slowly, usually resulting in selective loss of highly specialized cells and tissues.

SCOPE OF EMBRYOLOGY

The term *embryology* can be misleading; literally, it means the study of an embryo (second to eighth week, inclusive). However, embryology refers to the study of both the embryo and the fetus, that is, the *study of prenatal development*.

The term developmental anatomy refers to both prenatal and postnatal periods of development. There are no essential differences between these two stages of development. Prenatal development is more rapid and results in more striking changes, but the developmental mechanisms of the two periods are very similar.

The term *ontogeny* describes the series of successive stages of development that occur during the *complete life period of a person*.

The study of abnormal development (congenital malformations) is called *dysmorphology* or *teratology*. This branch of embryology is concerned with the various genetic and environmental factors that disturb normal development (see Chapter 8).

SIGNIFICANCE OF EMBRYOLOGY

The study of human embryology is important because it develops knowledge concerning the *beginnings of human life* and the changes occurring during development. Knowledge of the developing human is of practical value in helping us to understand the normal relationships of body structures and the causes of congenital malformations.

Prior to 1940, little was known about the *causes of human congenital malformations*. Now it is realized that many malformed infants have chromosomal abnormalities (e.g., those with Down syndrome, Fig. 8–4), and that the embryo is extremely vulnerable during the first eight weeks to large amounts of radiation, viruses, and certain drugs (e.g., thalidomide). Knowledge that physicians, especially obstetricians, have of normal development and of the causes of congenital malformations is necessary for giving the embryo the greatest possible chance of developing normally.

Much of the modern practice of *obstetrics* involves what might be called "applied developmental biology." Embryological topics of special interest to obstetricians are ovulation, oocyte and sperm transport, fertilization, implantation, fetal-maternal relations, fetal circulation, critical periods of development, and causes of congenital malformations. In addition to caring for the mother, obstetricians must guard the health of the embryo during its critical period of development.

The significance of embryology is readily apparent to *pediatricians* because many of their patients have disorders resulting from maldevelopment, e.g., diaphragmatic hernia, spina bifida, and congenital heart disease. *Developmental abnormalities are among the ten leading causes of death in infancy.* Knowledge of the development of structure and function is essential to understanding the physiological changes that occur during the newborn period and for helping babies in distress.

Progress in *surgery,* especially in the pediatric age group, has made knowledge of human development more clinically significant. The understanding and correction of most congenital malformations (e.g., cleft palate and cardiac defects) depend upon knowledge of normal development and of the deviations that have occurred. An understanding of common congenital malformations and their causes also enables physicians, dentists, and others to explain the developmental basis of abnormalities, often dispelling parental guilt feelings.

Physicians who are aware of common abnormalities and their embryological basis approach unusual situations with confidence rather than surprise. For example, when it is realized that the renal artery represents only one of several vessels originally supplying the kidney during development, the frequent variations in number and arrangement of renal vessels are understandable and not unexpected. Many anatomical relationships, such as those of the lesser sac of peritoneum, the course of the left recurrent laryngeal nerve, and the innervation of the diaphragm, make more sense when developmental events are understood. When one knows the "raison d'être" for such relationships and arrangements, their complexity also becomes more impressive.

The achievements of experimental embryologists are of fundamental significance to the science of *pathology*. In the healing of wounds, tissues are restored to normal by processes that characterize embryonic differentiation. Because of the apparent simplicity of cell relations during early stages of development, pathologists often turn to embryology when developing a basis of classification of new growths (tumors).

HISTORICAL GLEANINGS

If I have seen further, it is by standing on the shoulders of giants.
– Sir Isaac Newton
  English mathematician, 1643–1727

This statement, made about 300 years ago, emphasizes that each new study of a problem rests on a base of knowledge established by earlier investigators. The theories of every age offer explanations based on the knowledge and the experience of the investigators of the period; although we should not consider them final, we should appreciate, rather than scorn, these ideas. People have always been interested in knowing how they originated, how they were born, and why some people develop abnormally. Ancient people, filled with curiosity, developed many answers to these questions.

THE GREEKS

Although some of their ideas were later shown to be incorrect, the Greeks made important contributions to the science of embryology. The first recorded embryological studies are in the books of *Hippocrates* (Fig. 1–3), the famous Greek physician of the fifth century B.C. He wrote,

> Take twenty or more eggs and let them be incubated by two or more hens. Then each day from the second to that of hatching, remove an egg, break it, and examine it. You will find exactly as I say, for the nature of the bird can be likened to that of man.

In the fourth century B.C., *Aristotle* wrote the first known treatise on embryology, in which he described development of the chick and other embryos. Many embryologists regard Aristotle as "The Founder of Embryology," despite the fact that he promoted the idea that the embryo devel-

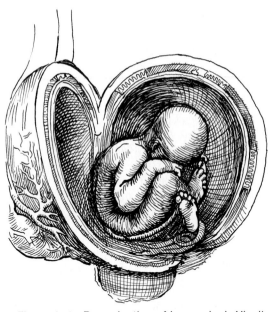

Figure 1–4   Reproduction of Leonardo da Vinci's drawing showing a fetus in an opened uterus.

oped from a formless mass that resulted from the union of semen with menstrual blood.

Galen (second century A.D.) wrote a book entitled *On the Formation of the Foetus,* in which he described the development and nutrition of fetuses and the structures we now call the *allantois,* the *amnion,* and the *placenta.*

THE MIDDLE AGES

Growth of science was slow during the medieval period, and few high points of embryological investigation undertaken during this age are known to us.

It is cited in the *Koran,* The Holy Book of the Muslims, that human beings are produced from a *mixture of secretions* from the male and the female. Several references are made to the creation of a human being from a *sperm drop,* and it is also suggested that the resulting organism settles in the woman like a seed, six days after its beginning. (The human blastocyst begins to implant about six days after fertilization.) The Koran also states that the sperm drop develops "into a clot of congealed blood." (An implanted blastocyst or a spontaneously aborted conceptus would resemble a blood clot.) Reference is also made to the leech-like appearance of the embryo. (The embryo shown in Fig. 5–3*A* is not unlike a leech, or bloodsucker, in appearance.) The embryo is also said to resemble "a chewed piece of substance" like gum or wood. (The somites shown in Figure 5–10 somewhat resemble the teethmarks in a chewed substance.)

Figure 1–3   Copy of a drawing of Hippocrates, the "Father of Medicine" (460–377 B.C.). He placed medicine on a scientific foundation.

The developing embryo was considered to become human at 40 to 42 days and to no longer resemble an animal embryo at this stage. (The human embryo begins to acquire human characteristics at this stage, as shown in Figure 5–14*C*.) The Koran also states that the embryo develops within "three veils of darkness." This probably refers to (1) the maternal anterior abdominal wall, (2) the uterine wall, and (3) the amniochorionic membrane. Space does not permit discussion of several other interesting references to human prenatal development that appear in the Koran.

THE RENAISSANCE

During the fifteenth century, *Leonardo da Vinci* made accurate drawings of dissections of the pregnant uterus and associated fetal membranes (Fig. 1–4) and introduced the quantitative approach by making measurements of embryonic growth.

THE MICROSCOPE

In 1651, *Harvey* studied chick embryos with simple lenses and made new observations, espe-

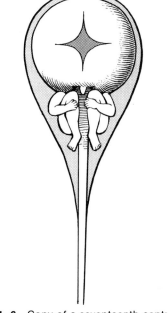

Figure 1–6 Copy of a seventeenth century drawing by Hartsoeker of a sperm. The miniature human being within it was thought to enlarge after the sperm entered an ovum.

cially on the circulation of blood. He also studied development of the fallow deer, but, when unable to observe early stages, concluded that the embryos were "secreted by the uterus."

Early microscopes were simple (Fig. 1–5), but they opened a new field of observation. In 1672 *de Graaf* observed little chambers in the rabbit's uterus and concluded that they could not have been secreted by the uterus, but must have come from the organs that he called *ovaries*. Undoubtedly, these little chambers were what we now call blastocysts. He described vesicular ovarian follicles, which are still sometimes called *graafian follicles* in his honor.

Malpighi, in 1675, studying what he believed were unfertilized hen's eggs, observed early embryos. As a result, he thought the egg contained a miniature chick. In 1677 *Hamm* and *Leeuwenhoek*, using an improved microscope, first observed human spermatozoa, but they misunderstood the sperm's role in fertilization: They thought it contained a *miniature human being* (Fig. 1–6).

Wolff, in 1759, refuted both versions of the *preformation theory* after observing parts of the embryo develop from "globules" (probably blastocysts). He examined unincubated eggs and could not see the embryos described by Malpighi. He proposed the *layer concept*, whereby division of the zygote produces layers of cells from which the embryo develops. His ideas formed the basis of the theory of *epigenesis*, which states that

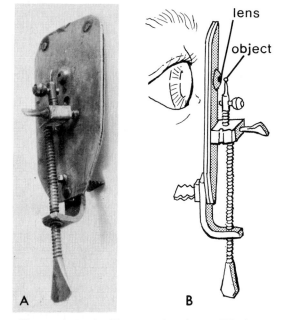

lens

object

A   **B**

Figure 1–5 *A,* Photograph of a *1673 Leeuwenhoek microscope. B,* Drawing of a lateral view illustrating its use. The object was held in front of the lens on the point of the short rod, and the screw arrangement was used to adjust the object under the lens.

development results from growth and differentiation of specialized cells.

The preformation controversy finally ended around 1775, when *Spallanzani* showed that both the ovum and the sperm were necessary for development of a new individual. From his experiments, including artifical insemination in dogs, he concluded that the sperm was the fertilizing agent.

In 1818, *Saint Hilaire* and his son made the first significant studies of congenital malformations. They did experiments in animals that were designed to produce developmental abnormalities, initiating what is now known as the science of *teratology*.

In 1827, about 150 years after the discovery of the sperm, *von Baer* described the oocyte in the ovarian follicle of a dog. He also observed dividing zygotes in the uterine tube and blastocysts in the uterus, and contributed much knowledge about the origin of tissues and organs from the layers described by Malpighi. His significant and far-reaching contributions resulted in his later being regarded as "The Father of Modern Embryology."

THE CELL THEORY

Great advances were made in embryology when the cell theory, stating that the body was composed of cells and cell products, was established in 1839 by *Schleiden and Schwann*. This concept soon led to the realization that the embryo developed from a single cell, the *zygote*.

THE CHROMOSOMES

In 1859, *Charles Darwin* published *On the Origin of Species,* in which he emphasized the *hereditary nature of variability* among members of a species as an important factor in evolution.

The *principles of heredity* were developed in 1865 by an Austrian monk named *Gregor Mendel,* but medical scientists and biologists did not understand the significance of these principles in the study of mammalian development for many years.

Flemming observed chromosomes in 1878 and suggested their probable role in fertilization. In 1883, *von Beneden* observed that mature germ cells have a reduced number of chromosomes. He also described some features of meiosis, the process whereby the chromosome number is reduced. In 1902 *Sutton* and *Boveri* declared independently that the behavior of the chromosomes during germ cell formation and fertilization agreed with Mendel's principles of inheritance. In the same year, *Garrod* reported alcaptonuria as the *first example of mendelian inheritance in human beings.* Many consider him the "Father of Medical Genetics."

The first significant observations on *human chromosomes* were made in 1912, when *von Winiwarter* reported that there were 47 chromosomes in the cells of the body. In 1923, *Painter* concluded that 48 was the correct number; this conclusion was widely accepted until 1956, when *Tjio and Levan* reported finding only 46 chromosomes in embryonic cells. Their descriptions and photomicrographs were so superior to those of previous workers that few cytologists doubted the accuracy of their counts.

Once the normal chromosomal pattern was firmly established, it soon became evident that some persons with congenital abnormalities had an abnormal number of chromosomes. *A new era in medical genetics* resulted from the demonstration by Lejeune et al. in 1959 that infants with mongolism (now called Down syndrome) have 47 chromosomes instead of the usual 46 in their body cells. It is now known that chromosomal aberrations are a significant cause of congenital malformations and embryonic death. About 8 per cent of all conceptuses are known to be chromosomally abnormal.

DESCRIPTIVE TERMS

In descriptive anatomy and embryology, several terms relating to position and direction are used, and various planes of the body are referred to in sections. Before reading this book, it would be helpful to become familiar with the language of anatomy, which forms the basis of medical language.

All descriptions of the adult are based on the assumption that the body is erect, with the upper limbs by the sides and the palms directed anteriorly (Fig. 1–7*A*). This is called the *anatomical position*. The terms *anterior,* or *ventral,* and *posterior,* or *dorsal,* are used to describe the front or back of the body or limbs, and the relations of structures within the body to one another. For embryos, the terms dorsal and ventral are commonly used (Fig. 1–7*B*).

Superior and *inferior* are used to indicate the relative levels of different structures (Fig. 1–7*A*). For embryos, the terms *cranial* (cephalic) and *caudal* are commonly used to denote relationships to the head and tail ends, respectively (Fig. 1–7*B*). When the suffix *-ad* is affixed to either of these terms, it implies movement toward the region concerned, e.g., *craniad* (cephalad) means toward the head, and *caudad* refers to movement toward the tail region.

The term *rostral* is used to indicate the relationships of structures to the nose (Fig.

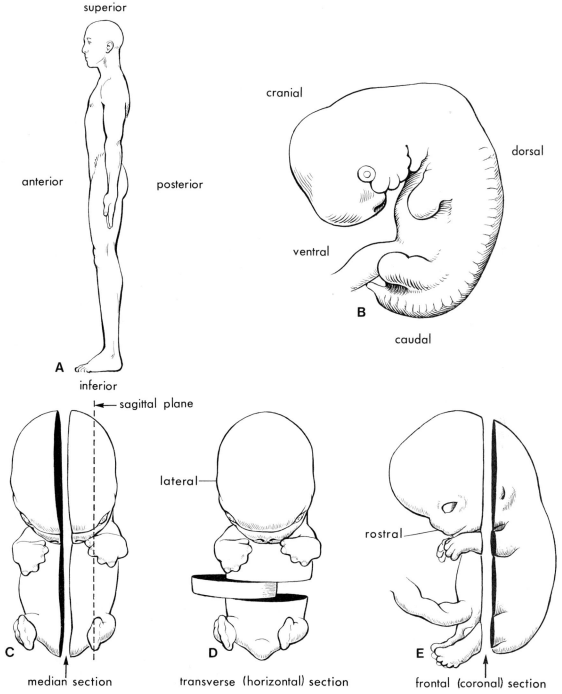

superior

cranial

dorsal

anterior

posterior

ventral

B

caudal

inferior

sagittal plane

lateral

rostral

C

D

E

median section

transverse (horizontal) section

frontal (coronal) section

Figure 1–7 Drawing illustrating descriptive terms of position, direction, and planes of the body. *A*, Lateral view of an adult in the anatomical position. *B*, Lateral view of a five-week embryo. *C* and *D*, Ventral views of six-week embryos. *E*, Lateral view of a seven-week embryo.

1–7E). Distances from the source of attachment of a structure are designated as *proximal* or *distal*; e.g., in the lower limb the knee is proximal to the ankle and the ankle is distal to the knee. Again, the suffix *-ad* implies movement toward a region, e.g., a nerve is traced distad, or distally, toward its ending.

The *median plane* is a vertical plane passing through the center of the body. Median sections divide the body into right and left halves (Fig. 1–7C). The terms *lateral* and *medial* refer to structures that are respectively farther from or nearer to the median plane of the body. A *sagittal plane* is any vertical plane passing through the body parallel to the median plane (Fig. 1–7 C).

A *transverse (horizontal) plane* refers to any plane that is at right angles to both the median and the frontal planes. A *frontal (coronal) plane* is any vertical plane that intersects the median plane at a right angle; it divides the body into front (anterior or ventral) and back (posterior or dorsal) parts.

Various terms are used to describe sections of embryos made through the aforementioned planes. A *median (midsagittal) section* is one cut through the median plane. Longitudinal sections parallel to the median plane, but not through it, are called *sagittal sections*. A vertical section through the frontal (coronal) plane is known as a *frontal (coronal) section*. Sections through the transverse plane are called *transverse (horizontal) sections,* or simply *cross sections* (Fig. 1–7D). *Oblique sections* are neither perpendicular nor horizontal, but are slanted or inclined.

SUGGESTIONS FOR ADDITIONAL READING

Persons wishing further information on introductory concepts and the history of embryology should consult the following books and some of the other references listed at the end of this chapter.

Arey, L. B.: *Developmental Anatomy: A Textbook and Laboratory Manual of Embryology*. Revised 7th ed. Philadelphia, W. B. Saunders Co., 1974, pp. 1–6. This classic text gives interesting accounts of the nature and scope of embryology, the historical background, and the roles of heredity and environment in development.

Balinsky, B. I.: *An Introduction To Embryology*. 4th ed. Philadelphia, W. B. Saunders Co., 1975. pp. 3–15. An excellent introductory text with a good balance between classic descriptive and experimental embryology. As it is oriented to undergraduate biology students, it should be very helpful to beginning medical and dental students.

Hamilton, W. J., and Mossman, H. W.: *Human Embryology — Prenatal Development of Form and Function*. 4th ed. Baltimore, The Williams & Wilkins Co., 1972, pp. 1–12. Many people consider this to be the finest textbook of embryology that has ever been written; however, it contains too much information for most beginning students.

CLINICALLY ORIENTED PROBLEMS

1. When does a new human being begin to develop?
2. What is a new human organism called at the beginning of its development?
3. Differentiate between the terms *conceptus* and *abortus*.
4. What sequence of events occurs during puberty? What are the respective ages of presumptive puberty in males and females?
5. Differentiate between the terms *embryology, ontogeny,* and *teratology*.

The answers to these questions are given at the back of the book.

REFERENCES

Corliss, C. E.: *Patten's Human Embryology. Elements of Clinical Development*. New York, McGraw-Hill Company, 1976.

Corner, G. W.: *Ourselves Unborn*. New Haven, Yale University Press, 1944.

Gasser, R.: *Atlas of Human Embryos*. Hagerstown, Harper & Row, Publishers, 1975.

Lejeune, J., Gautier, M., and Turpin, R.: Étude des chromosomes somatiques de neuf enfants mongoliens. *C.R. Acad. Sci.* (Paris) *248*:1721, 1959.

Meyer, A. W.: *The Rise of Embryology*. Stanford, Stanford University Press, 1939.

Moore, K. L.: *The Sex Chromatin*. Philadelphia, W. B. Saunders Co., 1966.

Needham, J.: *A History of Embryology*. 2nd ed. Cambridge, Cambridge University Press, 1959.

Oppenheimer, J. M.: Problems, concepts and their history; in B. H. Willier, P. A. Weiss, and V. Hamburger (Eds.): *Analysis of Development*. Philadelphia, W. B. Saunders Co., 1965, pp. 1–24.

O'Rahilly, R.: *Developmental Stages in Human Embryos, Part A: Embryos of the First Three Weeks*. Washington, D.C., Carnegie Institution of Washington, 1973.

Persaud, T. V. N.: *Teratogenesis: Experimental Aspects and Clinical Implications*. Germany, VEB Gustav Fischer Verlag, 1979, pp. 8–12.

Rigatto, H.: Common neonatal problems; *in* Persaud, T. V. N. (Ed.): *Prenatal Pathology. Fetal Medicine*. Springfield, Charles C Thomas, 1979, pp. 151–158.

Streeter, G. L.: Developmental horizons in human embryos. Description of age group XI, 13 to 20 somites, and age group XII, 21 to 29 somites. *Contrib. Embryol. Carnegie Inst. 30*:211, 1942.

Tanner, J. M.: Growth and Endocrinology of the Adolescent; *in* L. I. Gardner (Ed.): *Endocrine and Genetic Diseases of Childhood and Adolescence*. 2nd ed. Philadelphia, W. B. Saunders Co., 1975, pp. 14–63.

Willis, R. A.: *The Borderland of Embryology and Pathology*. 2nd ed. London, Butterworth & Co. Publishers, Ltd., 1962, pp. 17–53.

2

THE BEGINNING OF DEVELOPMENT

The First Week

He who sees things grow from the beginning will have
the finest view of them.

— Aristotle
Greek philosopher, 384–322 B.C.

Human development begins at fertilization, when a sperm unites with an ovum to form a unicellular organism called a *zygote* (Gr. *zygotos,* yoked together). This cell marks the *beginnings* of each of us as a unique individual.

Although it is a very large cell, the zygote is just visible to the unaided eye as a tiny speck. It contains genes (units of genetic information) in duplicate that are derived from the mother and the father. The unicellular organism, or zygote, becomes progressively transformed into a multicellular human being through the division, migration, growth, and differentiation of cells.

GAMETOGENESIS

Gametogenesis is the process of formation and development of specialized generative cells or gametes (germ cells). This process, which involves the chromosomes and the cytoplasm of the gametes, prepares these cells for *fertilization* (union of the male and female gametes). During gametogenesis, the chromosome number is reduced by half, and the shape of the cells is altered.

The *sperm* and the *ovum* (the male and female germ cells or gametes) are highly specialized *sex cells* (Fig. 2–1) that contain half the usual number of chromosomes. The number is reduced during *meiosis,* a special type of cell division that occurs during *gametogenesis.* This maturation process is called *spermatogenesis* in males and *oogenesis* in females (Fig. 2–2).

There are *two* successive meiotic divisions. In the *first meiotic division* (the re-

duction division), *homologous chromosomes* (a "matched pair," one from each parent) pair during prophase and then separate during anaphase, with *one representative of each pair going to each pole.* By the end of the first meiotic division, each new cell formed (secondary spermatocyte or secondary oocyte) has the *haploid chromosome number,* i.e., half the original number of chromosomes of the preceding cell (primary spermatocyte or primary oocyte). This *disjunction* of paired homologous chromosomes is the physical basis of *segregation,* the separation of allelic genes at meiosis.

The second meiotic division follows the first division without DNA replication and without a normal interphase. Each chromosome (which has two parallel strands, or chromatids) divides, and each half, or *chromatid,* is drawn to a different pole. Thus, the haploid number of chromosomes is retained, and each daughter cell formed by meiosis has this reduced number, with one representative of each chromosome pair. For more details and easy-to-follow illustrations of meiosis, see Thompson and Thompson (1980).

The significance of meiosis is that it provides for constancy of the chromosome number from generation to generation by producing *haploid sex cells.* Meiosis also allows independent assortment of maternal and paternal chromosomes among the gametes. *Crossing over,* by relocating segments of the maternal and paternal chromosomes, serves to "shuffle" the genes and thereby produce a recombination of genetic material.

Some embryologists consider gametogen-

14

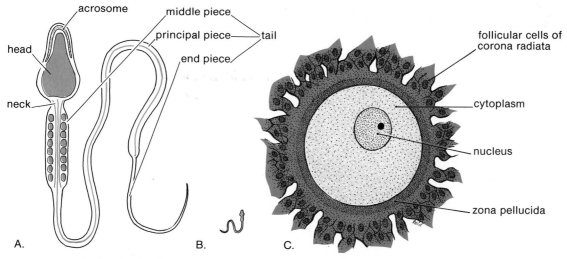

Figure 2–1 *A,* Drawing showing the main parts of a human sperm (× 1250). The head, composed mostly of the nucleus, is covered by the acrosome (acrosomal cap), an organelle containing lytic enzymes that are believed to have an important role during fertilization (see Fig. 2–13). The sperm tail consists of three regions: the middle piece, the principal piece, and the end piece. The mitochondria in the middle piece are believed to generate the energy for sperm motility. *B,* A sperm drawn to about the same scale as the oocyte. Because sperms were at one time regarded as parasites, they were given the name spermatozoa ("semen animals"). *C,* Drawing of a human secondary oocyte (×200), surrounded by the zona pellucida and the corona radiata.

esis the first phase of development, because oogenesis and spermatogenesis create the conditions from which subsequent embryogenesis arises. Disturbances during gametogenesis, e.g., *nondisjunction* (Fig. 2–3), result in abnormal gametes that cause abnormal development, such as occurs with trisomy 21, or Down syndrome (see Fig. 8–4).

SPERMATOGENESIS

The term *spermatogenesis* refers to the entire sequence of events by which spermatogonia are transformed into spermatozoa, or sperms. This maturation process begins at puberty (about 14 years) and continues into old age.

The *spermatogonia,* which have been dormant in the seminiferous tubules of the testes since the fetal period, begin to *increase in number at puberty.* After several mitotic divisions, the spermatogonia grow and undergo gradual changes that transform them into *primary spermatocytes* (Fig. 2–2), the largest germ cells in the tubules. Each primary spermatocyte subsequently undergoes a reduction division, called the *first meiotic division,* to form two haploid *secondary spermatocytes* that are about half the size of

primary spermatocytes. Subsequently, these secondary spermatocytes undergo a *second meiotic division* to form four haploid *spermatids* that are about half the size of secondary spermatocytes. The spermatids are gradually transformed into four *mature sperms* by an extensive process of differentiation known as *spermiogenesis.* Spermatogenesis, the process of transformation of a spermatogonium into a mature sperm, takes about 64 days (Clermont, 1963).

The mature sperm (Figs. 2–1*A* and 2–12) is a free-swimming, actively motile cell consisting of a *head* and a *tail,* or flagellum. The head, forming most of the bulk of the sperm, consists of the nucleus, whose chromatin is greatly condensed. The anterior two thirds of the nucleus is covered by the *acrosome,* a membrane-limited organelle containing enzymes that are believed to facilitate sperm penetration of the corona radiata and zona pellucida during fertilization (Fig. 2–13).

The tail of the sperm consists of three segments: the *middle piece,* the *principal piece,* and the *end piece.* The tail provides the motility of the sperm, which assists in its transport to the site of fertilization. The middle piece of the tail contains the energy-producing cytoplasmic and mitochondrial

NORMAL GAMETOGENESIS

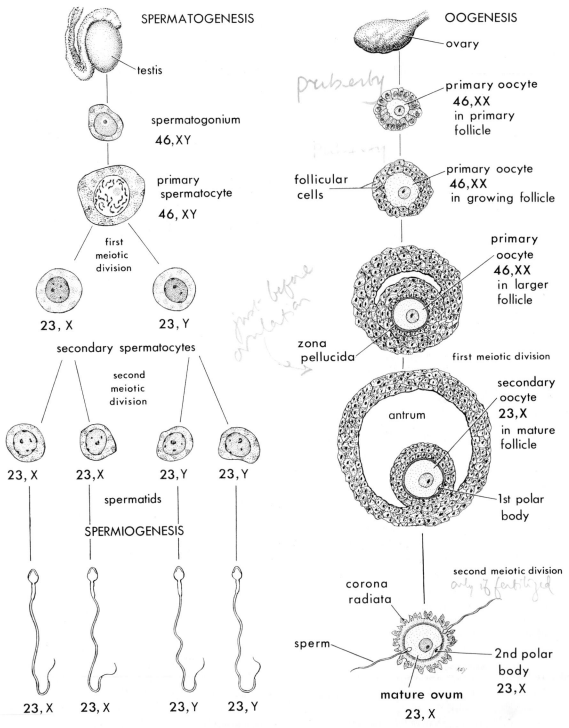

Figure 2–2 Drawings comparing spermatogenesis and oogenesis. Oogonia are not shown in this figure because all oogonia differentiate into primary oocytes before birth. The chromosome complement of the germ cells is shown at each stage. The number designates the total number of chromosomes, including the sex chromosome(s) shown after the comma. Note that (1) following the two meiotic divisions, the diploid number of chromosomes, 46, is reduced to the haploid number, 23; (2) *four sperms* form from one primary spermatocyte, whereas only *one* mature ovum results from maturation of a primary oocyte; and (3) the cytoplasm is conserved during oogenesis to form one large cell, the mature oocyte or ovum.

apparatus. The junction between the head and the tail is called the *neck* (Fig. 2–1).

OOGENESIS

The term *oogenesis* refers to the entire sequence of events by which oogonia are transformed into ova. This maturation process begins before birth, but it is not completed (if it occurs at all) until after puberty.

Prenatal Maturation. During early fetal life, the *oogonia* proliferate by mitotic division. All these oogonia enlarge to form *primary oocytes* before birth; for this reason, oogonia are not shown in Figures 2–2 and 2–3. As the primary oocyte forms, ovarian stromal cells surround it and form a single layer of flattened follicular cells. The primary oocyte enclosed by this layer of flattened follicular cells constitutes a *primordial follicle,* or unilaminar follicle (Fig. 2–8A). When the primary oocyte enlarges at puberty, the single layer of flattened, follicular cells surrounding it becomes cuboidal and then columnar, forming a *primary follicle* (Fig. 2–2). When the primary follicle has more than one layer of cuboidal follicular cells, it is often called a *growing follicle* (Fig. 2–8B).

Primary oocytes begin the first meiotic division before birth, but the completion of prophase does not occur until after puberty. *The primary oocytes remain in suspended prophase* (dictyotene) for several years, until sexual maturity is reached and the reproductive cycles begin at puberty. Apparently, the follicular cells surrounding the primary oocyte secrete a substance called *oocyte maturation inhibitor* (OMI), which acts to keep the meiotic process of the oocyte arrested (Page et al., 1981). This long duration of the first meiotic division may account in part for the relatively high frequency of meiotic errors such as *nondisjunction* (failure of paired chromosomes to disjoin) that occur with increasing maternal age.

It must be stressed that *no primary oocytes form after birth,* in contrast to the continuous production of primary spermatocytes in the male after puberty.

Postnatal Maturation. The primary oocytes remain dormant in the ovaries until puberty. As a follicle matures, the primary oocyte increases in size and a deeply staining membrane, the *zona pellucida,* forms around it (Figs. 2–1C and 2–8B). Shortly before ovulation (36 to 48 hours), the primary oocyte completes the *first meiotic division.* Unlike the corresponding stage of spermatogenesis, however, the division of cytoplasm is unequal. The *secondary oocyte* receives almost all the cytoplasm (Fig. 2–2), and the *first polar body* receives hardly any. The first polar body is a small, nonfunctional cell that soon degenerates. At ovulation the nucleus of the secondary oocyte begins the *second meiotic division,* but progresses only to metaphase, when division is arrested. *If fertilization occurs,* the second meiotic division is completed and most cytoplasm is again retained by one cell, the *mature oocyte* (Fig. 2–2). The other cell, the *second polar body,* is small and soon degenerates. As soon as the second polar body is extruded, maturation of the ovum is complete.

The secondary oocyte released at ovulation is surrounded by the *zona pellucida* and a layer of follicular cells called the *corona radiata* (Fig. 2–1C). Compared with ordinary cells it is truly large, but it is just visible to the unaided eye as a tiny speck.

It is generally believed that about 2 million primary oocytes are usually present in the ovaries of a newborn female infant, but many regress during childhood so that by puberty only 30 to 40 thousand remain. Of these, only about 400 become secondary oocytes and are expelled at ovulation during the reproductive period. The number of oocytes ovulated is greatly reduced in women who take contraceptive (birth control) pills because the hormones in these pills prevent ovulation.

COMPARISON OF THE SPERM AND THE OVUM

The sperm and ovum are dissimilar in several ways because of their adaptation to specialized roles. The mature ovum is massive compared with the sperm (Fig. 2–1) and is immotile, whereas the microscopic sperm is highly motile. The ovum has an abundance of cytoplasm containing *yolk granules,* which provide nutrition during the first few days of development. The sperm bears little resemblance to an ovum or any other cell because of its sparse cytoplasm and specialization for motility.

With respect to sex chromosome constitution, there are *two kinds of normal sperm* (Fig. 2–2): 23, X and 23, Y, whereas there is only *one kind of normal ovum:* 23, X.

ABNORMAL GAMETOGENESIS

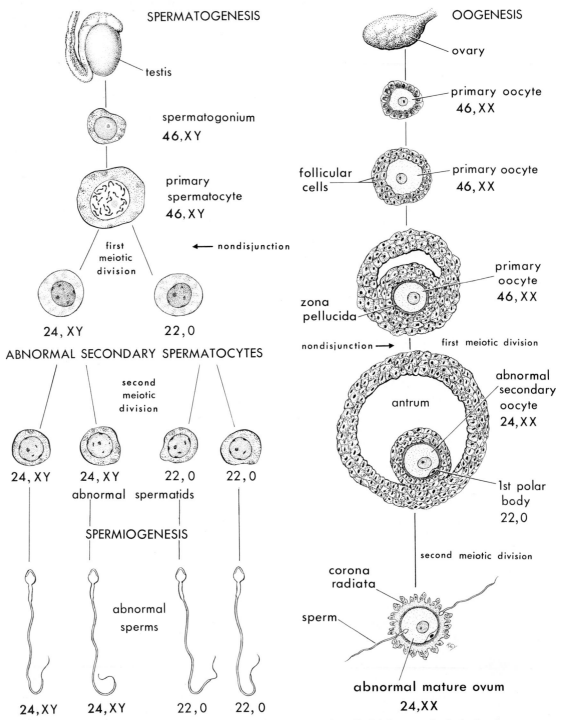

SPERMATOGENESIS

OOGENESIS

ovary

testis

primary oocyte
46,XX

spermatogonium
46,XY

primary oocyte
46,XX

follicular cells

primary spermatocyte
46,XY

primary oocyte
46,XX

first meiotic division

← nondisjunction

zona pellucida

24,XY   22,0

nondisjunction →

first meiotic division

ABNORMAL SECONDARY SPERMATOCYTES

abnormal secondary oocyte
24,XX

second meiotic division

antrum

24,XY   24,XY   22,0   22,0

1st polar body
22,0

abnormal spermatids

SPERMIOGENESIS

second meiotic division

corona radiata

abnormal sperms

sperm

24,XY   24,XY   22,0   22,0

abnormal mature ovum
24,XX

Figure 2–3 Drawings showing how *nondisjunction,* an error in cell division, results in faulty chromosome distribution in germ cells. (Although nondisjunction of sex chromosomes is illustrated, a similiar defect may occur during division of autosomes.) When nondisjunction occurs during the first meiotic division of spermatogenesis, one secondary spermatocyte contains 22 autosomes plus an X and a Y chromosome, and the other one contains 22 autosomes and no sex chromosome. Similarly, nondisjunction during oogenesis may give rise to an oocyte with 22 autosomes and two X chromosomes (as shown), or to one with 22 autosomes and no sex chromosomes. (See Fig. 8–3 for other drawings illustrating nondisjunction.)

In the foregoing descriptions, and in Figures 2–2 and 2–3, the number 23 indicates the *total* number of chromosomes in the complement, including the sex chromosome. This number is followed by a comma and an X or Y to indicate the sex chromosome constitution; e.g., 23, X indicates that there are 23 chromosomes in the complement, made up of 22 autosomes and 1 sex chromosome (an X in this case).

ABNORMAL GERM CELLS

The ideal maternal age for reproduction appears to be from 18 to 30 years of age (Smith et al., 1978). The likelihood of a severe chromosomal abnormality in the embryo increases significantly after the age of 35. It is also undesirable for the father to be older than this because the likelihood of a fresh gene mutation (alteration) increases with paternal age. The older the father at the time of conception, the more likely he is to have accumulated mutations that the embryo might inherit (Thompson and Thompson, 1980). This relationship does not hold for all dominant mutations and is not an important consideration in older mothers.

Chromosomal Abnormalities. During meiosis, homologous chromosomes sometimes fail to separate and go to opposite poles of the cell. As a result of this error of cell division, known as *nondisjunction,* some germ cells have 24 chromosomes and others have only 22 (Fig. 2–3).

If a germ cell with 24 chromosomes fuses with a normal one during fertilization, a zygote with 47 chromosomes forms. This condition is called *tri-somy* because of the presence of three representatives of a particular chromosome, instead of the usual two. If a germ cell with only 22 chromosomes fuses with a normal one, a zygote with 45 chromosomes forms. This condition is known as *monosomy* because only one representative of a particular chromosome is present, instead of the usual pair. For a description of the clinical conditions associated with numerical disorders of chromosomes, see Chapter 8.

Morphological Abnormalities. Up to 10 per cent of the sperms in an ejaculate may be grossly abnormal (Fig. 2–4B), but it is generally believed that they do not fertilize oocytes owing to their lack of normal motility. Most, if not all, morphologically abnormal sperms are unable to pass through the mucus in the cervical canal. X-rays, severe allergic reactions, and certain antispermatogenic agents have been reported to increase the percentage of abnormally shaped sperms in man. Such sperms are not believed to affect fertility unless their number exceeds 20 per cent.

Although some oocytes may have two or three nuclei, they probably never mature. Some abnormal follicles containing two (Fig. 2–4A) or more oocytes may develop, but this phenomenon is infrequent in human females. Although such compound follicles could result in multiple births, it is believed most of them never mature and never expel their oocytes at ovulation.

STRUCTURE OF THE UTERUS

A brief description of the histological structure of the uterus is given here as a basis for understanding reproductive cycles, implantation of the blastocyst, and placentation.

The wall of the uterus consists of three

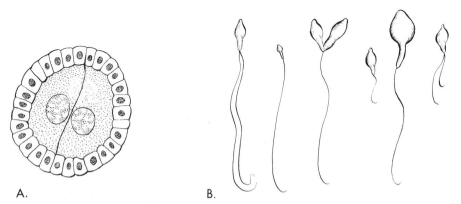

A.   B.

Figure 2–4 Drawings of germ cells. *A,* Primary follicle containing two primary oocytes. If this follicle matured and the oocytes were expelled and fertilized, they would give rise to dizygotic (nonidentical) twins. *B,* Various types of abnormal sperms. Although 10 per cent of the sperms may be abnormal, they are not believed to fertilize oocytes or to affect fertility unless 20 per cent or more are abnormally formed. Defective and immobile sperms are unable to penetrate the cervical mucus.

layers (Fig. 2–5*A*): (1) a very thin outer serosa, or *perimetrium;* (2) a thick middle smooth muscle layer, or *myometrium*; and (3) a thin inner layer, or *endometrium.*

During the secretory phase of the menstrual cycle (Fig. 2–6), *three layers of the endometrium* can be distinguished (Fig. 2–5*B*): (1) a thin superficial *compact layer* consisting of densely packed, stromal cells around the straight necks of the glands; (2) a thick *spongy layer* composed of edematous stroma containing the dilated, tortuous bodies of the glands; and (3) a thin *basal layer* containing the blind ends of the glands. This deep layer has its own blood supply and is not sloughed off during menstruation. The other two, the compact and spongy layers, disintegrate and are shed at menstruation and parturition (de-

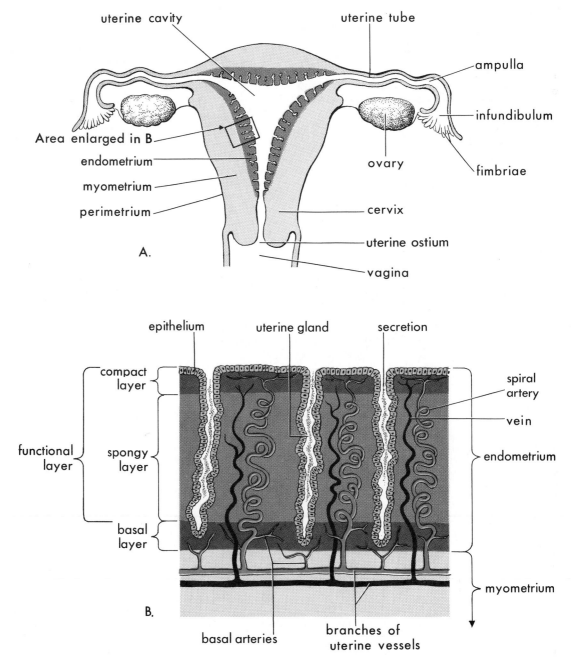

Figure 2–5 *A*, Diagrammatic frontal section of the uterus and uterine tubes. The ovaries and vagina are also indicated. *B*, Enlargement of the area outlined in *A*.

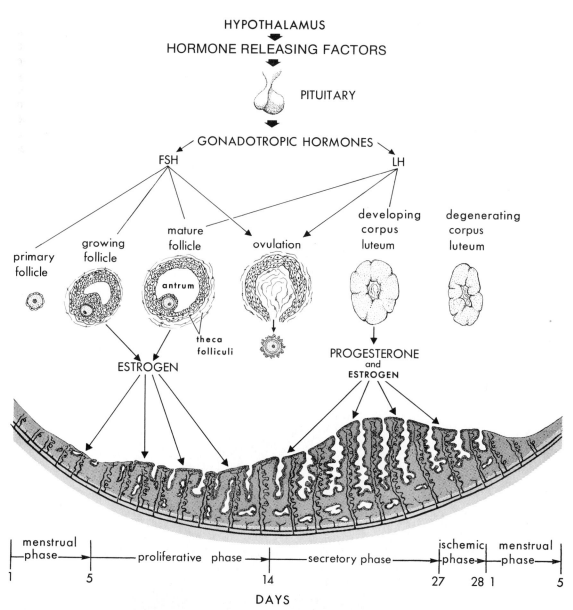

Figure 2–6 Schematic drawing illustrating the interrelations of the hypothalamus, hypophysis (pituitary gland), ovaries, and endometrium. One complete menstrual cycle and the beginning of another are shown. Changes in the ovaries, called the ovarian cycle, are promoted by the gonadotropic hormones (FSH and LH). Hormones from the ovaries (estrogens and progesterone) then promote changes in the structure and function of the endometrium, called the uterine cycle. Thus, the cyclical activity of the ovary is intimately linked with changes in the uterus.

livery of a baby), and so are commonly called the *functional layer* (Fig. 2–5B).

REPRODUCTIVE CYCLES

Commencing at puberty and normally continuing throughout the reproductive years, *human females undergo monthly reproduc-*tive (sexual) cycles involving activities of the hypothalamus, hypophysis (pituitary gland), ovaries, uterus, uterine tubes, vagina, and mammary glands. These cycles prepare the reproductive system for pregnancy.

A *hormone-releasing factor* (GnRH) synthesized by cells in the hypothalamus is carried by the *hypophyseal portal system*

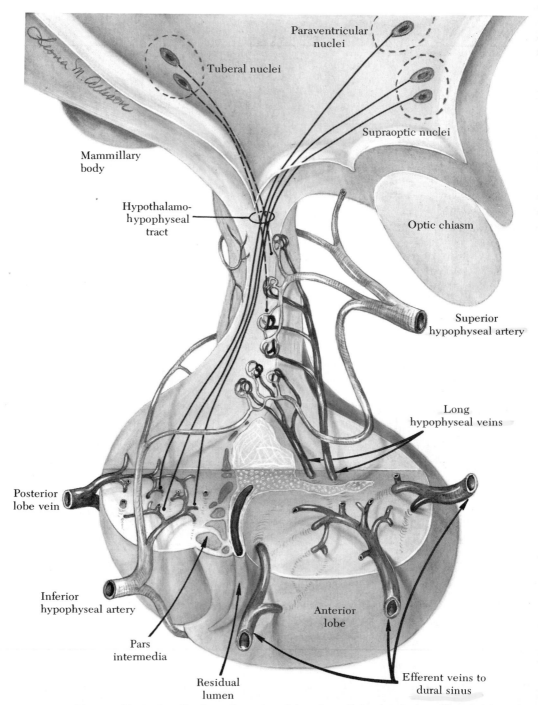

Figure 2–7 Diagram illustrating the hypophyseal portal system. Releasing factors liberated from hypothalamic cells are carried in the blood to the anterior lobe of the hypophysis (pituitary gland) via the capillaries, portal veins, and sinusoids of the hypophyseoportal system. (From Leeson, C. R., and Leeson, T. S.: *Histology.* 4th ed. Philadelphia, W. B. Saunders Co., 1981.)

(Fig. 2–7) to the anterior lobe of the hypophysis (pituitary). This causes the cyclic release of the gonadotropic hormones, follicle-stimulating hormone (FSH), and luteinizing hormone (LH).

THE OVARIAN CYCLE

The gonadotropins (FSH and LH) produce *cyclic changes in the ovaries* (development of follicles, ovulation, and corpus luteum formation) known as the ovarian cycle. In each cycle, FSH promotes growth of 5 to 12 primary follicles; however, usually only one of them develops into a mature follicle and ruptures through the surface of the ovary, expelling its oocyte (Fig. 2–10). Hence, 4 to 11 follicles degenerate and never mature. As the oocyte and follicle degenerate, they are replaced by connective tissue, forming a *corpus atreticum.*

Follicular Development. Development of a follicle is characterized by (1) growth and differentiation of the primary oocyte, (2) proliferation of follicular cells, (3) formation of the zona pellucida, and (4) development of a connective tissue capsule, the *theca folliculi,* from the ovarian stroma (Fig. 2–6). The follicular cells divide actively, producing a stratified layer around the oocyte (Fig. 2–8*B*). The follicle soon becomes oval in shape and the oocyte eccentric in position, because proliferation of the follicular cells occurs more rapidly on one side. Subsequently, fluid-filled spaces appear around the cells; these spaces soon coalesce to form a single large cavity, the *follicular antrum* (Fig. 2–8*C).*

When the antrum forms, the ovarian follicle is called a *secondary or vesicular follicle.* The primary oocyte gets pushed to one side of the follicle, where it is surrounded by a mound of follicular cells, the *cumulus oophorus,* and projects into the antrum (Figs. 2–8*C* and 2–9).

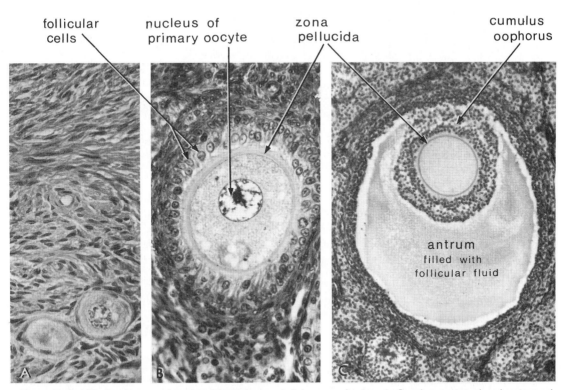

Figure 2–8 Photomicrographs of sections from adult human ovaries. *A,* Ovarian cortex showing two primordial follicles containing primary oocytes that have completed the prophase of the first meiotic division and have entered the dictyotene stage, a "resting" stage between prophase and metaphase (×250). *B,* Growing follicle containing a primary oocyte, surrounded by the zona pellucida and a stratified layer of follicular cells (×250). *C,* An almost mature follicle with a large antrum. The oocyte, embedded in the cumulus oophorus, does not show a nucleus because it has been sectioned tangentially (×100). (From Leeson, C. R., and Leeson, T. S.: *Histology.* 4th ed. Philadelphia, W. B. Saunders Co., 1981.)

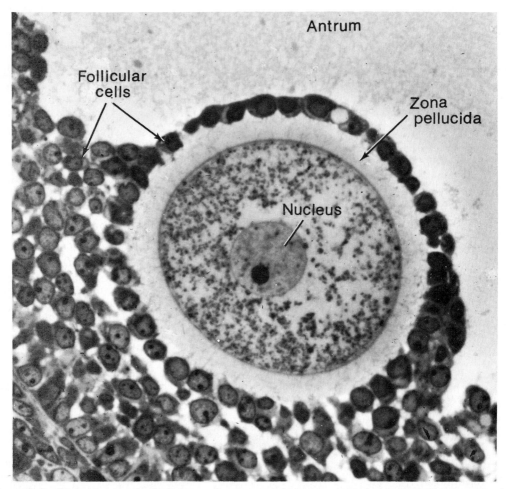

Figure 2–9 Photomicrograph of a human primary oocyte in a secondary or vesicular follicle, surrounded by follicular cells of the cumulus oophorus. The zona pellucida is a refractile, deeply staining layer of uniform thickness. It is a gel-like, neutral glycoprotein that protects the oocyte. (From Bloom, W., and Fawcett, D. W.: *A Textbook of Histology.* 10th ed. Philadelphia, W. B. Saunders Co., 1975. Courtesy of L. Zamboni.)

Early development of follicles is induced by FSH, but the final stages of maturation require LH as well. Growing follicles produce *estrogen,* a female sex hormone that regulates development and function of the reproductive organs. Estrogens are mainly formed by the inner layer of the theca folliculi (Fig. 2–6), known as the *theca interna*.

Ovulation. Around midcycle (14 days), under the influence of FSH and LH, the mature ovarian follicle undergoes a sudden growth spurt, producing a *cystic swelling* or bulge on the surface of the ovary. A small oval avascular spot, the *stigma,* soon appears on this swelling (Fig. 2–10A). Prior to ovulation, the secondary oocyte and some cells of the cumulus oophorus detach from the interior of the distended follicle (Fig.

2–10B). At ovulation there is a surge of LH release that appears to cause the stigma to balloon out, forming a vesicle. The stigma then ruptures, expelling the secondary oocyte with the follicular fluid (Figs. 2–10D and 2–11). The oocyte is surrounded by the *zona pellucida* and one or more layers of follicular cells that quickly become radially arranged as the *corona radiata* (Figs. 2–1C and 2–10C).

The surge in the release of luteinizing hormone also seems to induce resumption of the first meiotic division of the primary oocyte. Hence, *mature ovarian follicles contain secondary oocytes* (Fig. 2–10).

A variable amount of intermenstrual pain, called *mittelschmerz* (Ger. *mittel,* mid + *schmerz,*

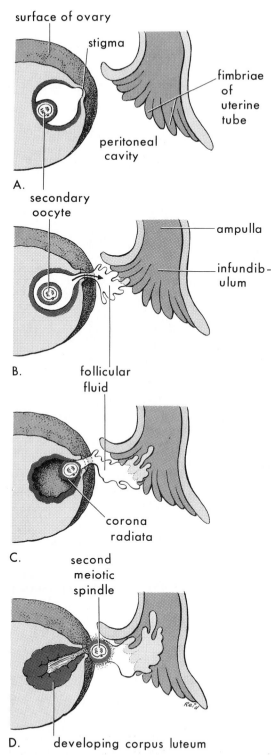

surface of ovary

stigma

fimbriae of uterine tube

peritoneal cavity

A.

secondary oocyte

ampulla

infundib-ulum

B.

follicular fluid

corona radiata

C.

second meiotic spindle

D. developing corpus luteum

Figure 2–10 Diagrams illustrating ovulation. The stigma ruptures and the secondary oocyte is expelled with the follicular fluid. The diameter of a secondary oocyte, freshly ovulated, is about 0.10 mm.

pain), accompanies ovulation in some women. This may be used as a sign of ovulation, but there are better ones; e.g., the basal body temperature usually shows a slight rise at ovulation.

Some women do not ovulate owing to an inadequate release of gonadotropins; as a result they are unable to become pregnant. In some of these patients *ovulation can be induced* by the administration of gonadotropins or by administration of an ovulatory agent (clomiphene citrate). This drug stimulates the release of pituitary gonadotropins (FSH and LH), which usually results in maturation of the ovarian follicle, ovulation, and development of the corpus luteum. The incidence of multiple pregnancy increases up to ten fold when ovulation is induced. Apparently, the fine control of FSH output is not present in these cases, and multiple ovulations occur, leading to multiple pregnancies and often spontaneous abortions (Hack et al., 1970).

The Corpus Luteum. Shortly after ovulation, the walls of the follicle and the theca folliculi collapse and are thrown into folds (Fig. 2–10D). Under LH influence, they develop into a glandular structure known as the corpus luteum (Fig. 2–6), which secretes mainly *progesterone,* but also produces some estrogen. These hormones, particularly progesterone, cause the endometrial glands to secrete and generally to prepare the endometrium for blastocyst implantation (see Fig. 3–1).

If the ovum is fertilized, the corpus luteum enlarges to form a *corpus luteum of pregnancy* (see Fig. 6–4B) and increases its hormone production. When pregnancy occurs, degeneration of the corpus luteum is prevented by *chorionic gonadotropin,* a hormone secreted by the trophoblast of the chorion (see Fig. 3–6).

If the ovum is not fertilized, the corpus lutuem begins to degenerate about 10 to 12 days after ovulation and is called a *corpus luteum of menstruation.* In each case the corpus luteum is subsequently transformed into a white scar called a *corpus albicans.*

THE MENSTRUAL CYCLE

The cyclic changes occurring in the endometrium constitute the *uterine (endometrial) cycle,* commonly referred to as the menstrual cycle because menstruation is an obvious event (Fig. 2–6).

The normal endometrium is a mirror of the ovarian cycle because it responds in a rather

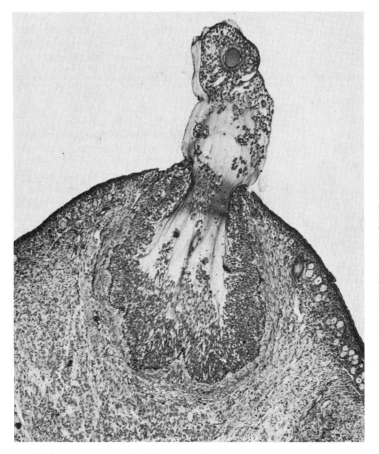

Figure 2–11 Photomicrograph of a section of a rabbit ovary taken just after rupture of a mature ovarian follicle during ovulation. The secondary oocyte, previously torn away from the cumulus oophorus (Fig. 2–10), has been carried with the gelatinous follicular fluid out of the follicle and the ovary into the peritoneal cavity. The follicular cells adhering to the secondary oocyte constitute the corona radiata. (From Page, E. W., Villee, C. A., and Villee, D. B.: *Human Reproduction. Essentials of Reproductive and Perinatal Medicine.* 3rd ed. Philadelphia, W. B. Saunders Co., 1981. Courtesy of Dr. Richard J. Blandau.)

precise manner to the fluctuating concentrations of ovarian steroids (Page et al., 1981).

The 28-day menstrual cycle shown in Figure 2–6 is by no means constant in length. In 90 per cent of healthy young women, the length of the endometrial cycles ranges between 23 and 35 days. Almost all these variations result from alterations in the duration of the proliferative phase (Page et al., 1981).

The typical cycles illustrated in Figure 2–6 are not always realized because the ovary may not produce a mature follicle. As a result, ovulation does not occur. In an *anovulatory cycle* the endometrial changes are minimal. The proliferative endometrium develops as usual; however, because there is no ovulation and no corpus luteum formation, the endometrium does not progress to the secretory phase but continues to be of the proliferative type until menstruation begins.

Anovulatory cycles may be produced by administering sex steroid hormones. The excess hormones act on the hypothalamus and hypophysis, resulting in an inhibition of secretion of the hypothalamic releasing factors and pituitary gonadotropic hormones essential for ovulation. Suppression of ovulation is the basis for the success of *birth control pills,* e.g., a combination of estrogen and progestin. These hormones act by suppression of the midcyle surge in LH, which normally causes ovulation. In 90 per cent of cases, the interval between cessation of oral contraception and the occurrence of pregnancy is 12 months, when no other method of contraception is used.

Ovarian hormones cause cyclic changes in the structure of the reproductive tract, notably the endometrium. Although divided into *three phases* for descriptive purposes (Fig. 2–6), it must be stressed that *the menstrual cycle is a continuous process;* each phase gradually passes into the next one.

The Menstrual Phase. The first day of menstruation is counted as the beginning of the menstrual cycle. The functional layer of the uterine wall is sloughed off and discarded during menstruation, which typically occurs at 28-day intervals and lasts for three to six days.

The Proliferative (Follicular) Phase.

This phase (days 6 to 14) coincides with the growth of ovarian follicles and is controlled by estrogens secreted by the theca interna surrounding the follicles. There is a two- to threefold increase in the thickness of the endometrium during this phase of repair and proliferation. Early during this phase, a continuous surface epithelium covers the endometrium; the glands increase in number and in length, and the spiral arteries elongate but do not reach the surface during this phase.

The Secretory (Luteal) Phase. This phase (days 15 to 28) coincides with the formation and growth of the corpus luteum. The progesterone secreted by the corpus luteum stimulates the glandular epithelium to produce a material rich in glycogen. The glands become wide, tortuous, and saccular, and the endometrium thickens, partly as a result of the increased fluid in the stroma. As the spiral arteries grow into the superficial compact layer, they become increasingly coiled. If the oocyte released at ovulation is fertilized, the blastocyst normally begins to implant in the endometrium on about the sixth day of the secretory phase (Fig. 2–17), or the twentieth day of the menstrual cycle (Fig. 2–6).

When fertilization does not occur, the secretory endometrium enters into an *ischemic (premenstrual) phase* during the last day or two of the menstrual cycle (Fig. 2–6). The ischemic phase is usually considered to be the last part of the secretory phase. The ischemia (localized deficiency of blood) gives the endometrium a pale appearance and occurs as the spiral arteries constrict intermittently. This intermittent constriction of spiral arteries results from the decreasing secretion of hormones by the degenerating corpus luteum. In addition to vascular changes, the hormone withdrawal results in a stoppage of glandular secretion, a loss of interstitial fluid, and a marked shrinking of the endometrium.

Toward the end of the ischemic part of the secretory phase, the spiral arteries become constricted for longer periods. Eventually, blood begins to seep through the ruptured walls of the spiral arteries into the surrounding stroma. Small pools of blood soon form and break through the endometrial surface, resulting in bleeding into the uterine lumen and the beginning of another menstrual phase.

As small pieces of the endometrium become detached and pass into the uterine cavity, the torn ends of the arteries bleed into the uterine cavity, resulting in an average loss of 35 ml of blood. Eventually, over three to six days, the entire compact layer and most of the spongy layer are discarded in the menstrual flow. The remnants of the spongy layer and the basal layer remain to undergo regeneration during the subsequent proliferative phase of the endometrium. Consequently, *the cyclic activity of the ovary is intimately linked with cyclic changes in the endometrium* (Fig. 2–6).

If pregnancy does not occur, the reproductive or menstrual cycles normally continue until the end of a woman's reproductive life, usually between the ages of 47 and 52. If pregnancy occurs, the menstrual cycles stop and the endometrium passes into a pregnancy phase. With the termination of pregnancy, the ovarian and menstrual cycles resume after a variable period of time (usually 6 to 10 weeks if the woman is not breast-feeding her baby).

GERM CELL TRANSPORT AND VIABILITY

Oocyte Transport. At ovulation, the secondary oocyte is carried in a stream of peritoneal fluid produced by the sweeping movements of the *fimbriae* of the uterine tube (Fig. 2–10A). These finger-like processes move back and forth over the ovary and "sweep" the secondary oocyte into the *infundibulum* of the tube. The oocyte then passes into the *ampulla* of the uterine tube, largely as a result of the beating action of cilia on some tubal epithelial cells, but partly by muscular contraction of the tubal wall. It has been estimated that it takes the oocyte about 25 minutes to reach the site of fertilization in the ampulla of the uterine tube (Fig. 2–18).

Sperm Transport. Usually 200 to 500 million sperms are deposited on the cervix and in the posterior fornix of the vagina at intercourse (Fig. 2–5A). The sperms pass by movements of their tails through the cervical canal, but passage of sperms through the uterus and uterine tubes appears to be assisted by muscular contractions of the walls of these organs. The *prostaglandins* present in the seminal plasma may stimulate uterine motility at the time of intercourse and assist in the movement of sperms through the uter-

Figure 2–12   Scanning electron micrograph of several human sperms. Each of these male germ cells consists of a head and a long tail. The head is formed principally by the nucleus, which contains the genetic traits that are transmitted by the male to the zygote during fertilization. The tail provides the motility that assists in the transport of the sperm to the fertilization site. (From Page, E. W., Villee, C. A., and Villee, D. B.: *Human Reproduction. Essentials of Reproductive and Perinatal Medicine.* 3rd ed. Philadelphia, W. B. Saunders Co., 1981. Courtesy of J. E. Flechon and E. S. E. Hafez.)

us and the uterine tubes to the site of fertilization (Page et al., 1981). It is not known how long it takes sperms to reach the fertilization site, but the time of transport is probably short. Settlage et al. (1973) found a few motile sperms in the ampulla of the uterine tube 5 minutes after their deposition near the uterine ostium (external os), but some sperms took up to 45 minutes to travel to the fertilization site. Only 300 to 500 sperms reach the fertilization site.

Sperm Counts.   During evaluation of fertility in a man, a clinical analysis of the semen is made. The average volume of the ejaculate is about 3.5 ml, and the sperms account for less than 10 per cent of this fluid. The rest consists of *seminal plasma* secreted by the ducts and accessory glands of the male reproductive tract.

In normal males there are usually more than 100 million sperms per ml of semen (Fig. 2–12). Although there is much variation in individual cases, men whose semen contains 20 million sperms per ml, or 50 million in the total specimen, are probably fertile. Men with less than 20 million sperms per ml of semen are likely to be sterile, especially if the specimen contains immotile and abnormal sperms (Fig. 2–4B).

The reduction in the number of sperms during passage through the female reproductive tract is mainly the result of filtering of abnormal and poorly motile sperms by the cervical mucus. Many sperms also invade the endometrial glands, where they die and are ingested by phagocytes.

Fertilization Site.   The usual site of fertilization is the *ampulla of the uterine tube,* its longest and widest part (Figs. 2–5A and 2–18). It is generally believed that fertilization cannot be delayed until the oocyte reaches the uterus, since it becomes overripe and undergoes degeneration. The final fate of unfertilized oocytes is dissolution in the uterus.

GERM CELL VIABILITY

Ova.   Studies on early stages of development indicate that the ovum is usually fertilized within 12 hours after expulsion of the secondary oocyte at ovulation, and observations have shown that in vitro the unfertilized human secondary oocyte dies within 12 to 24 hours.

Sperms.   Most sperms probably do not survive for more than 24 hours in the female genital tract. However, there is suggestive evidence that some sperms may be able to fertilize an ovum for as long as three days after insemination. Many sperms are stored in the crypts or aggregations of cervical glands and within the cervical mucus. They obtain oxygen by diffusion through the cervical plasma (Elstein, 1978). These sperms are gradually released for three or four days into the body of the uterus and pass into the uterine tubes. This storage of sperm ensures a constant release of sperms and thereby increases the chances of fertilization.

After being frozen to low temperatures, semen may be kept for years (Hancock, 1970; Friedman, 1977). Children have been born after artificial insemination of women with semen that had been stored for several years. Thus, frozen-storage of semen could result in a man, long-dead, fathering a child by means of artificial insemination.

Capacitation of Sperms. Before a mature motile sperm can penetrate the corona radiata and zona pellucida surrounding a second-

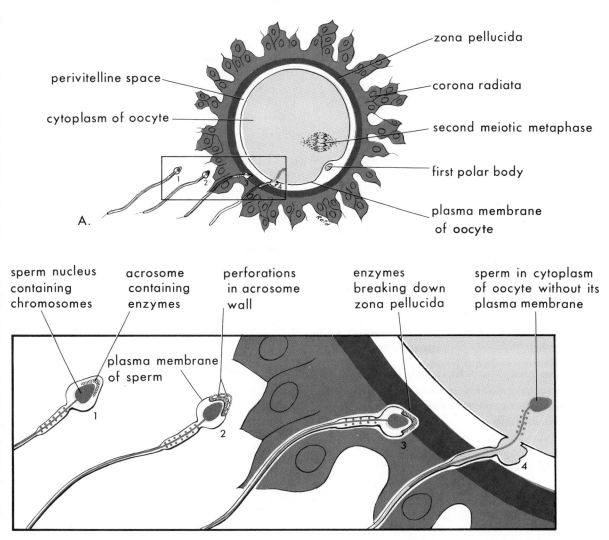

Figure 2-13 Diagrams illustrating the acrosome reaction and sperm penetration of an oocyte. The detail of the area outlined in *A* is given in *B. 1.* Sperm during capacitation. *2.* Sperm undergoing the acrosome reaction. *3,* Sperm digesting a path through the zona pellucida by the action of enzymes released from the acrosome. *4,* Sperm after entering the cytoplasm of the oocyte. Note that (1) the plasma membranes of the sperm and the oocyte have fused, and (2) the head and tail of the sperm enter the oocyte, leaving the sperm's plasma membrane attached to the oocyte's plasma membrane.

ary oocyte, it must undergo capacitation, an *activation process* that takes about seven hours. It is generally accepted that this process consists of enzymatic changes that result in the *removal of a glycoprotein coat* and seminal plasma proteins from the plasma membrane over the acrosome. No morphological changes are known to occur during the capacitation process.

Usually, sperms are capacitated in the uterus or the uterine tubes by substances in the secretions of the female genital tract. Follicular fluid is also known to have capacitating properties. During *in vitro fertilization*, capacitation is induced by artificial means using macromolecular components in the medium, such as gammaglobulin free serum, serum dialysate, follicular fluid, adrenal gland extracts, albumin, and dextran (Zaneveld, 1978). The process is enhanced when follicular fluid is added to the medium.

The Acrosome Reaction. An acrosome reaction may occur after capacitation of a sperm. This sequence of events, occurring during passage of the sperm through the corona radiata (Fig. 2–13*B*), consists of structural changes. The outer membrane of the acrosome fuses at many places with the overlying cell membrane of the sperm head, and the fused membranes then rupture, producing multiple perforations through which the enzymes in the acrosome escape. Progesterone seems to stimulate the acrosome reaction (Austin, 1975). It is present in large amounts in the follicular fluid released at ovulation (Figs. 2–10 and 2–11) and between the follicular cells of the corona radiata. Enzymes that are believed to facilitate passage of the sperm through the corona radiata and the zona pellucida are released from the acrosome during the acrosome reaction. *Hyaluronidase* enables the sperm to penetrate the corona radiata. *Trypsin-like substances* and a *zona lysin* digest a pathway for the sperm through the zona pellucida.

FERTILIZATION

Fertilization is the sequence of events that begins with contact between a sperm and a secondary oocyte, and ends with the fusion of the nuclei of the sperm and ovum and the intermingling of maternal and paternal chromosomes at the metaphase of the first mitotic division of the zygote (Fig. 2–14).

Embryonic development commences with fertilization; hence, the beginning of fertilization is the start of stage 1 of development. The fertilization process requires about 24 hours and occurs as follows (Fig. 2–13):

1. *The sperm passes through the corona radiata.* Dispersal of these follicular cells appears to result from the action of enzymes released from the acrosome, principally *hyaluronidase.* This enzyme is believed to effect removal of the corona radiata cells surrounding the secondary oocyte. Probably movements of the tail of the sperm also help it to penetrate the corona radiata.

2. *The sperm penetrates the zona pellucida,* digesting a path by the action of enzymes released from its acrosome. The enzyme *acrosin* appears to cause lysis of the zona pellucida, thereby forming a path for the sperm to follow. It has been shown experimentally that inhibition of acrosin prevents passage of sperms through the zona pellucida (Propping et al., 1978).

Once the first sperm passes through the zona pellucida, a *zona reaction* occurs that renders this layer impermeable to other sperms. Alterations in the physiochemical characteristics of the zona pellucida are associated with the zona reaction, but no morphological changes have been detected. The zona reaction is believed to be produced by cortical granules released from the secondary oocyte. They contain *lysosomal enzymes* that induce the zona reaction.

Although several sperms may penetrate the zona pellucida, usually only one sperm enters the ovum and fertilizes it. Two sperms may participate in fertilization during an abnormal process known as *dispermy;* this seems to be a relatively common occurrence in humans (Carr, 1971). The resulting triploid embryos (69 chromosomes) may appear quite normal, but they nearly always abort. A few triploid infants have been born, but they all died shortly after birth (Carr, 1971). There is some experimental evidence that *aged oocytes* do not release cortical granules after sperm penetration of the zona pellucida. As a result, the zona reaction does not take place, and multiple penetration of sperms occurs. *Polyspermy* is unlikely to produce a viable embryo.

3. *The sperm head attaches to the surface of the secondary oocyte;* the plasma membranes of the oocyte and sperm fuse and then break down at the point of contact. The head and tail of the sperm enter the cytoplasm of the oocyte, leaving the sperm's plasma membrane attached to the oocyte's plasma membrane.

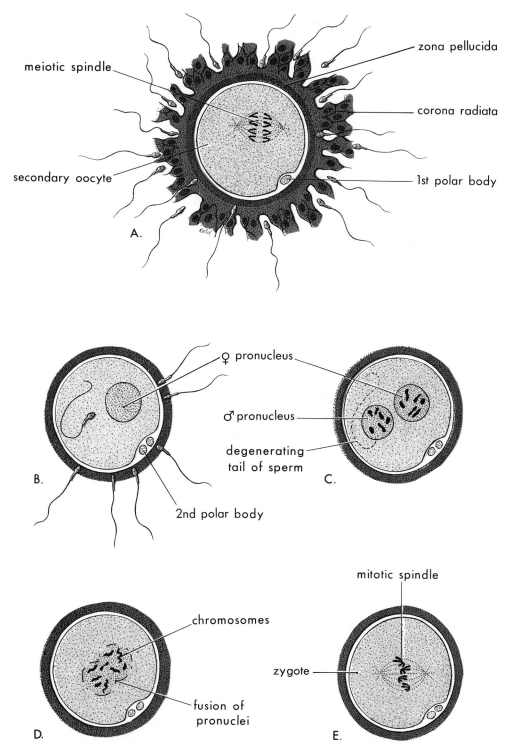

Figure 2–14 Diagrams illustrating fertilization (developmental stage 1), the procession of events beginning when the sperm contacts the secondary oocyte's plasma membrane and ending with the intermingling of maternal and paternal chromosomes at metaphase of the first mitotic division of the zygote. *A*, Secondary oocyte surrounded by several sperms. (Only four of the 23 chromosome pairs are shown.) *B*, The corona radiata has disappeared; a sperm has entered the oocyte, and the second meiotic division has occurred, forming a *mature ovum*. *C*, The sperm head has enlarged to form the male pronucleus. *D*, The pronuclei are fusing. *E*, The chromosomes of the zygote are arranged on a mitotic spindle in preparation for the first cleavage (mitotic) division.

4. *The secondary oocyte completes the second meiotic division,* forming a *mature ovum* and the second polar body (Fig. 2–14*B*). The nucleus of the ovum is known as the *female pronucleus.*

5. Once within the cytoplasm of the ovum, the tail of the sperm rapidly degenerates, and its head enlarges to form the *male pronucleus* (Fig. 2–14*C*).

6. *The male and female pronuclei approach each other* in the center of the ovum, where they come into contact and lose their nuclear membranes. Then the maternal and paternal chromosomes intermingle at metaphase of the first mitotic division of the zygote.

RESULTS OF FERTILIZATION

1. Restoration of Diploid Number. Fusion of the two haploid germ cells produces a zygote, a diploid cell with 46 chromosomes, the usual number for the human being.

Meiosis is the special form of cell division that results in the chromosome number being reduced by half. Thus, meiosis provides for the constancy of chromosome number from generation to generation by producing haploid germ cells.

2. Species Variation. Because half the chromosomes come from the mother and half from the father, *the zygote contains a new combination of chromosomes* that is different from those of the parents. This mechanism forms the basis of biparental inheritance and results in variation of the human species.

Meiosis allows independent assortment of maternal and paternal chromosomes among the germ cells. Crossing over of chromosomes, by relocating segments of the maternal and paternal chromosomes, serves to "shuffle" the genes, thereby producing a recombination of genetic material.

3. Sex Determination. The embryo's sex is determined at fertilization by the kind of sperm that fertilizes the ovum. Hence, *it is the father rather than the mother whose gamete determines the sex of their offspring.* Fertilization by an X-bearing sperm produces an XX zygote, which normally develops into a female, whereas fertilization by a Y-sperm produces an XY zygote, which normally develops into a male.

As discussed in Chapter 13, sexual development is a complex process that is suscepti-

ble to alteration by environmental factors, particularly sex hormones. However, the *chromosomal sex* established at fertilization is usually the same as the *phenotypic sex* that develops during the fetal period.

Control of the Sex of the Embryo. Because the sex of the embryo is determined by whether the sperm contributes an X or a Y chromosome to the zygote, and because X and Y sperms are formed in equal numbers, the expectation is that the sex ratio at fertilization *(primary sex ratio)* should be 1.00 (100 boys per 100 girls). However, there are more male babies than female babies born in all countries. In North America, for example, the sex ratio at birth *(secondary sex ratio)* is about 1.05 (105 boys per 100 girls).

The secondary sex ratio could be altered if parents were to request termination of pregnancies in which the fetus is not of the sex that they prefer. Although it is technically possible to determine the sex of the unborn fetus by examining cells in the amniotic fluid (see Chapter 6 and Figure 6–13), most physicians would not perform *amniocentesis* and abortion for the sole purpose of controlling the sex of the embryo.

Various *in vitro techniques* have been developed in an attempt to separate X- and Y- bearing sperms using (1) the differential swimming abilities of the two types of sperm, (2) different speeds of migration in an electric field, and (3) microscopic differences in the X and Y sperms.

Others claim that the timing and management of sexual intercourse can enable a couple to choose the sex of their child. However, no method of controlling the human embryo's sex has been shown to change the sex ratio consistently.

Most scientists are unconvinced that any of the current methods of controlling the embryo's sex are of any value in altering the human sex ratio.

4. Initiation of Cleavage. Fertilization initiates development by stimulating the zygote to undergo a series of rapid mitotic cell divisions called *cleavage.*

Cleaveage of a secondary oocyte may occur without fertilization as part of a process known as *parthenogenesis;* this may occur naturally or may be artifically induced.

Parthenogenesis is a normal event in some species. For example, some eggs laid by a queen bee are not fertilized but develop parthenogenetically into haploid males. In a few other species (e.g., rabbits), an unfertilized secondary oocyte can be experimentally induced to undergo parthenogenetic development. *No verified case of parthenogenesis has been reported in humans.* In a suspected case, an exhaustive battery of blood grouping, histocompatibility, and other tests would be required to provide proof. (For more information, see Karp, 1976.)

There is evidence that the human oocyte may start to undergo parthenogenetic cleavage, but this does not result in organized development (Beatty, 1957). It has been suggested that on rare occasions an oocyte in an ovarian follicle may undergo parthenogenetic cleavage and give rise to an ovarian teratoma (Simard, 1957).

Cloning in humans has not been reported in the scientific literature, but a clone of African clawed toads has been produced (Gurdon, 1962). A *clone* is a group of identical cells derived from one cell by fission (as in bacteria) or by mitotic cell division (as in cleavage). Experimental clonal reproduction resulting in live-born normal offspring has not been reported in the scientific literature. The possibility of experimental cloning in humans has been discussed widely since Rorvik (1978) wrote about the subject. In his supposedly factual story, the nucleus of a secondary oocyte was inactivated by an unspecified technique, and the nucleus from a somatic cell of an adult man was inserted into the oocyte. An embryo, or clone, that supposedly developed from the renucleated oocyte was transplanted to the uterus of a female farm laborer. *Rorvik's story is not accepted by the majority of the scientific community.*

CLEAVAGE

As the zygote, a single cell, passes down the uterine tube, it undergoes mitotic cell divisions known as *cleavage.* This phase of development begins with the first mitotic division of the zygote and ends with formation of the blastocyst (Fig. 2–15).

Division of the zygote into two daughter cells called *blastomeres* occurs during stage 2 of development (Fig. 2–15A), about 30 hours after fertilization. Subsequent divisions follow one another, forming progressively smaller blastomeres (Fig. 2–15D). The *morula* (L. *morus,* mulberry), a solid ball of 12 to 16 blastomeres, forms about three days after fertilization. The morula enters the uterus as it is forming.

If *nondisjunction* (failure of two members of a chromosome pair to disjoin during anaphase of cell division) happens at an *early cleavage* division of a zygote, an individual with two or more cell lines with different chromosome numbers is produced. Such individuals in whom *mosaicism* is present are termed *mosaics.* About 1 per cent of patients with *Down syndrome* (Fig. 8–4) are mosaics with a mixture of 46-chromosome and 47-chromosome tissues. These persons have relatively mild stigmata of the Down syndrome, and they have a higher risk than normal persons of having children with this syndrome if the mosaicism affects the germ cells.

BLASTOCYST FORMATION

During stage 3 of development (about four days), cavities appear inside the compact mass of cells forming the morula, and fluid soon passes into these cavities from the uterine cavity. As the fluid increases, it separates the cells into two parts: (1) an outer cell layer, the *trophoblast* (Gr. *trophe,* nutrition), which gives rise to part of the placenta, and (2) a group of centrally located cells, known as the *inner cell mass* (or embryoblast), which gives rise to the embryo. The fluid-filled spaces soon fuse to form a single large space, known as the *blastocyst cavity,* or *blastocoele.* At this stage of development, the conceptus is called a *blastocyst* (Figs. 2–15E and 2–16A). The inner cell mass now projects into the blastocyst cavity, and the trophoblast forms the wall of the blastocyst (Figs. 2–15F and 2–16B). The blastocyst lies free in the uterine secretions for about two days, then the *zona pellucida* degenerates and disappears (Figs. 2–15E and 2–16A).

During stage 4 of development (5 to 6 days), the blastocyst attaches to the endometrial epithelium (Fig. 2–17A). As soon as the trophoblast attaches to this epithelium, it starts proliferating rapidly and gradually differentiates into two layers: (1) an inner *cytotrophoblast* (cellular trophoblast), and (2) an outer *syncytiotrophoblast* (syntrophoblast, syncytial trophoblast) consisting of a multinucleated protoplasmic mass in which cell boundaries disappear and the mass of trophoblast becomes a syncytium (Fig. 2–17B). The finger-like processes of the syncytiotrophoblast grow into the endometrial epithelium and start to invade the endometrial stroma. By the end of the first week, the blastocyst is superficially implanted in the compact layer of the endometrium.

During stage 5 of development (7 to 12 days), as the blastocyst is implanting, early differentiation of the inner cell mass occurs. A flattened layer of cells, the *hypoblast* (primitive endoderm), appears on the surface of the inner cell mass facing the blastocyst cavity at about seven days (Fig. 2–17B). Recent evidence indicates that the hypoblast is probably displaced to extraembryonic regions (see Chapter 4 and Fig. 4–1).

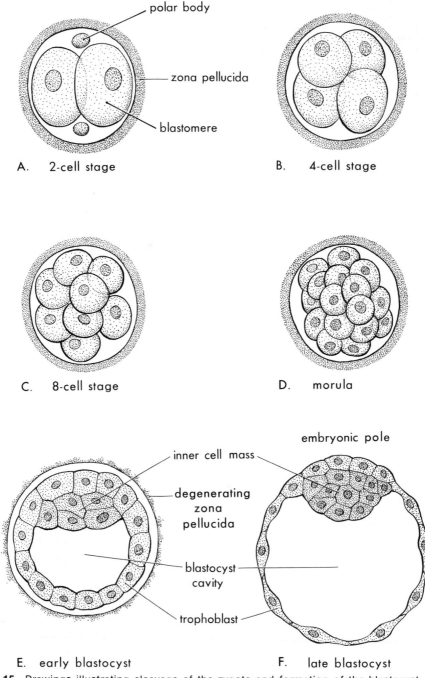

Figure 2–15 Drawings illustrating cleavage of the zygote and formation of the blastocyst. *A* to *D* show various stages of cleavage (developmental stage 2). The period of the morula begins at the 12- to 16-cell stage and ends when the blastocyst forms, which occurs when there are 50 to 60 blastomeres present. *E* and *F* are sections of blastocysts (developmental stage 3). The zona pellucida has disappeared by the late blastocyst stage (five days). The polar bodies shown in *A* are small, nonfunctional cells that soon degenerate.

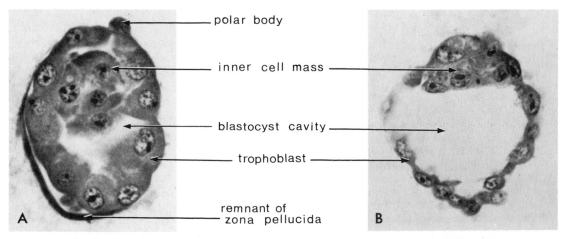

polar body

inner cell mass

blastocyst cavity

trophoblast

remnant of zona pellucida

A

B

Figure 2–16 Photomicrographs of sections of human blastocysts (developmental stage 3) recovered from the uterine cavity (×600). *A,* Four days; the blastocyst cavity is just beginning to form and the zona pellucida is deficient over part of the blastocyst. *B,* Four and a half days; the blastocyst cavity has enlarged and the inner cell mass and trophoblast are clearly defined. The zona pellucida has disappeared. (From Hertig, A. T., Rock, J., and Adams, E. C.: *Am. J. Anat. 98:*435, 1956. Courtesy of Carnegie Institution of Washington.)

Figure 2–17 Drawings of sections illustrating early stages of implantation. *A,* Six days; the trophoblast is attached to the endometrial epithelium at the embryonic pole of the blastocyst. *B,* Seven days; the syncytiotrophoblast has penetrated the epithelium and has started to invade the endometrial stroma.

Some students have difficulty interpreting illustrations such as these, because in histological studies it is conventional to draw the endometrial epithelium upward, whereas in embryological studies the embryo is usually shown with its dorsal surface upward. Because the embryo implants on its future dorsal surface, it would appear upside-down if the histological convention were followed. In this book, the histological convention is followed when the endometrium is the dominant consideration (e.g., Fig. 2–5*B*), and the embryological convention is used when the embryo is the center of interest, as in these illustrations.

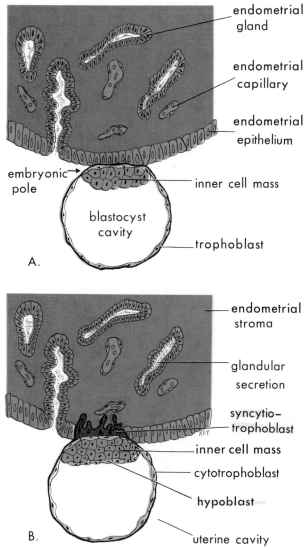

endometrial gland

endometrial capillary

endometrial epithelium

embryonic pole

inner cell mass

blastocyst cavity

trophoblast

A.

endometrial stroma

glandular secretion

syncytio-trophoblast

inner cell mass

cytotrophoblast

hypoblast

uterine cavity

B.

35

In vitro fertilization of human ova and cleavage of the zygote have been achieved by several investigators (Edwards, 1974; Steptoe and Edwards, 1976; Elliott, 1979; Grobstein, 1979). Most studies have been carried out on infertile women with occluded uterine tubes, with the intention of establishing pregnancy by transferring a morula (cultured in vitro) into the uterus. This technique offers hope to some infertile women who wish to have children. Synchronizing the endometrium and the cleavage stage of the zygote appears to present the major problem with this technique, because in vitro development is 20 to 30 per cent retarded as compared with normal in vivo development.

Abnormal Zygotes and Spontaneous Abortion. About 15 per cent of all zygotes result in detectable spontaneous abortion, but this estimate is undoubtedly low because the loss of zygotes during the first week is thought to be high. The actual rate is unknown because the women do not know they are pregnant at this early stage. Clinicians frequently have a patient who states that her last menstrual period was delayed by one or two weeks and that then her menstrual flow was unusually profuse. Very likely, such a patient has had an early spontaneous abortion.

Early abortion occurs for a variety of reasons, one being the presence of chromosomal abnormalities in the zygote. Carr and Gedeon (1977) estimate that about half of all known spontaneous abortions occur because of chromosomal abnormalities. Hertig et al. (1959), while examining specimens recovered from early pregnancies, found several clearly defective zygotes, some so abnormal that survival would not have been likely. This early loss of zygotes, once called *pregnancy wastage,* appears to represent a disposal of abnormal conceptuses that could not have developed normally, i.e., a *natural prenatal screening* of embryos. Without this screening, about 12 per cent instead of 2 to 3 per cent of infants would be congenitally malformed (Warkany, 1981).

SUMMARY OF FIRST WEEK

Some sperms deposited in the vagina pass through the cervical canal, the uterine cavity, and along the uterine tube to its ampulla, where fertilization usually occurs. When the secondary oocyte is contacted by a sperm, it completes the second meiotic division. As a result, a *mature ovum* and a second polar body are formed. The nucleus of the mature ovum constitutes the *female pronucleus.*

After the sperm enters the ovum's cyto-

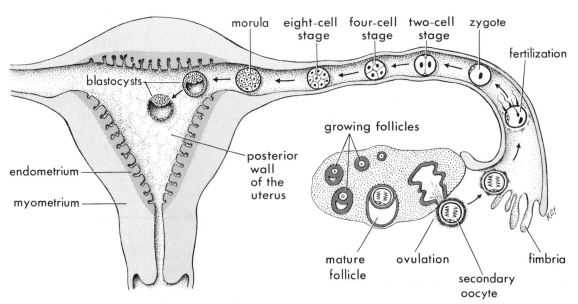

Figure 2–18 Diagrammatic summary of the ovarian cycle, fertilization, and human development during the first week. Developmental stage 1 begins with fertilization and ends when the zygote forms. Stage 2 (days 2 to 3) comprises the early stages of cleavage (from 2 to about 16 cells or the morula). Stage 3 (days 4 to 5) consists of the free unattached blastocyst. Stage 4 (days 5 to 6) is represented by the blastocyst attaching to the center of the posterior wall of the uterus, the usual site of implantation.

plasm, the head of the sperm separates from the tail and enlarges to become the *male pronucleus*. Fertilization is complete when the maternal and paternal chromosomes intermingle during metaphase of the first mitotic division of the *zygote,* the cell that gives rise to a human being.

As it passes along the uterine tube, the zygote undergoes *cleavage* (a series of mitotic divisions) into a number of small cells called *blastomeres* (Fig. 2–18). About three days after fertilization, a ball of 12 to 16 blastomeres, called the *morula,* enters the uterus. A cavity soon forms in the morula, converting it into a *blastocyst* consisting of (1) an *inner cell mass,* or embryoblast, which gives rise to the embryo, (2) a *blastocyst cavity,* or blastocoele, and (3) an outer layer of cells, the *trophoblast,* which encloses the inner cell mass and blastocyst cavity.

The zona pellucida disappears (days 4 to 5) and the blastocyst attaches to the endometrial epithelium (days 5 to 6). The trophoblastic cells then invade the epithelium and underlying endometrial stroma. Concurrently, the *hypoblast* (primitive endoderm) begins to form on the ventral surface of the inner cell mass. By the end of the first week, the blastocyst is superficially implanted in the endometrium.

At least 15 per cent of zygotes are lost during the first week of development, commonly as the result of chromosomal abnormalities.

SUGGESTIONS FOR ADDITIONAL READING

Allen, R. C.: The moment of fertilization. *Sci. Am. 201*:124, 1959. A well-written article with a particularly good discussion of fertilization by an expert in the field.

Edwards, R. G., and Fowler, R. E.: Human embryos in the laboratory. *Sci. Am. 233*:44, 1970. A description of techniques and an explanation of the way ova are fertilized in vitro.

Page, E. W., Villee, C. A., and Villee, D. B.: *Human Reproduction. Essentials of Reproductive and Perinatal Medicine.* 3rd ed. Philadelphia, W. B. Saunders Co., 1981. A good review of the reproductive processes, including a good account of the hormonal control of reproduction. *A clinical textbook.*

Shettles, L. B.: Fertilization and early development from the inner cell mass; *in* E. E. Philipp, J. Barnes, and M. Newton (Eds.): *Scientific Foundations of Obstetrics and Gynecology.* London, William Heinemann Ltd., 1970, pp. 132–158. A pioneer in the study of early embryos, Dr. Shettles tells how he obtains human oocytes at ovulation, fertilizes them in a Petri dish, and observes cleavage of the zygote.

CLINICALLY ORIENTED PROBLEMS

1. What is the main cause of numerical aberrations of chromosomes? Define this process. What is a common result of this defect?
2. All blastomeres of a morula were found to have an extra set of chromosomes. Explain how this could happen.
3. In about 50 per cent of *infertile couples,* the difficulty is attributable to some factor in the woman. What do you think would be a major source of female infertility?
4. Some people have a mixture of 46- and 47-chromosome cells (e.g., some Down syndrome patients are *mosaics*). *Read about mosaicism in Chapter 8.* How do mosaics form? Would children with mosaicism and the Down syndrome have the same stigmata as other infants with this condition? Would mosaics have a higher risk than normal people of producing children with chromosomal aberrations?

The answers to these questions are given at the back of the book.

REFERENCES

Arey, L. B.: *Developmental Anatomy: A Textbook and Laboratory Manual of Embryology.* Revised 7th ed. Philadelphia, W. B. Saunders Co., 1974, pp. 150–166.

Austin, C. R.: The egg and fertilization. *Sci. J.* (London) 6:37, 1970.

Austin, C. R.: Membrane fusion events in fertilization. *J. Reprod. Fertil. 44*:155, 1975.

Beatty, R. A.: *Parthenogenesis and Polyploidy in Mam-*

malian Development. Cambridge, Cambridge University Press, 1957.

Benirschke, K.: Gametogenesis and fertilization; *in* D. E. Reid, K. J. Ryan, and K. Benirschke: *Principles and Management of Human Reproduction*. Philadelphia, W. B. Saunders Co., 1972, pp. 166–179.

Biggers, J. D.: New observations on the nutrition of the mammalian oocyte and the preimplantation embryo; *in* R. J. Blandau (Ed.): *The Biology of the Blastocyst*. Chicago, University of Chicago Press, 1971.

Blandau, R. H. W, White, B. J., and Rumery, R. E.: Observations on the movements of the living primordial germ cells in the mouse. *Fertil. Steril. 14*:482, 1963.

Bloom, W., and Fawcett, D. W.: *A Textbook of Histology*. 10th ed. Philadelphia, W. B. Saunders Co., 1975, pp. 858–906.

Böving, B.: Anatomy of reproduction; *in* J. P. Greenhill (Ed.): *Obstetrics*. 14th ed. Philadelphia, W. B. Saunders Co., 1969.

Brackett, B. G., Seitz, H. M., Rocha, G., and Mastroianni, L.: The mammalian fertilization process; *in* K. S. Moghissi and E. S. E. Hafez (Eds.): *Biology of Mammalian Fertilization and Implantation*. Springfield, Charles C Thomas, Publisher, 1972, pp. 165–184.

Brown, R. L.: Rate of transport of sperma in the human uterus and the tubes. *Am. J. Obstet. Gynecol. 47*:153, 1955.

Carr, D. H.: Chromosome studies in spontaneous abortions. *Obstet. Gynecol. 26*:308, 1965.

Carr, D. H.: Chromosome studies on selected spontaneous abortions: Polyploidy in man. *J. Med. Genet. 8*:164, 1971.

Carr, D. H., and Gedeon, M.: Population cytogenetics of human abortuses; *in* E. B. Hook and I. H. Porter (Eds): *Population Cytogenetics: Studies in Humans*. New York, Academic Press, 1977.

Clermont, Y.: The cycle of the seminiferous epithelium of man. *Am. J. Anat. 112*:35, 1963.

Clermont, Y., and Trott, M.: Kinetics of spermatogenesis in mammals: seminiferous epithelium cycle and spermatogonial renewal. *Physiol. Rev. 52*:198, 1972.

Edwards, R. G.: Fertilization of human eggs in vitro: morals, ethics, and the law. *Q. Rev. Biol. 49*:3, 1974.

Edwards, R. G.: Advances in reproductive biology and their implications for studies on human congenital defects; *in* A. G. Motulsky and W. Lenz (Eds.): *Birth Defects*. Amsterdam, Excerpta Medica, 1974.

Ehrig, G., Mettler, L., and Rüping, L.: Mice in vitro fertilization model as prerequisite for the testing of possible factors interfering with human fertilization; *in* H. Ludwig and P. F. Tauber (Eds.): *Human Fertilization*. Stuttgart, Georg Thieme Publishers, 1978.

Elliott, J.: Finally, some details on *in vitro* fertilization. *J.A.M.A. 241*:9, 1979.

Elstein, M.: Cervix and cervical barrier; *in* H. Ludwig and P. F. Tauber (Eds.): *Human Fertilization*. Stuttgart, Georg Thieme Publishers, 1978.

Ferguson-Smith, M. A.: Sex chromatin, Klinefelter's syndrome and mental deficiency; *in* K. L. Moore (Ed.): *The Sex Chromatin*. Philadelphia, W. B. Saunders Co., 1966, pp. 277–315.

Friedman, S.: Artificial donor insemination with frozen human semen. *Fertil. and Steril. 28*:1230, 1977.

Gaddum-Rosse, P., and Blandau, R. J.: Comparative

observation on ciliary currents in mammalian oviducts. *Biol. Reprod. 14*:605, 1976.

Gamzell, C.: Ovulation; *in* E. E. Philipp, J. Barnes, and M. Newton (Eds.): *Scientific Foundations of Obstetrics and Gynecology*. London, William Heinemann, Ltd., 1970, pp. 131–134.

Grobstein, C.: External human fertilization. *Sci. Am. 240*:59, 1979.

Gurdon, J. B.: Multiple genetically identical toads. *J. Hered. 53*:4, 1962.

Hack, M., Brish, M., Serr, D. M., Insler, V., and Lunenfeld, B.: Outcome of pregnancy after induced ovulation: Follow-up of pregnancies and children born after gonadotropin therapy. *J.A.M.A. 211*:791, 1970.

Hafez, E. S.: *Human Ovulation. Mechanisms, Prediction, Detection and Induction*. Amsterdam, North Holland Publishing Co., 1979.

Hancock, J. L.: The sperm cell. *Sci. J.* (London) 6:31, 1970.

Hartman, C. J.: How do sperms get into the uterus? *Fertil. Steril. 8*:403, 1957.

Hartree, E. F.: Spermatozoa, eggs, and proteinases. *Biochem. Soc. Trans. 5*:375, 1977.

Hertig, A. T., Adams, E. C., and Mulligan, W. J.: On the preimplantation stages of the human ovum: A description of four normal and four abnormal specimens ranging from the second to the fifth day of development. *Contrib. Embryol. Carnegie Inst. 35*:199, 1954.

Hertig, A. T., and Rock, J.: Two human ova of the previllous stage, having a developmental age of about seven and nine days respectively. *Contrib. Embryol. Carnegie Inst. 31*:65, 1945.

Hertig, A. T., Rock, J., and Adams, E. C.: A description of 34 human ova within the first seventeen days of development. *Am. J. Anat. 98*:435, 1956.

Hertig, A. T., Rock, J., Adams, E. C., and Menkin, M. C.: Thirty-four fertilized human ova, good, bad, and indifferent, recovered from 210 women of known fertility. *Pediatrics 23*:202, 1959.

Hotchkiss, R. S. H.: *Fertility in Men*. Philadelphia, J. B. Lippincott Co., 1944.

Kaplan, S. A.: Pituitary disorders; *in* L. I. Gardner (Ed.): *Endocrine and Genetic Diseases of Childhood and Adolescence*. 2nd ed. Philadelphia, W. B. Saunders Co., 1975, pp. 107–108.

Karp, L. E.: *Genetic Engineering: Threat or Promise*. Chicago, Nelson-Hall, 1976.

Klopper, A.: The reproductive hormones. *Sci. J.* (London) 6:44, 1970.

Leeson, C. R., and Leeson, T. S.: *Histology*. 4th ed. Philadelphia, W. B. Saunders Co., 1981, pp. 484–573.

Ludwig, H., and Tauber, P. F. (Eds.); *Human Fertilization*. Stuttgart, Georg Thieme Publishers, 1978.

Mastroianni, L.: The tube; *in* E. E. Philipp, J. Barnes, and M. Newton (Eds.): *Scientific Foundations of Obstetrics and Gynecology*. London, William Heinemann, Ltd., 1970, pp. 81–87.

Morriss, G.: Growing embryos in vitro. *Nature 278:*402, 1979.

O'Rahilly, R.: *Developmental Stages in Human Embryos. Part A. Embryos of the First Three Weeks (Stages 1 to 9)*. Washington, D.C., Carnegie Institution of Washington, 1973, pp. 9–31.

Propping, D., Tauber, P. F., and Zaneveld, L. J. D.: Fertilization and implantation; *in* H. Ludwig and P. F.

Tauber (Eds.): *Human Fertilization*. Stuttgart, Georg Thieme Publishers, 1978.

Rafferty, K. A., Jr.: The beginning of development; *in* A. C. Barnes (Ed.): *Intra-Uterine Development*. Philadelphia, Lea & Febiger, 1968, pp. 1–25.

Reed, M.: Hypothalamic releasing factors; *in* E. E. Philipp, J. Barnes, and M. Newton (Eds.): *Scientific Foundations of Obstetrics and Gynecology*. London, William Heinemann, Ltd., 1970, pp. 489–495.

Rock, J., and Hertig, A. T.: The human conceptus during the first two weeks of gestation. *Am. J. Obstet. Gynecol. 55:*6, 1948.

Rorvik, D. M.: *In His Image: The Cloning of Man*. Philadelphia, J. B. Lippincott Company, 1978.

Settlage, D. S. F., Motoshima, M., and Tredway, D. R.: Sperm transport from the external cervical os to the Fallopian tubes in women. *Fertil. Steril. 24:* 655, 1973.

Shettles, L. B.: *Ovum Humanum*. New York, Hafner Publishing Co., Inc., 1960.

Simard, L. C.: Polyembryonic embryos of the ovary of parthenogenetic origin. *Cancer 10:*215, 1957.

Smith, D. W., Bierman, E. L., and Robinson, N. M.: *The Biologic Ages of Man: From Conception Through Old Age*. 2nd ed. Philadelphia, W. B. Saunders Co., 1978.

Steptoe, P. C., and Edwards, R. G.: Reimplantation of a human embryo with subsequent tubal pregnancy. *Lancet 79:*880, 1976.

Thompson, J. S., and Thompson, M. W.: *Medical Genetics*. 3rd ed. Philadelphia, W. B. Saunders Co., 1980, pp. 1–31.

Van Blerkom, J., and Motta, P.: *The Cellular Basis of Mammalian Reproduction*. Baltimore, Urban and Schwarzenberg, 1979.

Warkany, J.: Prevention of congenital malformations. *Teratology 23:*175, 1981.

Yanagimachi, R.: Specificity of sperm-egg interaction; *in* M. Edidin and M. M. Johnson (Eds.): *The Immunobiology of the Gametes*. Cambridge, England, Cambridge University Press, 1977.

Yen, S. S. C., Rankin, J., and Littel, A. S.: Hormonal relationships in the menstrual cycle. *J.A.M.A. 211:* 1513, 1970.

Zaneveld, L. J. D.: Capacitation of spermatozoa; *in* H. Ludwig and P. F. Tauber (Eds.): *Human Fertilization*. Stuttgart, Georg Thieme Publishers, 1978.

3

FORMATION OF THE BILAMINAR EMBRYO
The Second Week

As implantation of the blastocyst continues during the second week, morphological changes occur in the inner cell mass that produce a *bilaminar embryonic disc* composed of epiblast and hypoblast (Fig. 3–1*A*). As described in Chapter 4, the *epiblast* gives rise to all three germ layers of the embryo (ectoderm, mesoderm, and endoderm). The *hypoblast* represents the primitive embryonic endoderm, which is probably displaced to extraembryonic regions. It is now believed that most or all embryonic endoderm is derived from the epiblast (see Fig. 4–1). As the bilaminar embryonic disc forms, the *amniotic cavity,* the *yolk sac,* the *connecting stalk,* and the *chorion* develop.

Crowley (1974) refers to the second week of development as "the period of twos" because *two embryonic layers* form (epiblast and hypoblast) from the inner cell mass; *two cavities* develop (amniotic cavity and primary yolk sac); and *two layers of trophoblast* differentiate (cytotrophoblast and syncytiotrophoblast).

IMPLANTATION

STAGE 5 (7 TO 12 DAYS)*

As it would be a rare occurrence for a student to require the details of embryonic staging without being able to consult a reference book, it is sufficient for him or her to remember general statements (e.g., *implantation of the blastocyst is the main characteristic of stage 5*), and to refer to this book, or a similar one, for details if and when they are required.

Stage 5 comprises embryos that are implanted to a varying degree and are previllous, that is, chorionic villi have not started to develop. *Implantation of the blastocyst is the main characteristic of stage 5.* Invasion of the endometrium begins on day 7 (Fig. 2–17*B*) and the trophoblast differentiates into two types: cytotrophoblast and syncytiotrophoblast. Both embryonic and maternal tissues are involved in implantation.

The actively erosive *syncytiotrophoblast* invades the endometrial stroma containing capillaries and glands, and the blastocyst slowly sinks into the endometrium. Hertig (1968) describes the syncytiotrophoblast as "invasive, ingestive, and digestive." The stromal cells around the implantation site become laden with glycogen and lipids, and they become polyhedral in appearance. Some of these cells, called *decidual cells* (Fig. 7–1*B*), degenerate in the region of the penetrating syncytiotrophoblast and provide a rich source of material for embryonic nutrition. Later, the conceptus will receive nutrients directly from the maternal blood.

As more trophoblast contacts the endometrium, it proliferates and differentiates into two layers (Fig. 3–1*A*): (1) the *cytotrophoblast,* which is mitotically active and forms new cells that migrate into the increasing mass of syncytiotrophoblast; and (2) the *syncytiotrophoblast* at the embryonic pole (adjacent to the developing embryo), which rapidly becomes a large, thick, multinucleated protoplasmic mass in which no cell boundaries are discernible (Fig. 3–1*B*). The cells of the cytotrophoblast divide, and many of them migrate into the syncytiotrophoblast, where they fuse and lose their cell membranes to form a syncytium. *Mitoses can*

---

*Although early development is a continuous process, it is helpful to divide human embryonic development into stages for comparison of specimens in different laboratories and for utilization in teratological studies (descriptive and experimental work on abnormal embryos). The embryonic "stages are based on the apparent morphological state of development, and hence are not directly dependent on either chronological age or on size" (O'Rahilly, 1973). Stages 1 to 4 and the beginning of stage 5 occur during the first week.

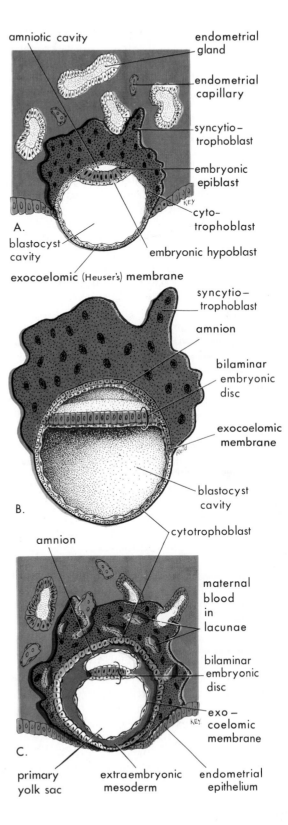

Figure 3–1 Drawings illustrating the implantation of a blastocyst into the endometrium (developmental stage 5). Actual size of the conceptus is about 0.1 mm. *A*, Drawing of a section through a blastocyst partially implanted in the endometrium (about 8 days). Note the slit-like amniotic cavity. *B*, An enlarged three-dimensional sketch of a slightly older blastocyst after removal from the endometrium. Note the extensive syncytiotrophoblast at the embryonic pole and the much larger amniotic cavity. *C*, Drawing of a section through a blastocyst of about 9 days implanted in the endometrium. (Based on Hertig and Rock, 1945). Note the spaces or lacunae appearing in the syncytiotrophoblast; These soon begin to communicate with the endometrial vessels.

be seen in the cytotrophoblast, but never in the syncytiotrophoblast.

While this trophoblastic development occurs, small spaces appear between the inner cell mass and the invading trophoblast. By day 8, these spaces have coalesced to form a slit-like *amniotic cavity* (Fig. 3–1*A*). Concurrently, morphological changes occur in the inner cell mass resulting in the formation of a flattened, essentially circular plate of cells called the *embryonic disc.* It consists of two layers: (1) the *epiblast,* consisting of high columnar cells related to the amniotic cavity; and (2) the *hypoblast,* consisting of cuboidal cells adjacent to the blastocyst cavity. The epiblast is a thick layer that is particularly active mitotically. *The epiblast gives rise to all or nearly all the cells of the embryo.* Most of or all the cells of the hypoblast are displaced laterally, where they contribute to the formation of the extraembryonic membranes.

As the amniotic cavity enlarges, it acquires a thin epithelial roof called the *amnion.* The cells forming the amnion, called *amnioblasts,* probably arise from cytotrophoblastic cells. The embryonic epiblast forms the floor of the amniotic cavity and is continuous peripherally with the amnion (Fig. 3–1*B*).

Concurrently, other cells delaminate from the cytotrophoblast and form a thin *exocoelomic membrane* (Fig. 3–1*B*). This membrane is continuous with the hypoblast of the embryonic disc and circumscribes a large cavity called the exocoelomic cavity, or the *primary (primitive) yolk sac.* Further delamination of trophoblastic cells gives rise to a layer of loosely arranged cells, the *extraembryonic mesoderm,* around the amnion and the primary yolk sac (Fig. 3–1*C*).

Isolated spaces, or cavities, called *lacunae* appear in the syncytiotrophoblast around day 9 (Fig. 3–1*C*) and soon become filled with maternal blood from ruptured capillaries and secretions from eroded endometrial glands. This nutritive fluid, called *embryotroph,* passes to the embryonic disc (future embryo) by diffusion.

The joining of uterine vessels with the lacunae in the syncytiotrophoblast represents the beginning of the uteroplacental circulation. When maternal blood flows into the *syncytiotrophoblastic lacunae,* its nutritive substances become available to the

embryonic tissues over the very large surface of the syncytiotrophoblast. As both arterial and venous branches of the maternal blood vessels come into communication with the syncytiotrophoblastic lacunae, blood

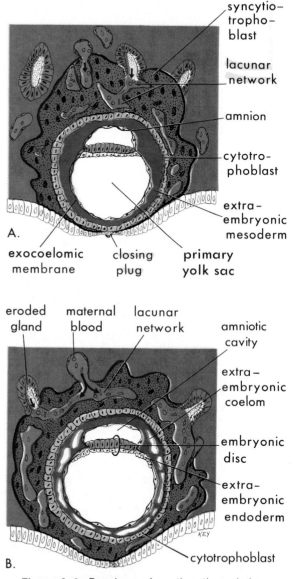

Figure 3–2 Drawings of section through implanted blastocysts (developmental stage 5c). *A,* 10 days; *B,* 12 days. (Based on Hertig and Rock, 1941.) This stage of development is characterized by the intercommunication of the lacunae filled with maternal blood. Note in *B* that large cavities have appeared in the extraembryonic mesoderm, forming the beginning of the extraembryonic coelom. Also note that extraembryonic endodermal cells have begun to form on the inside of the primary yolk sac.

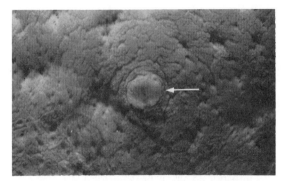

Figure 3–3 Photograph of the endometrial surface showing the implantation site of a human embryo of about 12 days; the implanted conceptus causes a small elevation (arrow) (×8). (From Hertig, A. T., and Rock, J.: *Contr. Embryol. Carneg. Instn.* 29:127, 1941. Courtesy of the Carnegie Institution of Washington.)

circulation is established. Oxygenated blood passes into the lacunae from the *spiral arteries,* and deoxygenated blood is removed from them via the veins of the uterus.

The 10-day conceptus is completely embedded in the endometrium (Fig. 3–2*A*). For a day or so, the defect in the surface epithelium is indicated by a *closing plug,* consisting of a blood clot and cellular debris. By day 12 the almost completely regenerated epithelium covers over the blastocyst (Fig. 3–2*B*); this produces a minute elevation on the endometrial surface (Figs. 3–3 and 3–4*A*). This type of implantation, during which the conceptus becomes completely embedded within the endometrium, is called *interstitial implantation* and occurs in humans and a few other species.

Adjacent syncytiotrophoblastic lacunae fuse to form intercommunicating *lacunar networks* (Fig. 3–2*B*), which give the syncytiotrophoblast a sponge-like structure (Figs. 3–4 and 3–5). These lacunar networks develop first at the *embryonic pole* and form the primordium of the *intervillous spaces* of the placenta. The endometrial capillaries around the implanted embryo first become congested and dilated to form *sinusoids,* and then the syncytiotrophoblast erodes them. Maternal blood now seeps directly into the lacunar networks and soon begins to flow slowly through the lacunar system, establishing a *primitive uteroplacental circulation.* The erosion of the endometrium continues until the *placenta* forms (see Fig. 4–12).

The failure of maternal tissue to reject the conceptus has puzzled embryologists and immunologists for some time. A current view is that the syncytiotrophoblast does not contain or exhibit *transplantation antigens,* and for this reason the conceptus is not rejected (Page et al., 1981).

The endometrial stromal cells around the conceptus enlarge and accumulate glycogen and lipid; these cellular changes, together with the vascular and the glandular alterations, are referred to as the *decidual reaction.* Although initially confined to the area immediately around the conceptus, the decidual reaction soon occurs throughout the endometrium.

As these changes occur in the trophoblast and the endometrium, the *extraembryonic mesoderm* increases (Fig. 3–2*A*), and by day 11 isolated *coelomic spaces* are visible within it. These spaces rapidly fuse to form large isolated cavities of *extraembryonic coelom* (Figs. 3–2*B* and 3–4*B*). The embryonic disc increases slightly in length, but it is still small (0.1 to 0.2 mm). The hypoblast (primitive embryonic endoderm) extends beyond the rim of the embryonic disc and along the inside of the dorsal part of the primary yolk sac (Fig. 3–2*B*).

STAGE 6 (13 TO 14 DAYS)

This stage is characterized by the first appearance of chorionic villi. Proliferation of the cytotrophoblast produces local masses, or clumps, that extend into the syncytiotrophoblast and indicate a first stage (*primary chorionic villi*) in the development of chorionic villi (Fig. 3–5*A*).

The isolated coelomic spaces in the extraembryonic mesoderm have fused to form a single large space called *extraembryonic coelom.* This fluid-filled cavity surrounds the amnion and the yolk sac, except where the amnion is attached to the trophoblast by the *connecting stalk* (Fig. 3–5*B*). As the extraembryonic coelom forms, the primary yolk sac decreases in size, and a smaller *secondary yolk sac* develops. It is formed by hypoblastic cells that grow out from the embryonic disc inside the primary yolk sac (Figs. 3–2*B* and 3–5*A*).

The coelom splits the extraembryonic mesoderm into two layers (Fig. 3–5*A* and *B*): (1) *extraembryonic somatic (somatopleuric) mesoderm,* lining the trophoblast and covering the amnion, and (2) *extraembryonic splanchnic (splanchnopleuric) mesoderm,*

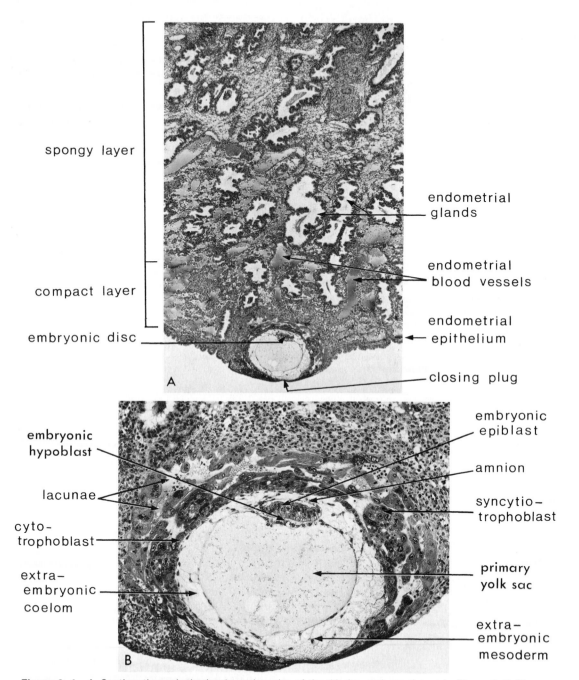

Figure 3–4   *A,* Section through the implantation site of the 12-day embryo shown in Figure 3–3. The embryo (developmental stage 5c) is embedded in the compact layer of the endometrium (×30). *B,* Higher magnification of the conceptus and surrounding endometrium (×100). Intercommunicating lacunae containing maternal blood are visible in the syncytiotrophoblast. (From Hertig, A. T., and Rock, J.: *Contr. Embryol. Carneg. Instn. 29:*127, 1941. Courtesy of the Carnegie Institution of Washington.)

Figure 3-5 Drawings of sections through implanted human embryos (developmental stage 6), based mainly on Hertig et al., 1956. In these drawings note that (1) the defect in the surface epithelium of the endometrium has disappeared; (2) a small secondary yolk sac has formed inside the primary yolk sac as it is "pinched off"; (3) a large cavity, the extraembryonic coelom, now surrounds the yolk sac and the amnion, except where the amnion is attached to the chorion by the connecting stalk; and (4) the extraembryonic coelom splits the extraembryonic mesoderm into two layers; extraembryonic somatic mesoderm lining the trophoblast and covering the amnion, and extraembryonic splanchnic mesoderm around the yolk sac. The trophoblast and extraembryonic somatic mesoderm together form the chorion, which eventually gives rise to the fetal part of the placenta. *A*, 13 days, illustrating the decrease in relative size of the primary yolk sac and the early appearance of primary chorionic villi at the embryonic pole. *B*, 14 days, showing the newly formed secondary yolk sac and the location of the prochordal plate (future site of mouth) in its roof. *C*, Detail of the prochordal plate area outlined in *B*.

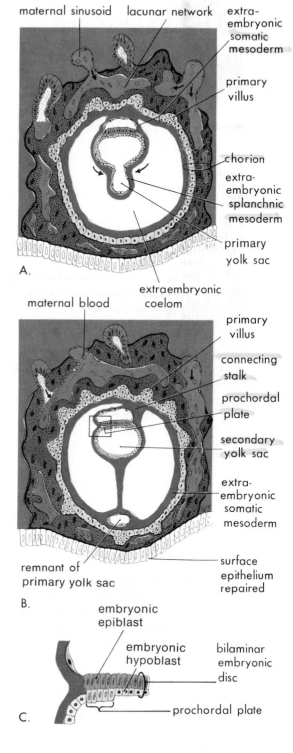

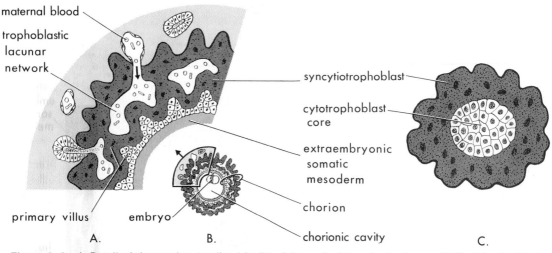

Figure 3-6   *A*, Detail of the section (outlined in *B*.) of the wall of the chorionic sac. *B*, Sketch of a 14-day conceptus illustrating the chorionic sac and the shaggy appearance created by the primary villi (×6). *C*, Drawing of a transverse section through a primary villus (×300). Developmental stage 6 is characterized by the appearance of primary chorionic villi.

around the yolk sac. The extraembryonic somatic mesoderm and the trophoblast together constitute the *chorion* (Fig. 3–5*A*). The chorion forms a sac, the *chorionic sac,* within which the embryo and its amnion and yolk sac are suspended by the connecting stalk (Fig. 3–6*B*). The extraembryonic coelom becomes the *chorionic cavity.*

The amniotic sac (with the embryonic epiblast forming its "floor") and the yolk sac (with the embryonic hypoblast forming its "roof") are analogous to two balloons pressed together (site of embryonic disc) and suspended by a cord (the connecting stalk) from the inside of a larger balloon (the chorionic sac).

The embryo is still in the form of a flat bilaminar embryonic disc, but the hypoblastic cells in a localized area have become columnar to form a thickened circular area, called the *prochordal plate* (Fig. 3–5*B* and *C*). The prochordal plate indicates the future site of the mouth and appears to serve as an important organizer of the head region. It is believed to give rise to mesenchyme in the head region and to the endodermal layer of the *oropharyngeal membrane* (see Fig. 4–6*C*).

IMPLANTATION SITES

Intrauterine Sites (See Figs. 2–18 and 3–7).   The blastocyst usually implants in the midportion of the body of the uterus, slightly more frequently on the posterior than on the anterior wall. Implantation al-

most always occurs close to a maternal capillary, suggesting some type of tropism toward maternal blood (Page et al., 1981).

Implantations in the lower segment of the uterus near the internal ostium (os) result in *placenta previa,* a placenta that covers or adjoins the internal ostium. Placenta previa may cause severe bleeding during pregnancy or at delivery.

Extrauterine Sites (Fig. 3–7).   Implantation often occurs *outside the cavity of the uterus.* It includes tubal, cervical, and interstitial types. *Over 90 per cent of ectopic implantations occur in the uterine tube* (Figs. 3–7, 3–8, and 3–9). About 60 per cent of tubal pregnancies are in the ampulla or infundibulum of the uterine tube (Figs. 3–7*A* and *F*). The incidence of tubal pregnancy varies from 1 in 80 to 1 in 250 pregnancies, depending on the socioeconomic level of the population studied (Page et al., 1981).

There are several causes of *ectopic tubal pregnancy,* but it is usually related to factors that delay or prevent transport of the dividing zygote to the uterus, for example, alterations resulting from *pelvic inflammatory disease.* In some cases, the blockage results from a previous tubal infection that has damaged the mucosa, causing adhesions between its folds.

Ectopic tubal pregnancies usually result in rupture of the uterine tube and hemorrhage during the first eight weeks, followed by death of the embryo (Figs. 3–8 and 3–9). Tubal rupture and subsequent hemorrhage

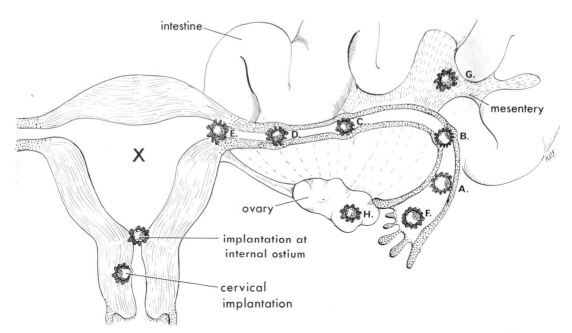

Figure 3–7 Drawing illustrating various implantation sites; the usual site in the posterior wall is indicated by an X. The approximate order of frequency of ectopic implantations is indicated alphabetically (*A,* most frequent to *H,* least frequent). *A* to *F,* Tubal pregnancies. *G,* Abdominal pregnancy. *H,* Ovarian pregnancy.

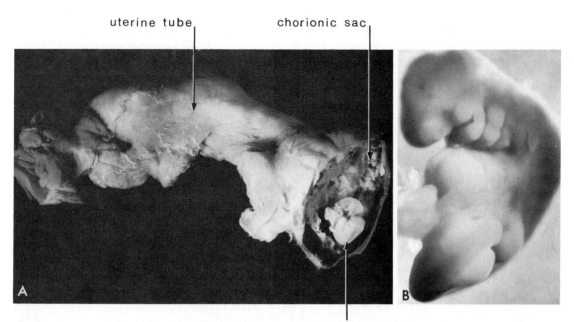

Figure 3–8 Photographs of a tubal pregnancy. *A,* The uterine tube has been sectioned to show the conceptus implanted in the mucous membrane (×3). *B,* Enlarged photograph of the normal-appearing four-week embryo (×13). Ectopic pregnancies occur most often in the ampulla of the uterine tube. This serious condition may be caused by a delay in the passage of the dividing zygote along the tube. Ectopic tubal pregnancy results in death of the embryo and usually sudden massive bleeding from the ruptured tube.

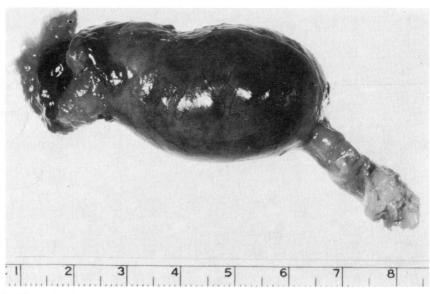

Figure 3–9   Photograph showing the gross appearance of an unruptured ectopic pregnancy located in the ampulla of the uterine tube. When the chorionic sac distends the tube, partial separation of the placenta and rupture of the tube often occur. Spurts of blood escape from the ruptured tube and its infundibulum (shown at the left). Tubal rupture and the associated hemorrhage constitute a threat to the mother's life. (From Page, E. W., Villee, C. A., and Villee, D. B.: *Human Reproduction. Essentials of Reproductive and Perinatal Medicine.* 3rd ed. Philadelphia, W. B. Saunders Co., 1981.)

constitute a threat to the mother's life and so are of major clinical importance.

When blastocysts implant in the isthmus of the uterine tube, the tube tends to rupture early, because this part of the tube is narrow and relatively unstretchable. Abortion of the embryo from this site often results in extensive bleeding, probably because of the rich anastomoses between ovarian and uterine vessels in this area. When blastocysts implant in the uterine, or intramural, part of the tube, they may develop into fetuses (12 to 16 weeks) before expulsion occurs. However, when such an *interstitial ectopic pregnancy* rup-

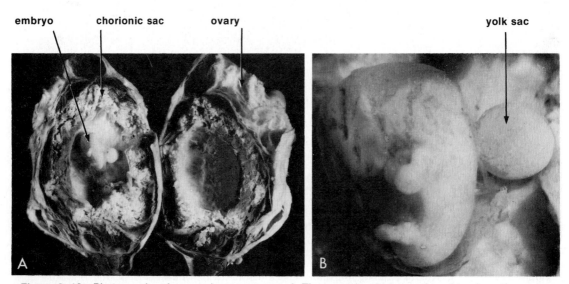

Figure 3–10   Photographs of an ovarian pregnancy. *A,* The ovary has been sectioned to show the embryo and its membranes, which occupy almost the entire ovary. *B,* Enlarged photograph of the normal-appearing five-week embryo (×6). Ovarian pregnancy is very rare and leads to death of the embryo and bleeding from the ovary.

embryo

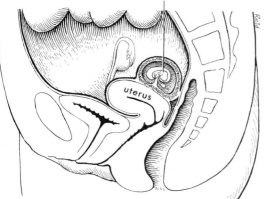

uterus

Figure 3–11 Drawing illustrating an abdominal pregnancy. Although a blastocyst expelled from the uterine tube may attach to any organ or to the mesentery of the intestines, it commonly attaches to the peritoneum of the rectouterine pouch.

tures, it opens up a large uterine sinus that bleeds profusely (Lawson and Stewart, 1967). Blastocysts that implant in the ampulla or on the fimbriae of the uterine tube are often expelled into the peritoneal cavity, where they commonly implant in the rectouterine pouch (Fig. 3–11).

Simultaneous intrauterine and extrauterine pregnancy is unusual. The ectopic pregnancy is masked initially by the presence of a pregnant uterus. Usually the extrauterine pregnancy can be terminated (e.g., by removal of the involved uterine tube) without interfering with the intrauterine pregnancy.

Cervical pregnancy is very rare (Fig. 3–7). Some of these pregnancies are not recognized because the conceptus is expelled early in the gestation. In other cases, the placenta of the embryo becomes firmly attached to the fibrous and muscular parts of the cervix, often resulting in bleeding and subsequent surgical intervention, e.g., *hysterectomy* (excision of the uterus).

Blastocysts may implant in the ovary (Fig. 3–10) or in the abdominal cavity (Fig. 3–11), but *ovarian and abdominal pregnancies are extremely rare.*

In exceptional cases, an abdominal pregnancy may progress to full term, and the fetus may be delivered alive. Usually an abdominal pregnancy creates a serious condition because the placenta often attaches to vital structures and causes considerable bleeding. Rarely, an abdominal pregnancy is not diagnosed, and the fetus dies and becomes calcified, forming a so-called stone fetus, or *lithopedion* (Gr. *lithos*, stone, + *paidion*, child).

EARLY ABORTIONS

Abortion is defined as the termination of pregnancy before 20 weeks' gestation, i.e., *before the period of viability.* Almost all abortions during the first three weeks occur spontaneously; that is, they are not induced. The frequency of early abortions is difficult to establish because they often occur before the woman is aware she is pregnant. An abortion just after the first missed period is very likely to be mistaken for a delayed menstruation.

Hertig et al. (1959) studied 34 embryos recovered from women of known fertility, and found 10 of them so abnormal that they probably would have aborted by the end of the second week. The incidence of chromosome abnormalities in early spontaneous abortions in one study was about 61 per cent (Boué et al., 1975). Summarizing the data of several studies, Carr and Gedeon (1977) estimated that about 50 per cent of all known spontaneous abortions result from chromosomal abnormalities. The higher incidence of early abortions in older women probably results from the increasing frequeny of nondisjunction during oogenesis (see Fig. 2–3).

Hertig (1967) estimated that of the 70 to 75 per cent of blastocysts that implant, only 58 per cent survive to the end of the second week. He further estimated that 16 per cent of this latter group would be abnormal and would soon abort.

Witschi (1970) estimated that from one third to one half of all zygotes never survive to implant. Failure to implant may result from a poorly developed endometrium, but probably in many cases there are chromosomal abnormalities in the zygote.

Exposure of embryos to *teratogens* (agents that produce congenital malformations or increase their incidence) during the first two weeks usually does not cause congenital malformations (see Fig. 8–14), but some agents kill the blastocyst or cause early abortions.

SUMMARY OF IMPLANTATION

Implantation of the blastocyst begins at the end of the first week and ends during the second week. The process may be summarized as follows:

1. *Zona pellucida degenerates* (days 4 to 5). Disappearance of the zona pellucida is brought about by enzymatic lysis. The lytic

enzymes are released from the acrosomes of the many sperms that partially penetrate the zona pellucida.

2. *Blastocyst attaches* to endometrial epithelium (days 5 to 6).

3. *Syncytiotrophoblast erodes* endometrial epithelium and stroma as it differentiates into an inner cellular layer, the cytotrophoblast, and an outer layer lacking cell boundaries, the syncytiotrophoblast (days 7 to 8). The erosion of the uterine mucosa results from *proteolytic enzymes* produced by the syncytiotrophoblast.

4. *Lacunae appear* in syncytiotrophoblast (day 9).

5. *Blastocyst sinks beneath surface* of the endometrial epithelium (days 9 to 10).

6. *Lacunar networks form* by fusion of adjacent lacunae (days 10 to 11).

7. *Syncytiotrophoblast invades endometrial sinusoids,* allowing maternal blood to seep into the lacunar networks and establish a *uteroplacental circulation* (days 11 to 12).

8. *Defect in endometrial epithelium disappears* (days 12 to 13).

9. *Marked decidual reaction* occurs in the endometrium around the conceptus (days 13 to 14).

Inhibition of Implantation.   The administration of relatively large doses of estrogen ("morning-after" pills) for several days after sexual intercourse will prevent pregnancy by inhibiting implantation of the blastocyst that may develop. It may also accelerate passage of the dividing zygote along the uterine tube (Seeman et al., 1980). Normally, the endometrium progresses to the secretory phase of the menstrual cycle as the zygote forms, undergoes cleavage, and enters the uterus.

The large amount of estrogen, usually administered as the synthetic estrogen *diethylstilbestrol (DES),* disturbs the normal balance between estrogen and progesterone that is necessary for preparation of the endometrium for implantation of the blastocyst (see Fig. 2–6). When the secretory phase does not occur, implantation cannot take place, and the blastocyst soon dies. Because *this treatment results in the death of the blastocyst* rather than in prevention of its formation, use of this method is largely restricted to special cases in which impregnation is not desired, e.g., after a rape or failure of a contraceptive method in a woman over 40. Another reason this method is not used routinely for birth control is that the treatment is associated with a relatively high frequency of nausea, vomiting, and other

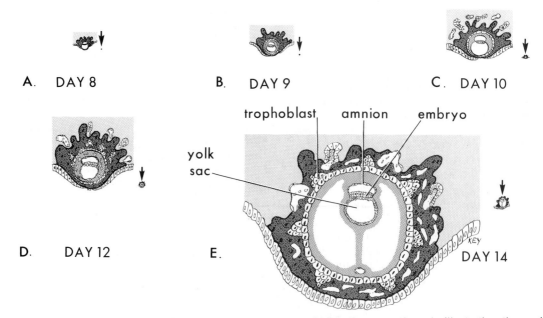

A.   DAY 8   B.   DAY 9   C.   DAY 10

D.   DAY 12   E.   DAY 14

trophoblast   amnion   embryo

yolk sac

Figure 3–12   Drawings of sections of human blastocysts during the second week, illustrating the rapid expansion of the trophoblast and the relatively minute size of the embryos (×25); the sketches indicated by the arrows show the actual size of the blastocysts.

adverse effects. *Contraception is preferable to contraimplantation.*

SUMMARY OF SECOND WEEK

Rapid proliferation and differentiation of the trophoblast occur as follows: (Fig. 3–12): (1) An inner cellular layer, the cytotrophoblast, and an outer syncytium, the syncytiotrophoblast, form; (2) lacunae develop in the syncytiotrophoblast and soon fuse to form *lacunar networks;* (3) the trophoblast erodes maternal sinusoids and endometrial glands; (4) a primitive *uteroplacental circulation* is established as blood seeps into the lacunar networks; (5) *primary chorionic villi form* on the external surface of the chorionic sac; and (6) implantation is completed as the conceptus is wholly embedded within the endometrium, and the defect in the surface epithelium is healed (Fig. 3–12*E*).

The various endometrial changes resulting from adaptation of the maternal tissues to implantation of the blastocyst are known collectively as the *decidual reaction.*

Concurrently, the primary yolk sac forms and *extraembryonic mesoderm* arises from the inner surface of the trophoblast. The *extraembryonic coelom* forms from spaces that develop in the extraembryonic mesoderm. The primary yolk sac becomes smaller and gradually disappears as the *secondary yolk sac* develops.

As these changes occur (1) the *amniotic cavity appears* as a slit-like space between the cytotrophoblast and the inner cell mass; (2) the inner cell mass differentiates into a *bilaminar embryonic disc* consisting of *epiblast* related to the amniotic cavity and *hypoblast* adjacent to the blastocyst cavity; and (3) the *prochordal plate* develops as a localized thickening of hypoblast, indicating the future cranial region of the embryo and the site of the future mouth.

SUGGESTIONS FOR ADDITIONAL READING

Gasser, R. F.: *Atlas of Human Embryos.* Hagerstown, Harper & Row, Publishers, 1975. An excellent collection of photomicrographs and drawings of specimens from the renowned Carnegie Collection. The important events in each week of development are presented in outline form.

Hamilton, W. J., and Mossman, H. W.: *Human Embryology. Prenatal Development of Form and Function.* 4th ed. Cambridge, W. Heffer and Sons, Ltd., 1972. pp. 83–117. A well-illustrated account of implantation of the blastocyst and development of the amniotic cavity, yolk sac, and embryonic disc.

Rugh, R., and Shettles, L. B.: *From Conception to Birth. The Drama of Life's Beginnings.* New York, Harper & Row, Publishers, 1971, pp. 1–28. A very readable account of events occurring during the first weeks of development. Although written for prospective parents, the descriptions should be helpful to beginning students.

Villee, D. B.: *Human Endocrinology. A Developmental Approach.* Philadelphia, W. B. Saunders Co., 1975, pp. 1–17. A summary of events occurring during the first weeks of life. You would probably find this less-exhaustive coverage of this complex subject helpful in getting an overview of embryogenesis.

CLINICALLY ORIENTED PROBLEMS

1. Is it advisable to make a radiological examination of a healthy 22-year-old female's chest during the last week of her menstrual cycle? Why or why not?

2. A woman who had been raped during her fertile period was given large doses of estrogen (25 mg of diethylstilbestrol (DES) twice daily for five days) to interrupt a possible pregnancy. If she happened to be pregnant, what do you think would be the mechanism of action of the DES? What do laypeople call this type of treatment?

3. A 23-year-old woman reported to her doctor with severe lower abdominal pain. She said that she had missed two menstrual periods. A diagnosis of *ectopic pregnancy* was made. What is the most likely site of the extrauterine gestation? Be specific. How do you think the doctor would treat the case?

4. A 30-year-old woman had an appendectomy toward the end of her menstrual cycle; 8½ months later she had a child with a congenital malformation of the brain. Could the surgery have produced the congenital malformation? Discuss briefly.

The answers to these questions are given at the back of the book.

REFERENCES

Balinsky, B. I.: *An Introduction to Embryology.* 4th ed., Philadelphia, W. B. Saunders Co., 1975.

Benirschke, K.: Implantation, placental development, uteroplacental blood flow; *in* D. E. Reid, K. J. Ryan, and K. Benirschke (Eds.): *Principles and Management of Human Reproduction.* Philadelphia, W. B. Saunders Co., 1972, pp. 179–196.

Billington, W. D.: Trophoblast; *in* E. E. Philipp, J. Barnes, and M. Newton (Eds.): *Scientific Foundations of Obstetrics and Gynecology.* London, William Heinemann, Ltd., 1970, pp. 159–167.

Blandau, R. J. (Ed.): *The Biology of the Blastocyst.* Chicago, University of Chicago Press, 1971.

Boronow, R. C., McElin, T. W., West, R. H., and Buckingham, J. C.: Ovarian pregnancy. *Am. J. Obstet. Gynecol.* 91:1095, 1965.

Boué, J., Boué, A., and Lazar, P.: Retrospective and prospective epidemiological studies of 1500 karyotyped spontaneous abortions. *Teratology 12*:11, 1975.

Böving, B. G.: Implantation mechanisms; *in* C. G. Hartmann (Ed.): *Mechanisms Concerned with Conception.* New York, Pergamon Press, 1963, pp. 321–396.

Böving, B. G.: Anatomy of reproduction; *in* J. P. Greenhill (Ed.): *Obstetrics.* 13th ed. Philadelphia, W. B. Saunders Co., 1965, pp. 3–101.

Boyd, J. D., and Hamilton, W. J.: *The Human Placenta.* Cambridge, W. Heffer and Sons, Ltd., 1970.

Carr, D. H.: Chromosome anomalies as a cause of spontaneous abortion. *Am. J. Obstet. Gynecol.* 97:283, 1967.

Carr, D. H.: Heredity and the embryo. *Sci. J.* (London) 6:75, 1970.

Carr, D. H.: Chromosome studies in selected spontaneous abortions. III. Early pregnancy loss. *Obstet. Gynecol.* 37:750, 1971.

Carr, D. H., and Gedeon, M.: Population cytogenetics of human abortuses; *in* E. B. Hook and I. H. Porter (Eds.): *Population Cytogenetics: Studies in Humans.* New York, Academic Press, 1977, pp. 1–9.

Carter, C.: The genetics of congenital malformations; *in* E. E. Philipp, J. Barnes, and M. Newton (Eds.): *Scientific Foundations of Obstetrics and Gynecology.* London, William Heinemann, Ltd., 1970, pp. 655–660.

Copenhaver, W. M., Kelly, D. E., and Wood, R. L.: *Bailey's Textbook of Histology.* 17th ed. Baltimore, The Williams & Wilkins Company, 1978, pp. 90–102.

Crowley, L. V.: *An Introduction To Clinical Embryology.* Chicago, Year Book Medical Publishers, Inc., 1974.

Fawcett, D. W., Wislocki, G. B., and Waldo, C. M.: The development of mouse ova in the anterior chamber of the eye and in the abdominal cavity. *Am. J. Anat. 81*:413, 1947.

Greenhill, J. P.: *Obstetrics.* 14th ed. Philadelphia, W. B. Saunders Co., 1969.

Greenhill, J. P., and Friedman, E. A.: *Biological Principles and Modern Practice of Obstetrics.* Philadelphia, W. B. Saunders Co., 1974, pp. 1–94.

Hamilton, W. J., and Boyd, J. D.: Development of the human placenta; *in* E. E. Philipp, J. Barnes, and M. Newton (Eds.): *Scientific Foundations of Obstetrics and Gynecology.* London, William Heinemann, Ltd., 1972, pp. 185–254.

Hertig, A. T.: The overall problem in man; *in* K. Benirschke (Ed.): *Comparative Aspects of Reproductive Failure.* New York, Springer Verlag, 1967.

Hertig, A. T.: *Human Trophoblast.* Springfield, Ill., Charles C Thomas, Publisher, 1968.

Hertig, A. T., and Rock, J.: Two human ova of the previllous stage, having a developmental age of about eleven and twelve days respectively. *Contrib. Embryol. Carnegie Inst. 29*:127, 1941.

Hertig, A. T., and Rock, J.: Two human ova of the previllous stage, having a developmental age of about seven and nine days respectively. *Contrib. Embryol. Carnegie Inst. 31*:65, 1945.

Hertig, A. T., and Rock, J.: Two human ova of the pre-villous stage, having a developmental age of about eight and nine days, respectively. *Contrib. Embryol. Carnegie Inst. 33*:169, 1949.

Hertig, A. T., Rock, J., and Adams, E. C.: A description of 34 human ova within the first seventeen days of development. *Am. J. Anat. 98*:435, 1956.

Hertig, A. T., Rock, J., Adams, E. C., and Menkin, M. C.: Thirty-four fertilized human ova, good, bad and indifferent, recovered from 210 women of known fertility. *Pediatrics 23*:202, 1959.

Lawson, J. B., and Stewart, D. B.: *Obstetrics and Gynecology in the Tropics.* London, Arnold Publishers, 1967, Chapter 21.

Luckett, W. P.: The origin of extraembryonic mesoderm in the early human and rhesus monkey embryos. *Anat. Rec., 169*:369, 1971.

Luckett, W. P.: Amniogenesis in the early human and rhesus monkey embryos. *Anat. Rec. 175*:375, 1973.

O'Rahilly, R.: *Developmental Stages in Human Embryos. Part A: Embryos of the First Three Weeks (Stages 1 to 9).* Washington, D.C., Carnegie Institution of Washington, 1973.

Page, E. W., Villee, C. A., and Villee, D. B.: *Human Reproduction. Essentials of Reproductive and Perinatal Medicine.* 3rd ed. Philadelphia, W. B. Saunders Co., 1981.

Peel, J., and Potts, M.: *Textbook of Contraceptive Practice.* New York, Cambridge University Press, 1969, Chapter 13.

Reid, D. E.: Obstetric hemorrhage in late pregnancy and post partum; *in* D. E. Reid, K. J. Ryan, and K. Benirschke: *Principles and Management of Human Reproduction.* Philadelphia, W. B. Saunders Co., 1972, pp. 308–318.

Reid, D. E., and Benirschke, K.: Ectopic pregnancy; *in* D. E. Reid, K. J. Ryan, and K. Benrischke: *Principles and Management of Human Reproduction.* Philadelphia, W. B. Saunders Co., 1972, pp. 277–285.

Saxén, L., and Rapola, J.: *Congenital Defects.* New York, Holt, Rinehart and Winston, Inc., 1969, pp. 112–116.

Seeman, P., Sellers, E. M., and Roschlau, W. H. E.: *Principles of Medical Pharmacology.* 3rd ed. Toronto, University of Toronto Press, 1980.

Stander, R. W.: Abdominal pregnancy. *Clin. Obstet. Gynecol. 5*:1065, 1962.

Streeter, G. L.: Developmental horizons in human embryos. Description of age group XI, 13 to 20 somites, and age group XII, 21 to 29 somites. *Contrib. Embryol. Carnegie Inst. 30*:211, 1942.

Tuchmann-Duplessis, H.: The effects of teratogenic drugs; *in* E. E. Philipp, J. Barnes, and M. Newton (Eds.): *Scientific Foundations of Obstetrics and Gynecology.* London, William Heinemann, Ltd., 1970, pp. 636–648.

Witschi, E.: Teratogenetic effects from overripeness of the egg; *in* F. C. Fraser and V. A. McKusick (Eds.): *Congenital Malformations.* Amsterdam, Excerpta Medica, 1970, p. 157.

FORMATION OF THE TRILAMINAR EMBRYO

The Third Week

The third week is characterized by the formation of the primitive streak (Fig. 4–1) and the three germ layers, from which all tissues and organs of the embryo develop. The third week is a period of rapid development of the conceptus, coinciding with the *first missed menstrual period* (see Fig. 1–1).

Cessation of menstruation is often the first sign that a woman may be pregnant. Relatively simple and rapid tests are now available for detecting pregnancy as early as the third week. These tests depend on the presence of *human chorionic gonadotropin* (hCG), a hormone produced by the trophoblast and excreted in the mother's urine. Bleeding at the expected time of menstruation does not rule out pregnancy because there may be hemorrhage from the implantation site in some cases. This *implantation bleeding* results from leakage of blood into the uterine lumen from disrupted blood vessels around the implantation site. When such bleeding is interpreted as menstruation, an error occurs in determining the expected delivery date of the baby.

There is no absolute sign of pregnancy during the early weeks because a gravid (pregnant) uterus may be mimicked by several other conditions, e.g., fibroid tumors of the uterus and ovarian tumors. Furthermore, a positive pregnancy test does not always indicate that an embryo is developing. After the death of an embryo, tumors such as *noninvasive hydatidiform mole* and highly malignant tumors, called *chorionepitheliomas,* may develop from the chorionic epithelium and produce hCG.

Early confirmation of pregnancy is possible with ultrasound techniques. Using gray-scale imaging, a pregnancy can be detected around the time of the first missed period (Carlson, 1977). Early confirmation of pregnancy is clinically important in patients with infertility who have *oligomenorrhea* (scanty menstruation), or in patients who have persistent bleeding during the first trimester (Depp, 1977).

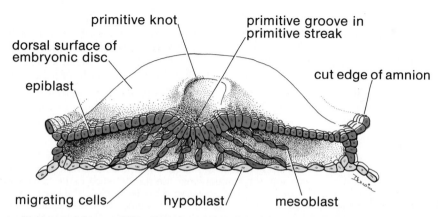

Figure 4–1 Drawing of the cranial half of the embryonic disc during the third week. The disc has been cut transversely to show the migration of mesenchymal cells from the primitive streak. This illustration also indicates that the definitive embryonic endoderm probably also arises from the epiblast. Presumably, the hypoblastic cells are displaced to extraembryonic regions.

The third week has been referred to as the "period of threes" (Crowley, 1974) because *three germ layers develop* and *three important structures form* (the primitive streak, the notochord, and the neural tube).

GASTRULATION

The process by which the bilaminar embryonic disc is converted into a trilaminar embryonic disc is called gastrulation. Many cells of the epiblast detach themselves from neighboring cells and migrate inwardly to form a loose network of tissue called *mesoblast,* which spreads laterally and cranially to form a layer between the epiblast and the hypoblast known as the *intraembryonic mesoderm.* Some mesoblastic cells invade the hypoblast and displace the hypoblastic cells laterally. This newly formed layer is known as the *embryonic endoderm.* The cells that remain in the epiblast form the layer called the *embryonic ectoderm.* Hence, the epiblast is the source of embryonic ectoderm, embryonic mesoderm, and most, if not all, embryonic endoderm. The cells of these three germ layers divide, migrate, aggregate, and differentiate into the tissues and the organs of the embryo (see Fig. 5–5).

The ectoderm gives rise to the outer epithelia (e.g., the epidermis) *and the nervous system. The endoderm is the source of the epithelial linings* of the respiratory passages and the digestive tract, including the glandular cells of associated organs such as the liver and pancreas. *The mesoderm gives rise to smooth muscular coats,* the connective tissues, and the vessels supplying these organs. Mesoderm is also the source of blood cells and bone marrow, the skeleton, striated muscles, and the reproductive and excretory organs.

Gastrulation begins on about day 14 and ends on about day 19. *Formation of the primitive streak and the notochord are the important processes occurring during gastrulation.* During this phase of development. the embryo is sometimes referred to as a *gastrula.* At the end of gastrulation, the embryo is said to be *triploblastic.*

THE PRIMITIVE STREAK

During developmental stage 6, usually on day 15, a thickened linear band of epiblast known as the *primitive streak* appears caudally in the midline of the dorsal aspect of the embryonic disc (Figs. 4–1 and 4–2*A*). The primitive streak results from the convergence of epiblastic cells toward the midline in the posterior part of the embryonic disc. As the primitive streak elongates by addition of cells at its caudal end, its cranial end thickens to form a *primitive knot* (Figs. 4–2*C* and 4–3*B*). Concurrently, a narrow *primitive groove* develops in the primitive streak, which is continuous with a depression in the primitive knot known as the *primitive pit* (Fig. 4–2*C* and *D*).

When the primitive streak appears, it is possible to identify the embryo's craniocaudal axis, its cranial and caudal ends, its dorsal and ventral surfaces, and its right and left sides.

By the end of stage 6 of embryonic development (about 16 days), *intraembryonic mesoderm* begins to appear (Figs. 4–1 and 4–2*B*). As soon as the primitive streak begins to produce mesoblastic cells destined to become intraembryonic mesoderm, the epiblast is referred to as the *embryonic ectoderm,* and the hypoblast is called the *embryonic endoderm.*

The intraembryonic mesoderm forms as follows: Cells of the epiblast move medially toward the primitive streak and enter the primitive groove. These cells lose their attachment to the cells of the epiblast and migrate inwardly between the epiblast and

Figure 4–2 Drawings illustrating formation of the trilaminar embryonic disc (developmental stage 6, days 15 to 16). The small sketch at the upper left is for orientation; the arrow indicates the dorsal aspect of the embryonic disc as shown in *A*. The arrows in all other drawings indicate migration of mesenchymal cells between the ectoderm and endoderm. *A, C,* and *E,* Dorsal views of the embryonic disc early in the third week, exposed by removal of the amnion. *B, D,* and *F,* Transverse sections through the embryonic disc at the levels indicated. The prochordal plate is indicated by a broken line because it is a thickening of endoderm that cannot be seen from the dorsal surface (see Fig. 4–5*C*).

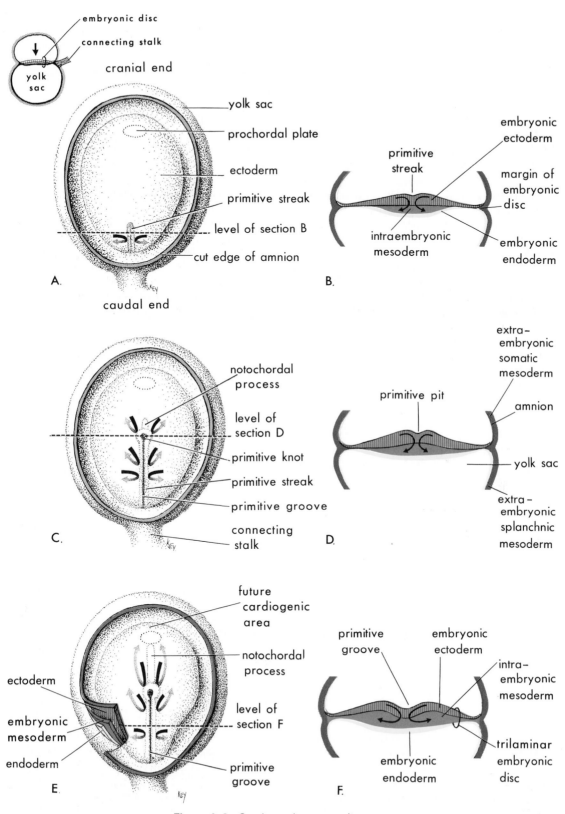

Figure 4–2 *See legend on opposite page.*

the hypoblast (Figs. 4–1 and 4–2). These wandering *mesoblastic cells* immediately begin to pass laterally and form a network of cells called the *mesoblast*. Some mesoblastic cells become organized into a layer called the *intraembryonic mesoderm*.

Mesoblastic cells are often called *mesenchymal cells*. These embryonic cells migrate and have the potential to proliferate and differentiate into diverse types of cells (fibroblasts, chondroblasts, and osteoblasts). Mesenchymal cells form a loosely woven tissue called *mesenchyme*, or embryonic connective tissue. It forms the supporting tissue of the embryo, e.g., most connective tissues of the body and the stromal components of all glands.

During developmental stage 7 (about 16 days), mesoblastic cells migrate cranially from the primitive knot and form a midline cord known as the *notochordal process* (Figs. 4–2C and 4–3B). This process grows between the ectoderm and the endoderm until it reaches the *prochordal plate*, a small circular area of columnar endodermal cells

(Figs. 3–5C and 4–2A). The rod-like notochordal process can extend no farther because the prochordal plate is firmly attached to the overlying ectoderm, forming the *oropharyngeal (buccopharyngeal) membrane* (Fig. 4–6C).

Other mesoblastic cells from the primitive streak and the notochordal process migrate laterally and cranially until they reach the margins of the embryonic disc. There they join the extraembryonic mesoderm covering the amnion and the yolk sac (Fig. 4–2F). Most extraembryonic mesoderm in human embryos is derived from the cytotrophoblast (see Chapter 3), but some of it may arise from the primitive streak, as in other mammals. Some cells from the primitive streak migrate cranially on each side of the notochordal process and around the prochordal plate, where they meet cranially in the *cardiogenic area* (Fig. 4–2E), the future site of the heart.

Caudal to the primitive streak there is a circular area known as the *cloacal membrane* (Figs. 4–3 and 4–5E). The embryonic disc

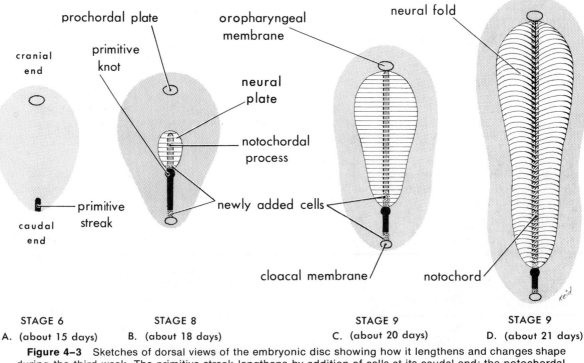

STAGE 6

A. (about 15 days)

STAGE 8

B. (about 18 days)

STAGE 9

C. (about 20 days)

STAGE 9

D. (about 21 days)

Figure 4–3 Sketches of dorsal views of the embryonic disc showing how it lengthens and changes shape during the third week. The primitive streak lengthens by addition of cells at its caudal end; the notochordal process lengthens by migration of cells from the primitive knot. The notochordal process and adjacent mesoderm induce the overlying embryonic ectoderm to form the neural plate, the primordium of the central nervous system (see also Fig. 4–8).

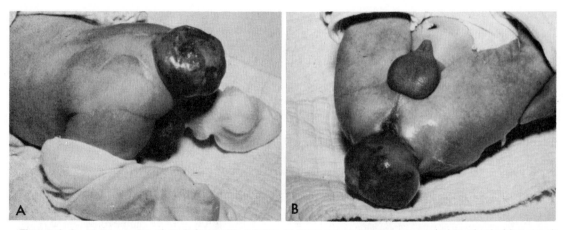

Figure 4–4 Photographs of an infant with a large sacrococcygeal teratoma, probably of primitive streak origin. These tumors are much more common in females than in males, and they often become malignant during infancy. (Courtesy of Dr. Jan Hoogstraten, Children's Centre, Winnipeg, Canada.)

remains bilaminar here and at the *oropharyngeal membrane* because the embryonic ectoderm and endoderm are fused at these sites, thus preventing migration of mesenchymal cells between them. By the middle of the third week, intraembryonic mesoderm separates the embryonic ectoderm and endoderm everywhere except (1) at the oropharyngeal membrane cranially; (2) in the midline cranial to the primitive knot where the notochordal process extends; and (3) at the cloacal membrane caudally.

Fate of the Primitive Streak. The primitive streak actively forms intraembryonic mesoderm until about the end of the fourth week; thereafter, production of mesoderm from this source slows down. The primitive streak quickly diminishes in relative size and becomes an insignificant structure in the sacrococcygeal region of the embryo (Fig. 4–3D). Normally it undergoes degenerative changes and disappears, but, rarely, primitive streak remnants persist and give rise to a tumor known as a *teratoma* (Fig. 4–4). As these cells are derived from pleuripotent primitive streak cells, the tumors often contain various types of tissue. Teratomas not derived from the primitive streak are usually located on or near the testes or the ovaries. In some of these tumors, structures have been reported that can be interpreted as disorganized fetal parts.

Changes in the Embryonic Disc (Fig. 4–3). Initially the embryonic disc is flat and essentially circular, but it soon becomes pear-shaped and then elongated as the notochordal process grows and the notochord forms. Expansion of the embryonic disc occurs mainly in the cranial region; the caudal end remains more or less unchanged. Much of the growth and elongation of the

embryonic disc results from the continuous migration of cells from the primitive streak (Figs. 4–1 and 4–2).

DEVELOPMENT OF NOTOCHORD

In the lower chordate *Amphioxus*, the notochord forms the only skeleton of the adult animal. In the human embryo, it forms a midline axis and the basis of the axial skeleton (the bones of the cranium, vertebral column, ribs, and sternum; see Chapter 15).

The notochord is a cellular rod that develops from the notochordal process and defines the primitive axis of the embryo (Figs. 4–5 and 4–6). As the notochordal process develops, the primitive pit extends into it to form a lumen known as the *notochordal canal* (Fig. 4–5B to E). The notochordal process is now a *tubular column of cells* that extends cranially from the primitive knot to the prochordal plate.

The notochord develops as follows: (1) The floor of the notochordal process fuses with the underlying embryonic endoderm; (2) degeneration of the fused regions occurs; (3) openings appear in the floor of the notochordal process, bringing the notochordal canal into communication with the yolk sac (Fig. 4–6B); (4) the openings rapidly become confluent and the floor of the notochordal canal disappears (Fig. 4–6C); (5) the remains of the notochordal process form a flattened grooved plate, known as the *notochordal plate* (Fig. 4–6D); (6) beginning at the cranial end, the notochordal plate infolds to form the *notochord* (Fig. 4–6F and G); (7) the embry-

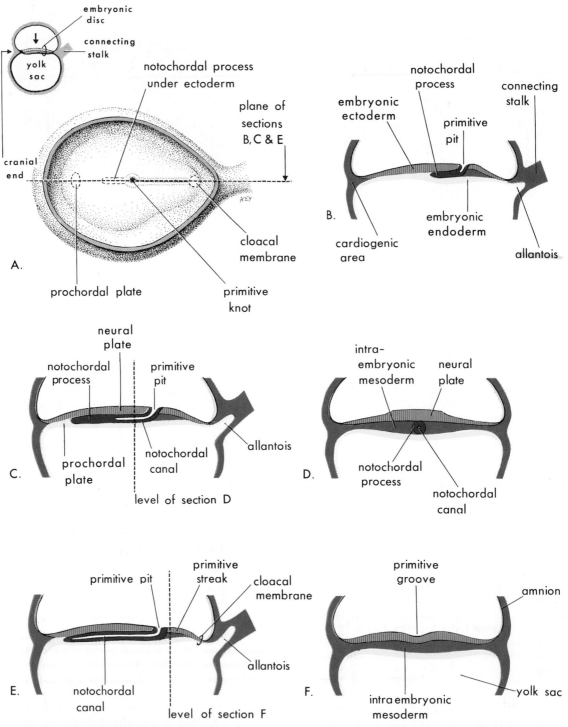

Figure 4–5   Drawings illustrating early stages of notochord development. The small sketch at the upper left is for orientation; the short arrow indicates the dorsal aspect of the embryonic disc. *A*, Dorsal view of the embryonic disc during stage 7 (about 16 days), exposed by removal of the amnion. The notochordal process is shown as if it were visible through the embryonic ectoderm. *B*, *C*, and *E*, Sagittal sections at the plane shown in *A*, illustrating successive stages in the development of the notochordal process and canal. Stages shown in *C* and *E* occur during stage 8 (about 18 days). *D* and *F*, Transverse sections through the embryonic disc at the levels shown.

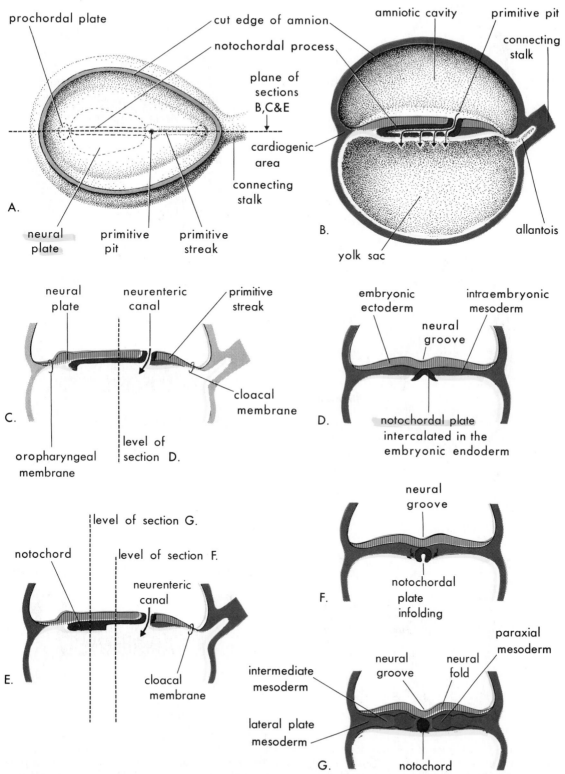

Figure 4–6   Drawings illustrating the final stages of notochord development. *A*, Dorsal view of the embryonic disc during stage 8 (about 18 days), exposed by removing the amnion. *B*, Three-dimensional sagittal section of the embryo. *C* and *E*, Sagittal sections of slightly older stage 8 embryos. *D*, *F*, and *G*, Transverse sections of the embryonic disc.

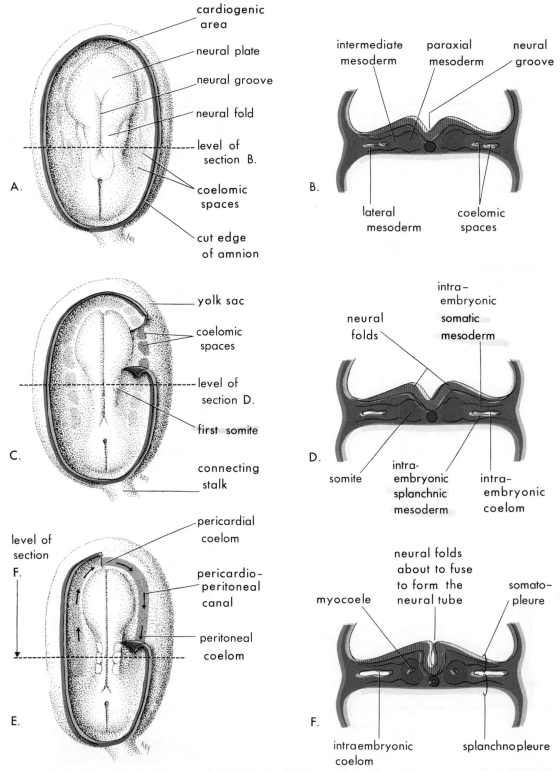

Figure 4–7 Drawings of embryos of 19 to 21 days (stages 8 and 9), illustrating development of the somites and intraembryonic coelom. *A, C,* and *E,* Dorsal view of the embryo, exposed by removal of the amnion. *B, D,* and *F,* Transverse sections through the embryonic disc at the levels shown. *A,* Presomite embryo of about 18 days (stage 8). *C,* An embryo of about 20 days (stage 9) showing the first pair of somites. A portion of the somatopleure on the right has been removed to show the isolated coelomic spaces in the lateral mesoderm. *E,* A three-somite embryo during stage 9 (about 21 days) showing the horseshoe-shaped intraembryonic coelom, exposed on the right by removal of a portion of the somatopleure.

onic endoderm again becomes a continuous layer ventral to the notochord (Fig. 4–6*E* and *G*).

The notochord is the structure around which the vertebral column forms (see Chapter 15). It degenerates and disappears where it is surrounded by the vertebral bodies, but persists as the <u>nucleus pulposus</u> of the intervertebral discs.

The notochord also induces the overlying ectoderm to form the neural plate, the primordium of the central nervous system.

A small passage, the *neurenteric canal,* temporarily connects the amniotic cavity and the yolk sac (Fig. 4–6*C* and *E*). By the end of the fourth week, the notochord is almost completely formed and extends from the oropharyngeal membrane cranially to the primitive knot caudally. When development of the notochord is completed, the neurenteric canal normally obliterates.

NEURULATION

The Neural Tube (Figs. 4–3 and 4–5 to 4–8).   The formation of the neural plate, the neural folds, and their closure to form the neural tube is called *neurulation*. This period ends during developmental stage 12 at about 26 days, when closure of the posterior neuropore occurs (see Fig. 5–8*D*). During neurulation, the embryo is sometimes referred to as a *neurula*.

As the notochord develops, the embryonic ectoderm over it thickens to form the *neural plate* (Figs. 4–3*B* and 4–5*C* and *D*). Experimental evidence (see Chapter 5) indicates that neural plate formation is induced by the developing notochord and the paraxial mesoderm on each side of it. *The ectoderm of the neural plate, called neuroectoderm, gives rise to the central nervous system, consisting of the brain and the spinal cord.*

The neural plate first appears cranial to the primitive knot, and dorsal to the notochordal process and the mesoderm adjacent to it (Fig. 4–3*B*). As the notochord forms and elongates, the neural plate broadens and eventually extends cranially as far as the *oropharyngeal membrane* (Fig. 4–3*C*). On about the eighteenth day, the neural plate invaginates along its central axis to form a *neural groove* with *neural folds* on each side (Figs. 4–3*D* and 4–6 to 4–8). By the end of the third week, the neural folds have begun to move together and fuse, converting the

neural plate into a *neural tube* (Figs. 4–7*F* and 4–8*D*).

The neural tube separates itself from the surface ectoderm, and the free edges of the ectoderm fuse so that this layer becomes continuous over the back of the embryo. Subsequently, the surface ectoderm differentiates into the epidermis of the skin.

The Neural Crest (Figs. 4–8 and 18–8). As the neural folds fuse, some ectodermal cells lying along the crest of each neural fold lose their epithelial affinities and attachments to the neighboring cells. As the neural tube separates from the surface ectoderm, these mesoblastic neural crest cells migrate inwardly and invade the mesoblast on each side of the neural tube. They appear as an irregular flattened mass, called the *neural crest,* between the neural tube and the overlying surface ectoderm. Initially continuous across the midline, it soon separates into right and left parts that migrate to the dorsolateral aspect of the neural tube, where they give rise to the *sensory ganglia* of the spinal and cranial nerves. Many mesoblastic neural crest cells begin to migrate in lateral and ventral directions and disperse. Although these cells are difficult to identify, special tracer techniques have revealed that they disseminate widely and have important derivatives.

Neural crest cells give rise to the spinal ganglia and the ganglia of the autonomic nervous system. The ganglia of the cranial nerves (V, VII, IX, and X) are also partly derived from the neural crest. In addition to forming ganglion cells, neural crest cells form the sheaths of nerves (Schwann cells) and the meningeal covering of the brain and the spinal cord (at least the pia mater and arachnoid). They also contribute to the formation of pigment cells, the suprarenal (adrenal) medulla, and several skeletal and muscular components in the head (see Chapter 10 and Fig. 18–8).

Congenital Malformations of the Central Nervous System.   Because of the primordium of the central nervous system (neural plate) appears during the third week and gives rise to the neural folds, disturbance of neurulation may result in severe abnormalities of the brain and spinal cord (see Fig. 15–10).

Available evidence suggests that the primary disturbances (e.g., a teratogenic drug) affect the neural epithelium itself and result in *failure of closure of the neural tube* in

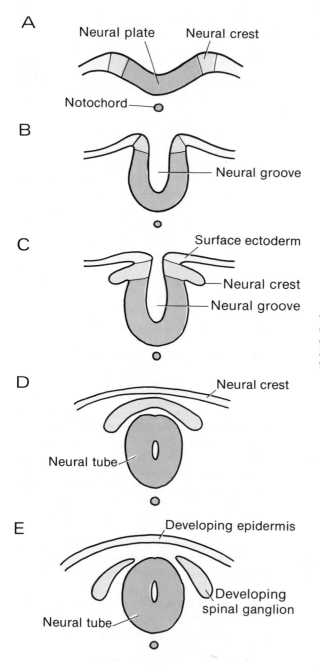

A

Neural plate   Neural crest

Notochord

B

Neural groove

C

Surface ectoderm

Neural crest

Neural groove

Figure 4–8 Diagrammatic transverse sections through progressively older embryos, illustrating formation of the neural groove, the neural tube, and the neural crest up to the end of the fourth week.

D

Neural crest

Neural tube

E

Developing epidermis

Developing spinal ganglion

Neural tube

the brain and/or the spinal cord regions. Extroversion of the neural tissue then occurs, and the exposed tissue degenerates. In *anencephaly,* the brain is represented by a mass of degenerated neural tissue exposed on the surface of the head (see Fig. 18–19).

Allantois. The allantois (Gr. *allantos,* sausage) appears on about day 16 as a rela-

tively tiny, finger-like outpouching or diverticulum from the caudal wall of the yolk sac (Fig. 4–5B, C, and E). The allantois in embryos of reptiles, birds, and some mammals has a respiratory function and/or acts as a storage place for urine during embryonic life. The allantois remains very small in the human embryo, but it is involved with early blood formation and is associated with development of the urinary bladder. As the bladder enlarges, the allantois becomes the *urachus* (see Chapter 13).

DEVELOPMENT OF SOMITES

As the notochord and the neural tube form, the intraembryonic mesoderm on each side of them thickens to form longitudinal columns of *paraxial mesoderm* (Fig. 4–7B). Each paraxial column is continuous laterally with the *intermediate mesoderm,* which gradually thins laterally into the *lateral mesoderm.* The lateral mesoderm is continuous with the extraembryonic mesoderm covering the yolk sac and the amnion.

During stage 9 (about 20 days), the paraxial mesoderm begins to divide into paired cuboidal bodies called *somites* (Gr. *soma,* a body). This series of *mesodermal tissue blocks* is located on each side of the developing neural tube (Fig. 4–7C and D). The somites give rise to most of the *axial skeleton* (the skeleton of the head and trunk) and associated musculature, as well as to much of the dermis of the skin (see Chapters 15 and 20).

The first pair of somites develops a short distance caudal to the cranial end of the notochord, and subsequent pairs form in a craniocaudal sequence. About 38 pairs of somites form during the so-called *somite period* (days 20 to 30); eventually, 42 to 44 pairs develop (4 occipital, 8 cervical, 12 thoracic, 5 lumbar, 5 sacral, and 8 to 10 coccygeal pairs). The first occipital and the caudal 5 to 7 coccygeal somites disappear. The others give rise to the axial skeleton. The somites form distinct surface elevations (Fig. 4–7E) and are somewhat triangular in transverse section. A slit-like cavity, the *myocoele,* appears within each somite (Fig. 4–7F), but it soon becomes occluded.

During the somite period, the somites are used as one of the criteria for determining the embryo's age (see Table 5–1).

DEVELOPMENT OF INTRAEMBRYONIC COELOM

The intraembryonic coelom first appears as a number of small, isolated *coelomic spaces* within the lateral mesoderm and the cardiogenic mesoderm (Fig. 4–7A and B). These spaces soon coalesce to form a horseshoe-shaped cavity, the *intraembryonic coelom* (Fig. 4–7E), which is lined by flattened epithelial cells. These cells will form the *mesothelium* lining the peritoneal cavity.

The intraembryonic coelom divides the lateral mesoderm into two layers (Fig. 4–7D): a *somatic (parietal) layer* continuous with the extraembryonic mesoderm covering the amnion; and a *splanchnic (visceral) layer* continuous with the extraembryonic mesoderm covering the yolk sac. The somatic mesoderm and the overlying embryonic ectoderm form the body wall, or *somatopleure* (Fig. 4–7F), whereas the splanchnic mesoderm and the embryonic endoderm form the *splanchnopleure,* or wall, of the primitive gut.

During the second month, the intraembryonic coelom is divided into the body cavities: (1) the *pericardial cavity*, (2) the *pleural cavities,* and (3) the *peritoneal cavity.* For a description of the division of the intraembryonic coelom, see Chapter 9.

PRIMITIVE CARDIOVASCULAR SYSTEM

Angiogenesis (Gr. *angeion,* vessel + *genesis,* production), or blood vessel formation, begins in the extraembryonic mesoderm of the yolk sac, the connecting stalk, and the chorion during stage 6 (13 to 15 days); embryonic vessels begin to develop about two days later. The early formation of the cardiovascular system is correlated with the absence of a significant amount of yolk in the ovum and the yolk sac and the consequent urgent need for vessels to bring nourishment and oxygen from the maternal circulation.

Blood and blood vessel formation may be summarized as follows: (1) Mesenchymal cells, known as *angioblasts,* aggregate to form isolated masses and cords known as *blood islands* (Fig. 4–9A to C); (2) spaces appear within these islands (Fig. 4–9D); (3) angioblasts arrange themselves around the cavity to form the primitive *endothelium* (Fig. 4–9E); (4) isolated vessels fuse to form networks of endothelial channels (Fig. 4–9F); and (5) vessels extend into adjacent areas by *endothelial budding* and by fusion with other vessels formed independently.

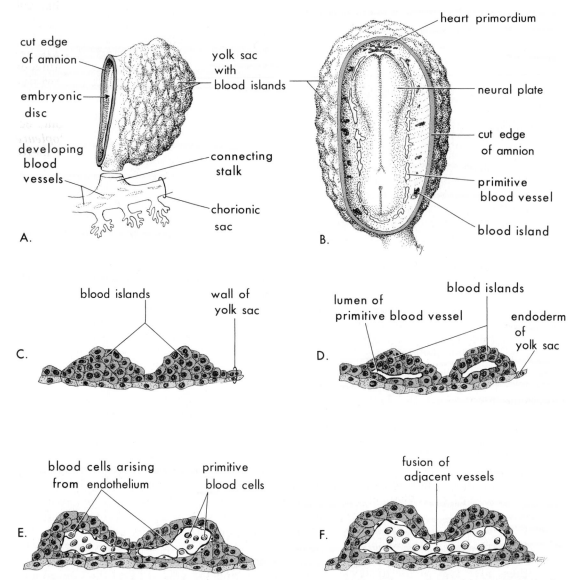

Figure 4–9   Successive stages in the development of blood and blood vessels. *A*, The yolk sac and a portion of the chorionic sac during stage 8 (about 18 days). *B*, Dorsal view showing the embryo exposed by removing the amnion. *C* to *F*, Sections of blood islands showing progressive stages in the development of blood and blood vessels.

Primitive plasma and *blood cells* develop from the endothelial cells as the vessels develop on the *yolk sac* and *allantois* (Fig. 4–9*E*). Blood formation does not begin within the embryo until the fifth week, when it first occurs in various parts of the embryonic mesenchyme, chiefly in the liver and later in the spleen, the bone marrow, and the lymph nodes. Mesenchymal cells surrounding the primitive endothelial vessels differentiate into the muscular and the connective tissue elements of the vessels.

The *primitive heart* forms in a similar manner from mesenchymal cells in the *cardiogenic area* (Figs. 4–6*B* and 4–9*B*). Paired, longitudinal endothelial channels, called *endocardial heart tubes*, develop before the end of the third week and begin to fuse into the primitive heart tube. By the twenty-first day, the fusing heart tubes have linked up with blood vessels in the embryo, the connecting stalk, the chorion, and the yolk sac to form a primitive cardiovascular system (Fig. 4–10). The circulation of blood has almost certainly started by the end of the third week. Thus, *the cardiovascular system is the first organ system to reach a functional state.*

DEVELOPMENT OF CHORIONIC VILLI

Shortly after the primary chorionic villi appear during stage 6 (see Fig. 3–6), they begin to branch. By the end of this stage (about 15 days), mesenchyme has grown into the villi, forming a core of loose connective tissue. The villi at this stage, called *secondary chorionic villi,* cover the entire surface of the chorion (Figs. 4–11 and 4–12*A* and *B*). Soon mesenchymal cells within the villi begin to differentiate into blood capillaries, forming an *arteriocapillary venous network* between stages 7 and 9 (embryos 15 to 20

days old). After blood vessels have developed in the villi, they are called *tertiary chorionic villi* (Figs. 4–10 and 4–12*D*). Vessels in these villi soon become connected with the embryonic heart via vessels that differentiate in the mesenchyme of the chorion and in the connecting stalk (Fig. 4–10). By the end of developmental stage 9 (about 21 days), embryonic blood begins to circulate through the capillaries of the chorionic villi. The villi absorb nutriments from the maternal blood in the intervillous spaces and excrete wastes from the embryo into them.

Concurrently, the cytotrophoblast cells of the chorionic villi penetrate the syncytiotrophoblastic layer and join to form a *cytotrophoblastic shell* (Fig. 4–12*C*), which attaches the chorionic sac to the endometrial tissues. Villi that are attached to the mater-

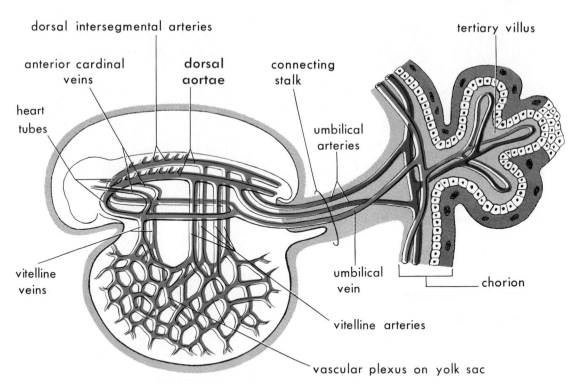

Figure 4–10 Diagram of the primitive cardiovascular system in an embryo during stage 9 of development (about 20 days), viewed from the left side, showing the transitory stage of paired symmetrical vessels. Each heart tube continues dorsally into a *dorsal aorta* that passes caudally. Branches of the aortae are (1) *umbilical arteries,* establishing connections with vessels in the chorion; (2) *vitelline arteries* to the yolk sac; and (3) *dorsal intersegmental arteries* to the body of the embryo. An *umbilical vein* returns blood from the chorion and divides into right and left umbilical veins within the embryo. Vessels on the yolk sac form a *vascular plexus* that is connected to the heart by *vitelline veins.* The *anterior cardinal veins* return blood from the head region. The umbilical vein is shown in red because it carries oxygenated blood and nutrients from the chorion (embryonic part of the placenta) to the embryo. The arteries are colored medium red to indicate that they are carrying partially deoxygenated blood and waste products to the chorion.

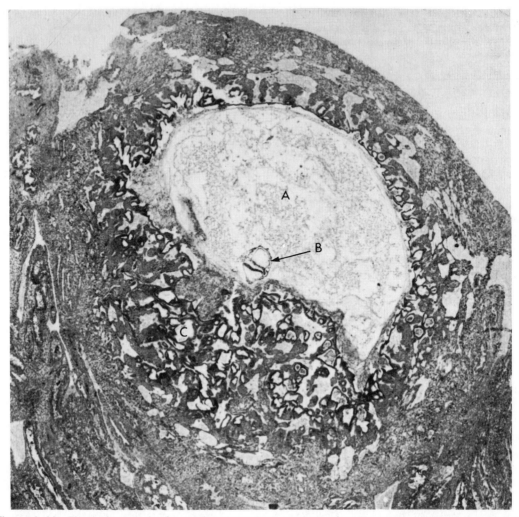

Figure 4–11   Photomicrograph of a section through the endometrium showing an implanted embryo (*B*), during stage 6 of development (about 15 days) (×15). Note the large chorionic cavity (*A*) and many secondary chorionic villi, especially at the embryonic pole (*C*). (Modified slightly from Leeson, C. R., and Leeson, T. S.: *Histology.* 4th ed. Philadelphia, W. B. Saunders Co., 1981.)

nal tissues via the cytotrophoblastic shell are called *stem* or *anchoring villi*. The villi that grow from the sides of the stem villi are called *branch villi*, and it is through them that the main exchange of material between the blood of the mother and the embryo takes place.

SUMMARY OF THIRD WEEK

Major changes occur as the bilaminar embryonic disc is converted into a *trilaminar disc*, composed of *three germ layers*. This process is called *gastrulation*.

The Primitive Streak and Intraembryonic Mesoderm (Figs. 4–1 and 4–2). The primitive streak appears towards the end of developmental stage 6 (about 15 days) as a midline thickening of the embryonic epiblast. It gives rise to cells that migrate ventrally, laterally, and cranially between the epiblast and the hypoblast. As soon as the primitive streak has begun to produce these mesenchymal cells, the epiblast layer is known as the *embryonic ectoderm*, and the hypoblast is known as the *embryonic endoderm*. The cells produced by the primitive streak soon organize into a *third germ layer*, the *intraembryonic mesoderm*.

Notochord Formation (Figs. 4–2, 4–3, 4–5, and 4–6). The *primitive knot* gives

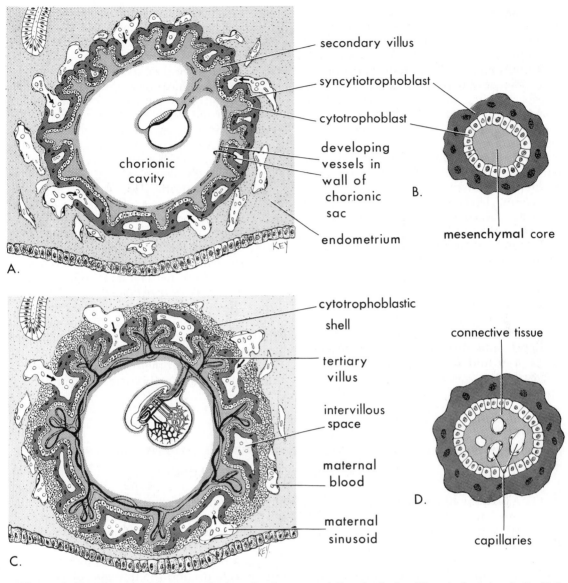

Figure 4–12 Diagrams illustrating further development of the chorionic villi and placenta. *A,* Sagittal section of an embryo during developmental stage 7 (about 16 days). *B,* Section of a secondary chorionic villus. *C,* Section of an implanted embryo during developmental stage 9 (about 21 days). *D,* Section of a tertiary chorionic villus. The fetal blood in the capillaries is separated from the maternal blood surrounding the villus by the placental membrane, composed of the endothelium of the capillary, mesenchyme, cytotrophoblast, and syncytiotrophoblast.

rise to the cells that form the *notochordal process* during developmental stage 7 (about 16 days). Cells from the primitive streak migrate to the edges of the embryonic disc, where they join the extraembryonic mesoderm covering the amnion and the yolk sac. By the end of the third week, mesoderm exists between the ectoderm and the endoderm everywhere except at the *oropharyngeal membrane,* in the midline occupied by

the *notochord* (a derivative of the notochordal process), and at the *cloacal membrane.* The primitive pit extends into the notochordal process to form the *notochordal canal.* Openings develop in the floor of the notochordal canal that soon coalesce, leaving the *notochordal plate.* The notochordal plate infolds to form the *notochord.*

Neural Tube and Neural Crest Formation (Figs. 4–3 and 4–5 to 4–8). The *neural plate* appears as a midline thickening of the embryonic ectoderm, cranial to the primitive knot. A longitudinal *neural groove* develops, which is flanked by *neural folds;* these folds meet and fuse to form the *neural tube.* The development of the neural plate and its infolding to form the neural tube is called *neurulation.* As this process occurs, some cells migrate ventrolaterally to form the *neural crest.*

Somite Formation (Fig. 4–7). The mesoderm on each side of the notochord thickens to form a longitudinal column of *paraxial mesoderm.* Division of the paraxial mesoderm into pairs of *somites* begins cranially by the end of the third week.

Coelom Formation (Fig. 4–7). The intraembryonic coelom arises as isolated spaces in the *lateral mesoderm* and in the *cardiogenic mesoderm.* These coelomic spaces subsequently coalesce to form a single, horseshoe-shaped cavity that eventually gives rise to the body cavities.

Blood and Blood Vessel Formation (Figs. 4–9 and 4–10). Blood vessels first appear on the yolk sac, around the allantois, and in the chorion. They develop within the embryo shortly thereafter. Spaces appear within aggregations of mesenchyme *(blood islands),* which soon become lined with endothelium and unite with other spaces to form a primitive cardiovascular system. At the end of the third week, the *primitive heart* is represented by paired *endocardial heart tubes* that are joined to blood vessels in the embryo and in the extraembryonic membranes. The primitive blood cells are derived mainly from the endothelial cells of blood vessels in the yolk sac and the allantois.

Villi Formation (Fig. 4–12). *Primary chorionic villi* become *secondary chorionic villi* as they acquire mesenchymal cores. Before the end of the third week, capillaries develop in the villi, transforming them into *tertiary chorionic villi.* Cytotrophoblastic extensions from the villi mushroom out and join to form a *cytotrophoblastic shell* that anchors the chorionic sac to the endometrium. The rapid development of chorionic villi during the third week greatly increases the surface area of the chorion for the exchange of nutrients and other substances between the maternal and embryonic circulations.

CLINICALLY ORIENTED PROBLEMS

1. A 30-year-old woman became pregnant two months after discontinuing use of the "pill" (oral contraception), and then had an *early spontaneous abortion.* What might have caused the abortion? What would the doctor likely have told the patient?

2. A 25-year-old woman with a history of regular menstrual cycles was five days *overdue on menses.* Owing to her mental condition and the undesirability of a possible pregnancy, the doctor decided to do a "menstrual extraction," or *uterine evacuation.* The tissue removed was examined for evidence of a pregnancy. What findings would indicate an early pregnancy? How old would the products of conception be?

3. What major system undergoes early development during the third week? What severe congenital malformation might result from teratological factors during this period of development?

+. A female infant was born with a large tumor situated between the rectum and the sacrum. A diagnosis of *protruding sacrococcygeal teratoma* was made, and the mass was surgically removed. What is the probable embryological origin of this tumor? Explain why these tumors often contain various types of tissue derived from all three germ layers. Does an infant's sex make him or her more susceptible to the development of one of these tumors?

The answers to these questions are given at the back of the book.

SUGGESTIONS FOR ADDITIONAL READING

Billington, W. D.: Trophoblast; *in* E. E. Philipp, J. Barnes, and M. Newton (Eds.): *Scientific Foundations of Obstretics and Gynecology*. London, William Heineman, Ltd., 1970, pp. 159–167. A very good account of the trophoblast, explaining its important functions. Also discussed is how knowledge about the trophoblast has clinical applications in early pregnancy, especially with regard to the treatment of patients with pathological disorders of the trophoblast.

REFERENCES

Arey, L. B.: *Development Anatomy: A Textbook and Laboratory Manual of Embryology*. Revised 7th ed. Philadelphia, W. B. Saunders Co., 1974, pp. 342–370.

Bessis, M.: The blood cells and their formation; *in* J. Brachet and A. E. Mirsky (Eds.): *The Cell*. Vol. 5. New York, Academic Press, 1961, pp. 163–217.

Boué J., Boué, A., and Lazar, P.: Retrospective and prospective epidemiological studies of 1500 karyotyped spontaneous abortions. *Teratology 12*:11, 1975.

Carlson, E. N.: Capabilities of gray-scale imaging in obstetrics and gynecology. *Clin. Obstet. Gynecol. 20*:235, 1977.

Carr, D. H.: Chromosomes and abortion. *Adv. Hum. Genet. 2*:201, 1971.

Crooke, A. C.: Human gonadotropins; *in* E. E. Philipp, J. Barnes, and M. Newton (Eds.): *Scientific Foundations of Obstetrics and Gynecology*. London, William Heinemann, Ltd., 1970, pp. 495–508.

Crowley, L. V.: *An Introduction to Clinical Embryology*. Chicago, Year Book Medical Publishers, Inc., 1974.

Dekaban, A. S.: Anencephaly in early human embryos. *J. Neuropathol. and Exp. Neurol. 22*:533, 1963.

Depp, R.: How ultrasound is used by the perinatologist. *Clin. Obstet. Gynecol. 20*:315, 1977.

Hamilton, W. J., and Boyd, J. D.: Development of the human placenta; *in* E. E. Philipp, J. Barnes, and M. Newton (Eds.): *Scientific Foundations of Obstetrics and Gynecology*. London, William Heinemann, Ltd., 1970, pp. 185–254.

Hay, E. D.: Organization and fine structure of epithelium and mesenchyme in the developing chick embryo; *in* R. Fleischmajer and R. E. Billingham (Eds.): *Epithelial-Mesenchymal Interactions*. Baltimore, The Williams & Wilkins Co., 1968, pp. 31–55.

Hamilton, W. J., and Mossman, H. W.: *Human Embryology: Prenatal Development of Form and Function*. 4th ed. Cambridge, W. Heffer and Sons, Ltd., 1972, pp. 69–81. An excellent description of the details of formation of the primitive streak, intraembryonic coelom, and notochord by world-renowned embryologists.

Tuchmann-Duplessis, H., and Haegel, P. (translated by L. S. Hurley): *Illustrated Human Embryology*. Vol. 1. New York, Springer-Verlag, 1972. An excellent aid to students who have difficulty visualizing early stages of development. The emphasis is on visual presentations.

Hertig, A. T.: Angiogenesis in the early human chorion and in the primary placenta of the macaque monkey. *Contrib. Embryol. Carnegie Inst. 25*:37, 1935.

Jolly, H.: *Diseases of Children*. 2nd ed. Oxford, Blackwell Scientific Publications, 1968, pp. 295–296.

Jones, H. O., and Brewer, J. I.: A human ovum in the primitive streak stage. *Contrib. Embryol. Carnegie Inst. 29*:157, 1941.

Leeson, C. R., and Leeson, T. S.: *Histology*. 4th ed. Philadelphia, W. B. Saunders Co., 1981.

O'Rahilly, R.: The manifestation of the axes of the human embryo. *Z. Anat. Entwicklungsgesch. 132*:50, 1970.

O'Rahilly, R.: *Development Stages in Human Embryos. Part A. Embryos of the First Three Weeks (Stages 1 to 9)*. Washington, D.C., Carnegie Institute of Washington, 1973.

Ramsey, E. M.: The placenta and fetal membranes; *in* J. P. Greenhill (Ed.): *Obstetrics*. 14th ed. Philadelphia, W. B. Saunders Co., 1969.

Springer, M.: Die Canalis neurentericus beim Menschen. *Z. Kinderchir. 11*:183, 1972.

Tavares, A.: Sex chromatin in tumors; *in* K. L. Moore (Ed.): *The Sex Chromatin*. Philadelphia, W. B. Saunders Co., 1966, pp. 405–433.

Wide, L.: Immunological determination of human gonadotropins; *in* E. E. Philipp, J. Barnes, and M. Newton (Eds.): *Scientific Foundations of Obstetrics and Gynecology*. London, William Heinemann, Ltd., 1970, pp. 509–514.

Willis, R. A.: *The Borderland of Embryology and Pathology*. 2nd ed. London, Butterworth & Co. Publishers, Ltd., 1962, pp. 442–466.

Wilson, K. M.: A normal human ovum of 16 days development, the Rochester ovum. *Contrib. Embryol. Carnegie Inst. 31*:103, 1945.

5

THE EMBRYONIC PERIOD

Fourth to Eighth Weeks

The embryonic period is a very important period of human development because the beginnings of all major external and internal structures develop during these five weeks. *By the end of the embryonic period, all the main organ systems have begun to develop,* but the function of most organs is minimal. As the organs develop, the shape of the embryo changes. Because the tissues and organs are developing, exposure of an embryo to teratogens during this period may cause major congenital malformations. *Morphogenesis,* the development of form, is an elaborate process during which many complex interactions occur in an orderly sequence.

A teratogen is an agent that produces or raises the incidence of congenital malformations. These agents (e.g., drugs and viruses) have their effect during the stage of active differentiation of an organ or a tissue. *Thalidomide,* the best-known teratogen in humans, caused major *limb malformations* and other congenital defects in infants whose mothers had ingested this drug during early pregnancy (also see Chapters 8 and 17).

FOLDING OF THE EMBRYO

The significant event in the establishment of general body form is folding of the flat trilaminar embryonic disc into a somewhat cylindrical embryo. This folding in both longitudinal and transverse planes is caused by rapid growth of the embryo, particularly of the neural tube. The rate of growth at the sides of the embryonic disc fails to keep pace with the rate of growth along the long axis as the embryo increases rapidly in length.

Although described separately in the following sections, it must be emphasized that *folding at the cranial and caudal ends and at the sides goes on simultaneously.* Concurrently, there is relative constriction at the junction of the embryo and the yolk sac.

FOLDING OF THE EMBRYO IN THE LONGITUDINAL PLANE

Folding of the ends of the embryo produces *head and tail folds* that result in the cranial and caudal regions moving, or "swinging," ventrally as if on hinges (Fig. 5–1A_2 to D_2).

The Head Fold (Fig. 5–2). At the end of the third week, the neural folds begin to develop into the brain and project dorsally into the amniotic cavity. Soon the forebrain grows cranially beyond the oropharyngeal membrane and overhangs the primitive heart. Concomitantly, the *septum transversum* (mesoderm cranial to the pericardial coelom), the *primitive heart,* the *pericardial coelom,* and the *oropharyngeal membrane* turn under onto the ventral surface.

During folding, part of the yolk sac is incorporated into the embryo as the *foregut;* it lies between the brain and the heart and

Figure 5–1 Drawings of embryos during the fourth week illustrating folding in both longitudinal and transverse planes. A_1, Dorsal view of an embryo at stage 10 (about 22 days). The continuity of the intraembryonic coelom and extraembryonic coelom is illustrated on the right side by removal of a portion of the embryonic ectoderm and mesoderm. B_1, C_1 and D_1, Lateral views of embryos at stages 11, 12, and 13 (about 24, 26, and 28 days, respectively). A_2 to D_2, Longitudinal sections at the plane shown in A_1. A_3 to D_3, Transverse sections at the levels indicated in A_1 to D_1. As the result of organ formation, the shape of the embryo changes, and major features of body form are established.

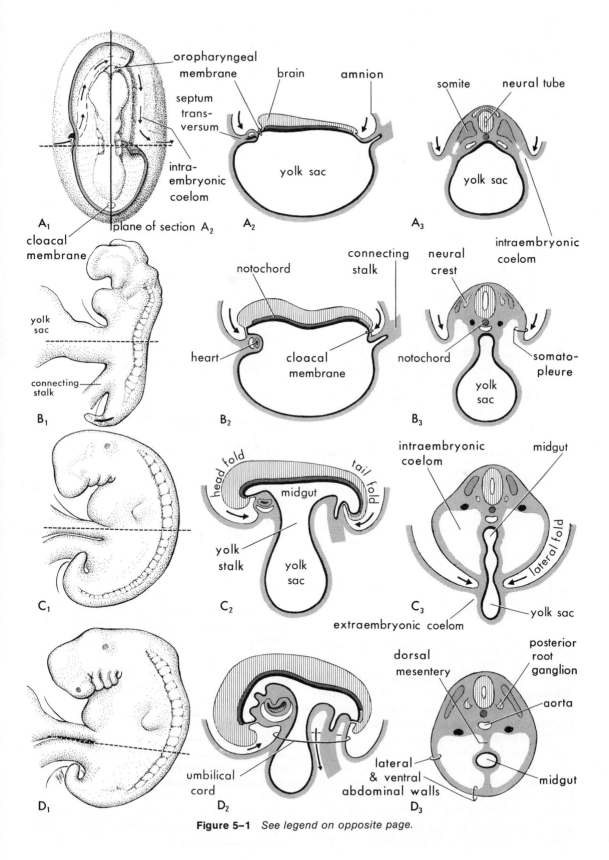

Figure 5–1 *See legend on opposite page.*

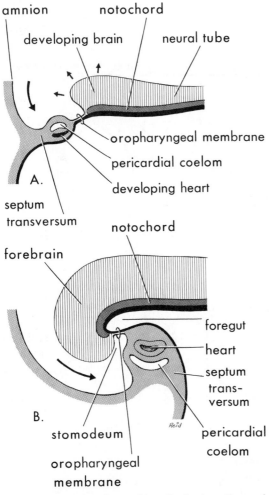

amnion   notochord

developing brain   neural tube

oropharyngeal membrane

pericardial coelom

A.

developing heart

septum
transversum

notochord

forebrain

foregut

heart

septum
trans-
versum

B.

Reid

stomodeum

pericardial
coelom

oropharyngeal
membrane

Figure 5–2 Drawings of longitudinal sections of the cranial region of four-week embryos showing the effect of the head fold on the position of the heart and other structures. *A*, Stage 10 (22 days). *B*, Stage 12 (26 days). Note that the septum transversum, heart, pericardial coelom, and oropharyngeal membrane turn under onto the ventral surface. Observe also that part of the yolk sac is incorporated into the embryo as the foregut.

ends blindly at the oropharyngeal membrane (Fig. 5–2*B*). This membrane separates the foregut from the *stomodeum,* or primitive mouth cavity. After folding, the septum transversum lies caudal to the heart, where it subsequently develops into the ventral part of the diaphragm (see Fig. 9–7).

The head fold also affects the arrangement of the intraembryonic coelom. Before folding, the coelom consists of a flattened horseshoe-shaped cavity (Fig. 5–1*A*₁); after folding, the pericardial coelom lies ventrally, and the pericardioperitoneal canals run dorsally over the septum transversum and join

the *peritoneal coelom* (Fig. 5–3*B*). At this stage the peritoneal coelom on each side communicates widely with the extraembryonic coelom (Fig. 5–3*A* and *B*).

The Tail Fold (Fig. 5–4). Folding of the caudal end of the embryo occurs a little later than that of the cranial end. The tail fold primarily results from dorsal and caudal growth of the neural tube. As the embryo grows, the tail region projects over the *cloacal membrane;* eventually this membrane lies ventrally (Fig. 5–4*B*).

During folding, part of the yolk sac is incorporated into the embryo as the *hindgut* (Fig. 5–4*B*). The terminal part of the hindgut soon dilates slightly to form the *cloaca;* it is separated from the amniotic cavity by the cloacal membrane. Before folding, the primitive streak lies cranial to the cloacal membrane (Fig. 5–4*A*); after folding, it lies caudal to this membrane (Fig. 5–4*B*). The *connecting stalk* now attaches to the ventral surface of the embryo, and the *allantois* is partially incorporated into the embryo.

FOLDING OF THE EMBRYO IN THE TRANSVERSE PLANE

Folding of the sides of the embryo produces right and left *lateral folds* (Fig. 5–1*A*₃ to *D*₃). Each lateral body wall, or somatopleure, folds toward the midline, rolling the edges of the embryonic disc ventrally and forming a roughly cylindrical embryo. As the lateral and ventral body walls form, part of the yolk sac is incorporated into the embryo as the *midgut* (Fig. 5–1*D*₂ and *D*₃). Concurrently, the connection of the midgut with the yolk sac is reduced to a narrow *yolk stalk*. After *folding,* the region of attachment of the amnion to the embryo is reduced to a relatively narrow region, *the umbilicus,* on the ventral surface.

As the midgut is separated from the yolk sac, it remains attached to the dorsal abdominal wall by a thick *dorsal mesentery* (Fig. 5–1*D*₃). As the umbilical cord forms, ventral fusion of the lateral folds reduces the region of communication between the intraembryonic and extraembryonic coeloms to a narrow communication that persists until about the tenth week (see long arrow in Fig. 5–1*D*₂). As the amniotic cavity expands and obliterates the extraembryonic coelom, *the amnion forms the epithelial covering for the umbilical cord* (Figs. 5–1*D*₂ and 7–17).

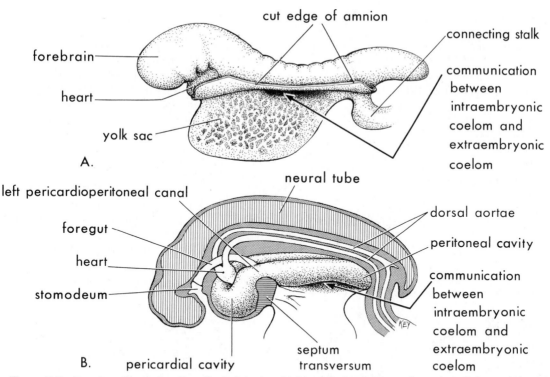

Figure 5–3 Drawings illustrating the effect of the head fold on the intraembryonic coelom. *A,* Lateral view of an embryo at stage 11 (24 to 25 days) during folding, showing (1) the large forebrain, (2) the ventral position of the heart, and (3) the communication between the intraembryonic coelom and the extraembryonic coelom. *B,* Schematic drawing of an embryo at stage 12 (26 to 27 days) after folding, showing (1) the pericardial cavity ventrally, (2) the pericardioperitoneal canals running dorsally on each side of the foregut, and (3) the peritoneal cavity in communication with the extraembryonic coelom.

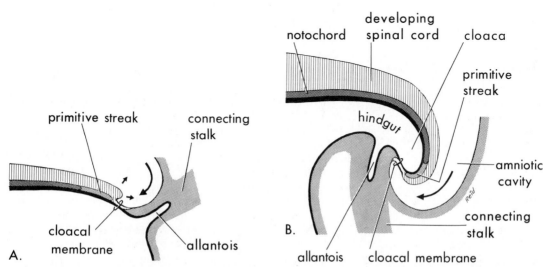

Figure 5–4 Drawings of longitudinal sections of the caudal area of four-week embryos, showing the effect of the tail fold on the position of the cloacal membrane and other structures. *A,* Stage 12 (26 to 27 days). Note that part of the yolk sac is incorporated into the embryo as the hindgut, and that the terminal part of the hindgut soon dilates to form the cloaca. Observe the change in position of the primitive streak, the allantois, the cloacal membrane, and the connecting stalk.

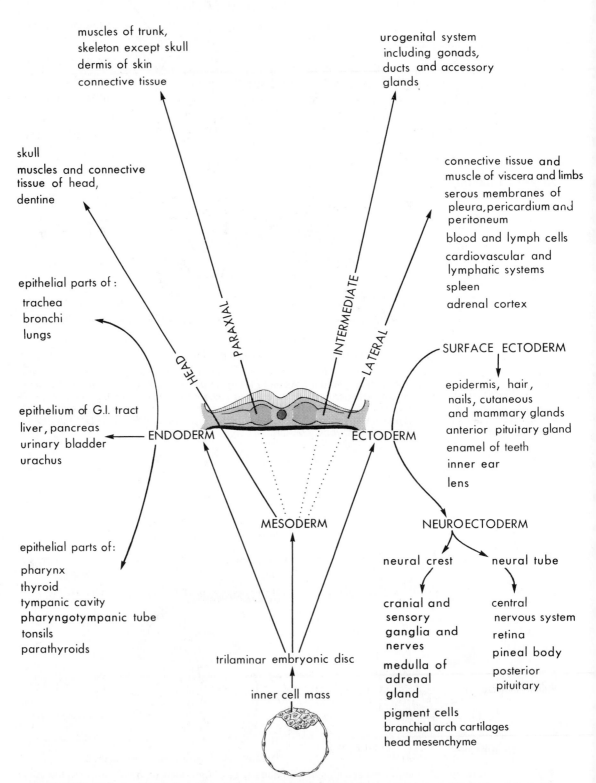

muscles of trunk,
skeleton except skull
dermis of skin
connective tissue

urogenital system
including gonads,
ducts and accessory
glands

skull
muscles and connective
tissue of head,
dentine

connective tissue and
muscle of viscera and limbs
serous membranes of
pleura, pericardium and
peritoneum
blood and lymph cells
cardiovascular and
lymphatic systems
spleen
adrenal cortex

epithelial parts of :
trachea
bronchi
lungs

PARAXIAL

INTERMEDIATE

LATERAL

HEAD

SURFACE ECTODERM

epidermis, hair,
nails, cutaneous
and mammary glands
anterior pituitary gland
enamel of teeth
inner ear
lens

epithelium of G.I. tract
liver, pancreas
urinary bladder
urachus

ENDODERM

ECTODERM

MESODERM

NEUROECTODERM

epithelial parts of:
pharynx
thyroid
tympanic cavity
pharyngotympanic tube
tonsils
parathyroids

neural crest

neural tube

cranial and
sensory
ganglia and
nerves
medulla of
adrenal
gland
pigment cells
branchial arch cartilages
head mesenchyme

central
nervous system
retina
pineal body
posterior
pituitary

trilaminar embryonic disc

inner cell mass

Figure 5–5 Scheme illustrating the origin and derivatives of the three germ layers. The cells of these layers make specific contributions to the formation of the different tissues and organs.

GERM LAYER DERIVATIVES

The three germ layers (embryonic ecto-derm, mesoderm, and endoderm) give rise to all the tissues and organs of the embryo. In addition to the derivatives shown in Figure 5–5, other tissues may originate from these germ layers under different normal or experimental influences; i.e., the specificity of the germ layers is not rigidly fixed. The cells of each germ layer divide, migrate, aggregate, and differentiate in rather precise patterns as they form the various organ systems. Tissues that develop from the different germ layers are commonly associated in the formation of an organ (*organogenesis*).

In general, the main germ layer derivatives are as follows:

ECTODERM. This layer gives rise to the central nervous system (brain and spinal cord), the peripheral nervous system, the sensory epithelia of the eye, the ear, and the nose, the epidermis and its appendages (hair and nails), the mammary glands, the hypophysis (pituitary gland), the subcutaneous glands, and the enamel of teeth.

Neural crest cells, derived from ectoderm (see Figs. 4–8, 5–5, and 18–8), give rise to the following: cells of the *spinal, cranial,* and *autonomic ganglia;* ensheathing cells of the peripheral nervous system; *pigment cells* of the dermis; muscle, connective tissues, and

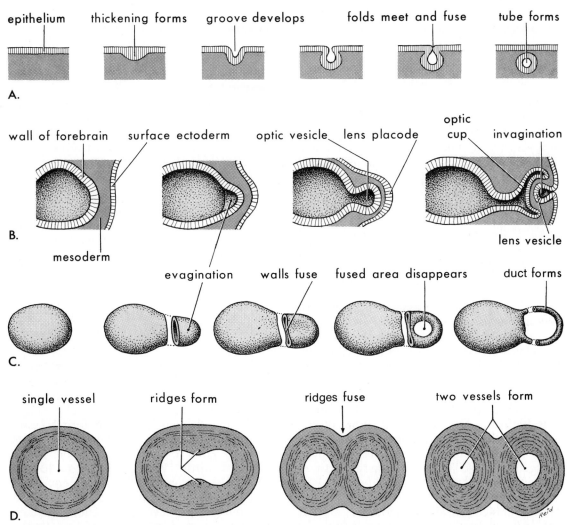

Figure 5–6 Diagrams illustrating various developmental processes used to form *A,* Tubes (e.g., neural tube), *B,* Vesicles (e.g., optic vesicle), and *C,* Ducts (e.g., semicircular duct); and to divide *D,* Tubular structures (e.g., large blood vessels).

bone of branchial arch origin (see Chapter 10); the *suprarenal (adrenal) medulla,* and the pia-arachnoid *(leptomeninges)*

MESODERM. This layer gives rise to cartilage, bone and connective tissue, striated and smooth muscles, the heart, blood and lymph vessels and cells, the kidneys, the gonads (ovaries and testes) and the genital ducts, the serous membranes lining the body cavities (pericardial, pleural, and peritoneal), the spleen, and the cortex of the suprarenal (adrenal) gland.

ENDODERM. This layer gives rise to the epithelial lining of the gastrointestinal and respiratory tracts, the parenchyma of the tonsils, the thyroid gland, the parathyroid glands, the thymus, the liver, and the pancreas, the epithelial lining of the urinary bladder and the urethra, and the epithelial lining of the tympanic cavity, the tympanic antrum, and the auditory tube.

CONTROL OF DEVELOPMENT

Development results from genetic plans contained in the chromosomes. The genetics of human development is a subject that is still very poorly understood (Thompson and Thompson, 1980). Much of our knowledge of human development has come from studies in animals, because there are ethical problems associated with the use of human embryos for laboratory studies.

Most developmental processes depend upon a precisely coordinated interaction of genetic and environmental factors. There are several control mechanisms that guide differentiation and ensure synchronized development, e.g., tissue interactions, regulated migrations of cells and cell colonies, controlled proliferations, and cell death. Each system of the body has its own developmental pattern, but most processes of morphogenesis are similar and are relatively simple (Fig. 5–6). Underlying all these changes are basic inducing and regulating mechanisms.

Induction. For a limited time during early development, certain embryonic tissues markedly influence the development of adjacent tissues. The tissues producing these influences or effects are called *inductors* or organizers. Induction involves two tissues: (1) the inducing tissue or inductor, and (2) the induced tissue. In order to induce, an inductor must be close to, but not necessarily in contact with, the tissue to be induced. In birds, and probably in humans, the primitive streak, the notochord, and the paraxial mesoderm act as primary organizers of the central nervous system (Fig. 5–7).

Once the basic embryonic plan has been established by primary inductors, a chain of *secondary inductions* occurs. Development of the eye is a good example (Fig. 5–6*B*): (1) development of the forebrain is induced by primary organizers, probably the notochordal process and the paraxial mesoderm; (2) the forebrain then reacts to the secondary inductive action of adjacent mesenchyme, producing an evagination called an *optic vesicle,* which soon becomes an *optic cup,* the primordium of the eye; (3) next the optic vesicle induces the adjacent surface ectoderm of the head to thicken into a *lens placode* (Figs. 5–6*B* and 5–8*E*), which soon invaginates to form a *lens vesicle,* the primordium of the lens, and (4) the lens induces the epidermis over it to become the *corneal epithelium* (see Chapter 19).

The nature of the inductive agents is not clearly understood at present, but it is generally accepted that some substance (protein in nature) passes from the inducing tissue to the induced tissue. Yamada (1967) states, "The possibility is not completely excluded that a group of small molecules are the real factors and are carried by some specific macromolecules." The observation that treatment of inducing substances with trypsin and pepsin destroys their inductive action has led to the conclusion that inducing substances are probably bound to proteins or ribonucleoproteins. For more information on the morphological and biochemical aspects of differentiation, see Balinsky (1975).

HIGHLIGHTS OF THE EMBRYONIC PERIOD

The details of organ formation are given with discussions of the various systems (see Chapters 11 to 20). The following descriptions summarize the main developmental events and changes in external form. Useful criteria for estimating developmental stages in human embryos are listed in Table 5–1.

The Fourth Week (Stages 10 to 13). During stage 10 of development (22 to 23 days), the embryo is almost straight and the *somites* produce conspicuous surface elevations (Figs. 5–8*A* and *B* and 5–9). By the end of stage 10, the neural tube is closed

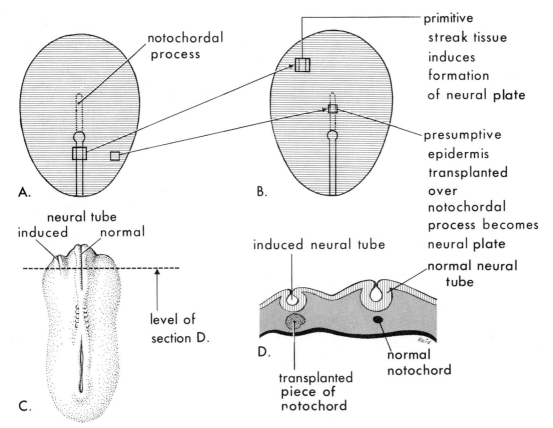

Figure 5–7 *A* and *B*, Dorsal views of avian embryonic discs illustrating experimental induction of the neural plate. *C*, Dorsal view of an embryo showing a normal and an induced neural tube. *D*, Transverse section through the embryo at the level shown, illustrating the secondary neural tube that was induced by the transplanted piece of notochord. (Based on experiments by Waddington, 1952.)

opposite the somites but is widely open at the rostral and caudal *neuropores* (Fig. 5–9*B*).

By stage 11 of development (about 24 days), the first (or mandibular) and the second (or hyoid) *branchial arches* are distinct (Figs. 5–8*C*, 5–10 and 5–11). The major portion, or mandibular prominence, of the first arch will give rise to the lower jaw, or *mandible,* and a rostral extension of it, the maxillary prominence, will contribute to the upper jaw, or *maxilla* (see Fig. 10–20). A slight curve is produced in the embryo by the head and tail folds, and the heart produces a large ventral prominence (Figs. 5–8*C* and 5–10).

Three pairs of branchial arches are visible by stage 12 of development (about 26 days) (Figs. 5–8*D* and 5–12), and the rostral neuropore is closed. The forebrain produces a prominent elevation of the head, and flexion,

or folding, of the embryo in the longitudinal plane has given the embryo a characteristic C-shaped curvature. Narrowing of the connection between the embryo and the yolk sac has been produced by folding in the transverse plane (Fig. 5–1*C₃*). The *upper limb buds* also become recognizable as small swellings on the ventrolateral body walls during stage 12 (Fig. 5–8*D*). The *otic pits,* the primordia of the inner ears, are also clearly visible.

The fourth pair of branchial arches and the *lower limb buds* are present by stage 13 of development, about 28 days (Figs. 5–8*E* and 5–13). Lens placodes, ectodermal thickenings indicating the future lenses of the eyes, are visible on the sides of the head. By the end of the fourth week, the *attenuated tail* with its somites is a characteristic feature (Fig. 5–13).

Text continued on page 85

TABLE 5-1 CRITERIA FOR ESTIMATING DEVELOPMENTAL STAGES IN HUMAN EMBRYOS

Age* (Days)	Figure Reference	Carnegie Stage	No. of Somites	Length (mm)†	Main Characteristics‡
20–21	5–1A_1	9	1–3	1.5–3.0	*Deep neural groove and first somites present.* Head fold evident.
22–23	5–8A 5–8B 5–9	10	4–12	2.0– 3.5	*Embryo straight or slightly curved.* Neural tube forming or formed opposite somites, but widely open at rostral and caudal neuropores. First and second pairs of branchial arches visible.
24–25	5–8C 5–10 5–11	11	13–20	2.5– 4.5	*Embryo curved owing to head and tail folds.* Rostral neuropore closing. Otic placodes present. Optic vesicles formed.
26–27	5–8D 5–12	12	21–29	3.0– 5.0	*Upper limb buds appear.* Caudal neuropore closing or closed. Three pairs of branchial arches visible. Heart prominence distinct. Otic pits present.
28–30	5–8E 5–13	13	30–35	4.0– 6.0	*Embryo has C-shaped curve. Upper limb buds are flipper-like.* Four pairs of branchial arches visible. Lower limb buds appear. *Otic vesicles* present. Lens placodes distinct. Attenuated *tail* present.
31–32	5–14A	14	§	5.0– 7.0	*Upper limbs are paddle-shaped.* Lens pits and nasal pits visible. Optic cups present.
33–36	5–14B	15		7.0– 9.0	*Hand plates formed.* Lens vesicles present. Nasal pits prominent. *Lower limbs are paddle-shaped.* Cervical sinus visible.
37–40		16		8.0–11.0	*Foot plates formed.* Pigment visible in retina. Auricular hillocks developing.
41–43	5–14C 5–15	17		11.0–14.0	*Digital, or finger, rays appear.* Auricular hillocks outline future auricle of external ear. Trunk beginning to straighten. Cerebral vesicles prominent.
44–46	5–16	18		13.0–17.0	*Digital, or toe, rays appearing.* Elbow region visible. Eyelids forming. Notches between finger rays. Nipples visible.
47–48	5–17A 5–18D	19		16.0–18.0	*Limbs extend ventrally.* Trunk elongating and straightening. Midgut herniation prominent.
49–51	5–17B 5–18B	20		18.0–22.0	*Upper limbs longer and bent at elbows. Fingers distinct but webbed.* Notches between toe rays. Scalp vascular plexus appears.
52–53	5–19A	21		22.0–24.0	*Hands and feet approach each other. Fingers are free and longer.* Toes *distinct* but webbed. Stubby tail present.
54–55		22		23.0–28.0	*Toes free and longer.* Eyelids and auricles of external ears are more developed.
56	5–19B 5–20	23		27.0–31.0	*Head more rounded and shows human characteristics.* External genitalia still have sexless appearance. Distinct bulge still present in umbilical cord; caused by herniation of intestines. *Tail has disappeared.*

See footnotes on facing page

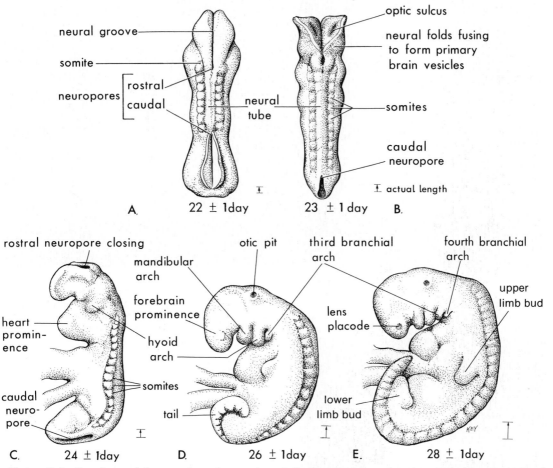

Figure 5–8 Drawings of four-week embryos. *A* and *B,* Dorsal views of embryos during stage 10 of development (22 to 23 days) showing 8 to 12 somites, respectively. *C, D,* and *E,* Lateral views of embryos during stages 11, 12, and 13 of development (24 to 28 days) showing 16, 27, and 33 somites, respectively.

*There is still uncertainty about the age of embryos in some embryonic stages. The ages given by Streeter (1951) were based on comparison with macaque embryos and are now known to be inaccurate for stages 14 to 23. For example, embryos at stage 23 are now generally believed to be at least 56 days and not 47 ± 1 days as described by Streeter.

†The embryonic lengths indicate the usual range, but do not indicate the full range within a given stage, especially when specimens of poor quality are included (O'Rahilly, 1973). In stages 10 and 11, the measurement is greatest length *(GL)*; in subsequent stages crown-rump *(CR)* measurements are given (Fig. 5–21).

‡Based on Streeter (1942, 1945, 1948, and 1951), O'Rahilly (1973), and Nishimura et al. (1974).

§At this and subsequent stages, the number of somites is difficult to determine and so is not a useful criterion.

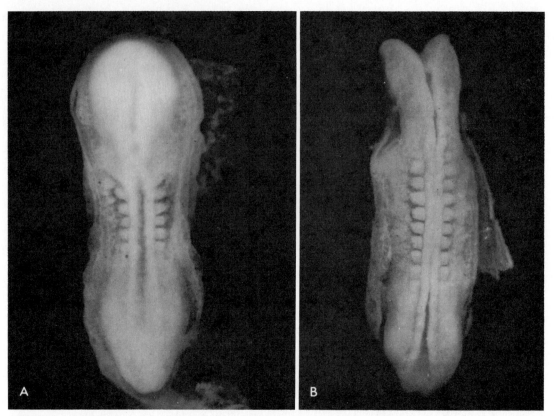

Figure 5–9 Photographs of embryos during stage 10 of development (22 to 23 days). In *A,* the embryo is essentially straight, whereas the embryo in *B* is slightly curved. In *A,* the neural groove is deep and is open throughout its entire extent. About one half of the longitudinal extent of the groove represents the future brain. In *B,* the neural tube has formed opposite the somites but is widely open at the rostral and caudal neuropores. Compare with Figure 5–8*A.* (Courtesy of Professor Hideo Nishimura, Kyoto University, Kyoto, Japan.)

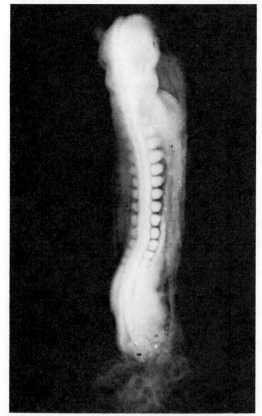

Figure 5–10 Photograph of an embryo during stage 11 of development (24 to 25 days). Ten of the 13 pairs of somites are easily recognized. The embryo is curved owing to folding at the cranial and caudal ends. The rostral neuropore is almost closed, and the caudal neuropore is closing. (Courtesy of Professor Hideo Nishimura, Kyoto University, Kyoto, Japan.)

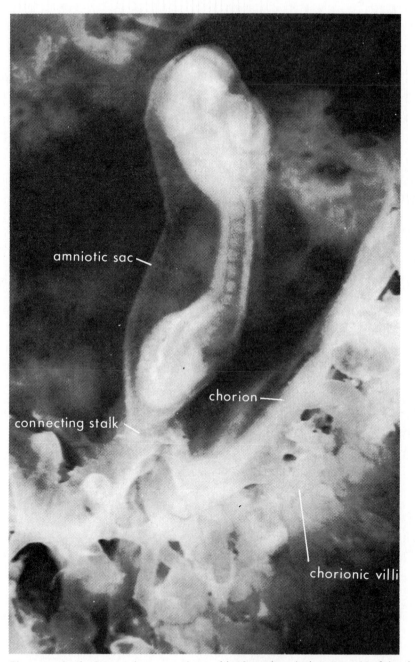

Figure 5–11   Photograph of a 3.1-mm human embryo with 13 somites during stage 11 of development (24 to 25 days). The embryo lies within its amniotic sac and is attached to the chorion by the connecting stalk. Note the well-developed chorionic villi. (Courtesy of Professor E. Blechschmidt, University of Göttingen, Göttingen, West Germany.)

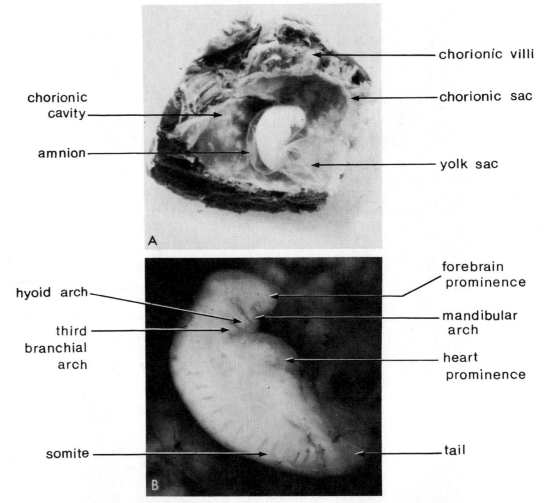

chorionic villi

chorionic sac

chorionic cavity

amnion

yolk sac

A

forebrain prominence

hyoid arch

mandibular arch

third branchial arch

heart prominence

somite

tail

B

Figure 5–12   *A,* Photograph of an embryo in its amniotic sac during stage 12 of development (26 to 27 days). The embryo was exposed by opening the chorionic sac (×5). *B,* Higher magnification of the 3.5-mm (crown-rump length) embryo (×18). Although present, the upper limb bud (Fig. 5–8*D*) is not visible in this photograph.

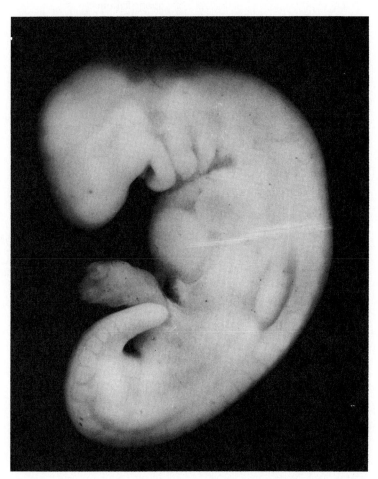

Figure 5–13 Photograph of an embryo during stage 13 (28 days). The embryo has a characteristic C-shaped curvature, four branchial arches, and upper and lower limb buds. The lower limb bud (Fig. 5–8E) is not recognizable in this photograph. The heart prominence is easily recognized. The ventrally-curled attenuated tail, with its somites, is a characteristic feature of this stage. (Courtesy of Professor Hideo Nishimura, Kyoto University, Kyoto, Japan.)

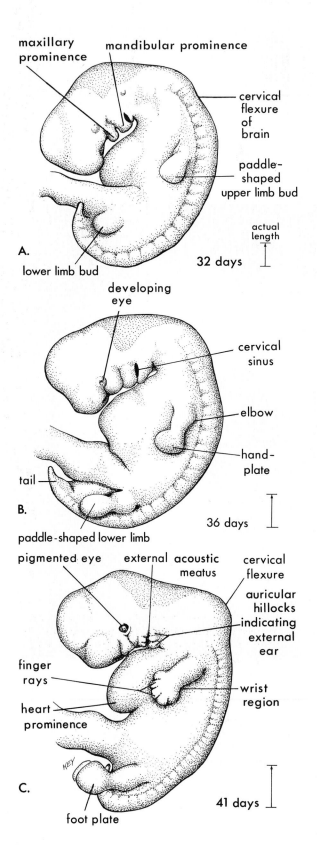

Figure 5–14 Drawings of lateral views of embryos during the fifth and sixth weeks. *A*, Stage 14, *B*, Stage 15. *C*, Stage 17. Note that the tail is undergoing retrogressive changes. Compare with Fig. 5–13.

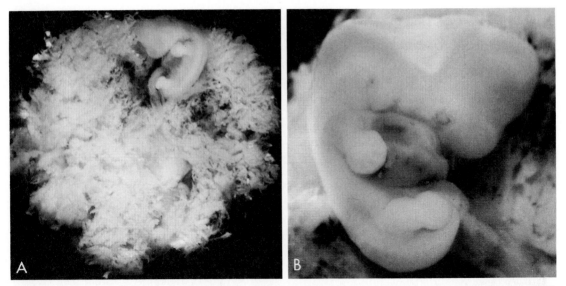

Figure 5–15   *A,* Photograph of an embryo in its amniotic sac, exposed by opening the chorionic sac (×2). *B,* Higher magnification of the 12-mm (crown-rump length) embryo during stage 17 (41 to 42 days) (×6). Note the large size of the head compared with the rest of the body, and the prominence of the cerebral vesicles, the primordia of the cerebral hemispheres.

The Fifth Week (Stages 14 and 15).   Changes in body form are minor compared with those that occur during the fourth week, but growth of the head exceeds that of other regions (Fig. 5–14*A* and *B*). This extensive head growth is caused mainly by the rapid development of the brain. The face soon contacts the heart prominence (Fig. 5–14*A*). The second, or hyoid, arch overgrows the third and fourth arches, forming an ectodermal depression known as the *cervical sinus* (Fig. 5–14*B*). The upper limbs begin to show some regional differentiation as the hand plates develop.

The Sixth Week (Stages 16 and 17).   The limb buds show considerable region differentiation, especially the upper limbs (Fig. 5–14*C*). The elbow and wrist regions become identifiable, and the paddle-shaped hand plates develop ridges, called *digital,* or *finger, rays,* indicating the future *digits* (fingers and thumb). Note that development of the lower limb occurs somewhat later than that of the upper limb.

Several small swellings develop around the branchial groove between the first two branchial arches (Fig. 5–14*C*); this groove becomes the *external acoustic meatus,* and the swellings eventually fuse to form the auricle of the external ear (see Fig. 19–17). Largely because retinal pigment begins to

appear, the eye becomes more obvious (Figs. 5–14*C* and 5–15*B*).

The head is now much larger relative to the trunk and is more bent over the *heart prominence* (Fig. 5–15*B*). This head position results from bending *(cervical flexure)* of the brain in the cervical region. The trunk and neck have begun to straighten. The somites are visible in the lumbosacral region until the middle of the week, but are not useful criteria for estimating age at this time.

The Seventh Week (Stages 18 and 19).   The communication between the primitive gut and the yolk sac has been reduced to a relatively small duct, the *yolk stalk* (Fig. 5–16). The intestines enter the extraembryonic coelom in the proximal portion of the umbilical cord; this is called *umbilical herniation.*

The limbs undergo considerable change during the seventh week. The upper limbs project over the heart (Figs. 5–16 and 5–17*A*). Notches appear between the digital rays in the hand plates, indicating the future digits (Fig. 5–17*A*).

The Eighth Week (Stages 20 to 23).   At the beginning of the final week of the embryonic period, the digits of the hand are short and noticeably webbed (Figs. 5–17*B* and 5–18*B*). Notches are visible between the digital, or toe, rays, and the tail is still visible,

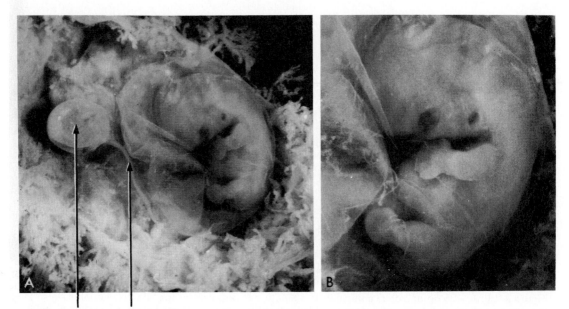

yolk sac   yolk stalk

Figure 5–16 *A,* Photograph of an embryo during stage 18 (44 to 46 days) in its amniotic sac, exposed by opening the chorionic sac (×2.8). *B,* Higher magnification of the 14-mm (crown-rump length) embryo (×5). Note the low position of the ear at this stage and the notches between the finger rays (Fig. 5–14*C*) in the hand plate.

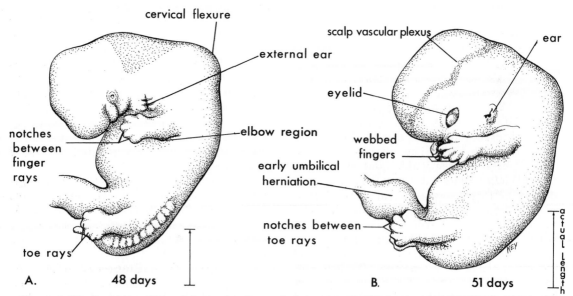

Figure 5–17 Drawings of lateral views of embryos during stages 19 (48 days) and 20 (51 days). *A,* Note that the limbs extend ventrally and that there are notches between the finger rays. *B,* Observe the short, stubby fingers with webbing between them. Beginning at stage 20, the position of the scalp vascular plexus (subcutaneous capillary plexus) is a good indicator of the stage of development.

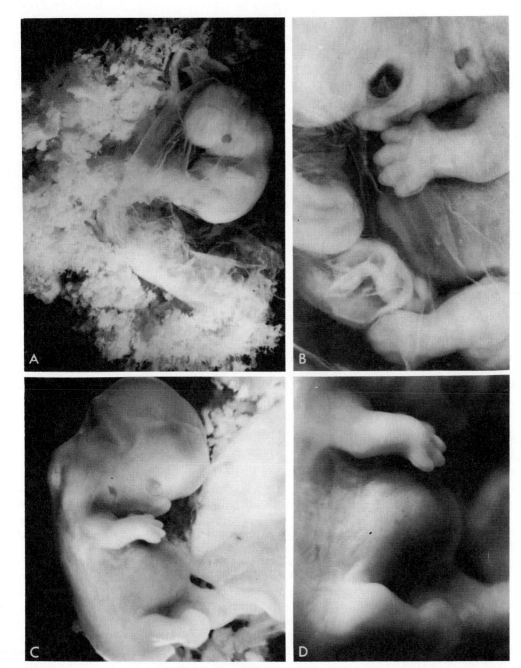

Figure 5–18 *A,* Photograph of an embryo in its amniotic sac, exposed by opening the chorionic sac (×2). *B,* Higher magnification of this 18.5-mm (crown-rump length) embryo during stage 20 (49 to 51 days) (×7). Note the webbed fingers and the notches between the toe rays. *C,* Photograph of a slightly younger six-week embryo, exposed by removal from the chorionic and amniotic sacs (×4). *D,* Higher magnification of this 18-mm (crown-rump length) embryo at stage 19 (47 to 48 days) (×7). The large ventral abdominal prominence is caused mainly by the liver; most of the intestine is in the proximal portion of the umbilical cord.

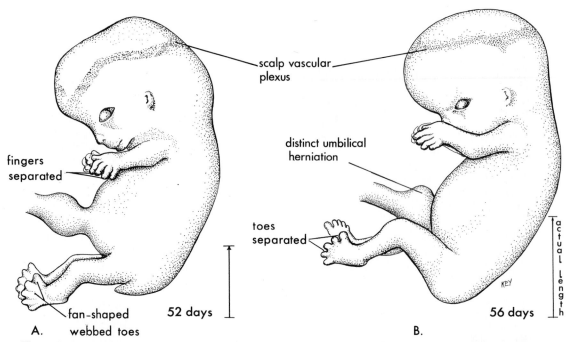

Figure 5–19 Drawings of lateral views of embryos during the eighth week. *A*, Stage 21 (52 days). *B*, Stage 23 (56 days). Note its human characteristics and that the scalp vascular plexus is near the vertex, or crown, of the head.

but it is stubby (Fig. 5–19*A*). The scalp vascular plexus has appeared and forms a characteristic band around the head (Fig. 5–17*B*).

By the end of the eighth week, the regions of the limbs are apparent, the fingers have lengthened, and the toes are distinct. All evidence of the tail disappears by the end of the eighth week (Figs. 5–19*B* and 5–20). The scalp vascular plexus now forms a band near the vertex (crown) of the head (Fig. 5–19*B*).

The embryo now has unquestionably human characteristics. The head is more round and erect, but is still disproportionately large, constituting almost half of the embryo. The neck region has become established, and the eyelids are more obvious. The abdomen is less protuberant, and the umbilical cord is relatively reduced in size. The intestines are still within the proximal portion of the cord (Fig. 5–20).

During the eighth week, the eyes are usually open, but toward the end of the week, the eyelids may begin to meet and become united by epithelial fusion. The auricles of the external ears begin to assume their final shape, but they are still low-set. Although sex differences exist in the appearance of the external genitalia, they are not distinct enough to permit accurate sexual identification to be made by lay persons.

DISORDERS OF EMBRYONIC DEVELOPMENT

As has been described, the embryonic period refers to the five weeks during which *morphogenesis* (development of shape) and *organogenesis* (formation of organs) occur. Consequently, it is the time during which the embryo is most susceptible to factors that may interfere with its development. Most congenital malformations originate during this critical period of embryonic development.

Dysmorphism (abnormality of shape) can result from a variety of mechanisms: (1) single-gene disorders such as *achondroplasia* (see Fig. 8–12); (2) chromosome disorders such as *Down syndrome* (see Fig. 8–4); (3) multifactorial inheritance, resulting from a *combination of genetic and environmental factors,* which accounts for much of the normal variation in families and many common congenital malformations such as cleft lip (see Fig. 10–26); and (4) *teratogene-*

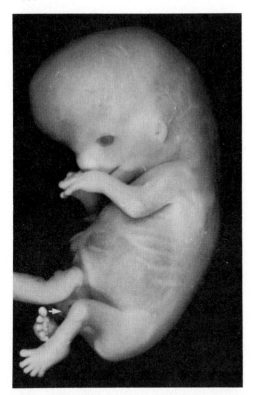

Figure 5–20 Photograph of a 29-mm (crown-rump length) embryo during stage 23 (56 days) (×2). The intestine is still in the umbilical cord (arrow). The digits are clearly defined. The regions of the limbs are apparent, and the tail has disappeared. Note the relatively large head.

sis (production of abnormalities by drugs, chemicals, and viruses known as *teratogens*). Exposure of an embryo to a teratogen (e.g., thalidomide) during the early embryonic period is likely to cause a congenital malformation (see Fig. 17–4).

Abuse of alcohol and heavy smoking during pregnancy are well-established causes of intrauterine growth retardation (Golbus, 1980). Many drugs and chemicals are teratogenic in rodents, but they have demonstrated no teratogenicity in humans. Unfortunately, drugs known to be nonteratogenic or "safe" in animals may cause human malformations if given during the embryonic period. Hence, obstetricians and other physicians who care for the embryo *expose the pregnant patient only to the most necessary therapeutic agents,* and only then after considering their possible effects on the embryo. For a complete discussion of the causes of congenital malformations, see Chapter 8.

ESTIMATION OF EMBRYONIC AGE

Determination of the starting date of pregnancy may be difficult, partly because it depends on the mother's memory. Two reference points are commonly used for estimating age: (1) the onset of the *last menstrual period* (LMP), and (2) the time of *fertilization.* The probability of error in establishing the last normal menses is highest in women who become pregnant after cessation of oral contraception; this is because the interval between discontinuance of the hormones and onset of ovulation is highly variable (Page et al., 1981). In addition, uterine bleeding (or "spotting") sometimes occurs after implantation of the blastocyst and may be incorrectly regarded as menstruation. In spite of these possible sources of error, LMP is commonly used by clinicians to estimate the age of embryos.

It must be emphasized that the zygote does not form until about two weeks after the onset of the last normal menstrual period. Consequently, 14 ± 2 days must be deducted from the so-called menstrual age to obtain the actual or *fertilization age* of an embryo. The day fertilization occurs is the most accurate reference point for estimating age; this is commonly calculated from the estimated time of ovulation because the ovum is usually fertilized within 12 hours after ovulation.

Because it may be important to know the fertilization age of an embryo, e.g., for determining its sensitivity to teratogenic agents (see Chapter 8), all statements about age should indicate the reference point used, i.e., weeks after LMP or days after ovulation, or the estimated time of fertilization.

Estimates of age of recovered embryos (e.g., after abortion) are determined from external characteristics and measurements of length (Table 5–1). Size alone may be an unreliable criterion because some embryos undergo a progressively slower rate of growth prior to death. The changing appearance of the developing limbs is a very useful criterion for estimating embryonic age.

Methods of Measurement (Fig. 5–21). Because embryos of the third and early fourth weeks (stages 6 to 10) are nearly straight, measurements indicate the *greatest length (GL).* The sitting height, or *crown-rump length (CR),* is most frequently used for older embryos. In those with greatly flexed heads, the CR length is actually a neck-rump measurement (Fig. 5–21C). Standing height, or *crown-heel length (CH),* is sometimes measured for eight-week-old embryos. CH measurements are often difficult to make on formalin-fixed embryos be-

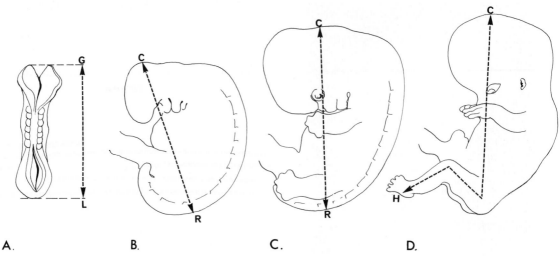

A.   B.   C.   D.

Figure 5–21 Sketches showing methods of measuring the length of embryos. *A*, Greatest length, *B* and *C*, Crown-rump length, *D*,, Crown-heel length.

cause such embryos are not easy to straighten.

Because the length of an embryo is only one criterion for establishing age, one must not refer to the 5-mm stage (O'Rahilly, 1973). If the Carnegie embryonic staging system (Table 5–1) is used internationally, detailed comparisons may be made between the findings of one person and those of another.

The size of the embryo in a pregnant woman can be estimated using gray-scale ultrasound measurements (Chilcote and Asokan, 1977). At 4 weeks (6 weeks after LMP), the embryo, the amnion, and the yolk sac compose a structure over 5 mm long that is detectable with careful scanning. After the fifth week (7 weeks after LMP), discrete embryonic structures can be visualized, and ''crown-rump'' measurements are predictive of embryonic age with an accuracy of ± 1 to 4 days. Furthermore, after the sixth week (8 weeks after LMP), dimensions of the head and trunk can be obtained and used for assessment of embryonic size (Chilcote and Asokan, 1977).

SUMMARY OF EMBRYONIC PERIOD

Early in the embryonic period, *longitudinal and transverse folding* converts the flat trilaminar embryonic disc into a C-shaped cylindrical embryo. The formation of head, tail, and lateral folds is a continuous sequence of events and results in a constriction between the embryo and the yolk sac. The dorsal part of the yolk sac is incorporated

into the embryo during folding and gives rise to the *primitive gut*. Transverse folding results in formation of the lateral and ventral body walls. The primitive gut becomes pinched off from the yolk sac but remains attached to it by a narrow *yolk stalk*. As the amnion expands, it forms an epithelial covering for the *umbilical cord*. The head fold results in the heart eventually lying ventrally and the brain becoming the most cranial part of the embryo. The tail fold causes the connecting stalk (primordium of the umbilical cord) and the allantois to move to the ventral surface of the embryo.

The three germ layers differentiate into various tissues and organs, so that by the end of the embryonic period, the beginnings of all the main organ systems have been established. The external appearance of the embryo is greatly affected by the formation of the brain, the heart, the liver, the somites, the limbs, the ears, the nose, and the eyes. As these structures develop, they affect the appearance of the embryo by forming characteristics that mark the embryo as unquestionably human. Because the beginnings of all essential external and internal structures are formed during the embryonic period, *the fourth to eighth weeks constitute the most critical period of development.* Developmental disturbances during this period may give rise to major congenital malformations.

Reasonable estimates of the age of embryos can be determined from (1) the day of onset of the last normal menstrual period, (2) the estimated time of fertilization, (3) measurements of length, and (4) external characteristics. The age of an embryo can also be estimated using gray-scale ultrasound measurements.

SUGGESTIONS FOR ADDITIONAL READING

Balinsky, B. I.: *An Introduction to Embryology,* 4th ed. Philadelphia, W. B. Saunders Co., 1975, pp. 208–220. A very good analysis of the nature of induction as well as a good description of the mechanism of action of inducing substances.

Gasser, F.: *Atlas of Human Embryos.* Hagerstown, Harper & Row, Publishers, 1975, pp. 25–297. An excellent atlas describing embryos in the Carnegie Embryological Collection. The important events are presented in outline form. There are many photomicrographs of representative levels through the embryos. You are urged to browse through this atlas.

Rugh, R., and Shettles, L. B.: *From Conception to Birth. The Drama of Life's Beginnings.* New York, Harper & Row, Publishers, 1971. Although this book was written for prospective mothers by a radiobiologist and an obstetrician, you are urged to peruse it because it is a well-written and instructive book. The photographs of embryos — many in color — are superb and will make you glad you examined this book.

CLINICALLY ORIENTED PROBLEMS

1. A 28-year-old woman, who has been a *heavy cigarette smoker* since her teens, was informed that she was in the second month of pregnancy. What would the doctor likely tell the patient about her smoking habit and the use of other drugs?
2. Why is the embryonic period such a critical stage of development?
3. Can one predict the possible harmful effects of drugs on the human embryo from studies performed in experimental animals?
4. Why may information about the starting date of a pregnancy provided by a patient be unreliable?

The answers to these questions are given at the back of the book.

REFERENCES

Benson, P.: *The Biochemistry of Development.* (Spastics International Medical Publications.) London, William Heinemann, Ltd., 1971, pp. 30–76.

Blechschmidt, E.: Die ersten drei Wochen mach der Befruchtung. *Image Roche* (Basel) 47:17, 1972.

Böving, B.: Anatomy of reproduction; *in* J. P. Greenhill (Ed.): *Obstetrics.* 13th ed. Philadelphia, W. B. Saunders Co., 1965, pp. 1–24.

Brachet, J.: *The Biochemistry of Development.* London, Pergamon Press, 1960.

Chilcote, W. S., and Asokan, S.: Evaluation of first-trimester pregnancy by ultrasound. *Clin. Obstet. Gynecol.* 20:253, 1977.

De Haan, R., and Ursprung, H.: *Organogenesis.* New York, Holt, Rinehart and Winston, Inc., 1965.

Golbus, M. S.: Teratology for the obstretician: Current status. *Obstet. Gynecol.* 55:269, 1980.

Grobstein, C.: Cell contact in relation to embryonic induction. *Exp. Cell Res.* (Suppl.) 8:234, 1961.

Iffy, L., Shepard, T. H., Jakobovits, A., Lemire, R. J., and Kerner, P.: The rate of growth in young human embryos of Streeter's horizons XIII and XXIII. *Acta Anat.* 66:178, 1967.

Nishimura, H., Takano, K., Tanimura, T., and Yasuda, M.: Normal and abnormal development of human embryos. *Teratology* 1:281, 1968.

Nishimura, H., Tanimura, T., Semba, R., and Uwabe, C.: Normal development of early human embryos: Observation of 90 specimens at Carnegie stages 7 to 13. *Teratology* 10:1, 1974.

Oppenheimer, J. M.: The non-specificity of the germ layers. *Q. Rev. Biol.* 15:1–27, 1940.

O'Rahilly, R.: Guide to the staging of human embryos. *Anat. Anz.* 130:556, 1972.

O'Rahilly, R.: *Developmental Stages in Human Embryos. Part A: Embryos of the First Three Weeks (Stages 1 to 9).* Washington, D. C., Carnegie Institution of Washington, 1973.

Page, E. W., Villee, C. A., and Villee, D. B.: *Human Reproduction: Essentials of Reproductive and Perinatal Medicine.* 3rd ed. Philadelphia, W. B. Saunders Co., 1981, pp. 357–360.

Reid, D. E., Ryan, K. J., and Benirschke, K.: *Principles and Management of Human Reproduction.* Philadelphia, W. B. Saunders Co., 1972, pp. 783–810.

Shepard, T. H.: Normal and abnormal growth patterns; *in* L. I. Gardner (Ed.): *Endocrine and Genetic Diseases of Childhood and Adolescence.* 2nd ed. Philadelphia, W. B. Saunders Co., 1975, pp. 1–8.

Shettles, L. B.: Fertilization and early development from the inner cell mass; *in* E. E. Philipp, J. Barnes, and M. Newton (Eds.): *Scientific Foundations of Obstetrics and Gynecology.* London. William Heinemann, Ltd., 1970, pp. 134–158.

Sirlin, J. L., and Brahma, S. L.: Studies on embryonic induction using radioactive tracers. II. The mobilization of protein components during induction of the lens. *Dev. Biol. 1*:234, 1959.

Streeter, G. L.: Developmental horizons in human embryos. Description of age group XI, 13 to 20 somites, and age group XII, 21 to 29 somites. *Contrib. Embryol. Carnegie Inst. 30*:211, 1942.

Streeter, G. L.: Developmental horizons in human embryos: Description of age group XIII, embryos of 4 or 5 millimeters long, and age group XIV, period of identification of the lens vesicle. *Contrib. Embryol. Carnegie Inst. 31*:27, 1945.

Streeter, G. L.: Developmental horizons in human embryos: Description of age groups XV, XVI, XVII, and XVIII. *Contrib. Embryol. Carnegie Inst. 32*:133, 1948.

Streeter, G. L., Heuser, C. H., and Corner, G. W.: Developmental horizons in human embryos: Descrip-

tion of age groups XIX, XX, XXI, XXII and XXIII. *Contrib. Embryol. Carnegie Inst. 34*:165, 1951.

Thompson, J. S., and Thompson, M. W.: *Genetics in Medicine*. 3rd ed. Philadelphia, W. B. Saunders Co., 1980.

Villee, C. A.: Biologic principles of growth; *in* F. Falkner (Ed.): *Human Development*. Philadelphia, W. B. Saunders Co., 1966, pp. 1–9.

Waddington, C. A.: *The Epigenetics of Birds*. Cambridge, Cambridge University Press, 1952.

Willier, B. H., Weiss, P. A., and Hamburger, V.: *Analysis of Development*. New York, Hafner Publishing Co., Inc., 1971.

Yamada, T.: Factors of embryonic induction; *in* M. Florkin and E. H. Stotz (Eds.): *Comprehensive Biochemistry*. Vol. 28. Amsterdam, Elsevier Publishing Co., 1967.

6

THE FETAL PERIOD
Ninth Week to Birth

The transition from embryo to fetus is not abrupt, but the name change is meaningful because it signifies that the embryo has developed from a single cell, the zygote, into a recognizable human being. *Development during the fetal period is primarily concerned with growth and differentiation* of tissues and organs that started to develop during the embryonic period. You should be aware that doctors, including obstetricians and radiologists using ultrasound, commonly refer to the developing human as a "fetus" throughout the entire gestation, and routinely date developmental events from the last menstrual period *(LMP).*

The rate of body growth during the fetal period is very rapid, especially between the ninth and twentieth weeks (Figs. 6–1 and 6–6), and weight gain is phenomenal during the terminal weeks (Table 6–2 and Fig. 6–12).

Fetuses weighing up to 500 gm at birth usually do not survive. The term *abortion* is applied to all pregnancies that terminate before the period of viability (Table 6–2). If given expert postnatal care, fetuses weighing 500 to 1000 gm may survive and are referred to as *immature infants.* Fetuses weighing between 1000 and 2500 gm are called *premature infants,* and most of them survive. However, *prematurity is one of the most common causes of perinatal death.*

During a pregnant woman's first prenatal visit to a doctor, the age of the embryo or fetus is estimated. The date of the *last menstrual period (LMP)* is a time-honored guide

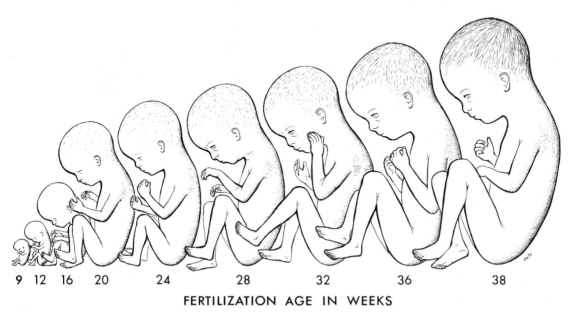

9 12 16   20   24   28   32   36   38

FERTILIZATION AGE IN WEEKS

Figure 6–1 Drawings of fetuses, about *one-fifth actual size.* Head hair begins to appear at about 20 weeks. Eyebrows and eyelashes are usually recognizable by 24 weeks, and the eyes reopen by 26 weeks. Fetuses born prematurely (22 weeks or more) may survive, but intensive care is required. The mean duration of pregnancy is 266 days (38 weeks) from fertilization, with a standard deviation of 12 days. *In clinical practice,* it is customary to refer to full term as 40 weeks from the first day of the last menstrual period (LMP), assuming that conception occurs two weeks after the onset of menses. Thus when a doctor refers to a pregnancy of 20 weeks, the true duration or actual age of the fetus is only 18 weeks.

to establishing gestational age, and it is reliable in most cases. To determine the actual age, or *fertilization age,* two weeks must be deducted from the *gestational age* because development does not begin until about two weeks after LMP. Page et al. (1981) state:

When, in the patient's opinion, the last menses was normal and the date is known with certainty, the probability of error is no more than 5 per cent. An exception would be those women who conceive shortly after discontinuing oral contraceptives or those women who do not have regular cycles.

ESTIMATION OF FETAL AGE

The gestational period may be divided into days, weeks, or months (Table 6–1), but confusion arises if it is not stated whether the time is calculated from (1) the onset of the last menstrual period (LMP), or (2) the estimated day of fertilization. Most uncertainty arises when months are used, particularly when it is not stated whether *calendar months* (28 to 31 days) or *lunar months* (28 days) are meant. Unless otherwise stated, age in this book is calculated from the estimated time of fertilization, and months refer to calendar months. *It is best to express fetal age in weeks* and to state whether the beginning or the end of a week is meant because statements such as "in the tenth week" are nonspecific.

Clinically, the gestational period is commonly divided into three parts or trimesters, each lasting three calendar months. By the end of the first trimester, all major systems are developed and the crown-rump length of the fetus is about the width of one's palm (Fig. 6–7). At the end of the second trimester (26 weeks after LMP, but only 24 weeks after the estimated day of fertilization), the fetus is usually too immature to survive if born prematurely, even though its length is now equal to about the span of one's hand.

Accurate determination of fetal age may not be possible for the reasons discussed in Chapter 5, but various measurements and external characteristics are useful in *estimating fetal age* (Table 6–2). *Crown-rump* (CR) is usually the most reliable measurement of length (see Fig. 5–21), but it must be emphasized that the length of fetuses, like infants, varies considerably for a given age. *Crown-heel (CH) length* is less useful because of the difficulty in straightening the fetus. *Foot length* correlates well with CR length and is particularly useful for estimating the age of incomplete or macerated fetuses. *Fetal weight* is often a useful criterion, but there may be a discrepancy between the fertilization age and the weight of a fetus, particularly when the mother has had metabolic disturbances during pregnancy, e.g., in diabetes mellitus, fetal weight often exceeds values considered normal for the CR length (Reid, 1972). Fresh fetal material has a shiny translucent appearance (Fig. 6–8), whereas fetuses that have been dead for several days prior to abortion have a tanned appearance and lack normal resilience (Shepard, 1975).

The fetal dimensions obtained from measurements of fetuses using *ultrasound techniques* closely approximate the crown-rump measurements obtained from aborted fetuses (Table 6–2). In addition, the *biparietal diameter* of the head and the dimension of the trunk may be obtained (Chilcote and Asokan, 1977). At 9 to 10 weeks (11 to 12 weeks after LMP), the head is still slightly larger (about 3 mm) than the trunk. Ultrasound crown-rump measurements of the fetus are predictive of fetal age with an accuracy of ± 1 to 4 days. Assessment of fetal size is enhanced when head and trunk dimensions are considered along with crown-rump measurements.

Determination of the size of the fetus, especially of its head, is of great value to the obstetrician for improving the management

TABLE 6–1   COMPARISON OF GESTATIONAL TIME UNITS

Reference Point	Days	Weeks	Calendar Months	Lunar Months
Fertilization*	266	38	8¾	9½
Last Menstrual Period	280	40	9¼	10

*The date of birth is calculated as about 266 days after the estimated day of fertilization, or 280 days after the onset of the last normal menstrual period. From fertilization to the end of the embryonic period, age is best expressed in days; thereafter age should be given in weeks. Because ovulation and fertilization are usually separated by not more than 12 hours, these events are more or less interchangeable in expressing prenatal age.

TABLE 6–2 CRITERIA FOR ESTIMATING FERTILIZATION AGE
DURING THE FETAL PERIOD

Age (weeks)	CR Length (mm)*	Foot Length (mm)*	Fetal Weight (gm)†	Main External Characteristics
PREVIABLE FETUSES				
9	50	7	8	*Eyes closing or closed.* Head more rounded. External genitalia still not distinguished as male or female. Intestines in umbilical cord.
10	61	9	14	*Intestine in abdomen.* Early fingernail development.
12	87	14	45	*Sex distinguishable externally.* Well-defined neck.
14	120	20	110	*Head erect.* Lower limbs well developed.
16	140	27	200	*Ears stand out* from head.
18	160	33	320	*Vernix caseosa present.* Early toenail development.
20	190	39	460	*Head and body hair (lanugo) visible.*
VIABLE FETUSES‡				
22	210	45	630	*Skin wrinkled* and red.
24	230	50	820	*Fingernails present.* Lean body.
26	250	55	1000	*Eyes partially open.* Eyelashes present.
28	270	59	1300	*Eyes open.* Good head of hair. Skin slightly wrinkled.
30	280	63	1700	*Toenails present.* Body filling out. Testes descending.
32	300	68	2100	*Fingernails reach finger tips.* Skin pink and smooth.
36	340	79	2900	*Body usually plump.* Lanugo hairs almost absent. Toenails reach toe tips. Flexed limbs; firm grasp.
38	360	83	3400	*Prominent chest;* breasts protrude. Testes in scrotum or palpable in inguinal canals. Fingernails extend beyond finger tips.

*These measurements are averages and so may not apply to specific cases; dimensional variations increase with age. The method for taking CR (crown-rump) measurements is illustrated in Figure 5–21.

†These weights refer to fetuses that have been fixed for about two weeks in 10 per cent formalin. Fresh specimens usually weigh about 5 per cent less (Böving, 1965).

‡There is no sharp limit of development, age, or weight at which a fetus automatically becomes viable or beyond which survival is assured, but experience has shown that it is rare for a baby to survive whose weight is less than 500 gm or whose fertilization age is less than 22 weeks. The term *abortion* refers to all pregnancies that terminate before the period of viability.

of patients (e.g., those women with small pelves and/or those fetuses with *intrauterine growth retardation and/or congenital abnormalities).*

HIGHLIGHTS OF THE FETAL PERIOD

No formal system of fetal staging using numbers has been proposed for the fetal period, as has been established for the embryonic period, but it is helpful to consider the changes in fetal development that occur in periods of four to five weeks.

Nine to Twelve Weeks. At the beginning of the ninth week, *the head constitutes almost half of the fetus* (Fig. 6–2). Subsequently, growth in body length accelerates

rapidly so that by the end of 12 weeks the CR length has more than doubled (Table 6–2). Growth of the head slows down considerably, however, compared with the rest of the body (Figs. 6–2, 6–3, and 6–4). The face is broad, the eyes widely separated, and the ears low-set. The eyelids are now closed (Fig. 6–3B).

At the beginning of the ninth week, the legs are short and the thighs are relatively small (Figs. 6–2 and 6–3). At the end of 12 weeks, the upper limbs have almost reached their final relative lengths, but the lower limbs are still not so well developed and are slightly shorter than their final relative length.

The external genitalia of males and fe-

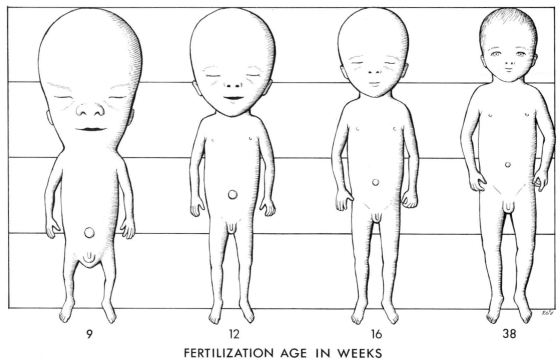

FERTILIZATION AGE IN WEEKS

Figure 6–2   Diagram illustrating the changing proportions of the body during the fetal period. By 36 weeks, the circumferences of the head and the abdomen are approximately equal. After this, the circumference of the abdomen may be greater. All stages are drawn to the same total height.

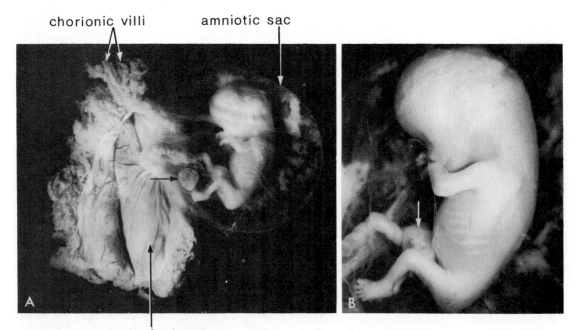

Figure 6–3   Photographs of a nine-week fetus in the amniotic sac exposed by removal from its chorionic sac. *A, actual size.* The remnant of the yolk sac is indicated by an arrow. *B,* enlarged photograph of the fetus (×2). Note the following features: (1) large head, (2) cartilaginous ribs, and (3) intestines in the umbilical cord (arrow). Ultrasound crown-rump measurements of the fetus can ascertain fetal age with an accuracy of ± one to four days. Determination of fetal size, especially of the head, allows the obstetrician to improve the management of patients (e.g., women whose fetuses may have congenital malformations such as hydro-cephalus; see Fig. 18–34).

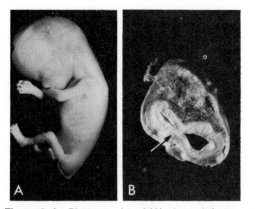

Figure 6–4   Photographs of *(A)* 10-week fetus and *(B)* cross section of the mother's ovary showing the corpus luteum of pregnancy (arrow). The corpus luteum secretes both estrogen and progesterone, important hormones for the maintenance of pregnancy. *Actual size.*

males appear somewhat similar until the end of the ninth week. Their mature fetal form is not established until the twelfth week. Intestinal coils are clearly visible within the proximal end of the umbilical cord (Fig. 6–3*B*) until the middle of the tenth week, when the intestines return to the abdomen (Fig. 6–5).

At the beginning of this period, the liver is the major site of *erythropoiesis*. By the end of the twelfth week, this activity decreases in the liver and begins in the spleen. *Urine starts to form* between the ninth and twelfth weeks and is excreted into the amniotic fluid.

Although this period is often called the stage of initial activity because the fetus reacts to stimuli (Hooker, 1936, and Rugh and Shettles, 1971), these movements by the relatively small fetus are too slight to be felt by the mother. By the end of the 12 weeks, stroking the lips causes the fetus to respond by sucking, and if the eyelids are stroked there is a reflex response. (These observations were made on spontaneously aborted fetuses.)

Thirteen to Sixteen Weeks.   *Growth is very rapid* during this period (Table 6–2 and Figs. 6–6 and 6–7). At the end of this period, the head is relatively small compared with that of the 12-week fetus (Fig. 6–2), and the lower limbs have lengthened. Ossification of the skeleton has progressed rapidly and shows clearly on x-ray films of the mother's abdomen by the beginning of the sixteenth week. Scalp hair patterning is determined during this period and gives a clue to early fetal brain development (Smith, 1973). By 16 weeks, the *ovaries are differentiated* and

have many primordial follicles containing *oogonia* (see Fig. 13–21).

Seventeen to Twenty Weeks.   Growth slows down during this period, but the fetus still increases its CR length by about 50 mm (Table 6–2). The lower limbs reach their final relative proportions (Fig. 6–8), and fetal movements, known as *quickening*, are commonly felt by the mother. The mean time that intervenes between a patient's first detection of fetal movements and delivery is 147 days, with a standard deviation of ± 15 days (Page et al., 1981).

The skin is covered with a greasy cheeselike material known as *vernix caseosa;* it consists of a mixture of a fatty secretion from the fetal sebaceous glands and dead epidermal cells. This vernix protects the fetus's delicate skin from abrasions, chapping, and hardening. The bodies of twenty-week fetuses are usually completely covered with

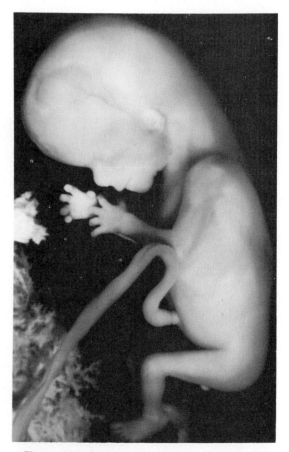

Figure 6–5   Photograph of an 11-week fetus exposed by removal from its chorionic and amniotic sacs (×1.5). Note the relatively large head and that the intestines are no longer in the umbilical cord.

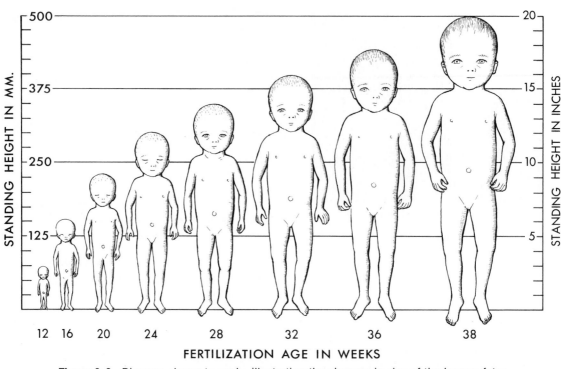

Figure 6–6   Diagram, drawn to scale, illustrating the changes in size of the human fetus.

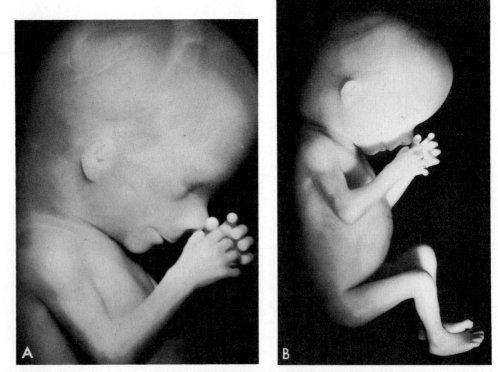

Figure 6–7   Photographs of a 13-week fetus. *A,* An enlarged photograph of the head and shoulders of this fetus that appears on the cover of this book (×2). *B, Actual size.*

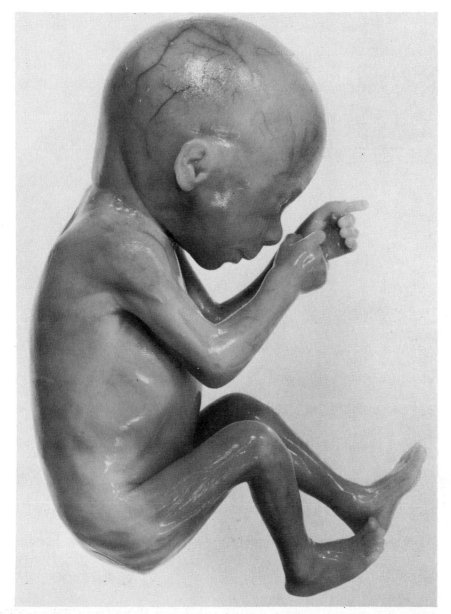

Figure 6–8 Photograph of a 17-week fetus. *Actual size.* Note that the ears stand out from the head and that no hair is visible. Because there is no subcutaneous fat and the skin is thin, the blood vessels of the scalp are visible. Fetuses at this age are unable to survive if born prematurely, mainly because their respiratory system is immature. The alveolar surface area is insufficient, and the vascularity of the lungs is underdeveloped.

fine downy hair called *lanugo;* this may help hold the vernix on the skin. Eyebrows and head hair are also visible at the end of the period.

Brown fat forms during this time and is the site of heat production, particularly in the newborn infant. This specialized adipose tissue produces heat by oxidizing fatty acids (Reid, 1972). Brown fat is chiefly found (1)

on the floor of the anterior triangle of the neck surrounding the subclavian and carotid vessels, (2) posterior to the sternum, and (3) in the perirenal area (Young, 1968, and Page et al., 1981). Brown fat has a high content of mitochondria, giving it a definite brown hue.

By 18 weeks, the uterus of a female fetus is completely formed, and canalization of the

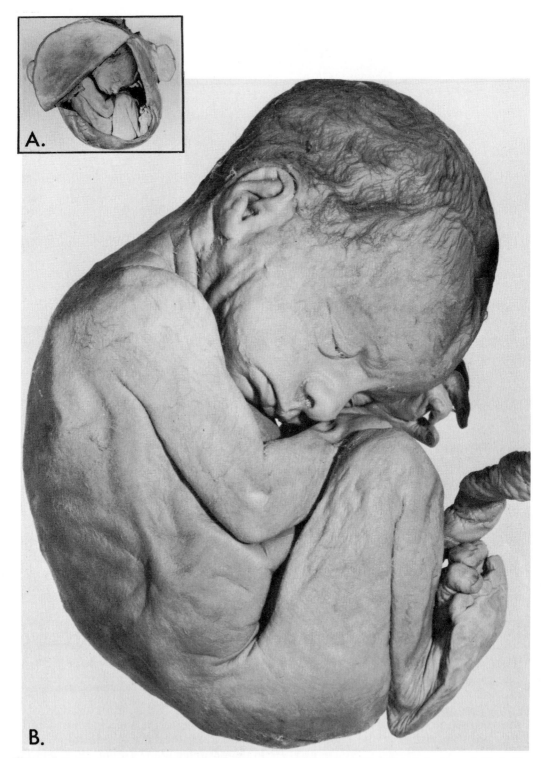

Figure 6–9   Photographs of a 25-week fetus. *A,* In the uterus. *B, Actual size.* Note the wrinkled skin and rather lean body caused by the scarcity of subcutaneous fat. Observe that the eyes are beginning to open. A fetus of this size might survive if born prematurely, hence it is considered a viable fetus. Termination of pregnancy is illegal after the period of viability (Table 6–2).

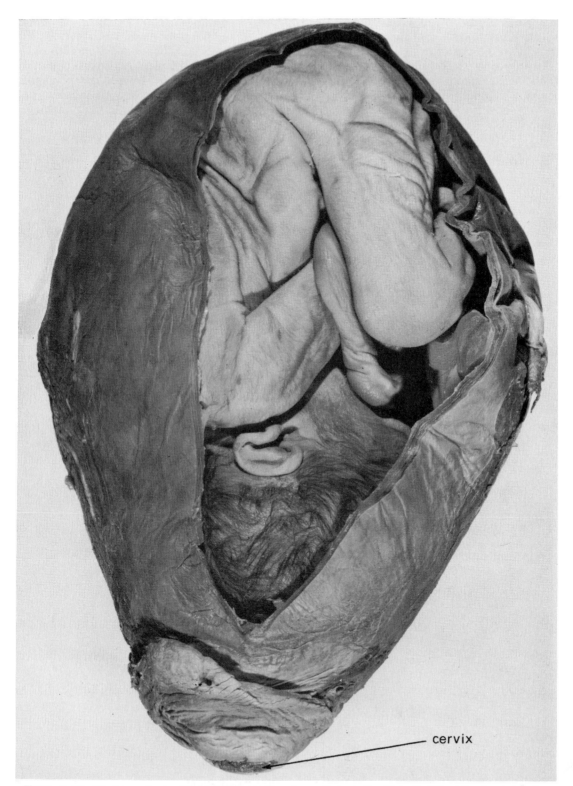

cervix

Figure 6–10 Photograph of a 29-week fetus in the uterus. *Actual size.* Note that the fetus is upside down; this is the normal presentation at this period of gestation. A portion of the wall of the uterus and parts of the chorion and amnion have been removed to show the fetus. Infants born at this stage usually survive because their lungs are capable of breathing air. This fetus and its mother died as the result of injuries sustained in an automobile accident.

vagina has begun. By 20 weeks, the testes of a male fetus have begun their descent, but are still on the posterior abdominal wall.

Twenty-one to Twenty-five Weeks. There is a substantial weight gain during this period. Although still somewhat lean, the body of the fetus is better proportioned (Fig. 6–9). The skin is usually wrinkled, particularly during the early part of this period. The skin is more translucent and is pink to red in color in fresh specimens because blood in the capillaries is visible. By 24 weeks, alveolar cells of the lung have begun to make *surfactant*, a surface-active lipid that maintains

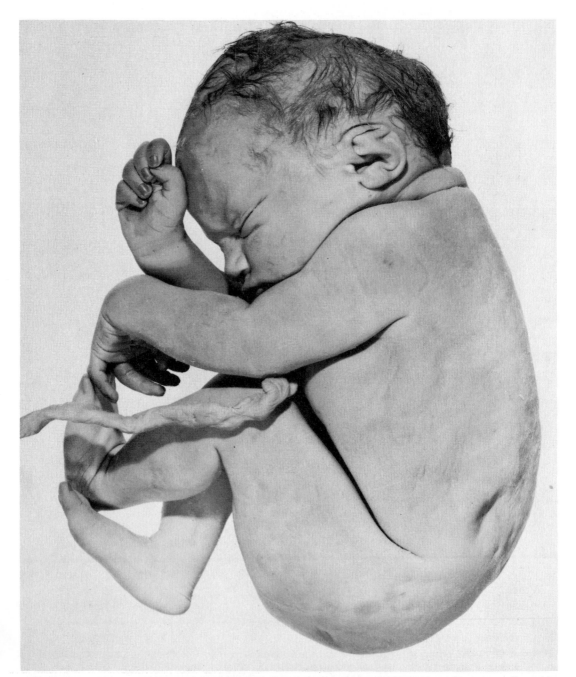

Figure 6–11   Photograph of a 36-week fetus. *Half actual size.* Fetuses at this size and age usually survive. Note the plump body resulting from the deposition of subcutaneous fat. This fetus' mother was killed in an automobile accident, and the fetus died before it could be delivered by cesarean section.

alveolar patency. Although all organs are rather well developed, a 22- to 25-week fetus may die within a few days if born prematurely, mainly because its respiratory system is still immature.

Twenty-six to Twenty-nine Weeks. A fetus may now survive if born prematurely (Table 6–2) because *the lungs are capable of breathing air*, and because the lungs and pulmonary vasculature have developed sufficiently to provide gas exchange (see Fig. 11–8). In addition, the central nervous system has matured to the stage at which it can direct rhythmic breathing movements and control body temperature. The eyes reopen during this period, and head and lanugo hair are well developed (Fig. 6–10). Considerable subcutaneous fat has now formed under the skin, smoothing out many of the wrinkles. During this period, the quantity of white fat in the body increases to about 3.5 per cent of body weight (Widdowson, 1974). *Erythropoiesis in the spleen* ends by 28 weeks, and the bone marrow becomes the major site of this process.

Thirty to Thirty-four Weeks. *The pupillary light reflex is present by 30 weeks.* Usually by the end of this period, the skin is pink and smooth, and the upper and lower limbs often have a chubby appearance. At this stage, the quantity of white fat in the body is about 7 to 8 per cent of body weight (Widdowson, 1974).

Thirty-five to Thirty-eight Weeks. Fetuses at 35 weeks have a firm grasp and exhibit a spontaneous orientation to light (Page et al., 1981). Most fetuses during this "finishing" period are plump (Fig. 6–11). *At 36 weeks, the circumference of the head and the abdomen are approximately equal.* After this, the circumference of the abdomen may be greater than that of the head.

There is a slowing of growth as the time of birth approaches (Fig. 6–12). Fetuses usually reach a CR length of 360 mm and weigh about 3400 gm. By full term, the amount of white fat in the body is about 16 per cent of body weight (Widdowson, 1974). The fetus lays down about 14 gm of fat a day during the last few weeks of gestation. In general, male fetuses grow faster than females, and male infants generally weigh more than female infants at birth. Succeeding pregnancies tend to last slightly longer, and result in larger babies.

By *full term* (38 weeks after fertilization, or 40 weeks after LMP), the skin is usually

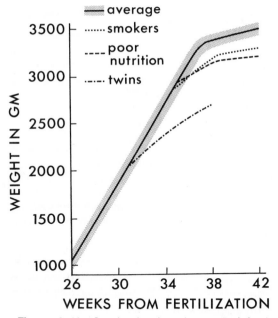

Figure 6–12 Graph showing the rate of fetal growth during the last trimester. (Adapted from Gruenwald, P.: *Am. J. Obstet. Gynecol.* 94:1112, 1966.) Average refers to babies born in the United States. After 36 weeks the growth rate deviates from the straight line. The decline, particularly after full term (38 weeks), probably reflects inadequate fetal nutrition caused by placental changes.

white or bluish-pink in color. The chest is prominent and the breasts protrude in both sexes. The testes are usually in the scrotum in full-term male infants; descent begins at about 28 to 32 weeks. Thus premature male infants commonly have undescended testes. Usually, the testes descend during early infancy.

Although the head at full term is much smaller in relation to the rest of the body than it was during early fetal life, it still is one of the largest parts of the fetus, an important consideration with regard to its passage through the cervix and vagina, the birth canal (see Figs. 7–10 and 7–11).

Not all low-weight babies are premature. About one third of "premature infants" with a birth weight of 2500 gm or less are actually small or undergrown for their gestational age (Lucey, 1972). This has become apparent only in the last few years since good intrauterine fetal growth charts have become available. These infants, sometimes called "small for dates," often result from placental insufficiency. It is important to distinguish between babies who are of low birth weight because of *intrauterine growth retardation* (Warkany, 1971) and those who are light because of a shortened gestation. Intrauterine growth retardation may be caused by placental insufficiency, but genetic and environmental factors are also known to cause such retardation (Warkany, 1971). These infants show a characteristic lack of subcutaneous fat, and the skin is wrinkled, suggesting that white fat has actually been lost. Brown fat is also reduced or absent.

Time of Birth. The expected time of birth is usually 266 days or 38 weeks after fertilization, or 280 days or 40 weeks after LMP (Table 6–1). Prolongation of pregnancy for two to three weeks beyond the expected time of birth is common (8 to 12 per cent). Some infants in prolonged pregnancy develop the *postmaturity syndrome*. These infants are usually thin and have dry "parchment-like" skin.

The common method of determining the expected date of delivery is to count back three calendar months from the first day of the last menstrual period and then add a year and one week. In women with typical 28-day menstrual cycles, this method is quite accurate. However, if the woman's cycles have been irregular, miscalculations of two to three weeks may occur. In addition, bleeding sometimes occurs in pregnant women at the time of the first missed period (about two weeks after fertilization). Should the woman interpret this *implantation bleeding* as menstruation, the estimated time of birth would be miscalculated by two or more weeks.

FACTORS INFLUENCING FETAL GROWTH

The fetus requires substrates for the production of energy and for growth. With the exception of a few metals, such as iron, all gases and nutrients pass freely to the fetus from the mother via the *placental membrane* (see Fig. 7–8).

Glucose, Insulin, and Amino Acids. Glucose is a primary source of energy for fetal metabolism and growth; amino acids are also required (Page et al., 1981). These substances are derived from the mother's blood via the placenta. The insulin required for the metabolism of glucose is secreted by the fetal pancreas; no significant quantities of maternal insulin reach the fetus because *the placental membrane is relatively impermeable to insulin*. Insulin is believed to stimulate fetal growth. The role of insulin as a "growth hormone" is fully discussed by Villee (1975). For a comprehensive account of fetal growth in humans, see Miller and Merritt (1979) and Page et al. (1981).

FACTORS CAUSING FETAL GROWTH RETARDATION

Many factors affect the rate of fetal growth. These may be maternal, fetal, or environmental. In general, factors operating throughout pregnancy (e.g., cigarette smoking) tend to produce *small infants,* whereas those operating during the last trimester (e.g., undernutrition of the mother) usually produce *underweight infants* with normal length and head size (Page et al., 1981).

Intrauterine growth retardation (IUGR) is usually defined as infant weight within the lowest ten percentile for gestational age.

Maternal Malnutrition. Severe maternal malnutrition resulting from a poor quality diet is known to cause reduced fetal growth (Fig. 6–12). Poor nutrition and faulty food habits are common and are not restricted to mothers belonging to poverty groups (Gruenwald, 1966).

Smoking. Maternal cigarette smoking is a well-established cause of intrauterine growth retardation (Golbus, 1980). The growth rate for fetuses of mothers who smoke cigarettes heavily is less than normal during the last six to eight weeks of pregnancy (Fig. 6–12). The effect is greater on fetuses whose mothers also receive inadequate nutrition; presumably, there is an additive effect of heavy smoking and poor quality diet.

Multiple Pregnancy. Individuals of twins, triplets, and other multiple births usually weigh considerably less than infants resulting from a single pregnancy. It is evident that the total requirements of twins (Fig. 6–12), triplets, and so forth exceed the nutritional supply available from the placenta during the third trimester.

Socially Used Drugs. Infants born to alcoholic mothers often exhibit intrauterine growth retardation (IUGR) as part of the *fetal alcohol syndrome* (see Chapter 8). Similarly, narcotic addiction can cause IUGR and other obstetrical complications (Golbus, 1980).

Impaired Uteroplacental Blood Flow. Maternal placental circulation may be reduced by a variety of conditions that decrease uterine blood flow (e.g., severe hypotension and renal disease). Chronic reduction of uterine blood flow can cause fetal starvation resulting in fetal growth retardation.

Placental Insufficiency. Placental dysfunction or defects (e.g., infarction or intervillous coagulation) can cause fetal growth retardation. The net effect of these placental changes is a reduction of the total area for exchange of nutrients between the fetal and maternal blood streams (see Chapter 7). It is very difficult to separate the effect of these changes from the effect of reduced maternal blood flow to the placenta. In some instances of chronic maternal disease, the maternal vascular changes in the uterus are primary and the placental defects are secondary (Page et al., 1981).

Genetic Factors and Chromosomal Aberrations. It is well established that genetic factors can cause retarded fetal growth. Repeated cases of intrauterine growth retardation in one family indicate that recessive genes are the cause of the abnormal growth in the family. In recent years, structural and numerical chromosomal aberrations have also been associated with cases of retarded fetal growth (Warkany, 1971). Intrauterine growth retardation is pronounced in Down syndrome and is very characteristic of trisomy 18 syndrome (see Chapter 8).

PERINATOLOGY

By accepting the shelter of the uterus, the fetus also takes the risk of maternal disease or malnutrition and of biochemical, immunological and hormonal adjustment.

–George W. Corner

Perinatology is the branch of medicine primarily concerned with the fetus and newborn infant, generally covering the period from about 26 weeks after fertilization to about 4 weeks after birth. The subspecialty known as *perinatal medicine* combines certain aspects of obstetrics and pediatrics.

The fetus is now commonly regarded as a patient on whom diagnostic and therapeutic procedures may be performed. Studies concerned with the fetus are sometimes called *fetology*. Several techniques are now available for assessing the status of the human fetus and for providing prenatal treatment.

AMNIOCENTESIS

Amniotic fluid is sampled by inserting a hollow needle through the mother's anterior abdominal wall and uterine wall into the amniotic cavity (Fig. 6–13). A syringe is then attached and amniotic fluid withdrawn. Because there is relatively little amniotic fluid, amniocentesis is difficult to perform prior to the fourteenth week.

Amniocentesis is relatively devoid of risk, especially when the procedure is performed

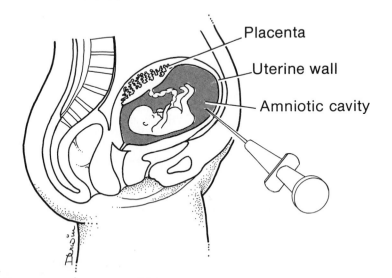

Placenta

Uterine wall

Amniotic cavity

Figure 6–13 Drawing illustrating the technique of amniocentesis. A needle is inserted through the lower abdominal wall and the uterine wall into the amniotic cavity. A syringe is attached and amniotic fluid is withdrawn for diagnostic purposes (e.g., for cell cultures or protein studies). Amniocentesis is relatively devoid of risk, especially when combined with ultrasonography for placental localization. The risk of injuring the fetus with the needle is also minimized by using ultrasound. The technique is usually performed at 15 to 16 weeks of gestation. Prior to this stage of development, there is relatively little amniotic fluid, and the difficulties in obtaining it without endangering the mother or the fetus are consequently greater. There is an excessive amount of amniotic fluid (polyhydramnios) in the case illustrated in this figure. This excess fluid is removed from some patients by amniocentesis.

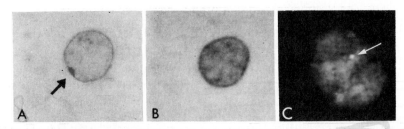

Figure 6–14　Nuclei of cells in amniotic fluid obtained by amniocentesis. *A,* Chromatin-positive nucleus indicating the presence of a female fetus; the sex chromatin is indicated by an arrow. *B,* Chromatin-negative nucleus indicating the presence of a male fetus. No sex chromatin is visible. Cresylecht violet stain (×1000). *C,* Y-chromatin-positive nucleus indicating the presence of a male fetus. The arrow indicates the Y-chromatin as an intensely fluorescent body obtained after staining the cell in quinacrine mustard. (*A* and *B*, From Riis, M., and Fuchs, F.: Sex chromatin and antenatal sex diagnosis; *in* K. L. Moore [Ed.]: *The Sex Chromatin.* Philadelphia, W. B. Saunders Co., 1966. *C,* Courtesy of Dr. M. Ray, Department of Pediatrics, Division of Genetics and Department of Anatomy, University of Manitoba and Health Sciences Centre, Winnipeg, Canada.)

by an experienced obstetrician who is guided by ultrasonography (Fig. 6–15) for placental localization. Amniocentesis is the most common technique for detecting genetic disorders and is usually performed at 15 to 16 weeks of gestation (i.e., after LMP).

The following are common indications for amniocentesis: (1) *late maternal age* (i.e., 40 years or older); (2) previous birth of a trisomic child (e.g., Down syndrome); (3) *chromosomal abnormality* in either parent (e.g., chromosome translocation); (4) women who are *carriers of X-linked recessive disorders* (e.g., hemophilia); (5) neural tube defects in family; and (6) carriers of inborn errors of metabolism.

Fetoprotein Measurements.　Chemical components are known to leak from skin defects of fetuses with neural tube defects into the amniotic fluid (see Chapter 18). The concentration of *α-fetoprotein (AFP)* in the amniotic fluid surrounding fetuses with spina bifida cystica and anencephaly is remarkably high (Brock, 1976). Thus, it is possible to detect the presence of these severe abnormalities by measuring the concentration of α-fetoprotein in amniotic fluid. Incidentally, these malformations also cause the α-fetoprotein to rise in the maternal serum; however, studies of maternal blood for α-fetoprotein are not so reliable at present as those on amniotic fluid. It is generally believed that *closed neural tube defects* (e.g., spina bifida with meningomyelocele, Fig. 18–16) cannot be detected through assay of α-fetoprotein. Closed neural tube defects represent about 5 to 10 per cent of the total number of neural tube defects (Brock, 1976).

Detection of an increased concentration of α-fetoprotein in amniotic fluid is likely to be a useful diagnostic tool for detecting the presence or absence of *open neural tube defects,* e.g., anencephaly (see Fig. 15–10) and severe types of spina bifida (see Fig. 18–17) in fetuses of mothers who have already had a child with a neural tube defect. The findings would help to decide whether the pregnancy should be terminated.

Spectrophotometric Studies.　Examination of amniotic fluid by this method may be used for assessing the degree of *erythroblastosis fetalis* (also called hemolytic disease of the fetus and newborn). This condition results from destruction of fetal red blood cells by maternal antibodies (see Chapter 7).

Sex Chromatin Patterns.　Fetal sex can be diagnosed by noting the presence or absence of sex chromatin in cells recovered from amniotic fluid (Fig. 6–14*A* and *B*). By use of a special staining technique, the Y chromosome can also be identified in the cells of male fetuses (Fig. 6–14*C*). Knowledge of fetal sex can be useful in diagnosing the presence of severe sex-linked hereditary diseases such as hemophilia or muscular dystrophy (Riis and Fuchs, 1966, and Page et al., 1981). These tests, however, are not routine and are not performed merely to satisfy the parents' curiosity about fetal sex.

Cell Cultures.　Fetal sex can also be determined by studying the sex chromosomes of cultured amniotic cells. These studies are more commonly done when an autosomal abnormality such as occurs in Down syndrome is suspected. *Inborn errors of metabolism* in fetuses can also be detected by studying cell cultures. Enzyme deficiencies can be determined by incubating cells recovered from amniotic fluid and then detecting the specific enzyme deficiencies in the cells. For a list of the hereditary metabolic disorders that can be diagnosed before birth, see Page et al., 1981. Amniocentesis and cell cultures permit prenatal diagnosis of severe diseases for which there is no effective treatment at present, and they afford the opportunity to terminate the pregnancy (Laurence and Gregory, 1976). Possible problems in the use of cultured amniotic fluid cells for biochemical diagnosis are discussed by Littlefield (1971) and Thompson and Thompson (1980).

INTRAUTERINE FETAL TRANSFUSION

Some fetuses with erythroblastosis fetalis can be saved by receiving intrauterine blood transfusions. The blood is injected through a needle inserted into the fetal peritoneal cavity (for technical details, see Allen and Umansky, 1972). Over a period of five to six days, most of the cells pass into the fetal circulation via the diaphragmatic lymphatics (Pritchard and Weisman, 1957). Intrauterine blood transfusions are seldom given nowadays, owing to the treatment of Rh negative mothers of Rh positive fetuses with *anti-Rh immunoglobulin*.

FETOSCOPY

Using fiberoptic lighting instruments, one may directly visualize parts of the fetal body. It is possible to scan the entire fetus, looking for congenital malformations such as cleft lip. The fetoscope is usually introduced through the anterior abdominal wall and the uterine wall into the amniotic cavity, similar to the way the needle is inserted during amniocentesis (Fig. 6–13). One not only can see the fetus, but also take biopsies of skin or blood samples. Probably the optimal time for doing this procedure is at 18 weeks of pregnancy, because the amniotic sac is large enough to allow easy entry of the trochar, and the fetus can be maneuvered for inspection (Kaback and Valenti, 1976, and Persaud, 1979).

The procedure takes more time than amniocentesis and creates a higher risk of infection or of spontaneous abortion. Such examinations may have a role to play in the future if the intrauterine diagnosis and treatment of conditions in the fetus are developed.

ULTRASONOGRAPHY

A chorionic sac may be visualized during the embryonic period by using ultrasound techniques; placental and fetal size, multiple births, and abnormal presentations can also be determined (Gosink, 1981, and MacVicar, 1976).

Ultrasonic scans give accurate measurements of the biparietal diameter of the fetal skull, from which close estimates of fetal length can be made (Page et al., 1981). Thus "small for date" fetuses can be detected. In most cases, the male genitalia can be visualized by ultrasound. Figure 6–15 illustrates

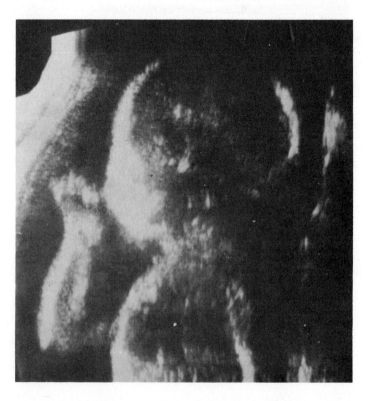

Figure 6–15 Ultrasound scan of a 30-week fetus that appears to be sucking its thumb. It reminds one of the cover photograph. The biparietal diameter of the head can be determined and compared with the abdominal diameter. Determination of these measurements facilitates estimation of the age and weight of the fetus. (From Thompson, J. S., and Thompson, M. W.: *Genetics In Medicine.* 3rd ed. Philadelphia, W. B. Saunders Co., 1980. Courtesy of Stuart Campbell.)

how details of the fetus can be observed in good ultrasound scans.

Ultrasound examinations are also helpful in diagnosing at a very early stage abnormal pregnancies, e.g., the so-called blighted embryo (a pregnancy in which there is a chorionic sac with no embryo, or one in which there is a *hydatidiform mole*).

Recent advances in ultrasonography have made this technique a major tool for prenatal diagnosis of fetal abnormalities such as the following: anencephaly, hydrocephaly, microcephaly, fetal ascites, and renal agenesis (Persaud, 1979; Sanders and James, 1980; Page et al., 1981). Because of the cost, ultrasonography for estimation of gestational age and for the determination of congenital malformations is not routine. It is indicated in *high-risk obstetrical patients* (e.g., when there is a medical indication for induction of labor).

AMNIOGRAPHY AND FETOGRAPHY

In amniography and fetography, a radiopaque substance is injected into the amniotic cavity in order to outline the amniotic sac and the external features of the fetus (Persaud, 1979). A water-soluble contrast medium is used in amniography, and in fetography an oil-soluble contrast medium is injected. The latter medium is apparently absorbed by the vernix caseosa.

The examinations are usually conducted 15 to 18 weeks after LMP on women who are, usually owing to evidence of elevated amniotic fluid fetoprotein, suspected of having fetuses with *open neural tube defects* or other severe congenital malformations.

Another benefit of amniography and fetography is that these methods allow one to study the gastrointestinal tract via observation of normal swallowing of the contrast-containing amniotic fluid (Balsam and Weiss, 1981). Fetuses with *esophageal atresia* (see Fig. 11–5A), *duodenal atresia* (see Fig. 12–5B), and *omphalocele* (see Fig. 12–16) can be diagnosed prenatally using these techniques.

It is likely that ultrasonography will ultimately replace amniography and fetography for prenatal diagnosis of neural tube defects because ultrasound techniques use no ionizing radiation and are noninvasive (i.e., they do not involve puncture of the fetal membranes).

SUMMARY OF FETAL PERIOD

The fetal period begins nine weeks after fertilization and ends at birth. It is primarily characterized by *rapid body growth* and *differentiation of organ systems*. An obvious change is the relative slowing of head growth compared with that of the rest of the body. Lanugo and head hair appear, and the skin is coated with vernix caseosa by the beginning of the twentieth week. The eyelids are closed during most of the fetal period but begin to reopen at about 26 weeks. Until this time, the fetus is usually incapable of extrauterine existence, mainly because of immaturity of the respiratory system.

Until about 30 weeks, the fetus appears reddish and wizened because of the thinness of the skin and the relative absence of subcutaneous fat. Fat usually develops rapidly during the last six to eight weeks, giving the fetus a smooth and plump appearance. This terminal period is devoted mainly to *building up of tissues* and to preparing systems involved in the transition from intrauterine to extrauterine environments, primarily the respiratory and cardiovascular systems.

Fetuses born prematurely during the 26- to 36-week period usually survive, but full-term fetuses have the best chance of survival.

Changes occurring during the fetal period are not so dramatic as those appearing in the embryonic period, but they are very important. The fetus is less vulnerable to the teratogenic effects of drugs, viruses, and radiation, but these agents may interfere with normal functional development, especially of the brain and the eyes (see Chapter 8).

Various techniques are available for *assessing the status of the fetus* and for diagnosing certain diseases and developmental abnormalities before birth. The physician can now determine whether or not a fetus has a particular disease or a congenital malformation by using amniocentesis and ultrasonography. The *prenatal diagnosis* can be made early enough to allow selective abortion of a defective fetus if this is the mother's decision and if the procedure is legal.

SUGGESTIONS FOR ADDITIONAL READING

Cheek, D. B., Graystone, J. E., and Niall, M.: Factors controlling fetal growth. *Clin. Obstet. Gynec. 20*:925, 1977. A comprehensive review of factors, especially hormones, that play a role in the growth of a fetus.

Gosink, B. B.: *Diagnostic Ultrasound*. 2nd ed. Philadelphia, W. B. Saunders Co., 1981. A complete introduction to applications of ultrasound through clinical exercises.

Mandelbaum, B., and Evans, T. N.: Life in the amniotic fluid. *Am. J. Obstet. Gynecol. 104*:365, 1969. Amniography, fetal radiography, and angiography are presented with respect to recent developments and the role that these techniques have in assessing the status of the fetus.

Page, E. W.: Human fetal nutrition and growth. *Am. J. Obstet. Gynecol., 104*:378, 1969. A review of intrauterine nutrition of the fetus in which fetal metabolism, alterations of maternal metabolism affecting the fetus, maternal nutrition, and the factors that regulate or interfere with fetal growth are discussed.

CLINICALLY ORIENTED PROBLEMS

1. A woman in the twentieth week of gestation was scheduled for a repeat cesarean section. Her doctor wanted to establish an estimated date of delivery. Can a doctor rely on the accuracy of the menstrual dates given by patients? How could an estimated delivery date (EDD) be established?
2. A 42-year-old pregnant woman was suspected of carrying a fetus with congenital malformations. How could confirmation of this be obtained? What chromosomal abnormality would most likely be found? How could the sex of the fetus be determined if this was of clinical interest (e.g., in a carrier of hemophilia)?
3. A woman in the second trimester of pregnancy asked her doctor whether her fetus was vulnerable to "over-the-counter drugs" and "street drugs." What would her doctor likely tell her?
4. Name at least four factors that may cause intrauterine growth retardation. Which factors can the mother eliminate?

The answers to these questions are given at the back of the book.

REFERENCES

Allen, F. H., Jr., and Umansky, I.: Erythroblastosis fetalis; *in* D. E. Reid, K. J. Ryan, and K. Benirschke (Eds.): *Principles and Management of Human Reproduction*. Philadelphia, W. B. Saunders Co., 1972, pp. 811–832.

Balsam, D., and Weiss, R. R.: Amniography in prenatal diagnosis. *Pediatr. Radiol. 141*:379, 1981.

Böving, B. G.: Anatomy of reproduction; *in* J. P. Greenhill (Ed.): *Obstetrics*. 13th ed. Philadelphia, W. B. Saunders Co., 1965, pp. 1–24.

Brock, D. J. H.: Prenatal diagnosis — chemical methods. *Br. Med. J. 32*:16, 1976.

Carlson, E. N.: Capabilities of gray-scale imaging in obstetrics and gynecology. *Clin. Obstet. Gynecol. 20*:235, 1977.

Chilcote, W. S., and Asokan, S.: Evaluation of first-trimester pregnancy by ultrasound. *Clin. Obstet. Gynecol. 20*:253, 1977.

Depp, R.: How ultrasound is used by the perinatologist. *Clin. Obstet. Gynecol., 20*:315, 1977.

Fraser, F. C., and Nora, J. J.: *Genetics of Man*. Philadelphia, Lea & Febiger, 1975.

Golbus, M. S.: Teratology for the obstetrician: Current status. *Obstet. Gynecol. 55*:269, 1980.

Greenhill, J. P., and Friedman, E. A. (Eds.): *Biological Principles and Modern Practice of Obstetrics*. Philadelphia, W. B. Saunders Co., 1974.

Gruenwald, R.: Growth of the human fetus. I. Normal growth and its variations. *Am. J. Obstet. Gynecol. 94*:1112, 1966.

Hellman, L. M., Kobayashi, M., Fillisti, L., and Cromb, E.: The sonographic depiction of the growth and development of the human fetus; *in* R. Caldeyro-Barcia (Ed.): *Perinatal Factors Affecting Human Development*. Washington, D.C., Pan American Health Organization, Scientific Publication No. 185, 1969, pp. 70–80.

Hooker, D.: Early fetal activity in mammals. *Yale J. Biol. Med. 8*:579, 1936.

Hull, D.: Brown adipose tissue in the new born; *in* E. E. Philipp, J. Barnes, and M. Newton (Eds.): *Scientific Foundations of Obstetrics and Gynecology*. London, William Heinemann, Ltd., 1970, pp. 407–411.

Kaback, M. M., and Valenti, C. (Eds.): *Intrauterine Fetal Visualization: A Multidisciplinary Approach*. New York, *Excerpta Medica,* American Elsevier Publishing Co., Inc., 1976.

Kumar, D.: Intrauterine diagnosis, indices of fetal jeopardy and intrauterine therapy; *in* A. C. Barnes (Ed.): *Intra-Uterine Development*. Philadelphia, Lea & Febiger, 1968, pp. 477–497.

Laurence, K. M., and Gregory, P.: Prenatal diagnosis of chromosome disorders. *Br. Med. J. 32*:9, 1976.

Liley, A. W.: The use of amniocentesis and fetal transfusion in erythroblastosis fetalis. *Pediatrics 35*:876, 1965.

Littlefield, J. W.: Problems in the use of cultured amniotic fluid cells for biochemical diagnosis; *in* D.

Bergsma, A. G. Motulsky, C. Jackson, and J. Sitter (Eds.): *Symposium on Intrauterine Diagnosis, Birth Defects* 7:10, 1971.

Lucey, J. F.: Conditions and diseases of the newborn; *in* D. E. Reid, K. J. Ryan, and K. Benirschke (Eds.): *Principles and Management of Human Reproduction.* Philadelphia, W. B. Saunders Co., 1972, pp. 848–861.

MacIntyre, M.: Chromosomal problems of intrauterine diagnosis; *in* D. Bergsma, A. G. Motulsky, C. Jackson, and J. Sitter (Eds.): *Symposium on Intrauterine Diagnosis, Birth Defects* 7:10, 1971.

MacVicar, J.: Antenatal detection of fetal abnormality—physical methods. Br. Med. J. *32*:4, 1976.

Miller, H. C., and Merritt, T. A.: *Fetal Growth in Humans.* Chicago, Year Book Medical Publishers, Inc., 1979.

Montague, A. C. W.: Hemolytic disease of the fetus; *in* A. C. Barnes (Ed.): *Intra-Uterine Development.* Philadelphia, Lea & Febiger, 1968, pp. 443–466.

Moore, K. L.: *The Sex Chromatin.* Philadelphia, W. B. Saunders Co., 1966.

Nesbitt, R. E. L., Jr.: Perinatal development; *in* F. Falkner (Ed.): *Human Development.* Philadelphia, W. B. Saunders Co., 1966, pp. 123–149.

Nyhan, W. L.: Intra-uterine diagnosis and the antenatal detection of inherited disease; *in* P. Benson (Ed.): *The Biochemistry of Development.* London, William Heinemann, Ltd., 1971, pp. 14–29.

Page, E. W., Villee, C. A., and Villee, D. B.: *Human Reproduction: Essentials of Reproductive and Perinatal Medicine.* 3rd ed. Philadelphia, W. B. Saunders Co., 1981.

Persaud, T. V. N.: *Prenatal pathology. Fetal Medicine.* Springfield, Illinois, Charles C Thomas, 1979.

Pritchard, J. A., and Weisman, R.: The absorption of labelled erythrocytes from the peritoneal cavity of humans. *J. Lab. Clin. Med. 49*:756, 1957.

Reid, D. E.: Fetal growth and physiology; *in* D. E. Reid, K. J. Ryan, and K. Benirschke (Eds.): *Principles and Management of Human Reproduction.* Philadelphia, W. B. Saunders Co., 1972, pp. 783–809.

Reid, D. E., Ryan, K. J., and Benirschke, K.: *Principles and Management of Human Reproduction.* Philadelphia, W. B. Saunders Co., 1972, pp. 404–447.

Riis, P., and Fuchs, F.: Sex chromatin and antenatal sex diagnosis: *in* K. L. Moore (Ed.): *The Sex Chromatin.* Philadelphia, W. B. Saunders Co., 1966, pp. 220–228.

Sanders, R. C., and James, A. E. (Eds.): *The Principles and Practice of Ultrasonography in Obstetrics and Gynecology.* 2nd ed. New York, Appleton-Century-Crofts, 1980.

Scammon, R. E., and Calkins, H. A.: *Development and Growth of the External Dimensions of the Human Body in the Fetal Period.* Minneapolis, University of Minnesota Press, 1929.

Scrimgeour, J. B.: Fetoscopy; *in* A. G. Motulsky and W. Lenz (Eds.): *Birth Defects: Proceedings of the Fourth International Conference.* Amsterdam, Excerpta Medica, 1974, pp. 234–239.

Shepard, T. H.: Normal and abnormal growth patterns; *in* L. I. Gardner (Ed.): *Endocrine and Genetic Diseases of Childhood and Adolescence.* 2nd ed. Philadelphia, W. B. Saunders Co., 1975, pp. 1–8.

Smith, D. W., and Gong, B. T.: Scalp hair patterning as a clue to early fetal brain development. J. Pediat. *83*:379, 1973.

Smyth, C. N.: Ultrasonics; *in* E. E. Philipp, J. Barnes, and M. Newton (Eds.): *Scientific Foundations of Obstetrics and Gynecology.* London, William Heinemann, Ltd., 1970, pp. 678–690.

Stevenson, R. E.: *The Fetus and Newly Born Infant. Influences of the Prenatal Environment.* Saint Louis, The C. V. Mosby Co., 1973, pp. 287–310.

Streeter, G. L.: Weight, sitting height, head size, foot length and menstrual age of the human embryo. *Contrib. Embryol. Carnegie Inst. 11*:143, 1920.

Thompson, J. S., and Thompson, M. W.: *Genetics In Medicine.* 3rd ed. Philadelphia, W. B. Saunders Co., 1980, pp. 344–347.

Tulchinsky, D., and Ryan, K. J.: *Maternal-Fetal Endocrinology.* Philadelphia, W. B. Saunders Co., 1980.

Usher, R. H., and McLean, F. H.: Normal fetal growth and the significance of fetal growth retardation; *in* J. A. Davis and J. Dobbing (Eds.): *Scientific Foundation of Paediatrics.* Philadelphia, W. B. Saunders Co., 1974, pp. 69–80.

Vaughan, V. C., McKay, R. J., and Behrman, R. E.: *Nelson Textbook of Pediatrics.* 11th ed. Philadelphia, W. B. Saunders Co., 1979, pp. 379–389.

Villee, D. B.: *Human Endocrinology. A Developmental Approach.* Philadelphia, W. B. Saunders Co., 1975, pp. 75–93.

Waisman, H. A., and Kerr, G. R.: *Fetal Growth and Development.* New York, McGraw-Hill Book Co., 1970.

Warkany, J.: *Congenital Malformations: Notes and Comments.* Chicago, Year Book Medical Publishers, Inc., 1971.

Widdowson, E. M.: Nutrition; *in* J. A. Davis and J. Dobbing (Eds.): *Scientific Foundations of Paediatrics.* Philadelphia, W. B. Saunders Co., 1974, pp. 44–55.

Young, I. M.: On being born; *in* R. Passmore and J. S. Robson (Eds.): *A Companion to Medical Studies. Vol. 1. Anatomy, Biochemistry, Physiology and Related Subjects.* Oxford, Blackwell Scientific Publications, 1968, pp. 39.1–39.11.

THE FETAL MEMBRANES AND PLACENTA

The *chorion,* the *amnion,* the *yolk sac,* and the *allantois* constitute the embryonic or fetal membranes. These membranes develop from the zygote but do not form parts of the embryo, with the exception of portions of the yolk sac and allantois. The dorsal part of the yolk sac is incorporated into the embryo as the primordium of the *primitive gut* (see Fig. 5–1). The allantois is represented in the adult as a fibrous cord, the *median umbilical ligament,* which extends from the apex of the urinary bladder to the umbilicus.

The placenta has two components: (1) a fetal portion developed from the chorion, and (2) a maternal portion formed by the endometrium. Before birth, the fetal membranes and placenta perform the following functions and activities: *protection, nutrition, respiration, excretion,* and *hormone production.* At birth, or *parturition,* the fetal membranes and placenta are expelled from the uterus as the *afterbirth* (Fig. 7–10*F*).

THE DECIDUA

The term *decidua* (L. *deciduus,* a falling off) is applied to the functional layer of the *gravid,* or pregnant, endometrium, indicating that it is shed at parturition. *Decidual cells* are a characteristic feature of the decidua (Fig. 7–1*B*); these large, pale-staining endometrial stromal cells contain large amounts of glycogen and lipids. The full significance of decidual cells is not understood at present, but it has been suggested that they may provide some nourishment for the embryo and protect the maternal tissue against uncontrolled invasion by the trophoblast (Ramsey, 1965). It has also been suggested that decidual cells may be involved in hormone production (Dallenbach-Hellweg and Nette, 1964).

Three regions of the decidua are identified according to their relation to the implantation site (Fig. 7–2): (1) The part underlying the conceptus and forming the maternal component of the placenta is the *decidua basalis;* (2) the superficial portion overlying the conceptus is the *decidua capsularis;* and (3) all the remaining uterine mucosa is the *decidua parietalis.*

PLACENTAL DEVELOPMENT AND STRUCTURE

Previous descriptions have traced the rapid proliferation of the trophoblast and development of the chorionic sac and its villi (see Chapters 3 and 4). By the beginning of

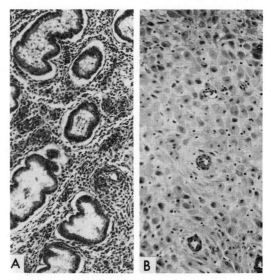

Figure 7–1 Photomicrographs of sections of the endometrium. *A,* During the secretory phase of the menstrual cycle (Chapter 2). Note the appearance of the stromal cells around the glands. *B,* The decidua during the second month of pregnancy, showing the characteristic pale-staining decidual cells. These glycogen- and lipid-laden cells are highly modified stromal cells (×80).

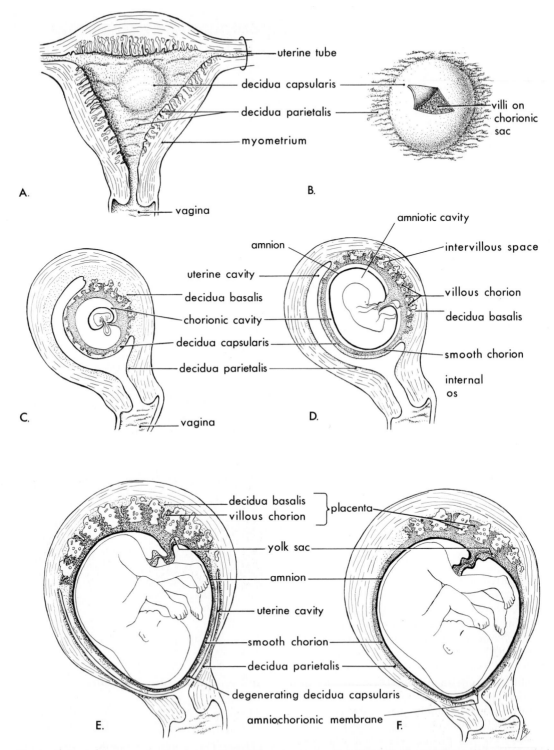

Figure 7–2 *A,* Drawing of a frontal section of the uterus showing the elevation of the decidua capsularis caused by the expanding chorionic sac of a four-week embryo, implanted in the endometrium on the posterior wall. *B,* Enlarged drawing of the implantation site. The chorionic villi have been exposed by cutting an opening in the decidua capsularis. *C* to *F,* Drawings of sagittal sections of the gravid uterus from the fourth to twenty-second weeks, showing the changing relations of the fetal membranes to the decidua. In *F,* the amnion and chorion are fused with each other and the decidua parietalis, thereby obliterating the uterine cavity. Note that the villi persist only where the chorion is associated with the decidua basalis; here they form the villous chorion (fetal portion of the placenta).

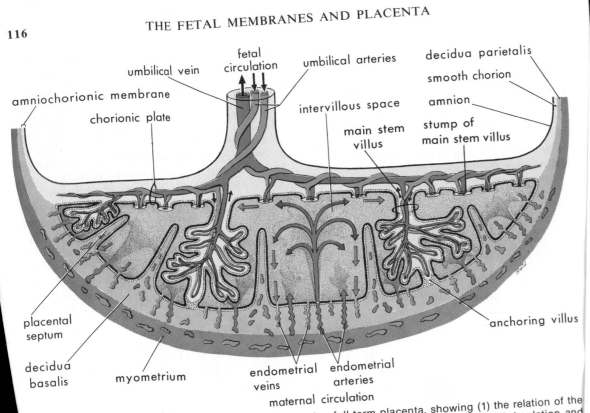

Figure 7–6   Schematic drawing of a section through a full-term placenta, showing (1) the relation of the villous chorion (fetal placenta) to the decidua basalis (maternal placenta), (2) the fetal placental circulation, and (3) the maternal placental circulation. Maternal blood flows into the intervillous spaces in funnel-shaped spurts, and exchanges occur with the fetal blood as the maternal blood flows around the villi. The inflowing arterial blood pushes venous blood out into the endometrial veins, which are scattered over the entire surface of the decidua basalis. Note that the umbilical arteries carry deoxygenated fetal blood (shown in blue) to the placenta and that the umbilical vein carries oxygenated blood (shown in red) to the fetus. Note that the cotyledons are separated from each other by decidual septa of the maternal portion of the placenta. Each cotyledon consists of two or more main stem villi and their many branches. In this drawing only one main stem villus is shown in each cotyledon, but the stumps of those that have been removed are indicated. (Based on Ramsey, 1965.)

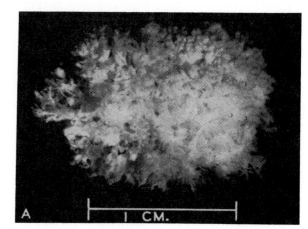

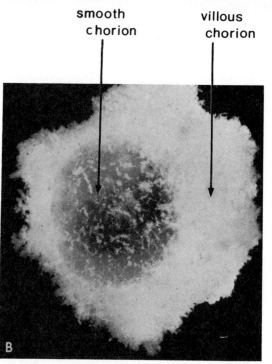

Figure 7–3   Photographs of human chorionic sacs. *A,* 21 days. The entire sac is covered with chorionic villi (×4). *B,* Eight weeks. *Actual size.* As the decidua capsularis becomes stretched and thin, the villi on the corresponding part of the chorionic sac gradually degenerate and disappear, leaving a smooth chorion (see Fig. 7–2C and D). The remaining villous chorion forms the fetal contribution to the placenta. (From Potter, E. L.: *Pathology of the Fetus and Infant.* 2nd ed. Copyright © 1961 by Year Book Medical Publishers, Inc., Chicago.

effects of oxygen, carbon dioxide, and drugs, see Martin (1968).

Maternal Placental Circulation (Fig. 7–

Blood in the intervillous space is temporarily outside the maternal circulatory system; it enters the intervillous space through to 100 *spiral arteries,* or endometrial ries (see Figs. 2–5B and 7–6). The flow m these vessels is pulsatile and is pro-ed in jet-like fountains or streams by the ernal blood pressure. The entering blood a considerably higher pressure than that intervillous space and so spurts toward chorionic plate, often called the "roof" e intervillous space. As the pressure pates, the blood slowly flows around ver the surface of the villi, allowing an nge of metabolic and gaseous products the fetal blood. The maternal blood ually reaches the decidual plate, often the "floor" of the intervillous space,

where it enters the endometrial veins (Fig. 7–6).

The welfare of the embryo and fetus de-pends more on the adequate bathing of t chorionic villi by maternal blood than on a other factor. Acute reductions of uterop cental circulation result in *fetal hypoxia,* fetal death. Chronic reductions of uterop cental circulation result in disturbances growth and development that constitut syndrome known as *intrauterine growth tardation.*

The intervillous space of the mature plac contains about 150 ml of blood, which is rep ished three or four times per minute. The ra uteroplacental blood flow increases during nancy from about 50 ml per minute at 10 wee 500 to 600 ml per minute by full term (M 1968). The intermittent contractions of the u during pregnancy decrease uteroplacental flow, but they do not squeeze significant am

the fourth week, the arrangements necessary for physiological exchanges between the mother and embryo are established.

Up to about the eighth week, *villi* cover the entire surface of the chorionic sac (Figs. 7–2C and 7–3A). As the sac grows, the villi associated with the decidua capsularis become compressed, and their blood supply reduced; subsequently, these villi begin to degenerate (Figs. 7–2D, and 7–3B), produc-ing a relatively avascular bare area known as the *smooth chorion,* or *chorion laeve* (L. *levis,* smooth). As this occurs, the villi asso-ciated with the decidua basalis rapidly in-crease in number, branch profusely, and

enlarge. This portion of the chorionic sac is known as the *villous chorion,* or *chorion frondosum* (L. *frondosus,* leafy).

The fetal component of the placenta con-sists of the chorionic plate (Fig. 7–5) and the chorionic villi that arise from it and project into the intervillous spaces containing mater-nal blood (Fig. 7–6).

The maternal component of the placenta is formed by the decidua basalis (Figs. 7–2 and 7–5). This comprises all the endometrium beneath the fetal component of the placenta, except the deepest part, which is often called the *decidual plate.* This layer remains after *parturition* (Fig. 7–10) and is involved in the regeneration of the endometrium during the subsequent uterine cycle, or *menstrual cycle* (see Fig. 2–6).

The shape of the placenta is determined by the form of the persistent area of chorionic villi (Fig. 7–2*F*); usually this is circular, giving the placenta a discoid shape (Fig. 7–4). As the villi invade the decidua basalis, they leave several wedge-shaped areas of decidual tissue called *placental septa* (Fig. 7–6). The placental septa divide the fetal part of the placenta into 10 to 38 irregular convex areas, composed of lobes and lobules, called *cotyledons* (Fig. 7–12*A*). Each cotyledon consists of two or more main stem villi and their many branches.

The Intervillous Spaces (Figs. 7–2D and 7–5). The blood-filled intervillous spaces are derived mainly from the lacunae that developed in the syncytiotrophoblast during the second week (Fig. 3–1*C*). During subsequent invasion by the syncytiotrophoblast, these spaces enlarge at the expense of the decidua basalis. Collectively, the spaces

form a large blood sinus, the *intervillous space,* which is bounded by the chorionic plate and decidua basalis (Figs. 7–5 and 7–6). The intervillous space is divided into compartments by the placental septa, but, because the septa do not reach the *chorionic plate,* there is communication between the intervillous spaces of different compartments. The intervillous space is drained by endometrial veins that are found over the entire surface of the decidua basalis (Fig. 7–6).

Maternal blood circulates through the intervillous spaces, bringing nutritive and other substances necessary for embryonic and fetal development, and taking away the waste products of fetal metabolism.

Fate of the Decidua Capsularis (Fig. 7–2). The decidua capsularis is the part of the endometrium that is superficial to the embryo, forming a *capsule* over it. As the conceptus enlarges, the decidua capsularis bulges into the uterine cavity and soon becomes greatly attenuated. Eventually, the

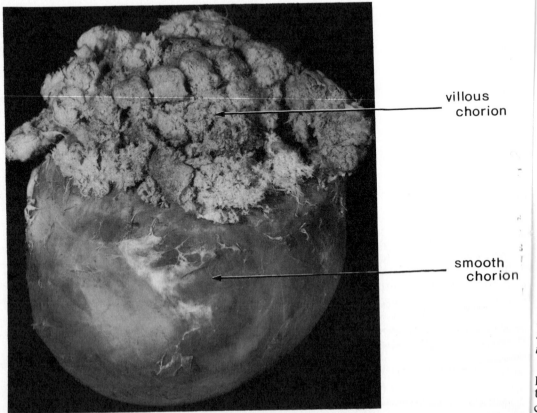

villous chorion

smooth chorion

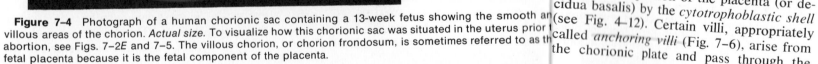

Figure 7–4 Photograph of a human chorionic sac containing a 13-week fetus showing the smooth and villous areas of the chorion. *Actual size.* To visualize how this chorionic sac was situated in the uterus prior to abortion, see Figs. 7–2*E* and 7–5. The villous chorion, or chorion frondosum, is sometimes referred to as the fetal placenta because it is the fetal component of the placenta.

maternal blood in intervillous space

amniochorionic membrane

decidua basalis

chorionic plate

amnion

smooth chorion

decidua parietalis

vagina

Figure 7–5 Drawing of a sagittal section of a gravid uterus at 22 weeks showing the relation of the fetal membranes to each other and to the decidua. The fetus has been removed, and the amnion and smooth chorion have been cut and reflected.

decidua capsularis fuses with the decidua parietalis, thereby obliterating the uterine cavity (Fig. 7–2*F*). By about 22 weeks, reduced blood supply causes the decidua capsularis to degenerate and disappear.

The Amniochorionic Membrane (Figs. 7–2, 7–5, and 7–6). The amniotic sac enlarges somewhat faster than the chorionic sac, and their walls soon fuse to form the amniochorionic membrane. This membrane fuses with the decidua capsularis and then, after disappearance of the decidua capsularis, fuses with the decidua parietalis.

It is the amniochorionic membrane, composed of the amnion and the smooth chorion, that ruptures or is ruptured artificially during labor (Fig. 7–10). *Premature rupture of the membranes is the most common event leading to premature labor* (Page et al., 1981).

The Fetomaternal Junction. The fetal placenta (or villous chorion) is anchored to the maternal portion of the placenta (or decidua basalis) by the *cytotrophoblastic shell* (see Fig. 4–12). Certain villi, appropriately called *anchoring villi* (Fig. 7–6), arise from the chorionic plate and pass through the

intervillous space to *attach firmly to the decidua basalis by way of the cytotrophoblastic shell.* In addition to anchoring the chorionic plate to the decidua basalis, the anchoring villi give origin to branches called *free, or floating, villi* because they float in the blood-filled intervillous spaces.

In placentas such as those in humans, the syncytiotrophoblast is highly invasive; as a result, there is considerable necrosis of the decidua basalis and deposition of *fibrinoid material* (Fig. 7–7*C*). A more or less continuous layer of fibrinoid material, often called *Nitabuch's layer,* forms and separates the fetal and maternal parts of the placenta. *Placental separation* occurs deep to this layer during the third stage of labor (Fig. 7–10*F*); thus it is shed with the placenta.

PLACENTAL CIRCULATION

The placenta essentially provides an area where materials may be exchanged across the placental membrane between the fetal and maternal circulations (Figs. 7–7 and 7–8). Although the circulations of the fetus and the mother are separated by the placental membrane of the fetal tissues, they are contiguous (*contiguus,* in contact).

Fetal Placental Circulation. Deoxygenated blood leaves the fetus and passes through the umbilical arteries to the placenta. Where the cord attaches to the placenta, these arteries divide into a number of radially disposed vessels that branch freely in the chorionic plate before entering the villi.

The blood vessels form an extensive arterio-capillary-venous system within the villus (Fig. 7–7*A*), bringing the fetal blood very close to the maternal blood (Figs. 7–7 and 7–8). *There is normally no intermingling of fetal and maternal blood.* The oxygenated fetal blood passes into thin-walled veins that follow the chorionic arteries back to the site of attachment of the umbilical cord, where they converge to form the umbilical vein. This large vessel carries oxygenated blood to the fetus. (For further details of the fetal circulation, see Chapter 14 and Figure 14–26).

The rate of blood flow through the fetal circulation is relatively slow. It is controlled by the interaction of the fetal arterial pressure and the fetal placental vascular resistance, which is probably under control of fetal prostaglandins.

of blood out of the intervillous space. Consequently, oxygen transfer to the fetus is decreased during uterine contractions, but it does not stop.

The Placental Membrane (Figs. 7–7 and 7–8). This membrane consists of the fetal tissues separating the maternal and fetal blood. Until about 20 weeks, it consists of four layers: (1) syncytiotrophoblast, (2) cytotrophoblast, (3) the connective tissue core of the villus, and (4) the endothelium of the fetal capillary.

The placental membrane is often called the placental "barrier", but this term is inappropriate because there are few compounds, endogenous or exogenous, that are unable to cross the placental membrane in detectable amounts (Page et al., 1981). The placental membrane acts as a true barrier only when the molecule has a certain size, configuration, and charge (e.g., heparin). *Most drugs or other chemicals present in the maternal plasma will also be found in the fetal plasma.*

Electron micrographs of the syncytiotrophoblast show that its free surface has many microvilli that increase the surface area for exchange between the maternal and fetal circulations. After 20 weeks, (1) the cytotrophoblast no longer forms a continuous layer, (2) the relative amount of connective tissue is reduced, and (3) the number and size of the fetal capillaries increase.

As pregnancy advances, the placental membrane becomes progressively thinner, and many capillaries come to lie very close to the syncytiotrophoblast (Fig. 7–7C). At some sites, the syncytiotrophoblastic nuclei form *nuclear aggregations,* or *syncytial knots* (Fig. 7–7C). These knots continually break off and are carried into the maternal

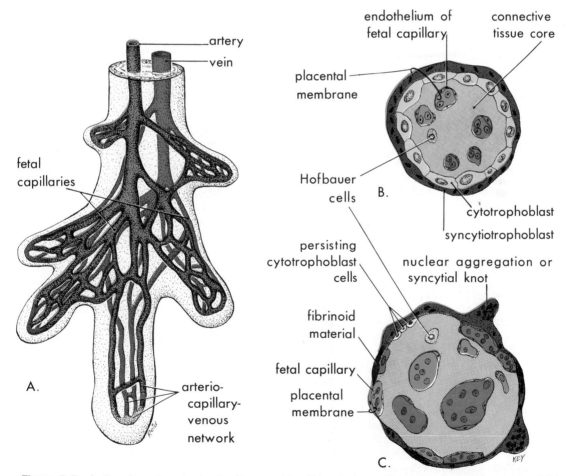

Figure 7–7 *A,* Drawing of a chorionic villus showing the arterio-capillary-venous system carrying fetal blood. The artery carries deoxygenated blood and waste products from the fetus, whereas the vein carries oxygenated blood and nutrients to the fetus. *B* and *C,* Drawings of sections through a chorionic villus at 10 weeks and at full term, respectively. The villi are bathed externally in maternal blood. The placental membrane, composed of fetal tissues, separates the maternal blood from the fetal blood (see also Fig. 7–8).

circulation. Some may lodge in capillaries of the maternal lung, where they all soon die and disappear (Benirschke, 1972). Toward the end of pregnancy, *fibrinoid material* forms on the surfaces of villi (Fig. 7–7C); it consists of fibrin and other, unidentified substances that stain intensely with eosin. All the aforementioned changes result mainly from aging and appear to cause reduced placental function.

Hofbauer Cells (Fig. 7–7B). These cells with large spherical nuclei and vacuolated cytoplasm appear in the cores of villi, particularly during the first half of pregnancy. Histochemical tests indicate that their vacuoles contain mucopolysaccharides, mucoproteins, and lipids (Boyd and Hamilton, 1970). The complete role of these cells is uncertain, but they have the general qualities of macrophages (Benirschke, 1972). Possibly, they are *phagocytic* (Ham and Cormack, 1979).

PLACENTAL ACTIVITIES

The placenta has three main activities: (1) metabolism, (2) transfer, and (3) endocrine

secretion; all are essential for maintaining pregnancy and promoting normal embryonic development.

PLACENTAL METABOLISM

The placenta, particularly during early pregnancy, synthesizes glycogen, cholesterol, and fatty acids, and serves as a source of nutrients and energy for the embryo. Many of its metabolic activities are undoubtedly critical for the major placental activities of transfer and endocrine secretion.

PLACENTAL TRANSFER

Almost all materials are transported across the placental membrane by one of the following four mechanisms: (1) *simple diffusion,* (2) *facilitated diffusion,* (3) *active transport,* and (4) *pinocytosis* (Page et al., 1981). Three other modes of transfer are sometimes used. (1) Fetal red blood cells pass into the maternal circulation, particularly during labor, presumably through microscopic breaks in

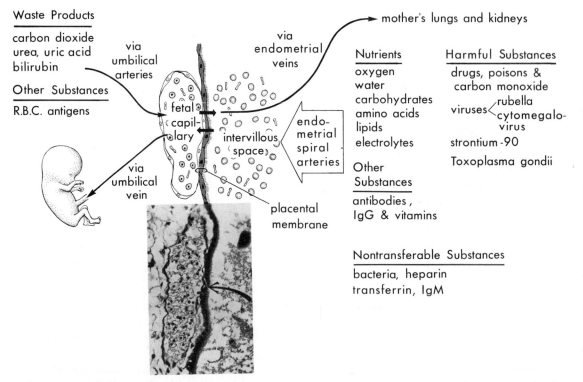

Waste Products

carbon dioxide
urea, uric acid
bilirubin

Other Substances

R.B.C. antigens

mother's lungs and kidneys

Nutrients

oxygen
water
carbohydrates
amino acids
lipids
electrolytes

Other Substances

antibodies,
IgG & vitamins

Harmful Substances

drugs, poisons &
carbon monoxide
viruses ⟨ rubella
cytomegalo-
virus
strontium-90

Toxoplasma gondii

Nontransferable Substances

bacteria, heparin
transferrin, IgM

Figure 7–8. Diagrammatic illustration of placental transfer. The tissues across which transport of substances between the mother and fetus occurs collectively constitute the placental membrane. This composite membrane is composed entirely of tissues of fetal origin: syncytiotrophoblast, stroma in the villus, and the endothelium of the fetus capillary (see also Fig. 7–7). (Inset photomicrograph from Javert, C. T.: *Spontaneous and Habitual Abortion.* 1957. Courtesy of The Blakiston Division, McGraw-Hill Book Co. Copyright ©1957 by McGraw-Hill, Inc. Used by permission of McGraw-Hill Book Company.)

the placental membrane (Fig. 7–8). Labeled maternal red blood cells have also been found in the fetal circulation (Page et al., 1981). Thus, *red blood cells may pass in either direction through breaks in the placental membrane.* (2) Other cells cross the placenta under their own power, e.g., maternal leucocytes and *Treponema pallidum,* the organism that causes syphilis. Some bacteria and protozoa infect the placenta by creating lesions and then enter the fetal blood.

Gases. Oxygen, carbon dioxide, and carbon monoxide cross the placental membrane by simple diffusion. Interruption of oxygen transport for even a few minutes will endanger fetal survival (Boyd and Hamilton, 1970). The uterus near term extracts 20 to 35 ml of oxygen per minute from the maternal blood (Page et al., 1981). Over half of this oxygen is transferred to the fetus by simple diffusion alone. The placental membrane approaches the efficiency of the lung for gas exchange. The quantity of oxygen reaching the fetus is primarily flow-limited rather than diffusion-limited. Hence, *fetal hypoxia* results primarily from factors that diminish either the uterine blood flow or the fetal blood flow.

Nutrients. Water is rapidly and freely exchanged between mother and fetus, and in increasing amounts as pregnancy advances. There is little or no transfer of maternal cholesterol, triglycerides, or phospholipids (Page et al., 1981). There is transport of free fatty acids, but the amount transferred is probably relatively small. Vitamins cross the placenta and are essential to normal development. Water-soluble vitamins cross the placental membrane more quickly than fat-soluble ones. Glucose is quickly transferred.

Hormones. Protein hormones do not reach the fetus in significant amounts, except for a slow transfer of thyroxine and triiodothyronine (Page et al., 1981). Unconjugated steroid hormones pass the placental membrane rather freely. Testosterone and certain synthetic progestins cross the placenta and may cause masculinization of female fetuses (see Fig. 8–16 and Table 8–3).

Electrolytes. These are freely exchanged across the placenta in significant quantities, each at its own rate. *When a mother receives intravenous fluids, they also pass to the fetus and affect its water and electrolyte status.*

Antibodies. Some passive immunity is conferred upon the fetus by placental transfer of maternal antibodies. The alpha and beta globulins reach the fetus in very small quantities, but many of the gamma globulins, notably the IgG (7S) class, are readily transported to the fetus. Maternal antibodies confer on the fetus immunity to such diseases as diphtheria, smallpox, and measles, but no immunity is acquired to pertussis (whooping cough) or chickenpox. *Maternal antibodies* are taken up by *pinocytosis* (Gr. *pinein,* to drink) of the syncytiotrophoblast, a process involving the ingestion of fluids and their contained solutes. The antibodies subsequently enter the fetal capillaries.

As previously stated, small amounts of blood may pass from the fetus to the mother through microscopic breaks in the placental membrane. If the fetus is Rh-positive and the mother Rh-negative, the fetal cells may stimulate the formation of anti-Rh antibody by the mother. This passes to the fetal blood stream and causes hemolysis of fetal Rh-positive blood cells and anemia in the fetus. Some fetuses with this condition, known as *hemolytic disease of the newborn (HDN),* or erythroblastosis fetalis, fail to make a satisfactory intrauterine adjustment and may die unless delivered early or given intrauterine blood transfusions (discussed in Chapter 6).

Exchange transfusions are also performed after birth, using the umbilical vein. Most of the infant's blood is replaced with Rh-negative donor blood. This technique prevents death of erythroblastotic babies who are very anemic. It also avoids brain damage by preventing or controlling hyperbilirubinemia (Reid et al., 1972).

When the placenta separates at birth (Fig. 7–10F), the mother often receives a small transfusion of fetal blood into her circulation from ruptured fetal chorionic vessels. If she is Rh-negative and the infant Rh-positive, the fetal red cells can stimulate a permanent antibody response in the mother. These fetal red blood cells can be destroyed rapidly by giving the mother high-titer anti-Rh antibody. In this way, she does not become sensitized. For more information on hemolytic disease of the newborn, see Thompson and Thompson (1980) and Page et al. (1981).

Wastes. The major waste product, carbon dioxide, diffuses across the placenta even more rapidly than oxygen. Urea and uric acid pass the placental membrane by simple diffusion, and bilirubin is quickly cleared.

Drugs. Most drugs and drug metabolites cross the placenta freely by simple diffusion, the exception being those with a structural similarity to amino acids, e.g., methyldopa and the antimetabolites. Some drugs, e.g., thalidomide, cause congenital malformations (see Chapter 8). *Fetal drug addiction* may occur after maternal use of drugs such as heroin. Except for the muscle relaxants, such as succinylcholine and curare, most agents used for the management of labor readily cross the placenta. These drugs may cause respiratory depression of the newborn infant, depending on the dose and timing in relation to delivery (Ryan, 1972). All sedatives and analgesics affect the fetus to some degree (Page et al., 1981).

Drugs taken by the mother can affect the embryo or fetus directly or indirectly by interfering with maternal or placental metabolism (Gill and Davis, 1974). The amount of drug or metabolite reaching the placenta is controlled by the maternal blood level and by the blood flow through the placenta from both the maternal and the fetal circulations.

Infectious Agents. Cytomegalovirus, rubella and Coxsackie viruses, and those viruses associated with variola, varicella, measles, and poliomyelitis may pass through the placental membrane and cause fetal infection. In some cases (e.g., rubella virus), congenital malformations may be produced (see Chapter 8). Organisms such as *Treponema pallidium,* which causes syphilis, may cross the placental membrane. The *spirochetes* (spiral microorganisms) penetrate the placental membrane and enter the fetal blood, often causing death and/or malformation of the fetus.

PLACENTAL ENDOCRINE SECRETION

Using precursors derived from the fetus, the mother, or both, the *syncytiotrophoblast* synthesizes various hormones.

Protein Hormones. The well-documented protein hormone products of the placenta are (1) *human chorionic gonadotropin* (hCG), and (2) *human chorionic somatomammotropin* (hCS), or human placental lactogen (hPL).

The glycoprotein hCG, similar to luteinizing hormone (LH), is secreted by the syncy- *tiotrophoblast during the third week,* a few days after the completion of implantation. The concentration of hCG in the maternal blood and urine rises to a maximum by the eighth week and then declines to a lower level.

In addition to hCG and hPL, human chorionic *thyrotropin* (hCT) and human chorionic *corticotropin* (hCACTH) are formed by the placenta. The placenta has a great potential for the synthesis of biologically active peptide and protein hormones (Page et al., 1981).

Steroid Hormones. The placenta plays a major role in the production of *progesterone* and *estrogens.* Progesterone can be obtained from the placenta at all stages of gestation, indicating that this hormone is necessary for the maintenance of pregnancy. The placenta forms progesterone from maternal cholesterol or pregnenolone.

The ovaries of a pregnant woman can be removed after the first trimester without causing an abortion. Large amounts of progesterone are produced during the first weeks of pregnancy by the *corpus luteum of pregnancy* in the ovary (see Fig. 6–4), but the placenta soon takes over this activity.

Estrogens are also produced in large quantities by the syncytiotrophoblast. The placenta forms them from 19 carbon precursors, many of which are supplied by the fetus (Villee, 1975).

UTERINE GROWTH DURING PREGNANCY

The uterus of a nonpregnant female lies in the pelvis minor (Fig. 7–9A). Naturally, the uterus increases in size during pregnancy to accommodate the needs of the growing fetus (Fig. 7–9B and C). While the uterus is increasing in size, it also increases in weight and its walls become thinner. During the first trimester, the uterus rises out of the pelvic cavity and usually reaches the level of the umbilicus by 20 weeks (Fig. 7–9B). By 28 to 30 weeks, it reaches the epigastric region (Fig. 7–9C). The increase in size of the uterus largely results from hypertrophy of preexisting muscular fibers, and partly from the development of new fibers.

PARTURITION, OR LABOR

The *birth process* by which the fetus, placenta, and fetal membranes are expelled

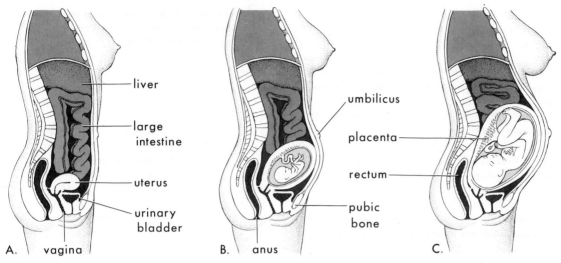

Figure 7–9 Diagrammatic drawings of sagittal sections of a female. *A,* Not pregnant. *B,* 20 weeks of pregnancy. *C,* 30 weeks of pregnancy. Note that as the fetus enlarges, the uterus increases in size to accommodate the rapidly growing fetus. By 20 weeks the uterus and fetus reach the level of the umbilicus; by 30 weeks they reach the epigastric region. The mother's abdominal viscera are displaced, and the skin and muscle of her anterior abdominal wall are greatly stretched.

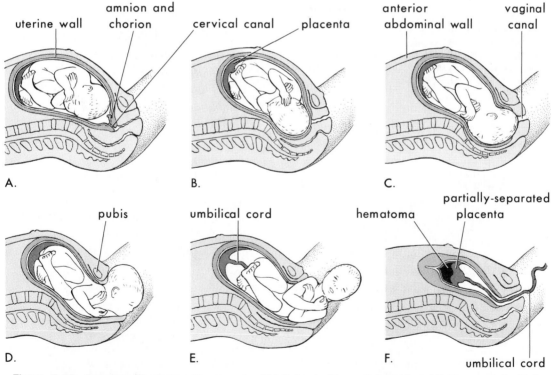

Figure 7–10 Drawings illustrating the process of birth (parturition or labor). *A* and *B,* The cervix is dilating during the first stage of labor. Note that the fused amnion and chorion are being forced into the cervical canal. *C* to *E,* The fetus passes through the cervix and the vagina during the second stage of labor. *F,* As the uterus contracts during the third stage of labor, the placenta folds up and pulls away from the uterine wall. Separation of the placenta results in bleeding and the formation of a large hematoma.

from the mother's reproductive tract (Fig. 7–10) is called *parturition* (labor, or childbirth). The factors that trigger labor are not clearly understood. For many years, it was thought that the release of oxytocin from the maternal neurohypophysis most likely initiated labor. There is now evidence that the quantity of steroids produced by the fetal adrenal cortex may be important in triggering some other event in the placenta or myometrium that culminates in the onset of labor. One such event may be the release of substances with oxytocic properties, i.e., substances that stimulate uterine contractions. Current opinion is that the intermittent release of oxytocin from the maternal neurohypophysis is important in determining the strength and duration of uterine contractions once labor is established (Page et al., 1981).

THE STAGES OF LABOR

Labor (L., toil, suffering) is the process of giving birth to a child. The word is synonymous with the term *parturition* (L. *parturitio*, childbirth). There are three stages:

The *first stage (dilatation stage)* begins when there is objective evidence of progressive dilatation of the cervix. This occurs with the onset of regular contractions of the uterus (less than 10 minutes apart and painful). The first stage ends with complete dilatation

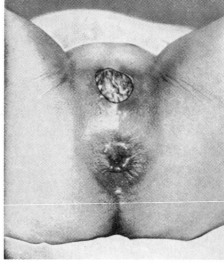

Figure 7–11 Photographs illustrating delivery of the baby's head during the second stage of labor. *A*, The head distends the mother's perineum, and part of the scalp becomes visible; this is called "crowning." *B*, The perineum slips back over the face. *C*, The head is delivered; subsequently the body of the fetus is expelled. (From Greenhill, J. B., and Friedman, E. A.; *Biological Principles and Modern Practice of Obstetrics.* Philadelphia, W. B. Saunders Co., 1974.)

A

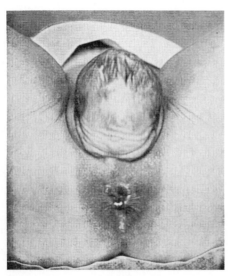

B

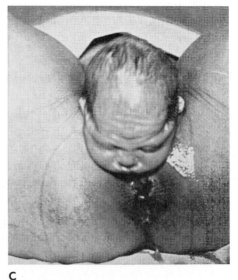

C

of the cervix. The average duration of the first stage is about 12 hours for first pregnancies (nulliparous patients, or primigravidas), and about 7 hours for women who have had a child previously (multiparous patients, or multiparas), but there are wide variations.

The *second stage (expulsion stage)* begins when the cervix is fully dilated and ends with the delivery of the baby (Figs. 7–10 and 7–11). The average duration for primigravidas is 50 minutes, and for multiparas, 20 minutes.

The *third stage (placental stage)* begins when the baby is born and ends when the placenta and membranes are delivered. The duration of this stage is ordinarily less than 10 minutes. The placenta separates through the spongy layer of the decidua basalis. After delivery of the baby, the uterus continues to contract. As a result, a hematoma forms behind the placenta (Fig. 7–10F) and separates it from the decidua basalis. After delivery of the placenta, the myometrial contractions constrict the spiral arteries that formerly supplied the intervillous spaces. These persistent contractions, which are almost "tonic" in nature, prevent excessive bleeding from the placental site.

THE FULL-TERM PLACENTA

After birth, the placenta, the umbilical cord, and the associated fetal membranes (amnion and smooth chorion) are expelled from the uterus. The placenta (Gr. *plakuos,* a flat cake) commonly has a discoid shape (Fig. 7–12), with a diameter of 15 to 20 cm and a thickness of 2 to 3 cm. The placenta weighs 500 to 600 gm, usually about one sixth the weight of the fetus. The margins of the placenta are continuous with the ruptured amniotic and chorionic sacs (Figs. 7–6 and 7–12A and C).

Variations in Placental Shape. Usually villi persist only where the chorion is in contact with the decidua basalis, giving rise to a discoid placenta (Fig. 7–12). When villi persist elsewhere, several variations in placental shape occur: accessory placenta (Fig. 7–13), bidiscoid placenta, diffuse placenta, and horseshoe placenta. Although there are innumerable variations in the size and shape of the placenta, most of them are of little physiological or clinical significance.

Examination of the placenta may provide information about the causes of (1) placental dysfunction, (2) fetal growth retardation, (3) neonatal illness, and (4) infant death. Placental studies can also determine if the placenta and membranes are complete. Retention of a cotyledon or an accessory placenta in the uterus may lead to late puerperal hemorrhage (bleeding after the first 24 hours post partum), but retained placental tissue is more often present in cases of delayed hemorrhage after abortion than after a term delivery.

Maternal Surface (Fig. 7–12A). The characteristic cobblestone appearance of this surface is caused by the 10 to 38 *cotyledons,* separated by grooves formerly occupied by the *placental septa* (Fig. 7–6). The surface of the cotyledons is covered by thin grayish shreds of decidua basalis, which are recognizable in sections of placenta examined under a microscope. Most of the decidua basalis, however, is temporarily retained in the uterus and shed with subsequent uterine bleeding.

Fetal Surface (Fig. 7–12B). The umbilical cord attaches to this surface, and its amniotic covering is continuous with the amnion adherent to this surface of the placenta. The vessels radiating from the umbilical cord are clearly visible through the smooth transparent amnion. The *umbilical cord vessels* subdivide on the fetal surface to form the *chorionic vessels,* which supply the chorionic villi (Fig. 7–6).

The Umbilical Cord. The attachment of the umbilical cord is usually near the center of the placenta (Fig. 7–12B), but it may be found at any point (e.g., at the edge producing a *battledore placenta,* Fig. 7–12D, or to the membranes forming a *velamentous insertion,* Fig. 7–14).

As the amniotic sac enlarges, the amnion sheathes the umbilical cord, forming the cord's epithelial covering (Figs. 7–16 and 7–17). The umbilical cord is usually 1 to 2 cm in diameter and 30 to 90 cm in length (average 55 cm). Excessively long or short cords are uncommon. Long cords have a tendency to prolapse and/or to coil around the fetus. Coiling of the cord around the fetus's neck may make the cord relatively short, interfering with delivery of the baby.

The umbilical cord usually contains two arteries and one vein surrounded by mucoid connective tissue, often called *Wharton's jelly* (Fig. 7–16A). Because the umbilical vessels are longer than the cord, twisting and bending of the vessels is common. The vessels frequently form loops, producing so-called *false knots* that are of no significance. In about 1 per cent of deliveries, there are *true knots* in the cord that may cause fetal death (Fig. 7–15). In most cases, the knots form during labor as a result of the fetus

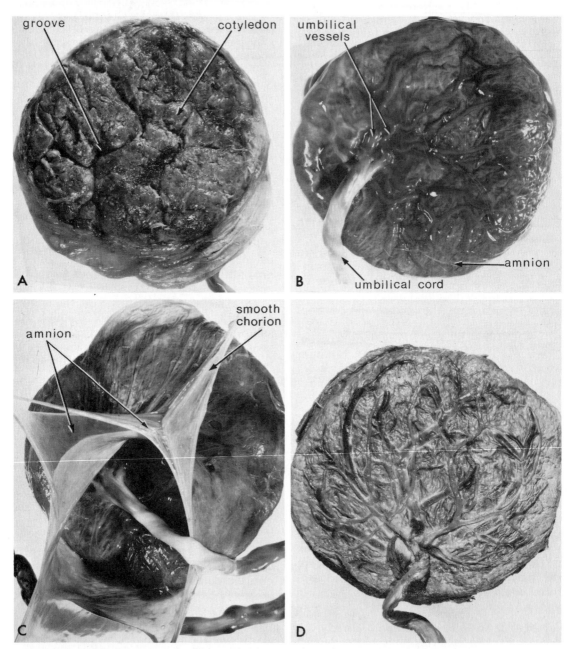

Figure 7–12 Photographs of full-term placentas. *About one-third actual size.* A, Maternal, or uterine, surface, showing cotyledons and grooves. Each convex cotyledon consists of a number of main stem villi with their many branches. The grooves were occupied by the placental septa when the maternal fetal portions of the placenta were together (see Fig. 7–6). B, Fetal, or amniotic, surface showing the blood vessels running in the chorionic plate (Fig. 7–6) under the amnion and converging to form the umbilical vessels at the attachment of the umbilical cord. C, The amnion and smooth chorion are arranged to show that they are (1) fused and (2) continuous with the margins of the placenta. See Figures 7–5 and 7–6 also for visualizing the relationship between the amnion and the smooth chorion. D, Placenta with a marginal attachment of the cord, often called a battledore placenta because of its resemblance to the bat used in the medieval game of battledore and shuttlecock.

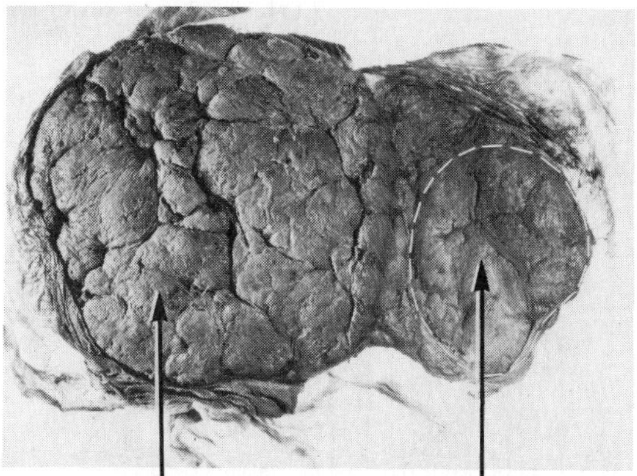

Figure 7–13 Photograph of the maternal surface of a full-term placenta with an accessory placenta. The accessory placental tissue developed from chorionic villi that persisted a short distance from the placenta. Retention of pieces of an accessory placenta in the uterus may prevent adequate uterine contraction and lead to delayed postpartum hemorrhage. *About one-quarter actual size.*

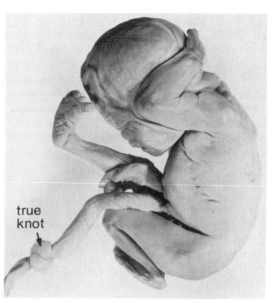

main
placenta

accessory
placenta

margin of placenta

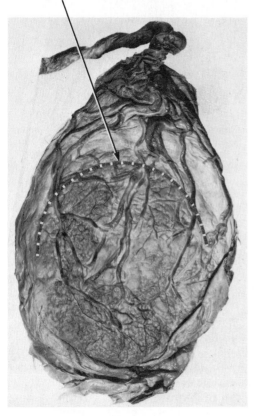

Figure 7–14 Photograph of a placenta with a velamentous insertion of the umbilical cord. The cord is attached to the membranes (amnion and chorion), not to the placenta. The umbilical vessels leave the cord and run between the amnion and smooth chorion before reaching the placenta. The vessels are more easily torn in this location, especially when they course over the lower uterine segment; this latter condition is known as vasa previa. If the vessels rupture before birth, the fetus loses blood and could be near exsanguination at birth.

Figure 7–15 Photograph of a 20-week fetus with a true knot (arrow) in the umbilical cord. *Half actual size.* The diameter of the cord is greater in the portion closest to the fetus, indicating that there was an obstruction of blood flow in the umbilical arteries. Undoubtedly, this knot was a major cause of the death of the fetus.

true
knot

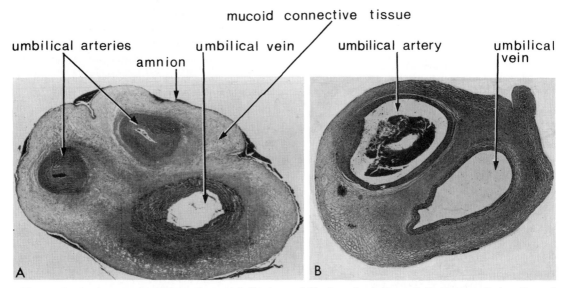

Figure 7–16 Transverse sections through full-term umbilical cords. *A,* Normal. *B,* Abnormal, showing only one artery (×3.) The umbilical cord of one in 200 fetuses has only one umbilical artery; and about 15 per cent of fetuses with this condition have cardiovascular malformations. (From Javert, C. T.: *Spontaneous and Habitual Abortion. 1957.* Courtesy of The Blakiston Division, McGraw-Hill Book Co. Copyright © 1957 by McGraw-Hill, Inc. Used by permission of McGraw-Hill Book Company.)

passing through a loop of the cord. Because the knots are usually loose, they have no clinical significance. If a true knot forms early in pregnancy and tightens owing to fetal movements, the knot may interfere with the fetal circulation and cause death and abortion of the fetus (Fig. 7–15).

Simple looping of the cord around the fetus occasionally occurs (Fig. 7–18). In about one fifth of all deliveries, the cord is looped once around the neck without increased fetal risk (Page et al., 1981).

In about 1 in 200 newborns, only one umbilical artery is present (Fig. 17–16*B*), a condition that may be associated with fetal abnormalities, particularly of the cardiovascular system (Benirschke and Driscoll, 1967). Page et al. (1981) state that absence of an umbilical artery is accompanied by a 15 to 20 per cent incidence of cardiovascular malformations in the fetus. Absence of an umbilical artery results from either agenesis or degeneration of this vessel early in development.

AMNION AND AMNIOTIC FLUID

The amnion is a *membranous sac* that surrounds the embryo (Fig. 7–17*A* and *B*) and, later, the fetus (Figs. 7–17*C* and *D* and 7–18). Formation of the amniotic cavity and the amnion was described in Chapter 3.

Because the amnion is attached to the margins of the embryonic disc (Fig. 7–17*A*), its junction with the embryo is located on the ventral surface after folding of the embryo (Fig. 7–17*B*). As the amnion enlarges, it gradually obliterates the chorionic cavity and sheathes the umbilical cord, forming its epithelial covering (Fig. 7–17*C* and *D*). The epithelial cells of the amnion possess microvilli that probably play a role in fluid transfer, e.g., the exchange of water and dissolved substances via vessels in the decidua parietalis.

Origin of Amniotic Fluid. Initially some fluid may be secreted by the amniotic cells, but most amniotic fluid is undoubtedly derived from the maternal blood by transport across the amnion. The fetus also makes a contribution by excreting urine into the amniotic fluid; by late pregnancy, about a half-liter of urine is added daily.

Volume of Amniotic Fluid. The volume of amniotic fluid normally increases slowly, reaching about 30 ml at 10 weeks, 350 ml at 20 weeks, and 1000 ml by 37 weeks. The volume then decreases sharply. Low volumes of amniotic fluid (e.g., 400 ml in the third trimester), a condition called *oligohydramnios,* is believed to result in most cases from placental insufficiency with diminished placental blood flow. In *renal agenesis* (absence of kidneys), the absence of the fetal

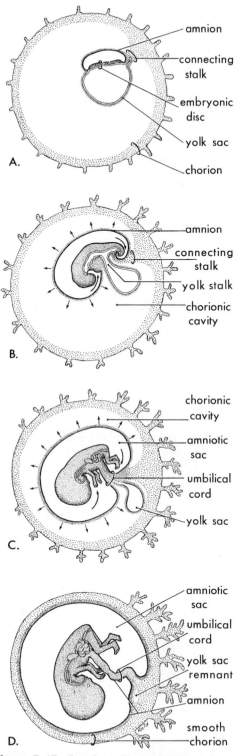

A.

- amnion
- connecting stalk
- embryonic disc
- yolk sac
- chorion

B.

- amnion
- connecting stalk
- yolk stalk
- chorionic cavity

C.

- chorionic cavity
- amniotic sac
- umbilical cord
- yolk sac

D.

- amniotic sac
- umbilical cord
- yolk sac remnant
- amnion
- smooth chorion

Figure 7–17 Drawings illustrating how the amnion becomes the outer covering of the umbilical cord and how the yolk sac is partially incorporated into the embryo as the primitive gut. Formation of the placenta and the degeneration of chorionic villi are also shown. *A*, three weeks. *B*, four weeks. *C*, 10 weeks. *D*, 20 weeks.

urine contribution to the amniotic fluid causes oligohydramnios.

An accumulation of amniotic fluid in excess of 2000 ml *(polyhydramnios)* may occur when the fetus does not drink the usual amount of fluid. This condition is often associated with malformations of the central nervous system, e.g., *anencephaly* (see Fig. 18–19). In other malformations, such as *esophageal atresia* (see Fig. 11–5A), amniotic fluid accumulates because it is unable to pass to the stomach and the intestine for absorption. *Multiple pregnancy* (e.g., twinning) is also a predisposing cause of polyhydramnios.

Exchange of Amniotic Fluid. Studies using radioactive isotopes have shown that water in the amniotic fluid changes every three hours (Gadd, 1970). Large volumes of fluid move in both directions between the fetal and maternal circulations, mainly via the placental membrane. *Fetal swallowing of amniotic fluid is a normal occurrence.* Most of the fluid passes into the gastrointestinal tract, but some of it also passes into the lungs. In either case, the fluid is absorbed into the fetal circulation and then passes into the maternal circulation via the placental membrane. In the final stages of pregnancy, *the fetus swallows up to 400 ml of amniotic fluid per day.* Some fluid also passes from the amniotic cavity into the maternal blood through the amniochorionic membrane (Fig. 7–5).

Composition of Amniotic Fluid. The fluid in the amniotic cavity is a solution in which undissolved material is suspended. It consists of desquamated fetal epithelial cells and approximately equal portions of organic and inorganic salts, in 98 to 99 per cent water. Half the organic constituents are protein; the other half consists of carbohydrates, fats, enzymes, hormones, and pigments. As pregnancy advances, the composition of the amniotic fluid changes as fetal excreta are added. Because fetal urine is added to amniotic fluid, studies of fetal enzyme systems, amino acids, hormones, and other substances can be conducted on fluid removed by amniocentesis (see Fig. 6–13). Studies of cells in the amniotic fluid permit diagnosis of the sex of the fetus and the detection of fetuses with chromosomal abnormalities (e.g., *trisomy 21* resulting in the Down syndrome).

Significance of Amniotic Fluid. The embryo, suspended by the umbilical cord, floats freely in amniotic fluid. This buoyant medium

1. *permits symmetrical external growth* of the embryo;

2. *prevents adherence of the amnion* to the embryo;

3. *cushions the embryo against injuries* by distributing impacts the mother may receive;

4. *helps control the embryo's body temperature* by maintaining a relatively constant temperature; and

5. *enables the fetus to move freely,* thus aiding musculoskeletal development.

THE YOLK SAC

Early development of the yolk sac is described in Chapters 3 and 5. By nine weeks, the yolk sac has shrunk to a pear-shaped remnant, about 5 mm in diameter, which is connected to the midgut by the narrow *yolk stalk* (Figs. 6–3*A* and 7–17*C*). By 20 weeks, the yolk sac is very small (see Fig. 8–17*D*); thereafter, it is usually not visible.

Significance of the Yolk Sac.   Although the human yolk sac is nonfunctional as far as yolk storage is concerned, it is essential for several reasons:

1. It appears to have a role in the *transfer of nutrients* to the embryo during the second and third weeks while the uteroplacental circulation is being established.

2. *Blood development* occurs in the wall of the yolk sac beginning in the third week (see Fig. 4–9) and continues to form there until hemopoietic activity begins in the liver during the sixth week.

3. During the fourth week, the dorsal part of the yolk sac is incorporated into the embryo as the *primitive gut* (see Fig. 5–1), which gives rise to the epithelium of the trachea, the bronchi, and the lungs (see Chapter 11) and of the digestive tract (see Chapter 12).

4. The *primordial germ cells* appear in the wall of the yolk sac in the third week and subsequently migrate to the developing sex glands or gonads (see Fig. 13–20), where they become the germ cells (spermatogonia or oogonia).

Fate of the Yolk Sac.   At 10 weeks, the small yolk sac lies in the chorionic cavity between the amnion and the chorionic sac

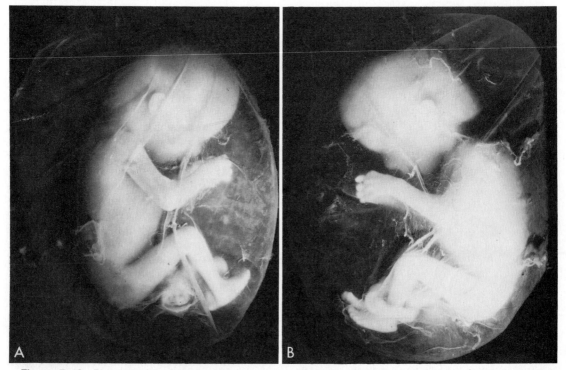

Figure 7–18 Photographs of a 12-week fetus within the amniotic sac. *Actual size.* In *B,* note that the umbilical cord is looped around the left ankle of the fetus. Coiling of the cord around parts of the fetus affects their development when the coils are so tight that the circulation in that part of the fetus is affected.

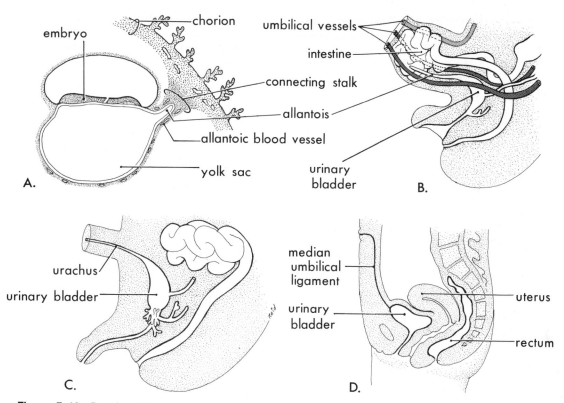

Figure 7-19 Drawings illustrating the development and usual fate of the allantois. *A,* Three weeks. *B,* Nine weeks. *C,* Three months. *D,* Adult.

(Fig. 7–17*C*). The yolk sac shrinks as pregnancy advances, eventually becoming very small and rather solid (Fig. 7–17*D*). It may persist throughout pregnancy and be recognizable on the fetal surface of the placenta beneath the amnion near the attachment of the umbilical cord, but this is extremely rare and of no significance.

The yolk stalk usually detaches from the midgut loop by the end of the sixth week (see Fig. 12–11). In about 2 per cent of adults, the proximal intra-abdominal part of the yolk stalk persists as a diverticulum of the ileum known as *Meckel's diverticulum* (see Figs. 12–17 and 12–18).

THE ALLANTOIS

The early development of the allantois is described in Chapter 4. During the second month, the extraembryonic portion of the allantois degenerates, but a remnant of it may be seen between the umbilical arteries in the proximal part of the umbilical cord for some time.

Significance of the Allantois. Al-
though the allantois does not function in human embryos, it is important for two reasons:

1. *Blood formation* occurs in its wall during the third to fifth weeks.

2. Its blood vessels become the umbilical vein and arteries (Fig. 7–19*A* and *B*).

Fate of the Allantois. The intraembryonic portion of the allantois runs from the umbilicus to the urinary bladder, with which it is continuous (Fig. 7–19*B*). As the bladder enlarges, the allantois involutes to form a thick tube called the *urachus* (Fig. 7–19 *C*). After birth, the urachus becomes a fibrous cord called the *median umbilical ligament,* which extends from the apex of the urinary bladder to the umbilicus (Fig. 7–19*D*).

MULTIPLE PREGNANCY

Multiple births are more common nowadays, owing to overstimulation of ovulation that occurs when human gonadotropins are administered to women with *ovulatory failure* (Fig. 7–30). In the United States, *twins* occur once in every 90 pregnancies. *Triplets* usually occur once in 90^2 pregnancies, *qua-druplets* once in 90^3, and *quintuplets* once in 90^4.

TWINS

Twins may originate from two zygoes (Fig. 7–20), in which case they are *dizygotic,* or fraternal, or from one zygote (Fig. 7–21), i.e., *monozygotic,* or identical. The fetal membranes and placenta(s) vary according to the derivation of the twins, and, in the case of monozygotic twinning, the type of membranes formed depends upon the time at which twinning occurs. If duplication of the *inner cell mass* occurs after the amniotic cavity forms (about 8 days), the embryos will be within the same amniotic and chorionic sacs (Fig. 7–25A).

As previously stated, twins occur about once in 90 pregnancies; about two thirds of these are dizygotic twins. The frequency of dizygotic twinning shows marked racial differences, but the incidence of monozygotic twinning is about the same in all populations. In addition, the rate of monozygotic twinning shows little variation with the mother's age, whereas the rate of dizygotic twinning increases with maternal age.

The tendency for dizygotic, but not monozygotic, twins to repeat in families is evidence of hereditary influence. Studies in a Mormon population showed that the genotype of the mother affects the frequency of dizygotic twins among her offspring, but the genotype of the father has no effect (Page et al., 1981). It has also been found that if the firstborn are twins, a repetition of twinning or some other form of multiple birth is about five times more likely to occur at the next pregnancy than it is in the general population.

Anastomoses between blood vessels of fused placentas of human dizygotic twins occasionally occur and result in erythrocyte mosaicism. The members of dizygotic twins have red cells of two different types because red cells were exchanged

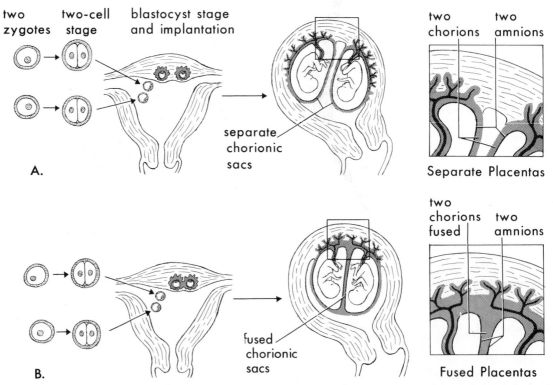

Figure 7–20 Diagrams illustrating how dizygotic twins develop from two zygotes. The relations of the fetal membranes and placentas are shown for instances in which *A,* the blastocysts implant separately, and *B,* the blastocysts implant close together. In both cases there are two amnions and two chorions, and the placentas may be separate or fused. *For convenience of illustration,* the blastocysts in these drawings (and in Figures 7–21, 7–24, and 7–25) are shown implanting in the fundic region of the uterus. As discussed in Chapter 3, blastocysts usually implant in the midportion of the body of the uterus, slightly more often on the posterior than on the anterior wall.

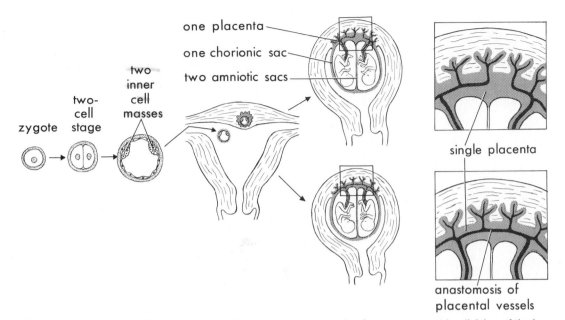

one placenta

one chorionic sac

two amniotic sacs

two inner cell masses

two-cell stage

zygote

single placenta

anastomosis of placental vessels

Figure 7-21 Diagrams illustrating how monozygotic twins develop from one zygote by division of the inner cell mass. This is the most common mechanism by which monozygotic (identical) twins develop. Such twins always have separate amnions, a single chorion, and a common placenta.

between the two circulations. This commonly occurs in cattle and causes *freemartinism* (Moore, 1966). Freemartins are intersexual female calves born as twins with male calves; they are intersexual because of male hormones that reach them through anastomosed placental vessels. If placental vascular anastomoses occur in human dizygotic twins, in cases in which one fetus is a male and the other one a female, masculinization of the female fetus does not occur, but the anastomotic condition may give rise to *chimeras,* i.e., persons with populations of blood cells of two genotypes that are from different zygotes.

Dizygotic Twins (Fig. 7-20). Because they result from the fertilization of two ova by two different sperms, dizygotic twins may be of the same sex or of different sexes. For the same reason, they are no more alike genetically than brothers or sisters born at different times. Dizygotic twins always have two amnions and two chorions, but the chorions and placentas may be fused.

Monozygotic Twins (Fig. 7-21). Because they result from the fertilization of one ovum, monozygotic twins are of the same sex, genetically identical, and very similar in physical appearance. Physical differences between *identical twins* are caused by environmental factors, e.g., anastomosis of placental vessels resulting in differences in blood supply from the placenta (Fig. 7-22).

Monozygotic twinning usually begins

around the end of the first week and results from division of the inner cell mass into two embryonic primordia. Subsequently, two embryos, each in its own amniotic sac, develop within one chorionic sac. The twins have a *common placenta* and often some

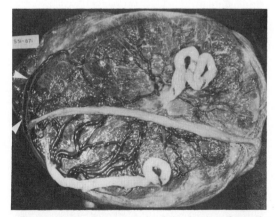

Figure 7-22 Photograph of the fetal surface of a placenta from monozygotic twins. The separate amnions are rolled at the center. There are several large vascular communications between the circulations of the placentas. A large arteriovenous shunt is visible at the left (arrows), and direct artery-to-artery anastomosis crosses the midline (under the amnions) from one placental half to the other. (Courtesy of Dr. K. Benirschke, University of California, San Diego.)

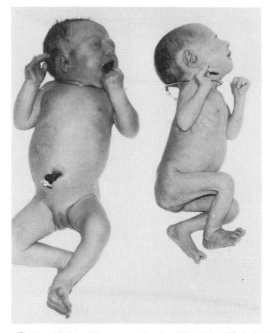

Figure 7-23 Monozygotic (or "identical") twins showing a wide discrepancy in size resulting from an uncompensated arteriovenous anastomosis of placental vessels. Blood was shunted from the smaller twin to the larger twin, producing the so-called fetal transfusion syndrome. (Courtesy of Dr. Harry Medovy, Children's Centre, Winnipeg, Canada.)

placental vessels join, but usually these anastomoses are well-balanced so that neither twin suffers (Benirschke, 1972). Occasionally, however, there exists a large *arteriovenous anastomosis* that causes a circulatory disturbance and results in physical differences in the twins (Figs. 7–22 and 7–23). In some cases, there is also a variety of congenital malformations.

Early division of embryonic cells (2 to 3 days) results in monozygotic twins who have two amnions, two chorions, and two placentas that may or may not be fused (Fig. 7–24). In such cases, it is impossible to determine from the membranes alone whether the twins are monozygotic or dizygotic. To determine the relationship of twins of the same sex and with similar blood groups, one must wait until other characteristics develop, e.g., eye color, fingerprints, and so forth (Smith and Penrose, 1955). The establishment of zygosity of twins has become important, particularly since the introduction of organ transplantation. For example, when a piece of skin is removed from one individual and grafted to another, it will continue to grow only if the donor and host are monozygotic twins. About 30 per

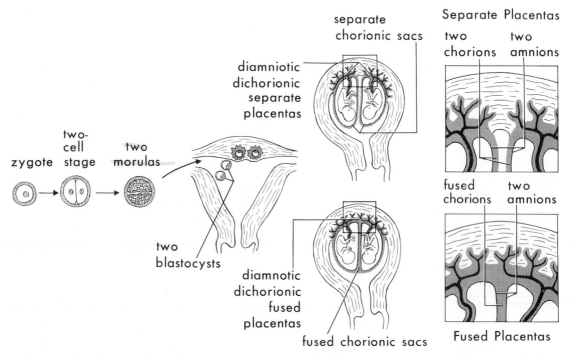

Figure 7-24 Diagrams illustrating how some monozygotic twins may develop from one zygote. Division may occur anywhere from the two-cell to the morula stage, producing two identical blastocysts. Each embryo subsequently develops its own amniotic and chorionic sacs. The placentas may be separate or fused.

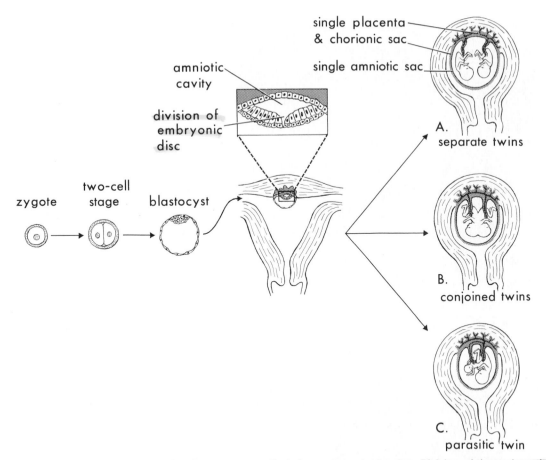

single placenta
& chorionic sac

single amniotic sac

amniotic
cavity

division of
embryonic
disc

zygote

two-cell
stage

blastocyst

A.
separate twins

B.
conjoined twins

C.
parasitic twin

Figure 7–25 Diagrams illustrating how monozygotic twins may rarely develop. Division of the embryonic disc results in two embryos within one amniotic sac. A, Complete division gives rise to separate twins. B and C, incomplete division results in various types of conjoined twins.

cent of monozygotic twins result from early division of the embryonic cells (Fig. 7–24).

Late division of embryonic cells (9 to 15 days) results in monozygotic twins that are in one amniotic sac and one chorionic sac (Fig. 7–25A). Such twins are rarely delivered alive because the umbilical cords are frequently so entangled that circulation ceases and one or both fetuses die. It has been estimated that the frequency of monoamniotic twins among monozygotic twins is about 4 per cent (Bulmer, 1970).

Conjoined Twins. If the inner cell mass, or the embryonic disc, does not divide completely, various types of conjoined twins may form (Figs. 7–25B and C, 7–26 and 7–27). These are named according to the regions that are attached, e.g., "thoracopagus" indicates that there is anterior union of the thoracic regions (Fig. 7–26). It has been estimated that about once in every 40 mon-

ozygotic twin pregnancies, the twinning is incomplete and conjoined ("Siamese") twins result. In some cases, the twins are connected to each other by skin only or by cutaneous and other tissues, e.g., liver (Fig. 7–26A). Some conjoined twins can be successfully separated by surgical procedures (de Vries, 1967). For descriptions of other possible types of conjoined twins, see Bergsma (1967).

OTHER MULTIPLE BIRTHS

Triplets occur once in about 8100 pregnancies and may be derived from (1) one zygote and be identical; (2) two zygotes and consist of identical twins and a single infant (Fig. 7–28); or (3) three zygotes and be of the same sex or of different sexes. In the last case, the infants are no more similar than siblings from three separate pregnancies.

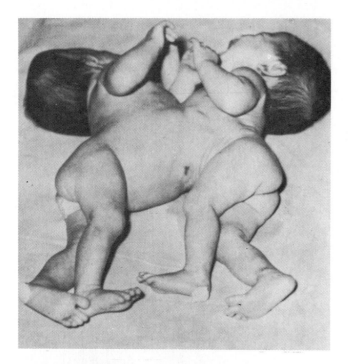

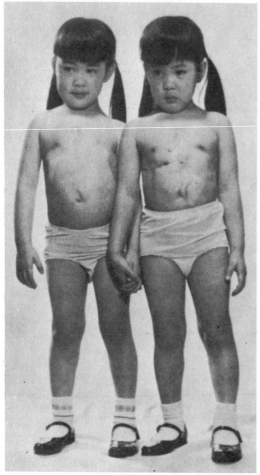

Figure 7-26 *A*, Photograph of newborn conjoined twins showing anterior union (thoracopagus). *B*, The twins about four years after separation. (From deVries, P. A.: Case history–the San Francisco twins; *in* D. Bergsma [Ed]: *Conjoined Twins. Birth Defects III*(1):141, 1967. © The National Foundation, New York.)

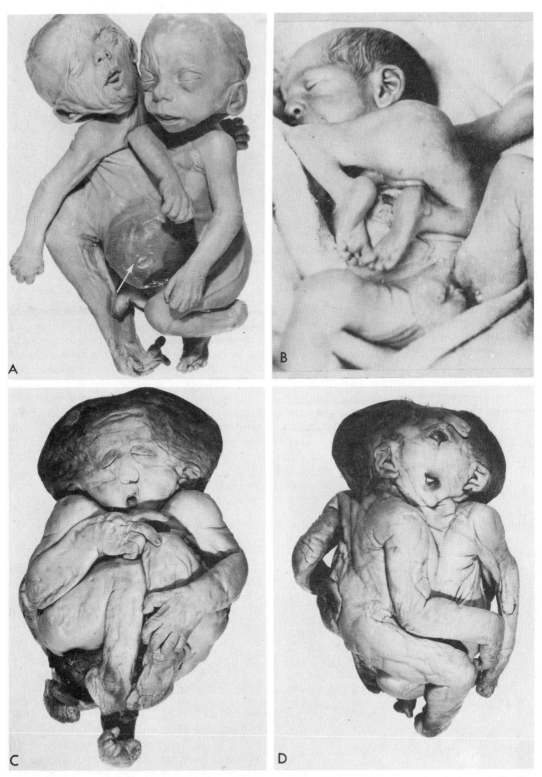

Figure 7–27 *A*, Conjoined twins showing extensive anterior fusion and an omphalocele (arrow). The lower limbs of the left fetus are fused, a condition known as sirenomelus. *B*, Parasitic fetus with well-developed lower limbs and pelvis attached to the thorax of an otherwise normal male infant. *C* and *D*, Two views of conjoined twins showing the so-called janiceps abnormality with two faces and fusion of the cranial and thoracic regions (cephalothoracopagus). Fortunately, this condition is extremely rare.

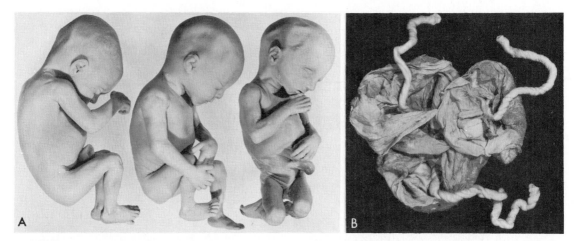

Figure 7–28 *A,* Photograph of 20-week triplets: monozygotic male twins (left) and a single female (right). *B,* Photograph of their fused placentas shows the twin placenta with two amnions (left) and the single placenta (upper right).

Similar combinations occur in *quadruplets* (Fig. 7–29), *quintuplets, sextuplets,* and *septuplets* (Fig. 7–30).

Superfetation is the implantation of one or more blastocysts in a uterus that already contains a developing embryo. This phenomenon occurs in some mammals, but evidence that it occurs in humans is inconclusive. The condition is most unlikely because ovulation is normally inhibited as soon as implantation of one blastocyst occurs (see Chapter 2).

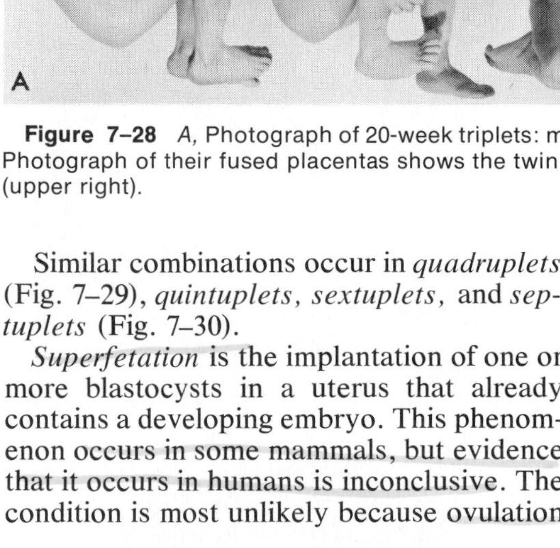

Superfecundation is the fertilization of two or more ova around the same time by sperms from different males. This phenomenon is known to occur in some mammals (e.g., cats and dogs), but has not been proved in humans (Greenhill and Friedman, 1974); however, in one case that occurred a few years ago in Bonn, Germany, a judge ruled on the basis of blood group studies that the dizygotic twins in question had different fathers. Obviously, both inseminations occurred within a few hours.

SUMMARY

In addition to the embryo, the fetal membranes and most of the placenta originate from the zygote. *The placenta consists of two parts:* (1) a fetal portion derived from the *villous chorion,* and (2) a maternal portion formed by the *decidua basalis.* The two parts are held together by *anchoring villi* and the cytotrophoblastic shell, and they work together in effecting placental transfer. The fetal circulation is separated from the maternal circulation by a thin layer of fetal tissues known as the *placental membrane* (placental barrier). It is a *permeable membrane* that allows water, oxygen, other nutritive substances, hormones, and noxious agents to pass from the mother to the embryo (fetus). Some products of excretion pass from the embryo (fetus) to the mother.

The principal activities of the placenta are

diamniotic dichorionic placenta

diamniotic monochorionic placenta

Figure 7–29 Photograph of the placentas from quadruplets. The upper two placentas (and the fetuses) were derived from two zygotes, whereas the lower, fused placenta was from monozygotic twins. (From Benirschke, K.: *Obstet. Gynecol. 18*:309, 1961.)

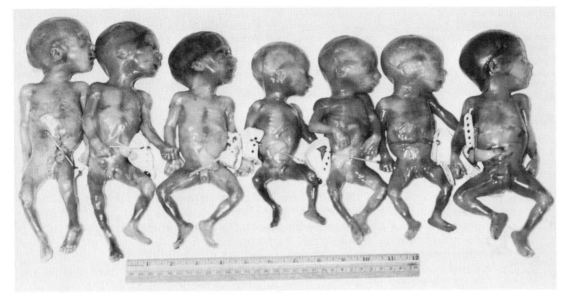

Figure 7-30 Photograph of septuplets, five females and two males, of about 17 weeks. Study of the fetal membranes indicated that they were probably derived from seven ova. (From Turksby, R. N., et al.: *Obstet. Gynecol. 30*:692, 1967.)

(1) metabolism, (2) transfer, and (3) endocrine secretion. All three activities are essential to maintaining pregnancy and to promoting normal embryonic development.

The fetal membranes and the placenta(s) in *multiple pregnancy* vary considerably, depending on the derivation of the embryos and the time at which division of the embryonic cells occurs. *The common type of twins is dizygotic,* with two amnions, two chorions, and two placentas that may or may not be fused. *Monozygotic twins,* the less common type, represent about a third of all twins, and they are derived from one zygote. These twins commonly have two amnions, one chorion, and one placenta. Twins with one amnion, one chorion, and one placenta are always monozygotic, and their umbilical cords are often entangled. Other types of multiple birth (triplets and so forth) may be derived from one or more zygotes.

The *yolk sac* and the *allantois* are vestigial structures, but their presence is essential to normal embryonic development. Both are *early sites of blood formation,* and the dorsal part of the yolk sac is incorporated into the embryo as the primitive gut.

The *amnion* forms a sac for amniotic fluid and provides a covering for the umbilical cord. The *amniotic fluid has three main functions;* it provides (1) a protective buffer for the embryo or fetus, (2) room for fetal movements, and (3) assistance in the regulation of fetal body temperature.

SUGGESTIONS FOR ADDITIONAL READING

Boyd, J. D., and Hamilton, W. J.: *The Human Placenta.* Cambridge, W. Heffer and Sons, Ltd., 1970.
An excellent reference text on the placenta, describing early development and maturation of the placenta. The electron micrographs and colored illustrations are superb.

Page, E. W., Villee, C. A., and Villee, D. B.: *Human Reproduction Essentials of Reproductive and Perinatal Medicine.* 3rd ed. Philadelphia, W. B. Saunders Co., 1981.
This book presents the essential foundations of obstetrics, gynecology, perinatal medicine, and the basic sciences of reproductive endocrinology, biochemistry, physiology, and genetics.

Phillipp, E. E., Barnes, J., and Newton, M. (Eds.): *Scientific Foundations of Obstetrics and Gynecology.* London, William Heinemann, Ltd., 1970.
This book describes the basic scientific information on which the clinical practice of obstetrics and gynecology depends. It contains contributions from many authors; subjects of special interest are "The Membranes" by G. Bourne; "Development of the Human Placenta" by W. J. Hamilton and J. D. Boyd; "The Liquor Amnii" by R. L. Gadd; "The Neurohypophysis and Labour" by G. W. Theobald; "The Endocrine Functions of the Placenta" by E. C. Amoroso and D. G. Porter; and "Ultrasonics" by C. N. Smyth.

Thompson, J. S., and Thompson, M. W.: *Genetics In Medicine.* 3rd ed. Philadelphia, W. B. Saunders Co., 1980, pp. 214–216.

A very good account of *hemolytic disease of the new-born,* including the factors influencing its development. This text gives an excellent, concise presentation of genetics and its application to medical problems.

Villee, D. B.: The placenta; *in* D. B. Villee (Ed.):

Human Endocrinology. A Developmental Approach. Philadelphia, W. B. Saunders Co., 1975, pp. 19–35.
A concise review of the development, metabolism, and hormone production of the placenta. Emphasis is given to the role of the placenta as an endocrine organ in pregnancy.

CLINICALLY ORIENTED PROBLEMS

1. How does a doctor calculate the estimated delivery date (EDD) of a baby? How could this EDD be confirmed in a high-risk obstetrical patient?
2. A pregnant woman was told that she had *polyhydramnios (hydramnios)* and asked you to explain what the term means. What would be your answer? What conditions are often associated with polyhydramnios? Explain why it occurs.
3. Does twinning run in families? Is maternal age a factor? How would you determine whether twins were monozygotic or dizygotic?
4. During the examination of an umbilical cord, you noticed that there was only one umbilical artery. How often does this malformation occur? What kind of fetal abnormalities might be associated with this condition?

The answers to these questions are given at the back of the book.

REFERENCES

Allen, F. H., Jr., and Umansky, I.: Erythroblastosis fetalis; *in* D. E. Reid, K. J. Ryan, and K. Benirschke (Eds.): *Principles and Management of Human Reproduction.* Philadelphia, W. B. Saunders Co., 1972, pp. 811–832.

Allen, M. S., and Turner, U. G.: Twin birth — identical or fraternal twins? *Obstet. Gynecol. 37*:538, 1971.

Beaconsfield, P., and Villee, C. (Eds.): *Placenta: A Neglected Experimental Animal.* Elmsford, New York, Pergamon Press, 1979.

Benirschke, K.: Implantation, placental development, uteroplacental blood flow; *in* D. E. Reid, K. J. Ryan, and K. Benirschke (Eds.): *Principles and Management of Human Reproduction.* Philadelphia, W. B. Saunders Co., 1972, 179–196.

Benirschke, K.: Incidence and prognostic implication of congenital absence of one umbilical artery. *Am. J. Obstet. Gynecol., 79*:251, 1960.

Benirschke, K.: Examination of the placenta. *Obstet. Gynecol. 18*:309, 1961.

Benirschke, K., and Driscoll, S. G.: *The Pathology of the Human Placenta.* New York, Springer-Verlag, 1967.

Benirschke, K., and Reid, D. E.: Multiple pregnancy; *in* D. E. Reid, K. J. Ryan, and K. Benirschke (Eds.): *Principles and Management of Human Reproduction.* Philadelphia, W. B. Saunders Co., 1972, pp. 197–210.

Bergsma, E. (Ed.): *Conjoined Twins. Birth Defects III* (1):1–147, 1967.

Billington, W. B.: Trophoblast; *in* E. E. Philipp, J. Barnes, and M. Newton (Eds.): *Scientific Foundations of Obstetrics and Gynecology.* London, William Heinemann, Ltd., 1970, pp. 159–167.

Bourne, G. L.: The membranes; *in* E. E. Philipp, J. Barnes, and M. Newton (Eds.): *Scientific Foundations of Obstetrics and Gynecology.* London, William Heinemann, Ltd., 1970, pp. 181–184.

Brock, D. J. H.: Prenatal diagnosis — chemical methods, *in* C. L. Berry (Ed.): *Human Malformations. Br. Med. Bull. 32*:16, 1976.

Bulmer, M. G.: *The Biology of Twinning in Man.* Oxford, Clarendon Press, 1970.

Chamberlain, G., and Wilkinson, A. (Eds.): *Placental Transfer.* Baltimore, University Park Press, 1979.

Dallenbach-Hellweg, G., and Nette, G.: Morphological and histochemical observations on trophoblast and decidua of the basal plate of the human placenta at term. *Am. J. Anat. 115*:309, 1964.

deVries, P. A.: Case history — the San Francisco twins; *in* D. Bergsma (Ed.): *Conjoined Twins. Birth Defects III*(1):141, 1967.

Fox, H.: *Pathology of the Placenta.* Philadelphia, W. B. Saunders Co., 1978.

Gadd, R. L.: The liquor amnii; *in* E. E. Philipp, J. Barnes, and M. Newton (Eds.): *Scientific Foundations of Obstetrics and Gynecology.* London, William Heinemann, Ltd., 1970, pp. 254–259.

Gill, S., and Davis, J. A.: The pharmacology of the fetus, baby, and growing child; *in* J. A. Davis and J. Dobbing (Eds.): *Scientific Foundations of Paediatrics.* Philadelphia, W. B. Saunders Co., 1974, pp. 801–818.

Glasser, S. R., and Bullock, D. W. (Eds.): *Cellular and Molecular Aspects of Implantation.* New York, Plenum Press, 1981.

Greenhill, J. P., and Friedman, E. A. (Eds.): *Biological Principles and Modern Practice of Obstetrics.* Philadelphia, W. B. Saunders Co., 1974.

Ham, A. W. and Cormack, D. H.: *Histology.* 8th ed. Philadelphia, J. B. Lippincott Company, 1979.

Harris, R.: Chromosomes in development; *in* J. A.

Davis and J. Dobbing (Eds.): *Scientific Foundations of Paediatrics*. Philadelphia, W. B. Saunders Co., 1974, pp. 22–28.

Hutchinson, D. L., Gray, M. J., Plentl, A. A., Alvarez, H., Caldeyro-Barcia, R., Kaplan, B., and Lind, J.: The role of the fetus in water exchange of the amniotic fluid of normal and hydramniotic patients. *J. Clin. Invest. 38*:971, 1959.

Javert, C. T.: *Spontaneous and Habitual Abortion*. New York, The Blakiston Division, McGraw-Hill Book Co., 1957.

Jones, G. S.: Endocrine functions of the placenta; *In* A. C. Barnes (Ed.): *Intra-Uterine Development*. Philadelphia, Lea & Febiger, 1968, pp. 68–94.

Klopper, A., and Chard, T. (Eds.): *Placental Proteins*. Berlin, Springer-Verlag, 1978.

Martin, C. B.: The anatomy and circulation of the placenta; *in* A. C. Barnes (Ed.): *Intra-Uterine Development*. Philadelphia, Lea & Febiger, 1968, pp. 35–67.

Moore, K. L.: The sex chromatin of freemartins and other animal intersexes; *in* K. L. Moore (Ed.): *The Sex Chromatin*, Philadelphia, W. B. Saunders Co., 1966, pp. 229–240.

Newman, H. H.: *Multiple Human Births*. Garden City, N.Y., Doubleday & Co., Inc., 1940.

Potter, E. L.: *Pathology of the Fetus and Infant*. 2nd ed. Chicago, Year Book Medical Publishers Inc., 1961, pp. 1–50.

Ramsey, E. M.: The placenta and fetal membranes; *in* J. P. Greenhill (Ed.): *Obstetrics*. 13th ed. Philadelphia, W. B. Saunders Co., 1965, pp. 101–136.

Reid, D. E., Ryan, K. J., and Benirschke, K. (Eds.): *Principles and Management of Human Reproduction*. Philadelphia, W. B. Saunders Co., 1972.

Ryan, K. J.: Endocrine organs of reproduction; *in* D. E. Reid, K. J. Ryan, and K. Benirschke (Eds.): *Principles and Management of Human Reproduction*. Philadelphia, W. B. Saunders Co., 1972, pp. 58–95.

Seeds, A. E., Jr.: Amniotic fluid and fetal water metabolism; *in* A. C. Barnes (Ed.): *Intra-Uterine Development*. Philadelphia, Lea & Febiger, 1968, pp. 129–144.

Smith, S. M., and Penrose, L. S.: Monozygotic and dizygotic twin diagnosis. *Ann. Hum. Genet. 19*:273, 1955.

Torpin, R.: *The Human Placenta*. Springfield, Ill., Charles C Thomas, Publisher, 1969.

Turksoy, R. N., Toy, B. L., Rogers, J., and Papageorge, W.: Birth of septuplets following human gonadotropin administration in Chiari-Frommel syndrome. *Obstet. Gynecol. 30*:692, 1967.

Vaughan, V. C., McKay, R. J., and Behrman, R. E.: *Nelson Textbook of Pediatrics*. 11th ed. Philadelphia, W. B. Saunders Co., 1979, pp. 379–388.

Villee, C. A. (Ed.): *The Placenta and Fetal Membranes*. Baltimore, The Williams & Wilkins Co., 1960.

Villee, D. B.: *Human Endocrinology. A Developmental Approach*. Philadelphia, W. B. Saunders Co., 1975.

Waisman, H. A., and Kerr, G.: *Fetal Growth and Development*. New York, McGraw-Hill Book Co., 1970.

Williams, E. L., and Warwick, R. (Eds.): *Gray's Anatomy*. 36th ed. Philadelphia, W. B. Saunders Co., 1980, pp. 120–138.

8

CAUSES OF CONGENITAL MALFORMATIONS
Human Teratology

We ought not to set them aside with idle thoughts or idle words about "curiosities" or "chances." Not one of them is without meaning; not one that might not become the beginning of excellent knowledge, if only we could answer the question — why is it rare or being rare, why did it in this instance happen?

— *James Paget, Lancet 2*:1017, 1882.

Congenital malformations are anatomical abnormalities present at birth (L. *congenitus*, born with). They may be macroscopic or microscopic, on the surface, or within the body. *Teratology* is the study of abnormal development and the causes of congenital malformations.

Until the early 1940's, it was generally accepted that human embryos were protected from environmental agents by their fetal membranes and their mother's abdominal walls and uterus. Gregg (1941) presented the first well-documented evidence that an environmental agent (rubella virus) could produce congenital abnormalities if present during the critical stages of development. However, it was the observations of Lenz (1961) and McBride (1961) that focused attention on the role of drugs in the etiology of human congenital malformations. It is now estimated that nearly 10 per cent of human developmental abnormalities result from the actions of drugs, viruses, and other environmental factors (Persaud, 1979).

About 20 per cent of deaths in the perinatal period are attributed to congenital malformations (MacVicar, 1976). Malformations are observed in about 2.7 per cent of newborn infants, and, during infancy, congenital abnormalities are detected in an additional 3 per cent (McKeown, 1976).

It is customary to divide the causes of congenital malformations into (1) *genetic factors* (chromosomal abnormalities or mutant genes) and (2) *environmental factors,* but many common congenital malformations are caused by a number of genetic and environmental factors acting together. This is called *multifactorial inheritance* and is discussed on page 161.

MALFORMATIONS CAUSED BY GENETIC FACTORS

Numerically, genetic factors are probably most important as the causes of congenital malformations (Warkany, 1981).

Any mechanism as complex as that underlying mitosis or meiosis may occasionally malfunction. Carr (1970) estimated that chromosomal abnormalities are present in about 1 of 200 newborn infants. Chromosome complements are subject to two kinds of change: (1) numerical, and (2) structural, and they may affect either the sex chromosomes or the autosomes. In rare cases, both kinds of chromosomes are affected.

Genetic factors initiate mechanisms of malformation by biochemical or other means at the subcellular, cellular, or tissue level. The mechanism initiated by the genetic factor may be identical with or similar to the

140

causal mechanism initiated by a teratogen (e.g., a drug). *A teratogen is any agent that can produce a congenital malformation or raise the incidence of the malformation in the population.*

Persons with chromosomal abnormalities usually have characteristic phenotypes (e.g., Down syndrome) and often look more like other persons with the same chromosomal abnormality than like their own siblings (brothers or sisters). The characteristic appearances result from a genetic imbalance that disrupts normal development.

NUMERICAL CHROMOSOMAL ABNORMALITIES

Numerical abnormalities of chromosomes usually arise as the result of nondisjunction. This is an error in cell division in which there is failure of the paired chromosomes or sister chromatids to separate or disjoin at anaphase. This error may occur during a mitotic division or during the first or second meiotic division (Fig. 8–3).

Normally, the chromosomes exist in pairs; the two chromosomes making up a pair are called *homologues*. Thus, human females have 22 pairs of autosomes plus two X chromosomes, and males have 22 pairs of autosomes plus one X and one Y chromosome (Fig. 8–1). One of the two X chromosomes in females forms a mass of X-chromatin, or *sex chromatin* (Fig. 8–9B), which is not present in cells of normal males (Fig. 8–9A) or in females lacking a sex chromosome (Fig. 8–2). Although sex chromatin studies, employing conventional methods, are useful in diagnosing errors of sex development (Moore, 1966), they provide no information about the presence or absence of the Y chromosome. However, Pearson et al. (1970) demonstrated that the Y chromosome can also be detected in interphase cells by means of quinacrine fluorescence staining. Each Y chromosome in a cell forms an intensely fluorescent body in interphase (Fig. 8–9D), and thus the number of Y-chromatin bodies observed indicates the number of Y chromosomes present. Sex chromatin studies alone do not provide any information about autosomal abnormalities. For this information, studies of chromosomes must be done.

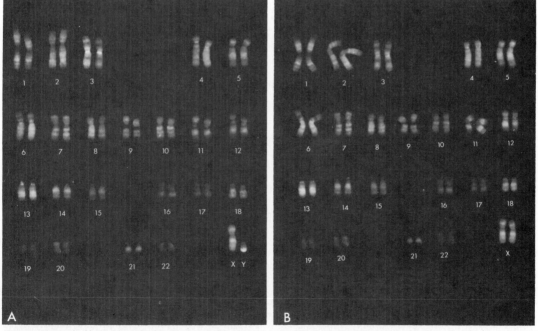

Figure 8–1 *A,* Normal male karyotype with Giemsa banding (G banding). *B,* Similar bands in a normal female karyotype. Caspersson et al. (1970) observed that when chromosomes are stained with quinacrine mustard or related compounds and examined by fluorescent microscopy, each pair of chromosomes stains in a distinctive pattern of bright and dim bands, called Q bands. They form the basis of the classification of chromosomes. (Courtesy of Dr. M. Ray, Department of Pediatrics, Division of Genetics, and Department of Anatomy, University of Manitoba and the Health Sciences Centre, Winnipeg, Canada.)

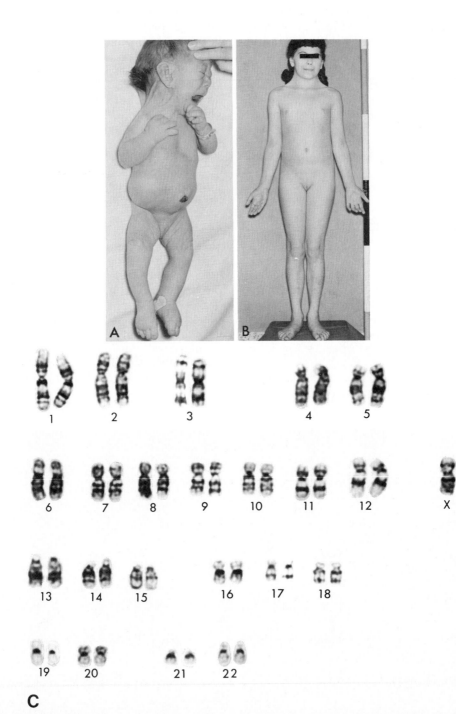

Figure 8–2 Females with Turner syndrome (XO sex chromosome complement). *A,* Newborn infant. Note the webbed neck and lymphedema of the hands and feet. *B,* 13-year-old girl showing the classic features: short stature, webbed neck, absence of sexual maturation, and broad, shield-like chest with widely spaced nipples. (From Moore, K. L.: *The Sex Chromatin,* Philadelphia, W. B. Saunders Co., 1966.) *C,* G-banding karyotype. Note the presence of only one X chromosome. (Courtesy of Dr. M. Ray, Department of Pediatrics, Division of Genetics, and Department of Anatomy, University of Manitoba and the Health Sciences Centre, Winnipeg, Canada.)

Changes in chromosome number represent either aneuploidy or polyploidy.

Aneuploidy. Any deviation from the diploid number of 46 chromosomes is called *aneuploidy*. An *aneuploid* is an individual or a cell that has a chromosome number that is not an exact multiple of the haploid number of 23 (e.g., 45 or 47). The principal cause of aneuploidy is nondisjunction during a meiotic cell division. This results in an unequal distribution of one pair of homologous chromosomes to the daughter cells. One cell has two chromosomes and the other has neither chromosome of the pair. As a result, the embryo's cells may be *hypodiploid* (usually 45, as in *Turner syndrome*, Fig. 8–2) or *hyperdiploid* (usually 47, as in *Down syndrome*, Fig. 8–4).

Monosomy. Embryos missing a chromosome usually die; hence, monosomy of an autosome is extremely rare in living persons (Thorburn and Johnson, 1966, and Challacombe and Taylor, 1969). About 97 per cent of embryos lacking a sex chromosome also

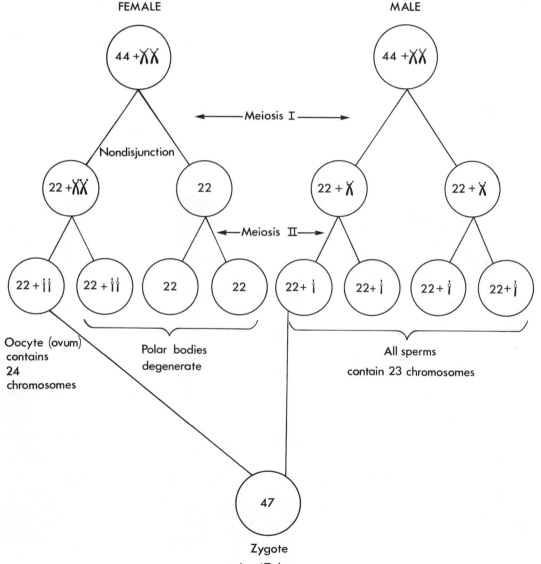

Figure 8–3 Diagram showing the first meiotic nondisjunction in a female resulting in an abnormal oocyte (ovum) with 24 chromosomes and how subsequent fertilization by a normal sperm produced a zygote with 47 chromosomes.

TABLE 8-1   TRISOMY OF THE AUTOSOMES

Disorder	Incidence	Usual Characteristics	Figures	References
Trisomy 21 or Down syndrome*	1:800	Mental deficiency; brachycephaly, flat nasal bridge; upward slant to palpebral fissures; protruding tongue; simian crease, clino-dactyly of 5th finger; congenital heart defects.	8–4	Breg (1975) Fraser and Nora (1975) Hook (1978) Smith (1976) Vaughan et al. (1979) Warkany (1971)
Trisomy 18 syndrome†	1:8000	Mental deficiency; growth retardation; prominent occiput; short sternum; ventricular septal defect; micrognathia; low-set malformed ears; flexed fingers, hypoplastic nails; rocker-bottom feet.	8–5	Breg (1975) Fraser and Nora (1975) Hook (1978) Smith (1976) Thompson and Thompson (1980)
Trisomy 13 syndrome†	1:7000	Mental deficiency; sloping forehead; malformed ears, scalp defects; microphthalmia; bilateral cleft lip and/or palate; polydactyly; posterior prominence of the heels.	8–6	Breg (1975) Fraser and Nora (1975) Hamerton (1971) Smith (1976) Thompson and Thompson (1980) Warkany (1971)

*The importance of this disorder in the overall problem of mental retardation is indicated by the fact that persons with Down syndrome represent 10 to 15 per cent of institutionalized mental defectives (Breg, 1975).

†Infants with this syndrome rarely survive beyond a few months.

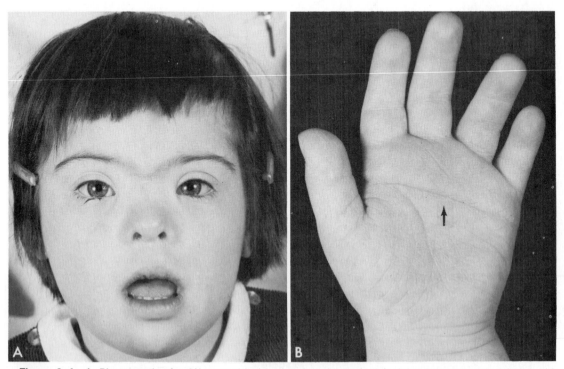

Figure 8–4   *A,* Photograph of a 3½-year-old girl, showing the typical facial appearance associated with Down syndrome. Note the flat, broad face, oblique palpebral fissures, epicanthus, speckling of the iris, and furrowed lower lip. *B,* The typical short, broad hand of this child shows the characteristic single transverse palmar or simian crease (arrow). (From Bartalos, M., and Baramki, T. A.: *Medical Cytogenetics.* 1967. Copyright © 1967, by The Williams and Wilkins Co., Baltimore.

die (Carr, 1970), but some survive and develop characteristics of *Turner syndrome* (Fig. 8–2). The incidence of XO Turner syndrome in the newborn population is approximately 1 in 10,000 (Hook and Hamerton, 1977). The phenotype of the 45,X Turner syndrome is illustrated in Figure 8–2. The Turner sex chromosome abnormality is the most common cytogenetic abnormality seen in fetuses that are spontaneously aborted, and it accounts for about 18 per cent of all abortions that are caused by chromosomal abnormalities. The error in gametogenesis *(nondisjunction)*, when it can be traced, is usually in the paternal gamete (i.e., the paternal X or Y chromosome is missing), and maternal age is not advanced (Uchida and Summitt, 1979).

Trisomy. If three chromosomes are present instead of the usual pair, the disorder is called *trisomy*. The usual cause of trisomy is *nondisjunction* (Fig. 8–3), resulting in a germ cell with 24 instead of 23 chromosomes and, subsequently, in a zygote with 47 chromosomes.

Trisomy of the autosomes is associated primarily with three syndromes (Table 8–1). The most common condition is *trisomy 21*, or *Down syndrome* (Fig. 8–4), in which three number 21 chromosomes are present. Trisomy 18 (Fig. 8–5) and trisomy 13 (Fig. 8–6)

are less common. Trisomy 22 is a very rare syndrome (Punnett et al., 1973).

Autosomal trisomies occur with increasing frequency as maternal age increases, particularly trisomy 21, which is present once in about 2000 births in mothers under 25, but once in about 100 in mothers over the age of 40 (Carter, 1970). Partial G-banding karyotypes showing trisomies for chromosome numbers 21, 18, and 13, are shown in Figure 8–7.

Trisomy of the sex chromosomes is a common condition (Table 8–2); however, because there are no characteristic physical findings in infants or children, it is rarely detected until adolescence. *Sex chromatin patterns are useful in detecting some types of trisomy of the sex chromosomes* because two masses of X-chromatin or sex chromatin are present in XXX females (Fig. 8–9C), and cells of XXY males (Fig. 8–8) show one mass of sex chromatin. The cells of XYY males show two fluorescing Y-chromatin bodies.

Tetrasomy and Pentasomy. Some persons, usually mentally retarded, have four or five sex chromosomes. The following sex chromosome complexes have been reported: *in females, XXXX* and *XXXXX*; and *in males, XXXY, XXYY, XXXYY,* and *XXXXY.* Usually, the greater the number of X chromosomes present,

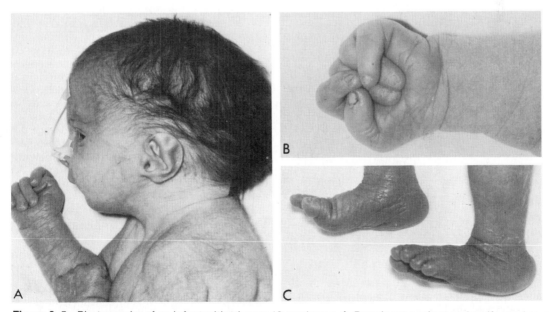

Figure 8–5 Photographs of an infant with trisomy 18 syndrome. *A,* Prominent occiput and malformed ears. *B,* Typical flexed fingers. *C,* So-called rocker-bottom feet, showing posterior prominences of the heels. (Courtesy of Dr. Harry Medovy, Children's Centre, Winnipeg, Canada.)

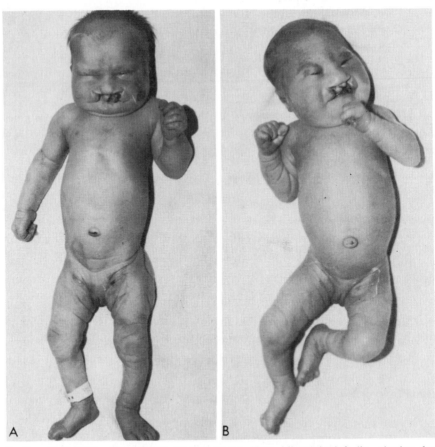

Figure 8–6 Female infants with trisomy 13 syndrome. Note bilateral cleft lip, sloping forehead, and rocker-bottom feet. (From Smith, D. W.: *Am. J. Obstet. Gynceol. 90*:1055, 1964.)

the greater the severity of the mental retardation and physical impairment (Neu and Gardner, 1969). The extra sex chromosomes do not accentuate male or female characteristics.

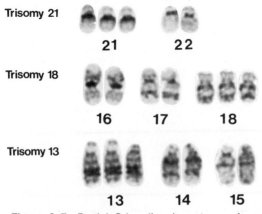

Figure 8–7 Partial G-banding karyotypes showing trisomies for chromosome numbers 21, 18, and 13. (Courtesy of Dr. M. Ray, Department of Pediatrics, Division of Genetics, and Department of Anatomy, University of Manitoba and the Health Sciences Centre, Winnipeg, Canada.)

Mosaicism. Persons with this condition have two or more cell lines with different karyotypes; either the autosomes or the sex chromosomes may be involved. Usually, the malformations are less serious than in persons with monosomy or trisomy, e.g., the features of Turner syndrome are not so evident in 45,X/46,XX mosaic females as in the usual 45,X females (Moore, 1966, and Neu and Gardner, 1975). Mosaicism usually arises by nondisjunction during early mitotic cleavage divisions. Mosaicism due to loss of a chromosome by so-called *anaphase lagging* is also known to occur. The chromosomes separate normally, but one chromosome is delayed in its migration and is eventually lost.

Polyploidy. Polyploid cells contain multiples of the haploid number of chromosomes (i.e., 69, 92, and so forth). Polyploidy is a significant cause of spontaneous abortion (Carr, 1971, and Carr et al., 1972).

The commonest type of polyploidy in human embryos is *triploidy* (69 chromosomes). This can result from the second polar body failing to separate from the oocyte (see Chapter 2) or by an ovum being fertilized by two sperms (dispermy)

TABLE 8–2   TRISOMY OF THE SEX CHROMOSOMES

Chromosome Complement*	Sex	Incidence	Usual Characteristics	References
47,XXX	Female	1:1000	Normal in appearance; usually fertile; may be mentally retarded.	Fraser and Nora (1975) Hook (1978) Neu and Gardner (1975) Miller (1964)
47,XXY	Male	1:1000	Klinefelter syndrome: small testes, hyalinization of seminiferous tubules; asperma-togenesis; often tall with dis-proportionately long lower limbs.	Ferguson-Smith (1966) Fraser and Nora (1975) Hamerton (1971) Neu and Gardner (1975) Thompson and Thompson (1980)
47,XYY	Male	1:1000	Normal in appearance; often tall; may have difficulty with impulse control.	Hamerton (1971) Mittwoch (1967) Neu and Gardner (1975) Nora and Fraser (1974)

*The number designates the total number of chromosomes, including the sex chromosomes shown after the comma.

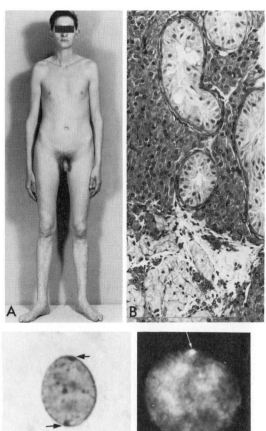

Figure 8–8 A, Adult male with XXY Klinefelter syndrome. Note the long lower limbs and normal trunk length. About 20 per cent of persons with Klinefelter syndrome have gynecomastia (excessive development of the male mammary glands). B, Section of a testicular biopsy showing some seminiferous tubules without germ cells and others that are hyalinized. (From Ferguson-Smith, M. A.; in Moore, K. L. [Ed.]: The Sex Chromatin. Philadelphia, W. B. Saunders Co., 1966.)

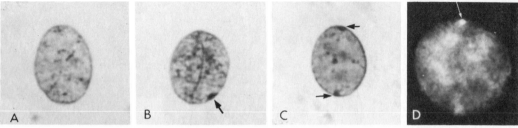

Figure 8–9 Oral epithelial nuclei stained with cresyl echt violet (A, B, and C) and quinacrine mustard (D) (×2000). A, from normal male. No sex chromatin is visible (chromatin negative). B, From normal female. The arrow indicates a typical mass of X-chromatin, or sex chromatin (chromatin positive). C, From female with XXX trisomy. The arrows indicate two masses of sex chromatin. D, From normal male. The arrow indicates the Y-chromatin as an intensely fluorescent body. (A and B are from Moore, K. L., and Barr, M. L.: Lancet 2:57, 1955. D, Courtesy of Dr. M. Ray, Department of Pediatrics, Division of Genetics, and Department of Anatomy, University of Manitoba and the Health Sciences Centre, Winnipeg, Canada.)

almost simultaneously (Carr, 1970). Although some fetuses with triploidy have been born alive, they all died within a few days.

Doubling of the diploid chromosome number to 92, or *tetraploidy*, probably occurs during preparation for the first cleavage division (see Fig. 2–14). Normally, each chromosome replicates, then divides into two; consequently, when the

zygote divides, each blastomere contains 46 chromosomes. If the chromosomes divide but the zygote does not undergo cleavage at this stage, the zygote will contain 92 chromosomes. Division of this zygote would subsequently result in an embryo with cells containing 92 chromosomes. Tetraploid embryos abort very early, and often all that is recovered is an empty chorionic sac (Carr,

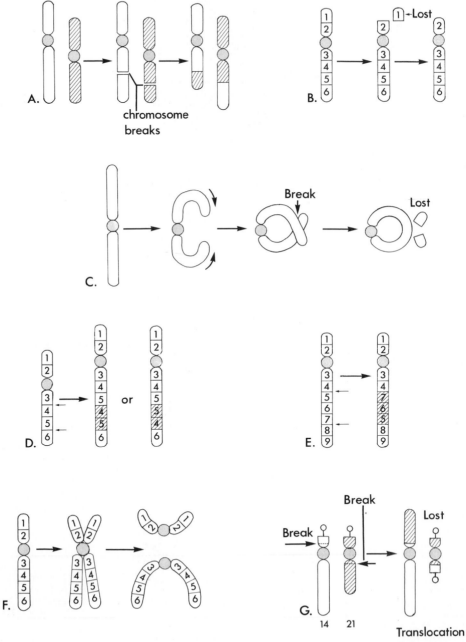

Figure 8–10   Diagrams illustrating structural abnormalities of chromosomes. *A*, Reciprocal translocation. *B*, Terminal deletion. *C*, Ring. *D*, Duplication. *E*, Paracentric inversion. *F*, Isochromosome. *G*, Robertsonian translocation.

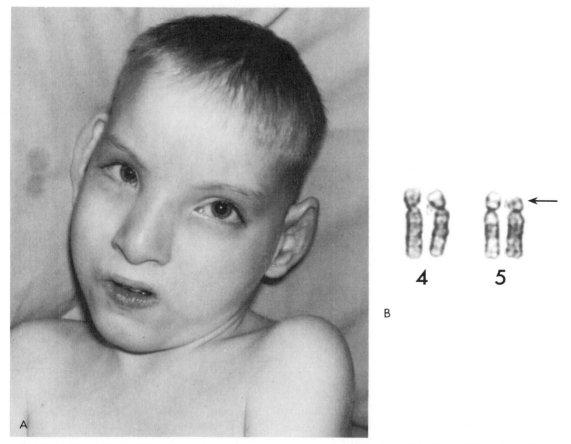

Figure 8–11 *A,* Male child with cri du chat syndrome. (From Gardner, E. J.: *Principles of Genetics,* 5th ed. New York, John Wiley & Sons, Inc., 1975.) *B,* A partial karyotype of this child showing a terminal deletion from the short arm end of chromosome number 5. The arrow indicates the normal short arm. (Courtesy of Dr. M. Ray, Department of Pediatrics, Division of Genetics, and Department of Anatomy, University of Manitoba and the Health Sciences Centre, Winnipeg, Canada.)

1971). Because this sac is derived from the zygote, chromosome analysis of chorionic cells reveals the chromosome complement of the embryo that died and degenerated.

STRUCTURAL ABNORMALITIES

Most structural abnormalities result from chromosome breaks induced by various environmental factors, e.g., radiation, drugs, and viruses (Bartalos and Baramki, 1967, and Saxén and Rapola, 1969). The type of abnormality that results depends upon what happens to the broken pieces (Fig. 8–10).

Translocation. This is the transfer of a piece of one chromosome to a nonhomologous chromosome. If two nonhomologous chromosomes exchange pieces, the translocation is *reciprocal* (Fig. 8–10*A* and *G*). Translocation does not necessarily lead to abnormal development. A person

with a translocation — for example, between a number 21 chromosome and a number 14 (Fig. 8–10*G*) — is phenotypically normal. Such persons are called *translocation carriers.* They have a tendency, independent of age, to produce germ cells with an abnormal translocation chromosome. About 3 to 4 per cent of persons with Down syndrome are translocation trisomies (Breg, 1975).

Deletion. When a chromosome breaks, a portion of the chromosome may be lost (Fig. 8–10*B*). A partial terminal deletion from the short arm end of a chromosome number 5 (Fig. 8–11*B*) causes the *cri du chat* syndrome (Fig. 8–11*A*). Affected infants have a weak cat-like cry, microcephaly, severe mental retardation, and congenital heart disease (Breg, 1975). The only invariable feature is mental retardation. About 1 per cent of persons with an I.Q. below 20 have a B-group chromosome deletion. For descriptions of several other conditions caused by deletions, see Bergsma (1975), and Uchida and Summitt (1979).

A *ring chromosome* is a type of deletion chromosome from which both ends have been lost and in which the broken ends have rejoined to form a ring-shaped chromosome (Fig. 8–10C). These abnormal chromosomes have been described in persons with Turner syndrome, trisomy 18, and other abnormalities (Ray, 1968, and Uchida and Summitt, 1979).

Duplication. This abnormality may be represented as a duplicated portion of a chromosome (1) within a chromosome (Fig. 8–10D), (2) attached to a chromosome, or (3) as a separate fragment. Duplications are more common than deletions, and, because there is no loss of genetic material, they are less harmful (Thompson and Thompson, 1980). Duplication may involve part of a gene, whole genes, or a series of genes.

Inversion. This is a chromosomal aberration in which a segment of a chromosome is reversed. Paracentric inversion (Fig. 8–10E) is confined to a single arm of the chromosome, whereas pericentric inversion involves both arms and includes the centromere. Pericentric inversion has been described in connection with Down syndrome and other abnormalities (Gray et al., 1962, and Allderdice et al., 1975).

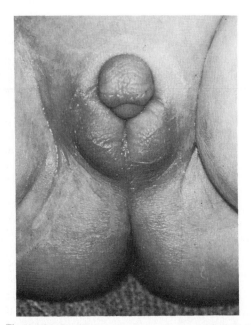

Figure 8–13 The masculinized external genitalia of a female infant with female pseudohermaphroditism caused by congenital virilizing adrenal hyperplasia. The 17-ketosteroid output was elevated. The virilization was caused by excessive androgens produced by the fetal adrenal glands.

Isochromosome. This abnormality results from the centromere dividing transversely instead of longitudinally (Fig. 8–10F) and appears to be the commonest structural abnormality of the X chromosome. Patients with this chromosomal abnormality often are short in stature and have other signs of Turner syndrome. These characteristics are related to the loss of a short arm of one X chromosome (Neu and Gardner, 1975).

MALFORMATIONS CAUSED BY MUTANT GENES

Probably 10 to 15 per cent of congenital malformations are caused by mutant genes (Gray and Skandalakis, 1973, and Vaughan et al., 1979). Because these malformations are inherited according to mendelian laws, predictions can be made about the probability of their occurrence in the affected person's children and in other relatives. Gene mutations causing malformations are much rarer than numerical and structural chromosomal abnormalities.

Although a great many genes mutate, most mutant genes do not cause congenital malformations. Examples of *dominantly inherited congenital malformations* are achondroplasia (Fig. 8–12) and polydactyly or extra digits (see Chapter 17). Other malformations are

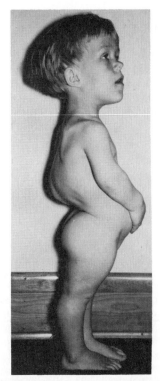

Figure 8–12 A child with achondroplasia showing short extremities, relatively large head, thoracic kyphosis, a sharply angled upper lumbar lordosis, and protrusion of the abdomen. Observe that the child's ribs rest on the iliac crests. (Courtesy of Dr. Harry Medovy, Children's Centre, Winnipeg, Canada.)

attributed to *autosomal recessive inheritance*, e.g., congenital adrenal hyperplasia (Fig. 8–13) and microcephaly (see Fig. 18–33). Autosomal recessive genes manifest themselves only when homozygous; as a consequence, many carriers of these genes (heterozygous persons) remain undetected. McKusick (1975) listed 1218 traits as dominantly inherited, 947 as recessive autosomal, and 171 as X-linked in humans. For details about the genetic aspects of congenital malformations, see Thompson and Thompson (1980).

MALFORMATIONS CAUSED BY ENVIRONMENTAL FACTORS

Although the human embryo is well protected in the uterus, certain agents, called *teratogens*, may induce congenital malformations when the tissues and organs are developing. The embryonic organs are most sensitive to noxious agents during periods of rapid differentiation. Damage to the primitive streak during stage 6 of development (about 15 days) could cause severe congenital malformations of the embryo, because this primitive structure produces intraembryonic mesoderm, the source of all types of connective tissue. Because biochemical differentiation precedes morphological differentiation, the period during which structures are sensitive to interference often precedes the stage of their visible development (Fig. 8–14).

Six mechanisms can cause congenital malformations: (1) too little growth, (2) too little resorption, (3) too much resorption, (4) resorption in the wrong locations, (5) normal growth in an abnormal position, and (6) local overgrowth of a tissue or structure (Patten, 1957).

CRITICAL PERIODS IN HUMAN DEVELOPMENT

The most critical period in the development of an embryo or in the growth of a particular tissue or organ is during the *time of most rapid cell division*. The critical period varies in accordance with the timing and duration of the period of increasing cell numbers for the tissue or organ concerned.

The critical period for brain growth and development extends into infancy. The brain is growing rapidly at birth and continues to do so throughout the first two years after birth. *Tooth development* also continues long after birth (see Chapter 20); hence, development of the permanent teeth may be affected by *tetracyclines* from 18 weeks (prenatal) to 16 years.

The skeletal system has a prolonged critical period of development, extending into adolescence and early adulthood. Hence, growth of skeletal tissues provides a very good gauge of general growth.

Environmental disturbances during the first two weeks after fertilization may interfere with implantation of the blastocyst and/or cause early death and abortion of the embryo, but they rarely cause congenital malformations in human embryos. Teratogens may, however, cause mitotic nondisjunction during cleavage, resulting in chromosomal abnormalities that subsequently cause congenital malformations.

Development of the embryo is most easily disturbed during the organogenetic period, particularly from day 15 to day 60. During this period, teratogenic agents may be lethal, but they are more likely to produce major morphological abnormalities. Physiological defects, minor morphological abnormalities, and functional disturbances are likely to result from disturbances during the fetal period. However, certain microorganisms are known to cause serious congenital malformations, particularly of the brain and eyes, when they infect the fetus (see Table 8–3).

Each organ has a critical period during which its development may be deranged (Fig. 8–14). The following examples illustrate the ways in which various teratogens may affect different organ systems that are developing at the same time: (1) Radiation tends to produce abnormalities of the central nervous system and eye as well as mental retardation; (2) the rubella virus mainly causes cataracts, deafness, and cardiac malformations; and (3) thalidomide induces skeletal and many other malformations.

Embryological timetables (such as Fig. 8–14) are helpful in studying the etiology of human abnormalities, but it is wrong to assume that malformations always result from a single event occurring during the sensitive period, or that one can determine from these tables the exact day on which the malformation was produced. All that one can reliably say is that the teratogen probably had its

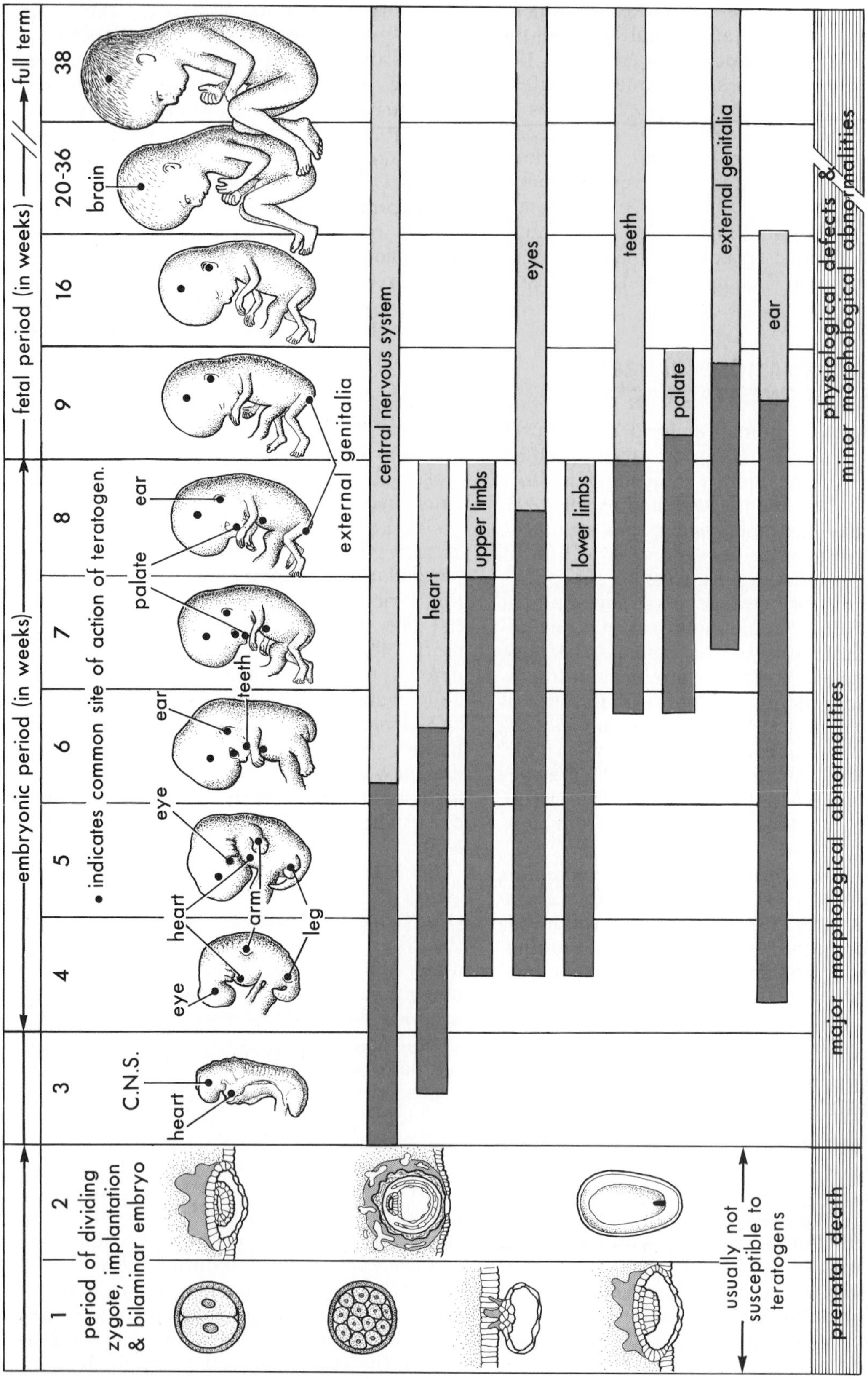

Figure 8-14   Schematic illustration of the critical periods in human development. During the first two weeks of development, the embryo is usually not susceptible to teratogens. During these predifferentiation stages, a substance either damages all or most of the cells of the embryo, resulting in its death, or it damages only a few cells, allowing the embryo to recover without developing defects. Red denotes highly sensitive periods; yellow indicates stages that are less sensitive to teratogens.

effect before the end of the organogenetic period of the structure or organ concerned as indicated by *red* on the chart.

TERATOGENS AND HUMAN MALFORMATIONS

A teratogen is any agent that can induce or increase the incidence of a congenital malformation. The general objective of teratogenicity testing of such chemicals as drugs, food additives, or pesticides is to attempt to identify agents that may be teratogenic during human development.

To prove that a given agent is teratogenic, one must show either that the frequency of malformations is increased above the "spontaneous" rate in pregnancies in which the mother is exposed to the agent (the prospective approach), or that malformed children have a history of maternal exposure to the agent more often than normal children (the retrospective approach). Both types of data are hard to get in an unbiased form. Individual case reports are not convincing unless both the agent and type of malformation are so rare that their association in several cases can be judged not coincidental.

— Fraser, 1967

Drug Testing in Animals. Although the testing of drugs in pregnant animals is important, it should be emphasized that the results are of limited value for predicting drug effects on human embryos. Animal experiments can only suggest similar effects in humans. However, if two or three species respond to a specific compound, then, even if the malformations differ among species, the probability of potential human hazard must be considered high.

DRUGS AND CHEMICALS AS TERATOGENS

Drugs vary considerably in their teratogenicity. Some cause severe malformations (e.g., thalidomide); other commonly-used drugs produce mental and growth retardation (e.g., alcohol). Pregnant women take an average of four drugs, excluding nutritional supplements and 40 per cent of these women take the drugs during the critical period of human development (Golbus, 1980). *Only 2 to 3 per cent of congenital malformations are caused by drugs and chemicals.* Few drugs have been positively implicated as teratogenic agents during human development (Table 8–3). Their use should be avoided in pregnant women and in those likely to conceive.

Discussions of these teratogenic drugs appear in regular print.

Several drugs are suspected of having teratogenic potential because of a few well-documented case reports (Wilson, 1973a). Still others must be regarded as possibly teratogenic on the basis of scattered evidence. Discussions of these drugs appear in small print.

It is best for women to avoid medication around the time of possible conception and throughout early pregnancy, unless there is a strong medical reason for its use and only if it is recognized as safe for the human embryo.

Alkaloids. Nicotine and caffeine do not produce congenital malformations in human embryos, but nicotine has an effect on fetal growth. *Maternal smoking is a well-established cause of intrauterine growth retardation (IUGR).* In heavy cigarette smokers (20 or more cigarettes per day), premature delivery is twice as frequent as in mothers who do not smoke, and their infants weight less than normal (see Fig. 6–12). Nicotine causes a decrease in uterine blood flow, thereby lowering the supply of oxygen in the intervillous space that is available to the embryo. The resulting oxygen deficiency in the embryo impairs cell growth and may have an adverse effect on mental development. Page et al. (1981) believe that the growth deficit results from the direct *toxic effects of smoking.* Because high levels of carboxyhemoglobin are present in the bloods of both the mother and embryo, this may alter the capacity of the blood to transport oxygen. As a result, *fetal hypoxia* occurs.

Although caffeine is not known to be a human teratogen, there is no assurance that excessive maternal consumption of it is safe for the embryo. For this reason, excessive drinking of coffee, tea, and colas that contain caffeine should be avoided.

Alcohol. Alcoholism is the most common drug abuse problem and affects 1 to 2 per cent of women of childbearing age (Golbus, 1980). Infants born to chronic alcoholic mothers exhibit prenatal and postnatal growth deficiency, mental retardation, and other malformations (Fig. 8–15 and Table 8–3). Short palpebral fissures, maxillary hypoplasia, abnormal palmar creases, joint anomalies, and congenital heart disease are present in most infants (Jones et al., 1974). This set of symptoms is known as the *fetal alcohol syndrome* (Mulvihill and Yeager, 1976). Even moderate maternal alcohol consumption (e.g., 2 to 3 ounces per day) may produce some symptoms of the syndrome

TABLE 8–3   TERATOGENS KNOWN TO CAUSE HUMAN MALFORMATIONS

Teratogens	Congenital Malformations
Androgenic Agents Ethisterone Norethisterone Testosterone	Varying degrees of masculinization of female fetuses: ambiguous external genitalia caused by labial fusion and clitoral hypertrophy.
Drugs and Chemicals Alcohol	*Fetal alcohol syndrome:* intrauterine growth retardation (IUGR); mental retardation; microcephaly; ocular anomalies; joint abnormalities; short palpebral fissures.
Aminopterin	Wide range of skeletal defects; IUGR; malformations of central nervous system, notably anencephaly.
Busulfan	Stunted growth; skeletal abnormalities; corneal opacities; cleft palate; hypoplasia of various organs.
Phenytoin (diphenylhydantoin)	*Fetal hydantoin syndrome:* IUGR; microcephaly; mental retardation; ridged metopic suture; inner epicanthal folds; eyelid ptosis; broad depressed nasal bridge; phalangeal hypoplasia.
Lithium carbonate	Various malformations, usually involving the heart and great vessels.
Methotrexate	Multiple malformations, especially skeletal, involving the face, skull, limbs, and vertebral column.
Thalidomide	Amelia, meromelia, and other limb deformities; external ear, cardiac, and gastrointestinal malformations.
Warfarin	Nasal hypoplasia; chondroplasia punctata; mental retardation; optic atrophy; microcephaly.
Infectious Agents Cytomegalovirus	Microcephaly; hydrocephaly; microphthalmia; microgyria; mental retardation; cerebral calcifications.
Herpes simplex virus	Microcephaly; microphthalmia; retinal dysplasia.
Rubella virus	Cataracts; glaucoma; chorioretinitis; deafness; microphthalmia; congenital heart defects.
Toxoplasma gondii	Microcephaly; microphthalmia; hydrocephaly; chorioretinitis; cerebral calcifications.
Treponema pallidum	Hydrocephalus; congenital deafness; mental retardation.
High level Radiation	Microcephaly; mental retardation; skeletal malformations.

(Streissguth et al., 1977), especially if the drinking is associated with malnutrition. "*Binge drinking*" (heavy consumption of alcohol for 1 to 3 days) during early pregnancy is very likely to harm the embryo.

Androgenic Agents.   *Any hormone that possesses masculinizing activities may also affect the embryo or fetus*, producing masculinization of female fetuses (Fig. 8–16). The incidence varies with the drug and the dosage. The preparations most frequently involved were the *progestins ethisterone* and *norethisterone* (Venning, 1965). Progestin exposure during the critical period of

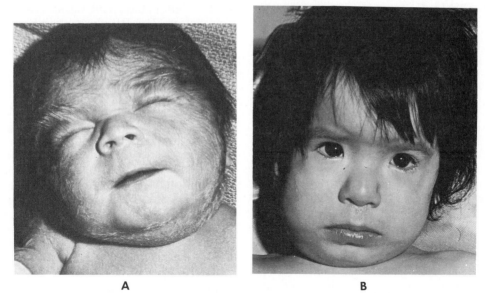

A   **B**

Figure 8–15   Photographs showing the facial appearance of an infant with the *fetal alcohol syndrome.* The characteristic triad of abnormalities include growth deficiency, mental retardation, and abnormal facial features. *A,* At birth. *B,* At one year. (*A* is from Jones, K. L., and Smith, D. W.; *Lancet 2*:999, 1973; *B* is from Jones, K. L., et al.: *Lancet 1*:1267, 1973.)

development is also associated with an increased prevalence of cardiovascular abnormalities (Heinonen et al., 1977). The administration of *testosterone* may produce similar masculinizing effects on female fetuses.

Oral contraceptives, containing progestin and estrogen, taken during the early stages of an unrecognized pregnancy, are strongly suspected of being teratogenic agents (Nora and Nora, 1975). The infants of 13 of 19 mothers who had taken progestogen-estrogen pills during the critical period of development exhibited the VACTERAL syndrome (in the acronym *vacteral*, the letters stand for *v*ertebral, *a*nal, *c*ardiac, *t*rachoesophageal, *r*enal, *a*nd *l*imb malformations).

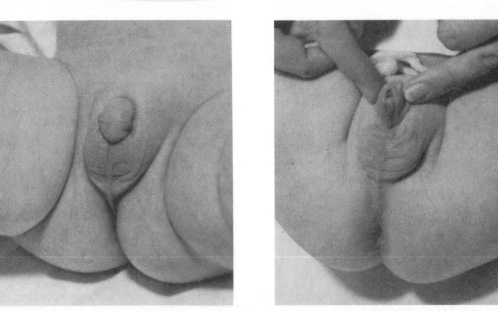

Figure 8–16   The external genitalia of a newborn female infant showing labial fusion and enlargement of the clitoris caused by an androgenic agent given to the infant's mother during the first trimester. The 17-ketosteroid output was normal. (From Jones, H. W., and Scott, W. W.: *Hermaphroditism, Genital Anomalies and Related Endocrine Disorders.* 1958. Copyright © 1958 by The William & Wilkins Co., Baltimore.)

There is not sufficient evidence to indicate that diethylstilbestrol (DES) taken during pregnancy is teratogenic, but it may produce tumors in female offspring. A number of young women aged 16 to 22 years have developed adenocarcinoma of the vagina after a common history of exposure to the synthetic estrogen in utero (Herbst et al., 1974, and Hart et al., 1976). However, the probability of cancer's developing at this early age in females exposed to DES in utero now appears to be low (Golbus, 1980).

Antibiotics. *Tetracyclines pass the placental membrane and are deposited in the embryo's bones and teeth* at sites of active calcification (Seeman et al., 1980). Hence, tetracycline therapy during the second and third trimesters of pregnancy may cause tooth defects (e.g., hypoplasia of enamel), yellow to brown discoloration of the teeth, and diminished growth of long bones (Rendle-Short, 1962, and Witkop et al., 1965). Therefore, if possible, *tetracyclines should not be administered to pregnant women or to prepubertal children.*

Deafness has been reported in infants of mothers who have been treated with high doses of streptomycin and dihydrostreptomycin as *antituberculosis agents* (Golbus, 1980). More than 30 cases of hearing deficit and VIII nerve damage have been reported in infants exposed to streptomycin derivatives in utero (Rasmussen, 1969, and Ganguin and Rempt, 1970).

Penicillin has been used extensively during pregnancy and appears to be harmless to the human embryo (Golbus, 1980).

Anticoagulants. All anticoagulants except heparin cross the placental membrane (see Fig. 7–8) and may cause hemorrhage in the fetus. *Warfarin* is now considered to be a teratogen. Warfarin derivatives are vitamin K antagonists and therefore are anticoagulants. There are several reports of infants born with hypoplasia of the nasal bones and other abnormalities whose mothers took this anticoagulant during the critical period of their embryo's development (Holzgreve et al., 1976). Second- and third-trimester exposure may result in mental retardation, optic atrophy, and microcephaly (Golbus, 1980).

As heparin does not cross the placental membrane, it is not a teratogen and does not affect the embryo or fetus.

Anticonvulsants. There is strong suggestive evidence that *trimethadione* (Tridione) and *paramethadione* (Paradione) may cause fetal facial dysmorphia, cardiac de-

fects, cleft palate, and intrauterine growth retardation (*IUGR*) when given to pregnant women (German et al., 1970). Loughnan et al. (1973) reported seven cases of hypoplasia of the terminal phalanges in infants of epileptics. All mothers had taken phenytoin and a barbiturate.

Phenytoin (diphenylhydantoin) is definitely a teratogen. A *fetal hydantoin syndrome* is now recognized (Hanson et al., 1976, and Golbus, 1980), consisting of the following abnormalities: IUGR, microcephaly, mental retardation, a ridged metopic suture, inner epicanthal folds, eyelid ptosis, a broad depressed nasal bridge, nail and/or distal phalangeal hypoplasia, and hernias. *Phenobarbital* appears to be a safe antiepileptic drug for use during pregnancy (Golbus, 1980).

Antineoplastic Agents. Tumor inhibiting chemicals are highly teratogenic. This is not surprising, because these agents inhibit rapidly dividing cells. Treatment with *folic acid antagonists* during the embryonic period often results in intrauterine death of the embryos, but the 20 to 30 per cent of those that survive are malformed (Wilson, 1973b). *Busulfan* and *6-mercaptopurine* administered in alternating courses throughout pregnancy have produced multiple severe abnormalities, but neither drug alone appears to cause major malformations (Sokal and Lessman, 1960). *Aminopterin is a potent teratogen that can produce major congenital malformations* (Fig. 8–17), especially of the central nervous system (Thiersch, 1952, and Warkany et al., 1960). Aminopterin, an antimetabolite, is an *antagonist of folic acid.* Milunsky et al. (1968) described multiple skeletal and other congenital malformations in an infant born to a mother who attempted to terminate her pregnancy by taking *methotrexate,* a derivative of aminopterin.

Corticosteroids. Cortisone causes cleft palate and cardiac defects in susceptible strains of mice and rabbits (Fraser, 1967). The scanty data concerning humans suggest that cortisone may be a weak teratogen (Karnofsky, 1965, and Fraser, 1967); however, there is no conclusive evidence that cortisone induces cleft palate or any other malformation in human embryos.

Environmental Chemicals. In recent years, there has been increasing concern over the possible teratogenicity of environmental chemicals, including industrial pollutants and food additives. At present, none of these chemicals has been positively implicat-

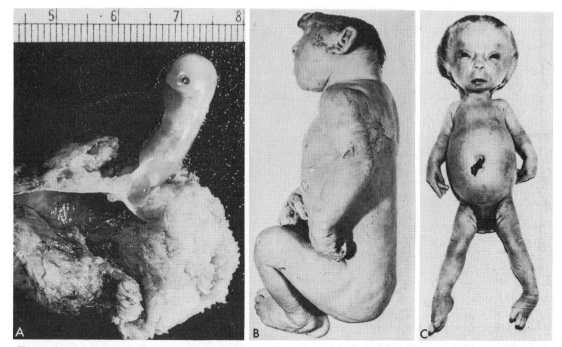

Figure 8–17   Aminopterin-induced congenital malformations. *A,* Grossly malformed embryo and its membranes. (Courtesy of Dr. J. B. Thiersch, Seattle, Washington.) *B,* Newborn infant with anencephaly or partial absence of the brain (From Thiersch, J. B.: *in* Wolstenholme, G. E. W., and O'Connor, C. M. [Eds.]: *Ciba Foundation Symposium on Congenital Malformations.* London, J. & A. Churchill, Ltd., 1960, pp. 152–154.) *C,* Newborn infant showing marked intrauterine growth retardation (2380 gm), a large head, a small mandible, deformed ears, club-hands, and clubfeet. (From Warkany, J., Beaudry, P. H., and Hornstein, S.: *Am. J. Dis. Child.* 97:274, 1960.)

ed as teratogenic in humans. However, infants of mothers whose main diet during pregnancy consisted of fish containing abnormally high levels of *organic mercury* acquire fetal Minamata disease and exhibit neurological and behavioral disturbances resembling cerebral palsy (Matsumoto et al., 1965, Bakir et al., 1973). In some cases, severe brain damage, mental retardation, and blindness are present (Amin-Zaki et al., 1974). Similar observations have been made in infants of mothers who ate pork that became contaminated when the pigs ate corn grown from seeds that had been sprayed with a mercury-containing fungicide (Snyder, 1971). Major morphological defects are seldom detected in infants whose mothers ingested methyl mercury (Persaud, 1979).

Insulin and Hypoglycemic Drugs.   Insulin is not teratogenic in human embryos, except possibly in maternal insulin coma therapy (Wickes, 1954). Hypoglycemic drugs (e.g., tolbutamide) have been implicated, but evidence for their teratogenicity is very weak. Consequently, in spite of their marked teratogenicity in rodents, there is no convincing evidence that oral hypoglycemic agents (particularly sulfonylureas) are teratogenic in human embryos. The incidence of congenital malformations (e.g., *sacral agenesis*) is somewhat increased in the offspring of diabetic mothers, but it is not clear whether it is the diabetes itself or the inadequate management of the condition that causes the malformations (Thompson and Thompson, 1980).

Lysergic Acid Diethylamide (LSD).   This socially used drug may be teratogenic. Long (1972) reviewed the literature on 161 infants born to women who ingested LSD before conception and/or during the pregnancy. Five infants had limb deficiency anomalies similar to those previously reported by Zellweger et al. (1967). Jacobson and Berlin (1972) also reported limb defects and noted a 9.6 per cent incidence of nervous system defects. These observations *suggest* that LSD may be teratogenic and that ingestion of it should be avoided during pregnancy (Golbus, 1980).

Marijuana.   There is no evidence that this drug is teratogenic in humans, but there is no assurance that heavy usage of it does not affect the embryo in some way.

Phencyclidine (PCP, "angel dust").   Golden et al. (1980) reported a case of an infant with several malformations and behavioral abnormalities whose mother used PCP throughout her pregnancy. This suggests, but does not prove, a causal association.

Salicylates.   There is some evidence that *aspirin*, the most commonly ingested drug during pregnancy, is potentially harmful to the embryo or fetus when administered to the mother in *large doses* (Corby, 1978).

Thyroid Drugs. *Potassium iodide* in cough mixtures and large doses of *radioactive iodine* may cause congenital goiter (Fraser, 1967). Iodides readily cross the placental membrane and interfere with thyroid production, and they may cause thyroid enlargement and *cretinism* (arrested physical and mental development and dystrophy of bones and soft parts).

Pregnant women are advised to avoid using douches or creams containing povidone-iodine because it is absorbed by the vagina (Vorherr et al., 1980). *Propylthiouracil* interferes with thyroxin formation in the fetus and may cause goiter. Warkany (1954) reported that maternal iodine deficiency may cause congenital cretinism.

Tranquilizers.   A mass of evidence has shown that thalidomide is a potent teratogen. This potent hypnotic agent was once widely used in Europe as a tranquilizer and a sedative. Lenz (1966) estimated that 7000 infants were malformed by thalidomide. The characteristic feature of the *thalidomide syndrome* is *meromelia* (e.g., phocomelia or seal-like limbs, Fig. 8–18), but malformations range from *amelia* (absence of limbs) through intermediate stages of development (rudimentary limbs) to *micromelia* (abnormally small and/or short limbs). Thalidomide may also cause absence of the external and internal ears, hemangioma on the forehead, heart defects, and malformations of the urinary and alimentary systems (Persaud, 1979). Thalidomide was withdrawn from the market in November 1961.

Lithium carbonate has been used widely for patients with manic-depressive psychosis (Golbus, 1980) and has caused congenital malformations, mainly of the heart and great vessels, in infants born to mothers given the drug early in pregnancy.

There appears to be a relationship between the use of *diazepam* during the first trimester of pregnancy and cleft lip with or without cleft palate (Golbus, 1980). Discretion suggests that this drug should be avoided in early pregnancy if possible.

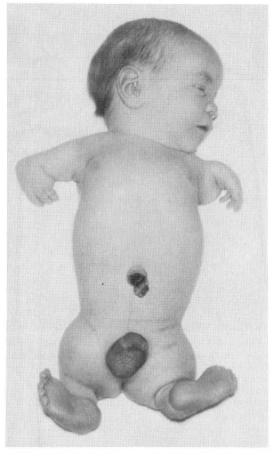

Figure 8–18   Newborn male infant showing typically malformed limbs (meromelia) caused by thalidomide ingested by the mother during early pregnancy. (From Moore, K. L.: *Manit. Med. Rev. 43*:306, 1963.)

INFECTIOUS AGENTS AS TERATOGENS

Throughout prenatal life, the embryo and fetus are endangered by a variety of microorganisms. In most cases, the assault is resisted; in some cases, an abortion or stillbirth occurs; and in others, the infants are born with congenital malformations or disease. The microorganisms cross the placental membrane (see Fig. 7–8) and enter the fetal blood stream. The fetal blood-brain barrier also appears to offer little resistance to microorganisms because there is a propensity for the central nervous system to be affected.

Three viruses are known to be teratogenic in humans: rubella virus, cytomegalovirus, and herpes simplex virus.

Rubella Virus (German Measles).   This

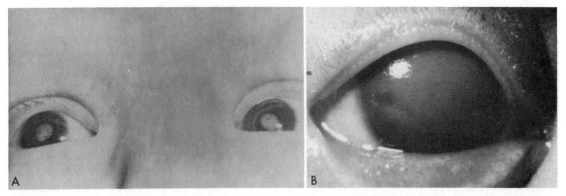

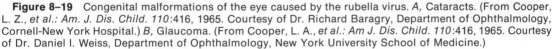

Figure 8–19 Congenital malformations of the eye caused by the rubella virus. *A*, Cataracts. (From Cooper, L. Z., *et al.: Am. J. Dis. Child. 110*:416, 1965. Courtesy of Dr. Richard Baragry, Department of Ophthalmology, Cornell-New York Hospital.) *B*, Glaucoma. (From Cooper, L. A., *et al.: Am J. Dis. Child. 110*:416, 1965. Courtesy of Dr. Daniel I. Weiss, Department of Ophthalmology, New York University School of Medicine.)

virus is the prime example of an *infective teratogen*. About 15 to 20 per cent of infants born to women who have had German measles during the first trimester of pregnancy are congenitally malformed (Sever, 1970). The usual triad of malformations is *cataract* (Fig. 8–19*A*), *cardiac malformation*, and *deafness* (Dudgeon, 1976), but the following abnormalities are occasionally observed: chorioretinitis, glaucoma (Fig. 8–19*B*), microcephaly, microphthalmia, and tooth defects (Cooper, 1975).

The earlier in pregnancy the maternal rubella infection occurs, the greater the danger of the embryo being malformed. Most infants have congenital malformations if the disease occurs during the first four to five weeks after fertilization; this is understandable, because this period of development includes the most susceptible organogenetic periods of the eye, the ear, the heart, and the brain (Fig. 8–14). The risk of malformations due to infections during the second and third trimesters is low, but functional defects of the central nervous system and the ear may result if infection occurs as late as the twenty-fifth week (Sever, 1970).

Although there is no conclusive information about the teratogenic potential of *live attenuated rubella virus*, Larson et al. (1971) state that vaccination with live attenuated virus during early pregnancy is contraindicated. They further believe that women and adolescent girls should be immunized with live attenuated virus only when pregnancy is not planned during the following two months.

Cytomegalovirus. Infection with cytomegalovirus (CMV) is probably the most common viral infection of the human fetus (Dudgeon, 1976). Because the disease seems to be fatal when it affects the embryo or young fetus, it is believed that most pregnancies end in abortion when the infection occurs during the first trimester. Later, infection may result in intrauterine growth retardation, microphthalmia, chorioretinitis, blindness, microcephaly, cerebral calcification, mental retardation, deafness, cerebral palsy, and hepatosplenomegaly (Persaud, 1979).

Herpes Simplex Virus. Infection of the fetus with this virus usually occurs late in pregnancy, probably most often during delivery. The congenital abnormalities that have been observed in fetuses infected several weeks before birth are microcephaly, microphthalmia, retinal dysplasia, and mental retardation (South et al., 1969, and Dudgeon, 1976).

Two other microorganisms are known to be teratogens in humans, *Toxoplasma gondii* and *Treponema pallidum*.

Toxoplasma Gondii. This protozoan is an *intracellular parasite*. Infection can be contracted from eating raw or poorly cooked meat (usually pork or mutton), by close contact with infected animals (usually cats), or from the soil. It is thought that the soil becomes contaminated with infected cat feces carrying the oocyst (Dudgeon, 1976). This organism probably crosses the placental membrane and infects the fetus, causing destructive changes in the brain and eye that result in microcephaly, microphthalmia, and hydrocephaly (White and Sever, 1967). The mothers of congenitally affected infants are usually unaware of having had toxoplasmosis, the disease caused by *Toxoplasma gondii*, during pregnancy. Because animals (cats, dogs, rabbits, and wild animals) may may be infected with this parasite, pregnant women should avoid them.

Syphilis. *Treponema pallidum*, the small spiral-shaped microorganism that causes syphilis, rapidly penetrates the placental membrane after the twentieth week of gestation, when the cytotrophoblast disappears (see Chapter 7). Untreated primary maternal infections (acquired during pregnancy) nearly always cause serious fetal infection, but adequate treatment of the mother before the sixteenth week kills the organism, thereby preventing it from crossing the placental membrane (see Fig. 7–8) and infecting the fetus (Holmes, 1977). Secondary maternal infections (acquired before pregnancy) seldom result in fetal disease and malformations. If the mother is untreated, stillbirths result in one fourth of cases (Vaughan et al., 1979). The tissues most often extensively involved in the dead fetuses are bone, bone marrow, lungs, liver, and spleen, but any organ system may be involved.

It was once believed that syphilis was a major cause of congenital malformations. While this is not true, owing to the onset of the infection after the twentieth week in most cases, it is well established that *Treponema pallidum* may produce *congenital deafness, hydrocephalus,* and *mental retardation* (Ingall and Norins, 1976, and Persaud, 1979). Late manifestations of untreated congenital syphilis are destructive lesions of the palate and nasal septum, dental abnormalities (centrally notched, widely spaced, peg-shaped upper central incisors called *Hutchinson's teeth*), and abnormal facies (frontal bossing, saddlenose, and poorly developed maxilla).

Treponema pallidum is therefore a teratogen that may produce congenital developmental abnormalities, and it may continue to produce developmental defects during postnatal development if the infant's infection is not treated.

RADIATION AS A TERATOGEN

Exposure to radiation may injure embryonic cells, resulting in cell death, chromosome injury, and retardation of growth. The severity of embryonic damage is related to the absorbed dose, the dose rate, and the stage of embryonic or fetal development during which the exposure occurs (Rennert, 1975, and Brent and Harris, 1976).

In the past, large amounts of ionizing radiation (hundreds to several thousand rads) were given inadvertently to embryos and fetuses of pregnant women who had cancer of the cervix of the uterus. In all cases, their embryos were severely malformed or killed (Bartalos and Baramki, 1967, Saxén and Rapola, 1969, and Persaud, 1979). Microcephaly, spina bifida cystica (see Fig. 18–15), cleft palate, skeletal and visceral abnormalities, and mental retardation were observed in the infants who survived. Development of the central nervous system was nearly always affected. An increase of 10 to 15 per cent in congenital malformations was also reported in infants of pregnant women who lived just outside the lethal range of radiation of the atomic bombs exploded in Japan in 1945 (Fraser, 1967, and Makino, 1975). It is evident therefore that *large amounts of ionizing radiation produce congenital malformations.* It is generally accepted that large doses of radiation (over 25,000 millirads) are harmful to the *developing* central nervous system. For this reason, therapeutic abortion may be recommended when radiation exposure to the embryo or young fetus exceeds 25,000 millirads.

There is no proof that human congenital malformations have been caused by diagnostic levels of radiation (Oppenheim et al., 1975, and Holmes, 1979). Scattered radiation from an x-ray examination of a part of the body that is not near the uterus (e.g., chest, sinuses, teeth) produces only a dose of a few millirads, which is not teratogenic to the embryo. For example, a single frontal radiograph of the chest of a pregnant woman in the first trimester results in a whole body dose to her embryo or fetus of approximately 1 millirad.

It is prudent to be cautious during diagnostic examinations of the pelvic region in pregnant women (x-ray examinations and medical diagnostic tests using radioisotopes) because they result in exposure of the embryo to 0.3 to 2 rads. Doses below 10 rads have not been shown to cause congenital malformations or intrauterine growth retardation, especially when the pelvic or abdominal x-ray examinations are undertaken after the twentieth week of pregnancy, but it is best to err on the side of caution. *The recommended limit of maternal exposure of the whole body to radiation from all sources is 500 millirads for the entire gestational period* (Holmes, 1979).

MECHANICAL FACTORS AS TERATOGENS

The significance of mechanical influences in the uterus on congenital postural deformities is still an open question (McKeown, 1976). The amniotic fluid absorbs mechanical pressures, thereby protecting the embryo from most external trauma. Consequently, it is generally accepted that congenital abnormalities caused by external injury to the mother are extremely rare but possible (Hinden, 1965, and Warkany, 1971). Dislocation of the hip and clubfoot may rarely be caused by mechanical forces, particularly in a malformed uterus (Browne, 1960), but genetic factors appear to be involved in most of these abnormalities.

A reduced quantity of amniotic fluid *(oligohydramnios)* may result in mechanically induced abnormalities of the fetal limbs (Dunn, 1976), e.g., congenital genu recurvatum or hyperextension of the knee. Intrauterine amputations or other malformations caused by local constriction during fetal growth may result from amniotic bands or fibrous rings, presumably formed as a result of rupture of the fetal membranes (amnion and smooth chorion) during early pregnancy (Vaughan et al., 1979).

MALFORMATIONS CAUSED BY MULTIFACTORIAL INHERITANCE

Most common congenital malformations have familial distributions consistent with multifactorial inheritance. Multifactorial inheritance may be represented by a model in which "liability" to a disorder is a continuous variable, determined by a combination of genetic and environmental factors, with a developmental "threshold" dividing individuals with the malformation from those without it (Fig. 8–20). Multifactorial traits are usually single major malformations such as *cleft lip* with or without cleft palate, *isolated cleft palate, neural tube defects* (anencephaly and spina bifida cystica), *pyloric stenosis,* and *congenital dislocation of the hip.* Some of these malformations may also occur as part of the phenotype in syndromes determined by single-gene inheritance, chromosomal abnormality or an environmental teratogen, or their etiology may be unknown.

The recurrence risks used for genetic counseling of families having congenital malformations determined by multifactorial inheritance are *empirical risks* based on the frequency of the malformation in the general population and in different categories of relatives. In individual families, such estimates may be inaccurate because they are usually averages for the population rather than precise probabilities for the individual family. For further discussion of multifactorial inheritance, see Chapter 12 in Thompson and Thompson's *Genetics in Medicine* (1980).

Figure 8–20 Quasicontinuous variation. Liability to a given trait is normally distributed, but the population is divided by a threshold into normal and abnormal classes. (From Thompson, J. S., and Thompson, M. W.: *Genetics in Medicine.* 3rd ed. Philadelphia, W. B. Saunders Co., 1980.)

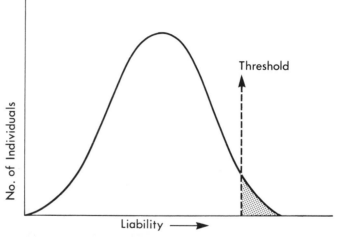

SUMMARY

A congenital malformation is an anatomical abnormality present at birth. It may be the result of either genetic factors or some environmental insult that occurred during prenatal development, but most common congenital malformations have been shown to fit the family patterns expected of *multifactorial inheritance* with a threshold and are determined by a combination of genetic and environmental factors. Developmental abnormalities may be macroscopic or microscopic, on the surface or within the body.

Some congenital malformations are caused by *genetic factors* (chromosomal abnormalities and mutant genes). A few congenital abnormalities are caused by *environmental factors* (infectious agents and teratogenic drugs), but most common malformations probably result from a complex *interaction between genetic and environmental factors* with arrest of a developmental process. Subsequent developmental stages therefore cannot occur, and a malformation results.

During the first two weeks of development, teratogenic agents usually kill the embryo rather than cause congenital malformations. During the *organogenetic period,* particularly from day 15 to day 60, teratogenic agents may cause *major congenital malformations.* During the fetal period, teratogens may produce morphological and functional abnormalities, particularly of the brain and the eyes. However, it must be stressed that some teratogenic agents (e.g., drugs and infections) may adversely affect the fetus without producing congenital malformations.

SUGGESTIONS FOR ADDITIONAL READING

Berry, C. L.: Human malformations. *Br. Med. J. 3*:1–94, 1976. A special issue of the *British Medical Bulletin* devoted exclusively to problems of congenital malformations in man. Of special interest are the three papers concerned with etiological factors that may lead to birth defects. The legal rights of the malformed infant are also discussed.

Persaud, T. V. N.: *Prenatal Pathology: Fetal Medicine.* Springfield, Illinois, Charles C Thomas, 1979. A book covering the major aspects of teratology and providing the essential information on epidemiology, causes, mechanisms, and prenatal detection of birth defects.

Thompson, J. S., and Thompson, M. W.: *Genetics in Medicine.* 3rd ed. Philadelphia, W. B. Saunders Co., 1980. This book has been written primarily to introduce medical students to the principles and language of human genetics and to indicate some of its clinical applications. There is a comprehensive account of the genetic causes of congenital malformations.

Warkany, J.: *Congenital Malformations: Notes and Comments.* Chicago, Year Book Medical Publishers, In., 1971. An excellent reference text on human malformations. Contains data on congenital malformations gathered over a 35-year-period by an outstanding professor of research pediatrics. It would be very difficult to find a congenital malformation that is not well covered in this book consisting of 1309 pages.

Warkany, J.: Prevention of congenital malformations. *Teratology 23*:175–189, 1981. A world authority on this subject, Dr. Warkany discusses the causes of congenital malformations and suggests ways of eliminating them as a source of mortality and morbidity.

CLINICALLY ORIENTED PROBLEMS

1. Women take an average of four drugs, excluding nutritional supplements, during the first trimester of pregnancy. What percentage of congenital malformations are caused by drugs and chemicals? Why may it be difficult for doctors to attribute specific congenital malformations to specific drugs?
2. Do women over the age of 35 have an increased risk of bearing abnormal children? If a 40-year-old woman becomes pregnant, what prenatal diagnostic test would likely be performed? What abnormality might be detected?
3. Are there any drugs that are considered safe during early pregnancy? If so, name some commonly prescribed ones.
4. A 12-year-old girl contracted German measles, and her mother was worried that the child might develop cataracts. What would the doctor likely tell the mother?

The answers to these questions are given at the back of the book.

REFERENCES

Allderice, P. W., Browne, N., and Murphy, D. P.: Chromosome 3 duplication q21-qter deletion p25-pter syndrome in children of carriers of a pericentric inversion inv (3) (p25 q21). *Am. J. Hum. Genet.* 27:699, 1975.

Amin-Zaki, L., Elhassani, S., Majeed, M. A., Clarkson, T. W., Doherty, R. A., and Greenwood, M.: Intrauterine methylmercury poisoning in Iraq. *Pediatrics* 54:587, 1974.

Anders, G. J.: Congenital malformations: genetic background and genetic counselling; in A. J. Huffstadt (Ed.): *Congenital Malformations.* Amsterdam, Excerpta Medica, 1980, pp. 18–55.

Bakir, F., Damluji, S. F., Amin-Zaki, L., Murtadha, M., Khalidi, A., Al-Rawi, N. Y., Tikriti, S., Dhahir, H. I., Clarkson, T. W., Smith, J. C., and Doherty, R. A.: Methylmercury poisoning in Iraq. *Science* 181:320, 1973.

Barr, M. L.: The sex chromosomes in evolution and medicine. *Can. Med. Assoc. J.* 95:1137, 1966.

Bartalos, M., and Baramki, T. A.: *Medical Cytogenetics.* Baltimore, The Williams & Wilkins Co., 1967.

Benirschke, K.: Teratology; in Reid, D. E., Ryan, K. J., and Benirschke, K. (Eds.): *Principles and Management of Human Reproduction.* Philadelphia, W. B. Saunders Co., 1972, pp. 392–402.

Bennett, R., Persaud, T. V. N., and Moore, K. L.: Experimental studies on the effects of aluminum on pregnancy and fetal development. *Anat. Anz.* 138:365, 1975.

Bergsma, D. (Ed.): The First Conference on the Clinical Delineation of Birth Defects. Part V: Phenotypic Aspects of Chromosomal Aberrations. The National Foundation — March of Dimes. Birth Defects: Original Article Series, Vol. V, No. 5, 1969.

Bergsma, D. (Ed.): *Birth Defects Atlas and Compendium.* The National Foundation — March of Dimes. Birth Defects: Original Article Series. Baltimore, The Williams & Wilkins Co., 1973.

Bergsma, D. (Ed.): *New Chromosomal and Malformation Syndromes.* The National Foundation — March of Dimes. Birth Defects: Original Article Series. Vol. XI, No. 5, 1975.

Berlin, C. M., and Jacobson, C. B.: Congenital anomalies associated with parental LSD ingestion. *Society for Pediatric Research Abstracts, Second Plenary Session,* 1970.

Breg, W. R.: Autosomal abnormalities; in L. I. Gardner (Ed.): *Endocrine and Genetic Diseases of Childhood and Adolescence.* 2nd ed. Philadelphia, W. B. Saunders Co., 1975, pp. 730–762.

Brent, R. L., and Harris, M. I.: *Prevention of Embryonic, Fetal, and Perinatal Disease.* Fogarty International Center Series on Preventive Medicine. Vol. 3, 1976.

Browne, D.: Congenital deformity. *Br. Med. J.* 2:1450, 1960.

Carakushansky, G., Neu, R. L., and Gardner, L. I.: Lysergide and cannabis as possible teratogens in man. *Lancet 1*:150, 1969.

Carr, D. H.: Heredity and the embryo. *Science J.* (London) 6:75, 1970.

Carr, D. H.: Chromosome studies in selected spontaneous abortions: Polyploidy in man. *J. Med. Genet.* 8:164, 1971.

Carr, D. H., Law, E. M., and Ekins, J. G.: Chromosome studies in selected spontaneous abortions. IV.

Unusual cytogenetic disorders. *Teratology* 5:49–56, 1972.

Carter, C. O.: The genetics of congenital malformations; in E. E. Philipp, J. Barnes, and M. Newton (Eds.): *Scientific Foundations of Obstetrics and Gynecology.* London, William Heinemann, Ltd., 1970, pp. 665–670.

Caspersson, T., Zech, L., Johansson, C., and Modest, E. J.: Identification of human chromosomes by DNA-binding fluorescent agents. *Chromosoma* 30:215, 1970.

Challacombe, D. N., and Taylor, A.: Monosomy for a G autosome. *Arch. Dis. Child.* 44:113, 1969.

Cooper, L. Z., Green, F. H., Krugman, S., Giles, J. P., and Mirick, G. S.: Neonatal thrombocytopenic purpura and other manifestations of rubella contracted in utero. *Am. J. Dis. Child. 110*:416, 1965.

Cooper, L. Z.: Congenital rubella in the United States; in S. Krugman and A. A. Gershon (Eds.): *Progress in Clinical and Biological Research, Infections of the Fetus and Newborn Infant.* New York, Alan R. Liss Inc., 1975, Vol. 3, p. 1.

Corby, D. G.: Aspirin in pregnancy: Maternal and fetal effects. *Pediatrics 62*:930, 1978.

Dishotsky, N. I., Loughman, W. D., Mogar, R. E., and Lipscomb, W. R.: LSD and genetic damage. *Science 172*:431–440, 1971.

Dudgeon, J. A.: Infective causes of human malformations. *Br. Med. J. 32*:77, 1976.

Dunn, P. M.: Congenital malformations and maternal diabetes. *Lancet 2*:644, 1964.

Dunn, P. M.: Congenital postural deformities. *Br. Med. J. 32*:71, 1976.

Epstein, S. S.: Environmental pathology. *Am. J. Pathol. 66*:352, 1972.

Federman, D. D.: *Abnormal Sexual Development: A Genetic and Endocrine Approach to Differential Diagnosis.* Philadelphia, W. B. Saunders Co., 1967, pp. 121–133.

Ferguson-Smith, M. A.: Sex chromatin, Klinefelter's syndrome and mental deficiency; in K. L. Moore (Ed.): *The Sex Chromatin.* Philadelphia, W. B. Saunders Co., 1966, pp. 277–315.

Fraser, F. C.: Some genetic aspects of teratology; in J. G. Wilson, and J. Warkany (Eds.) *Teratology: Principles and Techniques.* Chicago, University of Chicago Press, 1965, pp. 21–37.

Fraser, F. C.: Etiologic agents. II. Physical and chemical agents; in A. Rubin (Ed.): *Handbook of Congenital Malformations.* Philadelphia, W. B. Saunders Co., 1967, pp. 365–371.

Fraser, F. C., and Nora, J. J.: *Genetics of Man.* Philadelphia, Lea and Febiger, 1975.

Ganguin, G., and Rempt, E.: Streptomycinbehandlung in der Schwanger-schaft und ihre Answirkung auf des Gehör des Kindes. *Z. Laryngol. Rhinol. Otol.* 49:496, 1970.

Gardner, E. J.: *Principles of Genetics.* 5th ed. New York, John Wiley & Sons, Inc., 1975.

Gardner, L. I. *Endocrine and Genetic Diseases of Childhood and Adolescence.* 2nd ed. Philadelphia, W. B. Saunders Co., 1975.

German, J., Kowal, A., and Ehlers, K. H.: Trimethadione and human teratogenesis. *Teratology 3*:349, 1970.

Golbus, M. S.: Teratology for the obstetrician: Current status. *Obstet. Gynecol.* 55:269, 1980.

Golden, N. L., Sokol, R. J., and Rubin, I.: Angel dust: possible effects on the fetus. *Pediatrics 65*:18, 1980.

Gray, J. E., Mutton, D. E., and Ashby, D. W.: Pericentric inversion of chromosome 21. A possible further cytogenetic mechanism in mongolism. *Lancet 1*:21, 1962.

Gray, S. W., and Skandalakis, J. E.: *Embryology for Surgeons. The Embryological Basis for the Treatment of Congenital Defects.* Philadelphia, W. B. Saunders Co., 1972.

Gregg, N. M.: Congenital cataract following German measles in the mother. *Trans. Ophthalmol Soc. Aust. 3*:35, 1941.

Hall, J. G., Pauli, R. M., and Wilson, K. M.: Maternal and fetal sequelae of anticoagulation during pregnancy. *Am. J. Med. 68*:122, 1980.

Hamerton, J.: *Human Cytogenetics. Clinical Cytogenetics.* Vol. II. New York, Academic Press, Inc., 1971, pp. 169–195.

Hanson, J. W., Myrianthopoulos, N. C., Harvey, M. A. S., et al.: Risks to offspring of women treated with hydantoin anticonvulsants with emphasis on the fetal hydantoin syndrome. *J. Pediatr. 89*:662, 1976.

Harley, J. D., Farrar, J. F., Gray, J. B., and Dunlop, I. C.: Aromatic drugs and congenital cataracts. *Lancet 1*:472, 1964.

Harris, L. E., Stayura, L. A., Ramirez-Talavera, P. F., and Annegers, J. F.: Congenital and acquired abnormalities observed in live-born and stillborn neonates. *Mayo Clin. Proc. 50*:85, 1975.

Hart, W. R., Zaharrow, I., Kaplan, B. J., Townsend, D. E., Aldrich, J. O., Henderson, B. E., Roy, M., and Benton, B.: Cytologic findings in stilbestrol-exposed females with emphasis on detection of vaginal adenosis. *Acta Cytol. (Baltimore) 20*:7, 1976.

Hecht, F., Beals, R. K., Lees, M. H., Jolly, H., and Roberts, P.: Lysergic-acid-diethylamide and cannabis as possible teratogens in man. *Lancet 2*:1087, 1968.

Heinonen, O. P., Slone, D., and Shapiro, S.: *Birth Defects and Drugs in Pregnancy.* Littleton, Massachusetts, Publishing Sciences Group, 1977.

Herbst, A. L. H., Ulfelder, H., and Poskanzer, D. C.: Adenocarcinoma of the vagina. *N. Engl. J. Med. 284*:878, 1971.

Herbst, A. L., Robboy, S. J., Scully, R. E., and Poskanzer, D. C.: Clear-cell adenocarcinoma of the vagina and cervix in girls. Analysis of 170 registry cases. *Am. J. Obstet. Gynecol. 119*:713, 1974.

Hicks, S. P., and D'Amato, C. J.: Effects of ionizing radiations on mammalian development. *Adv. Teratol. 1*:195, 1966.

Hill, R. M., Craig, J. P., Chaney, M. D., Tennyson, L. M., and McCulley, L. B.: Utilization of over-the counter drugs during pregnancy. *Clin. Obstet. Gynecol. 20*:381, 1977.

Hinden, E.: External injury causing foetal deformity. *Arch. Dis. Child. 40*:80, 1965.

Holmes, K. K.: Syphilis; *in* Thorn, G. W., Adams, R. D., Braunwald, E., Isselbacher, K. J., and Petersdorf, R. G. (Eds.): *Harrison's Principles of Internal Medicine.* 8th ed. New York, McGraw-Hill Book Company, 1977.

Holmes, L. B.: Radiation; *in* V. C. Vaughan, R. J. McKay, and R. D. Behrman (Eds.): *Nelson Textbook of Pediatrics.* 11th ed. Philadelphia, W. B. Saunders Co., 1979.

Holzgreve, W., Carey, J. C., and Hall, B. D.: Warfarin-induced fetal abnormalities. *Lancet 2*:914, 1976.

Hook, E. B.: Rates of Down's syndrome in live births and at midtrimester amniocentesis. *Lancet 1*:1053, 1978.

Hook, E. B., and Hamerton, J. L.: The frequency of chromosome abnormalities detected in consecutive newborn studies — differences between studies —results by sex and by severity of phenotypic involvement; *in* E. B. Hook and I. H. Porter (Eds.): *Population Cytogenetics: Studies in Humans.* New York, Academic Press, 1977.

Ingall, D., and Norins, L.: Syphilis; *in* Remington, J. S., and Klein, J. O. (Eds.): *Diseases of the Fetus and Newborn Infant.* Philadelphia, W. B. Saunders Co., 1976.

Jacobson, C. B., and Berlin, C. M.: Possible reproductive detriment in LSD users. *J.A.M.A. 222*:1367, 1972.

Jones, H. W., Jr., and Scott, W. W.: *Hermaphroditism, Genital Anomalies and Related Endocrine Disorders.* Baltimore, The Williams & Wilkins Co., 1958.

Jones, K. L., Smith, D. W., Streissguth, A. P., and Myrianthopoulos, N. C.: Outcome in offspring of chronic alcoholic women. *Lancet 1*:1076, 1974.

Karnofsky, D. A.: Drugs as teratogens in animals and man. Ann. Rev. Pharmacol. 5:447, 1965.

Kucera, J., Lenz, W., and Maier, W.: Malformations of the lower limbs and the caudal part of the spinal column in children of diabetic mothers. *Ger. Med. Mon 10*:393, 1965.

Larson, H. E., Parkman, P. D., Davis, W. J., Hopps, H. E., and Meyer, H. M., Jr.: Inadvertent rubella virus vaccination during pregnancy. *N. Engl. J. Med. 284*:870, 1971.

Laurence, K. M., and Gregory, P.: Prenatal diagnosis of chromosome disorders. *Br. Med. J. 32*:9, 1976.

Lenz, W.: Kindliche Missbildungen nach Medikament während der Gravidität? *Dtsch. Med. Wochenschr. 86*:2555, 1961.

Lenz, W.: Malformations caused by drugs in pregnancy. *Am. J. Dis. Child. 112*:99, 1966.

Lenz, W., and Knapp, K.: Foetal malformations due to thalidomide. *Ger. Med. Mon. 7*:253, 1962.

Long, S. Y.: Does LSD induce chromosomal damage and malformations? A review of the literature. *Teratology 6*:75, 1972.

Longo, L. D.: Environmental pollution and pregnancy: Risks and uncertainties for the fetus and infant. *Am. J. Obstet. Gynecol. 137*:162, 1980.

MacVicar, J.: Antenatal detection of fetal abnormality; physical methods. *Br. Med. J. 32*:4, 1976.

Makino, S.: *Human Chromosomes.* Tokyo, Igaku Shoin Ltd., 1975.

Matsumoto, H. G., Goyo, L., and Takevchi, T.: Fetal minamata disease. A neuropathological study of two cases of intrauterine intoxication by a methyl mercury compound. *J. Neuropathol. Exp. Neurol. 24*:563, 1965.

McBride, W. G.: Thalidomide and congenital abnormalities. *Lancet 2*:1358, 1961.

McBride, W. G.: Studies of the etiology of thalidomide dysmorphogenesis. *Teratology 14*:71, 1976.

McKeown, T.: Human malformations: Introduction. *Br. Med. J. 32*:1, 1976.

McKusick, V. A.: *Mendelian Inheritance in Man. Catalogs of Autosomal Dominant, Autosomal Recessive, and X-linked Phenotypes.* Baltimore, The Johns Hopkins University Press, 1975.

Miller, O. J.: The sex chromosome anomalies. *Am. J. Obstet. Gynecol. 90*:1078, 1964.

Milunsky, A., Graef, J. W., and Gaynor, M. F., Jr.: Methotrexate-induced congenital malformations. *J. Pediatr. 72*:790, 1968.

Mims, C.: Comparative aspects of infective malformations. *Br. Med. J. 32*:84, 1976.

Mittwoch, U.: *Sex Chromosomes.* New York, Academic Press, Inc., 1967.

Mole, R. H.: Radiation effects on prenatal development and the radiological significance. *Br. J. Radiol. 52*:89, 1979.

Moore, K. L.: The vulnerable embryo: Causes of malformation in man. *Manitoba Med. Rev. 43*:306, 1963.

Moore, K. L.: *The Sex Chromatin.* Philadelphia, W. B. Saunders Co., 1966.

Moore, K. L., and Barr, M. L.: Smears from the oral mucosa in the detection of chromosomal sex. *Lancet 2*:57, 1955.

Mulvihill, J. J., and Yeager, A. M.: Fetal alcohol syndrome. *Teratology 13*:345, 1976.

Nahmias, A. J., Visintine, A. M., Reimer, C. B., Del Buono, I., Shore, S. L., and Starr, S. E.: Herpes simplex virus infection of the fetus and newborn; *in* S. Krugman, and A. A. Gershon (Eds.): *Progress in Clinical and Biological Research, Infections of the Fetus and Newborn.* New York, Alan R. Liss Inc., 1975, Vol. 3, p. 63.

Nelson, M. M., and Fofar, J. L.: Associations between drugs administered during pregnancy and congenital abnormalities of the fetus. *Br. Med. J. 1*:523, 1971.

Neu, R. L., and Gardner, L. I.: Abnormalities of the sex chromosomes; *in* L. I. Gardner (Ed.): *Endocrine and Genetic Diseases of Childhood and Adolescence.* 2nd ed. Philadelphia, W. B. Saunders Co., 1975, pp. 793–814.

Nishimura, H., and Tanimura. T.: *Clinical Aspects of the Teratogenicity of Drugs.* Amsterdam, Excerpta Media, 1975.

Nora, J. J., and Fraser, F. C.: *Medical Genetics: Principles and Practice.* Philadelphia, Lea and Febiger, 1974.

Nora, A. H., and Nora, J. J.: A syndrome of multiple congenital anomalies associated with teratogenic exposure. *Arch. Environ. Health 30*:17, 1975.

Oppenheim, B. E., Griem, M. L., and Meier, P.: The effects of diagnostic x-ray on the human fetus: An examination of evidence. *Radiology 114*:529, 1975.

Page, E. W., Villee, C. A., and Villee, D. B.: *Human Reproduction: Essentials of Reproductive and Perinatal Medicine.* 3rd ed. Philadelphia, W. B. Saunders Co., 1981.

Patten, B. M.: Varying developing mechanisms in teratology. *Pediatrics 19*:734, 1957.

Persaud, T. V. N.: *Problems of Birth Defects. From Hippocrates to Thalidomide and After.* Baltimore, University Park Press, 1977.

Persaud, T. V. N.: *Teratogenesis. Experimental Aspects and Clinical Implications.* Jena, Gustav Fischer Verlag, 1979.

Persaud, T. V. N., and Ellington, A. C.: Teratogenic activity of cannabis resin. *Lancet 2*:406, 1968.

Persaud, T. V. N., and Moore, K. L.: Causes and prenatal diagnosis of congenital abnormalities. *J. Obstet. Gynecol. Nursing 3*:40, 1974.

Poswillo, S.: Mechanisms and pathogenesis of malformation. *Br. Med. J. 32*:59, 1976.

Punnett, H. H., Kistenmacher, M. L., Toro-Sola, M. A., and Kohn, G.: Quinacrine fluorescence and giemsa banding in trisomy 22. *Theoret. Appl. Genet. 43*:134, 1973.

Rasmussen, F.: The oto-toxic effect of streptomycin and dihydrostreptomycin on the foetus. *Scand. J. Respir. Dis. 50*:61, 1969.

Ray, M.: Autosomal deficiencies in humans. *The Nucleus (Suppl.)*:275, 1968.

Rendle-Short, T. J.: Tetracycline in teeth and bone. *Lancet 1*:118, 1962.

Rennert, O. M.: Irradiation and radiation exposure. *Clin. Obstet. Gynecol. 18*:177, 1975.

Rubin, A.: *Handbook of Congenital Malformations.* Philadelphia, W. B. Saunders Co., 1967.

Saxén, L., and Rapola, J.: *Congenital Defects.* New York, Holt, Rinehart and Winston, Inc., 1969, pp. 35–75.

Seeman, P., Sellers, E. M., and Roschlan, W. H. (Eds.): *Principles of Medical Pharmacology.* 3rd ed. Toronto, Univ. of Toronto Press, 1980.

Sever, J. L.: Rubella and cytomegalovirus; *in* F. C. Fraser, V. A. McKusick, and R. Robinson (Eds.): *Congenital Malformations.* Proc. Third, Internat. Conf. New York, Excerpta Medica Foundation, 1970, p. 377.

Sever, J. L., Nelson, K. B., and Gilkeson, M. R.: Rubella epidemic 1964: Effect on 6000 pregnancies. *Am. J. Dis. Child. 110*:395, 1964.

Shaw, E. B., and Steinbach, H. L.: Aminopterin-induced fetal malformation: Survival of infant after attempted abortion. *Am. J. Dis. Child 115*:477, 1968.

Smith, D. W.: Autosomal abnormalities. *Am. J. Obstet. Gynecol. 90*:1055, 1964.

Smith, D. W.: The 18 trisomy and D₁ trisomy syndromes; *in* L. I. Gardner (Ed.): *Endocrine and Genetic Diseases of Childhood.* 2nd ed. Philadelphia, W. B. Saunders Co., 1975, pp. 715–729.

Smith, D. W.: *Recognizable Patterns of Human Malformation: Genetic, Embryologic, and Clinical Aspects.* 2nd ed. Philadelphia, W. B. Saunders Co., 1976.

Smith, D. W., Bierman, E. L., and Robinson, N. M. (Eds.): *The Biologic Ages of Man: From Conception Through Old Age.* 2nd ed. Philadelphia, W. B. Saunders Co., 1978.

Smithels, R. W.: Drugs and human malformations. *Adv. Teratol. 1*:251, 1966.

Smithels, R. W.: Environmental teratogens of man. *Br. Med. J. 32*:27, 1976.

Snyder, R. D.: Congenital mercury poisoning. *N. Engl. J. Med. 284*:1014, 1971.

Sokal, J. E., and Lessmann, E. M.: Effects of cancer chemotherapeutic agents on the human fetus. *J.A.M.A. 172*:1765, 1960.

South, M. A., Tompkins, W. A. F., Morris, C. R., and Rawls, W. E.: Congenital malformations of the central nervous system associated with genital type (type 2) herpesvirus. *J. Pediatr. 75*:13, 1969.

Speidel, B. D., and Meadow, S. R.: Maternal epilepsy and abnormalities of the fetus and newborn. *Lancet 2*:839, 1972.

Stempfel, R. S., Jr.: Disorders of sexual development; *in* L. I. Gardner (Ed.): *Endocrine and Genetic Diseases of Childhood.* 2nd ed. Philadelphia, W. B. Saunders Co., 1975, pp. 551–570.

Streissguth, A. P., Hanson, J. W., et al.: The effects of moderate alcohol consumption on fetal growth and morphogenesis. Fifth International Conference on Birth Defects. Amsterdam, Excerpta Medica, 1977.

Sutherland, J. S., and Light, I. J.: The effects of drugs upon the developing fetus. *Pediatr. Clin. North Am. 12*:781, 1965.

Thiersch, J. B.: Therapeutic abortions with a folic acid antagonist, 4-aminopteroylglutamic acid (4-Amino-PGA), administered by the oral route. *Am. J. Obstet. Gynecol.* 63:1298, 1952.

Thornburn, M. J., and Johnson, B. E.: Apparent monosomy of a G autosome in a Jamaican infant. *J. Med. Genet.* 3:290, 1966.

Tuchmann-Duplessis, H.: The effects of teratogenic drugs; *in* E. E. Philipp, J. Barnes, and M. Newton (Eds.): *Scientific Foundations of Obstetrics and Gynecology.* London, William Heinemann, Ltd., 1970, pp. 636–648.

Uchida, I. A.: Radiation-induced nondisjunction. *Environ. Health Perspect.* 31:13, 1979.

Uchida, I. A., and Summitt, R. L.: Chromosomes and their abnormalities; *in* V. C. Vaughan, R. J. McKay, and R. D. Behrman, (Eds.): *Nelson Textbook of Pediatrics.* 11th ed. Philadelphia, W. B. Saunders Co., 1979, pp. 344–369.

Uchida, I. A., Ray, M., McRae, K. N., and Besant, D. F.: Familial occurrence of trisomy 22. *Am. J. Hum. Genet.* 20:107, 1968.

Vaughan, V. C., McKay, R. J., and Behrman, R. E.: *Nelson Textbook of Pediatrics.* 11th ed. Philadelphia, W. B. Saunders Co., 1979.

Venning, G. R.: The problem of human foetal abnormalities with special reference to sex hormones; *in* J. M. Robson, F. M. Sullivan, and R. L. Smith (Eds.): *Embryopathic Activity of Drugs.* London, J. & A. Churchill, Ltd., 1965, pp. 94–104.

Vorherr, H., Vorherr, U. F., Mehta, P., Ulrich, J. A., and Messer, R. H.: Vaginal absorption of povidone-iodine. *J.A.M.A.* 244:2628, 1980.

Warkany, J.: Congenital malformations induced by maternal dietary deficiency; Experiments and their interpretations. *Harvey Lect.* 48:383, 1954.

Warkany, J.: *Congenital Malformations: Notes and Comments.* Chicago, Year Book Medical Publishers, Inc., 1971.

Warkany, J.: Warfarin embryopathy. *Teratology* 14:205, 1976.

Warkany, J., Beaudry, P. H., and Hornstein, S.: Attempted abortion with 4-aminopterolyglutamic acid (Aminopterin): Malformations of the child. *Am. J. Dis. Child.* 97:274, 1960.

Weiss, B., and Doherty, R. A.: Methylmercury poisoning. *Teratology.* 12:311, 1975.

White, L. R., and Sever, J. L.: Etiology agents. I. Infectious agents; *in* A. Rubin (Ed.): *Handbook of Congenital Malformations.* Philadelphia, W. B. Saunders Co., 1967, pp. 353–364.

Wickes, I. G.: Foetal defects following insulin coma therapy in early pregnancy. *Br. Med. J.* 2:1029, 1954.

Wilkins, L., Jones, H. W., Jr., Holman, G. H., and Stempfel, R. S., Jr.: Masculinization of the female fetus in association with administration of oral and intramuscular progestins during gestation; nonadrenal female pseudohermaphroditism. *J. Clin. Endocrinol. Metab.* 18:559, 1958.

Wilson, J. G.: *Environment and Birth Defects.* New York, Academic Press, Inc., 1973a.

Wilson, J. G.: Present status of drugs as teratogens in man. *Teratology.* 7:3, 1973b.

Wilson, J. G.: Mechanisms of teratogenesis. *Am. J. Anat.* 136:129, 1973c.

Witkop, C. J., Jr., Wolf, R. O., and Mehaffey, M. H.: The frequency of discolored teeth showing yellow fluorescence under ultraviolet light. *J. Oral Therap. Pharmacol.* 2:81, 1965.

Zellweger, H., McDonald, J. S., and Abbo, G.: Is lysergic acid diethylamide a teratogen? *Lancet* 2:1066, 1967.

9

BODY CAVITIES, PRIMITIVE MESENTERIES, AND DIAPHRAGM

Early development of the intraembryonic coelom, or body cavity, is described in Chapter 4 (see Fig. 4–7). By the fourth week, it appears as a horseshoe-shaped cavity in the cardiogenic and lateral mesoderm (Fig. 9–1). The curve, or bend, in this cavity represents the future *pericardial cavity,* and its limbs indicate the future *pleural and peritoneal cavities*. The greater part of each limb of the cavity opens into the extraembryonic coelom at the lateral edges of the embryonic disc (Fig. 9–1A and B). This communication is important, for, as described in Chapter 12, most of the midgut herniates through this communication into the umbilical cord (see Figs. 12–6 and 12–12), where it develops into most of the small intestine and part of the large intestine.

In embryos of lower forms of animals, the intraembryonic coelom provides short-term storage for excretory products. In the human embryo, the coelom primarily provides room for organ development and movement. The communication between the intraembryonic and extraembryonic coeloms, or cavities, may aid in the transfer of fluid and nutrients to the early embryo, particularly during establishment of the uteroplacental circulation.

During transverse folding of the embryo, the limbs or lateral parts of the intraembryonic coelom are brought together on the ventral aspect of the embryo (Fig. 9–2C and F). In the region of the future peritoneal cavity, the ventral mesentery degenerates, resulting in a large embryonic peritoneal cavity extending from inferior to the heart to the pelvic region (Fig. 9–2F).

Three well-defined coelomic or body cavities are now recognizable: (1) a large *pericardial cavity* (Fig. 9–2B and E); (2) two relatively small *pericardioperitoneal canals*

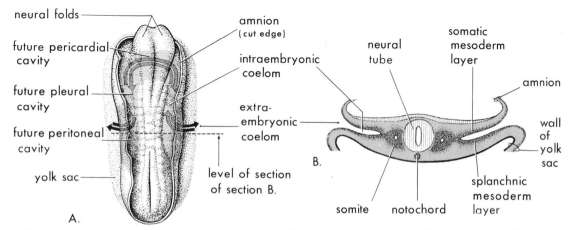

Figure 9–1 *A*, Diagram of embryo during stage 10 of development (about 22 days) showing the outline of the horseshoe-shaped intraembryonic coelom. The amnion has been removed and the coelom is shown as if the embryo were translucent. The continuity of the coelom, as well as the communication of its right and left extremities with the extraembryonic coelom, is indicated by arrows. *B*, Transverse section through the embryo at the level shown in *A*.

167

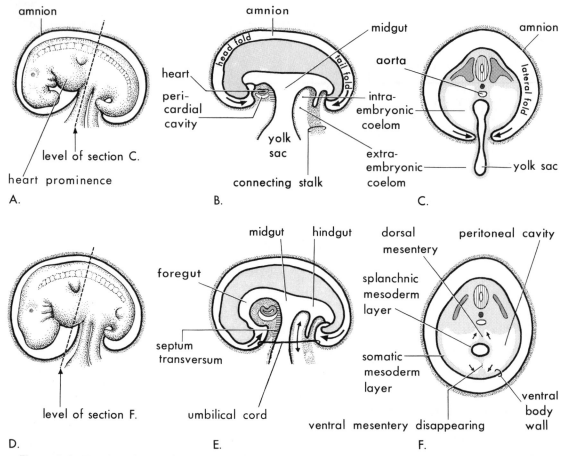

Figure 9–2 Drawings illustrating folding of the embryo and its effects on the intraembryonic coelom and other structures. *A*, Lateral view of an embryo during stage 12 of development (about 26 days). *B*, Schematic longitudinal section of this embryo showing the head and tail folds. *C*, Transverse section at the level shown in *A*, indicating how fusion of the lateral folds gives the embryo a cylindrical form. *D*, Lateral view of an embryo during stage 13 (about 28 days). *E*, Schematic longitudinal section of this embryo showing the reduced communication between the intraembryonic and extraembryonic coeloms (double-headed arrow). *F*, Transverse section as indicated in *D*, illustrating formation of the ventral body wall and disappearance of the ventral mesentery. The arrows indicate the junction of the somatic and splanchnic mesoderm layers. The somatic mesoderm layer will become the parietal peritoneum, which lines the abdominal wall, and the splanchnic mesoderm layer will become the visceral peritoneum covering an organ (e.g., the stomach).

connecting the pericardial and peritoneal cavities (Figs. 9–3*B* and 9–4); and (3) a large *peritoneal cavity* (Figs. 9–2*F* and 9–3*C* to *E*). These cavities possess a parietal wall lined by mesothelium derived from the somatic mesoderm, and a visceral wall covered by mesothelium derived from the splanchnic mesoderm (Fig. 9–3*E*). The peritoneal cavity becomes separated from the extraembryonic coelom at the umbilicus when the intestines return to the abdomen from the umbilical cord during the tenth week (see Chapter 12 and Fig. 12–11).

With formation of the head fold, the heart and pericardial cavity are carried ventrally

and caudally, anterior to the foregut (Figs. 9–2*B* and *E* and 9–3*A* and *B*). The pericardial cavity then opens dorsally into the pericardioperitoneal canals, which pass dorsal to the *septum transversum* on each side of the foregut (Fig. 9–4*B* and *D*).

After folding of the embryo, the caudal part of the foregut, midgut, and hindgut are suspended in the peritoneal cavity from the posterior abdominal wall by the *dorsal mesentery* (Figs. 9–2*F* and 9–3*C* to *E*). A mesentery is a *double fold of peritoneum*. Transiently, the dorsal and ventral mesenteries divide the peritoneal cavity into right and left halves, but the ventral mesentery soon dis-

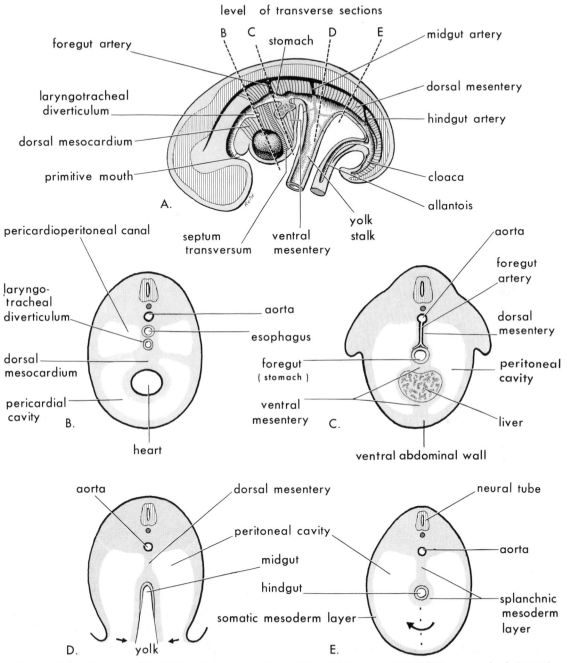

Figure 9–3 Diagrams illustrating the mesenteries at the beginning of the fifth week. *A,* Schematic longitudinal section. Note that the dorsal mesentery serves as a pathway for the arteries supplying the developing gut. Nerves and lymphatics also pass between the layers of the dorsal mesentery. *B to E,* Transverse sections through the embryo at the levels indicated in *A.* The ventral mesentery disappears, except in the region of the terminal esophagus, the stomach, and the first part of the duodenum.

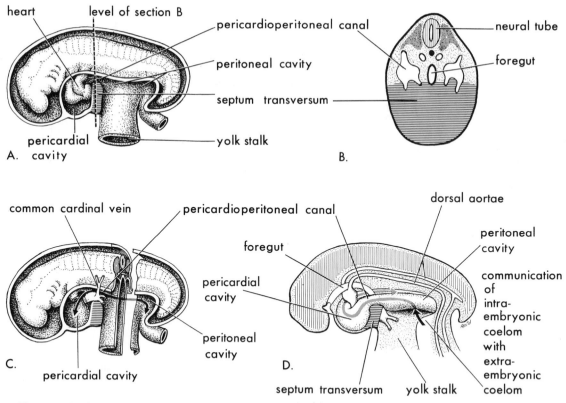

Figure 9–4 Schematic drawings of an embryo during stage 11 of development (about 24 days). *A*, The lateral wall of the pericardial cavity has been removed to show the primitive heart. *B*, Transverse section illustrating the relationship of the pericardioperitoneal canals to the septum transversum (primordium of diaphragm, in part) and the foregut. *C*, Lateral view with the heart removed. The embryo has been sectioned transversely to show the continuity of the intraembryonic and extraembryonic coeloms. *D*, Sketch showing the pericardioperitoneal canals arising from the dorsal wall of the pericardial cavity and passing on each side of the foregut to join the peritoneal cavity. The arrows show the communication of the extraembryonic coelom with the intraembryonic coelom and the continuity of the intraembryonic coelom at this stage.

appears, except where it is attached to the caudal part of the foregut. The peritoneal cavity then becomes a large continuous space (Figs. 9–3 and 9–4). The arteries supplying the primitive gut, i.e., the celiac (foregut), the superior mesenteric (midgut), and the inferior mesenteric (hindgut), pass between the layers of the dorsal mesentery (Fig. 9–3).

DIVISION OF THE COELOM

Each pericardioperitoneal canal lies lateral to the foregut (future esophagus) and dorsal to the septum transversum (Fig. 9–4B). Partitions form concurrently in each pericardioperitoneal canal and separate the pericardial cavity from the pleural cavities and the pleural cavities from the peritoneal cavity.

Owing to lateral growth of the *lung buds* (developing lungs) into the pericardioperi-

toneal canals (Fig. 9–5A), a pair of ridges is produced in the lateral wall of each canal. The cranial ridges, or *pleuropericardial membranes,* are located superior to the developing lungs, and the caudal ridges, or pleuroperitoneal membranes, are located inferior to them.

The Pleuropericardial Membranes (Fig. 9–5). *These cranial partitions gradually separate the pericardial cavity from the pleural cavities.* Initially, this pair of membranes appears as ridges or bulges of mesenchyme containing the *common cardinal veins,* which pass to the heart (Fig. 9–5A). These veins drain the primitive venous system into the sinus venosus of the primitive heart (see Figs. 14–1 and 14–2).

At this stage, the lung buds are very small relative to the heart and pericardial cavity. They grow laterally from the caudal end of the trachea into the corresponding pericardioperitoneal canal (pleural canal), which becomes the *primitive pleural cavity* (Fig. 9–5B and C). As these cavities expand ventrally around the heart, they extend into the body wall and split the mesenchyme into (1) an outer layer that becomes the thoracic wall and (2) an inner layer (the pleuropericardial membrane) that becomes the *fibrous pericardium,* the outer fibrous layer of the membranous sac enclosing the heart (Fig. 9–5C and D).

Initially, the pleuropericardial membranes project into the cranial ends of the pericardioperitoneal canals (Fig. 9–5A). With subsequent growth of the common cardinal veins, descent of the heart, and expansion of the pleural cavities, these membranes become mesentery-like folds extending from the lateral thoracic wall (Fig. 9–5B). By the seventh week, the pleuropericardial membranes fuse with the mesoderm ventral to the esophagus, or *primitive mediastinum,* separating the pericardial cavity from the pleural cavities (Fig. 9–5C). The mediastinum in the embryo is filled with a mass of mesenchyme (embryonic connective tissue). It extends from the sternum to the vertebral column and separates the developing lungs.

The right pleuropericardial opening closes slightly earlier than the one on the left,

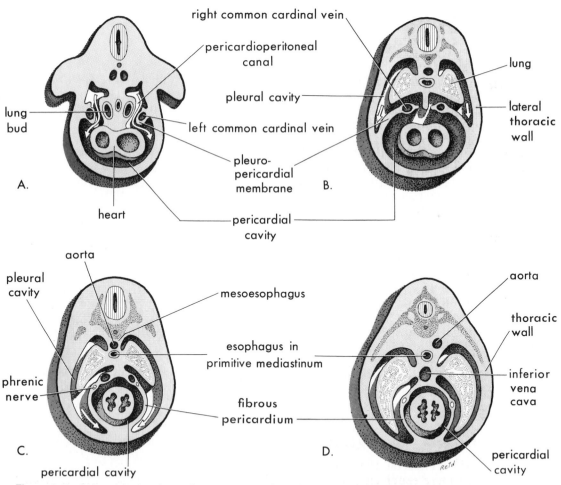

Figure 9–5 Schematic drawings of transverse sections through embryos cranial to the septum transversum, illustrating successive stages in the separation of the pleural cavities from the pericardial cavity. Growth and development of the lungs, expansion of the pleural cavities, and formation of the fibrous pericardium are also shown. *A,* Five weeks. The arrows indicate the communications between the pericardioperitoneal canals and the pericardial cavity. *B,* Six weeks. The arrows indicate development of the pleural cavities as expansions into the body wall. *C,* Seven weeks. Expansion of the pleural cavities ventrally around the heart is shown. The pleuropericardial membranes are now fused in the midline with each other and with the mesoderm ventral to the esophagus. *D,* Eight weeks. Continued expansion of the lungs and pleural cavities and formation of the fibrous pericardium and chest wall are illustrated.

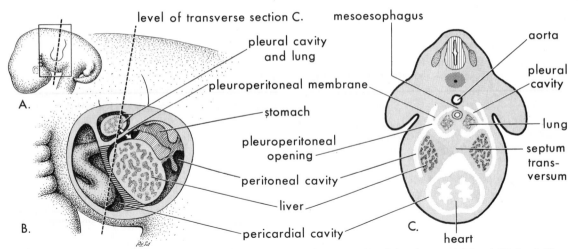

Figure 9–6 *A*, Sketch of a lateral view of an embryo during stage 15 of development (about 33 days). The rectangle indicates the area enlarged in the drawing below it. *B*, The primitive coelom or body cavities viewed from the left side after removal of the lateral body wall. *C*, Transverse section through the embryo at the level shown in *A* and *B*.

probably because the right common cardinal vein is larger than the left one and produces a larger pleuropericardial membrane.

The Pleuroperitoneal Membranes. (Figs. 9–6 and 9–7). *These caudal partitions gradually separate the pleural cavities from the peritoneal cavity.* This pair of membranes is mainly produced as the developing lungs and pleural cavities expand by invading the body wall. They are attached dorsolaterally to the body wall, and their crescentic free edges initially project into the caudal ends of the pericardioperitoneal canals. They become relatively more prominent as a result of the growth of the lungs cranial to them and the liver caudal to them.

During the sixth week, the pleuroperitoneal membranes extend medially and ventrally until their free edges fuse with the dorsal mesentery of the esophagus and with the septum transversum (Fig. 9–7C). This separates the pleural cavities from the peritoneal cavity. Closure of the pleuroperitoneal openings is assisted by growth of myoblasts (primitive muscle cells) into these membranes, which form the posterolateral parts of the diaphragm (Fig. 9–7). The opening on the right side closes slightly before the one on the left; the reason for this is uncertain, but it might be related to the large size of the liver at this stage of development.

DEVELOPMENT OF THE DIAPHRAGM

This dome-shaped, musculotendinous partition separating the thoracic and abdomino-

pelvic cavities mainly *develops from four structures* (Fig. 9–7):

1. The Septum Transversum. This transverse septum, composed of mesoderm, forms the *central tendon* of the diaphragm (Fig. 9–7E). The septum transversum is first identifiable at the end of the third week as a mass of mesoderm cranial to the pericardial cavity (see Fig. 5–1A). After the head fold forms during the fourth week, the septum transversum becomes a thick incomplete partition, or partial diaphragm, between the pericardial and abdominopelvic cavities (Fig. 9–4). The septum transversum fuses dorsally with the mesenchyme ventral to the esophagus (primitive mediastinum) and with the pleuroperitoneal membranes (Fig. 9–7C).

2. The Pleuroperitoneal Membranes. These membranes fuse with the dorsal mesentery of the esophagus and with the dorsal portion of the septum transversum (Fig. 9–7C). This completes the partition between the thoracic and abdominopelvic cavities and forms the primitive diaphragm.

Although the pleuroperitoneal membranes form large portions of the primitive diaphragm, they represent relatively small intermediate portions of the fully developed diaphragm (Fig. 9–7E).

3. Dorsal Mesentery of Esophagus. As previously described, the septum transversum and the pleuroperitoneal membranes fuse with the dorsal mesentery of the esophagus (dorsal mesoesophagus). This mesentery constitutes the median portion of the diaphragm. The *crura of the diaphragm*

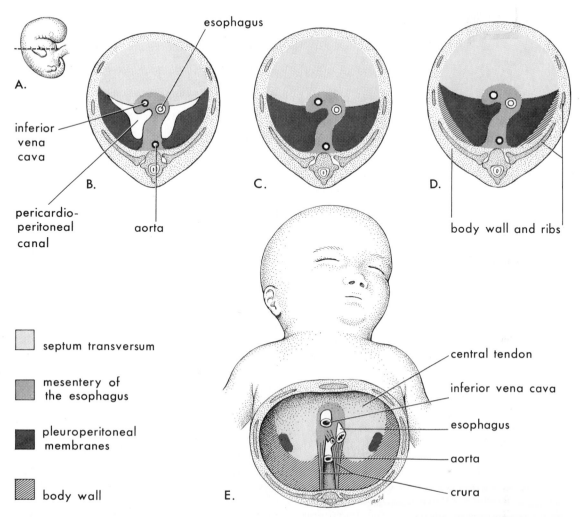

esophagus

A.

inferior
vena
cava

B.

pericardio-
peritoneal
canal

aorta

C.

D.

body wall and ribs

septum transversum

mesentery of
the esophagus

pleuroperitoneal
membranes

body wall

E.

central tendon

inferior vena cava

esophagus

aorta

crura

Figure 9–7   Drawings illustrating development of the diaphragm. *A,* Sketch of a lateral view of an embryo at the end of the fifth week (*actual size*) indicating the level of section. *B* to *E* show the diaphragm as viewed from below. *B,* Transverse section showing the unfused pleuroperitoneal membranes. *C,* Similar section at the end of the sixth week after fusion of the pleuroperitoneal membranes with the other two diaphragmatic components. *D,* Transverse section through a 12-week embryo after ingrowth of the fourth diaphragmatic component from the body wall. *E,* View of the diaphragm of a newborn infant, indicating the probable embryological origin of its components.

develop from muscle fibers that grow into the dorsal mesentery of the esophagus (Fig. 9–7*E*).

4. The Body Wall.   During the ninth to twelfth weeks, the lungs and pleural cavities enlarge and burrow into the lateral body walls (Fig. 9–5). During this "excavation" process, body-wall tissue is split into two layers: (1) an outer layer that will form part of the definite body wall, and (2) an inner layer that contributes to peripheral portions of the diaphragm, external to the portions derived from the pleuroperitoneal membranes (Fig. 9–7*D*).

Further extension of the pleural cavities into the body walls forms the right and left

costodiaphragmatic recesses and establishes the characteristic dome-shaped configuration of the adult diaphragm (Fig. 9–8). After birth, the costodiaphragmatic recesses become alternately smaller and larger as the lungs move in and out of them during inspiration and expiration.

POSITIONAL CHANGES AND INNERVATION OF THE DIAPHRAGM

During the fourth week, the septum transversum lies opposite the third, fourth, and fifth *cervical somites* (Fig. 9–9*A*). During the fifth week, myoblasts from the *myotomes* of these somites migrate into the developing

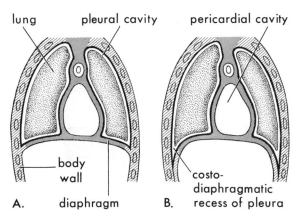

A.

B.

Figure 9–8 Diagrams illustrating the extension of the pleural cavities into the body walls to form the peripheral portions of the diaphragm, the costodiaphragmatic recesses, and the establishment of the characteristic dome-shaped configuration of the diaphragm. Note that the body wall tissue is added to the diaphragm peripherally as the lungs and pleural cavities enlarge.

diaphragm and bring their nerves with them. Thus, the supply of the diaphragm is from the *third, fourth, and fifth cervical nerves,* which are contained in the *phrenic nerves.* These nerves pass to the septum transversum via the pleuropericardial membranes. This explains why the phrenic nerves subsequently come to lie on the fibrous pericardium (Fig. 9–5C and D).

Rapid growth of the dorsal part of the embryo's body compared with the ventral part results in an apparent migration or descent of the diaphragm. By the sixth week, the developing diaphragm is at the level of the thoracic somites (Fig. 9–9B). The phrenic nerves now take a descending course, and, as the diaphragm "moves" relatively farther caudally in the body, these nerves are correspondingly lengthened. By the beginning of the eighth week, the dorsal part of the diaphragm lies at the level of the first lumbar vertebra (Fig. 9–9C).

As the parts of the diaphragm fuse (Fig.

9–7), mesenchyme of the septum transversum extends into the other parts and forms myoblasts that differentiate into the muscle of the diaphragm. Hence, *the motor nerve supply to the diaphragm is via the phrenic nerves (ventral rami of C3, 4, 5).*

The phrenic nerve is also sensory to the central region of the diaphragm, but the peripheral region, which develops from the body wall (Fig. 9–7E), receives sensory nerves from the lower six or seven intercostal nerves.

CONGENITAL MALFORMATIONS

Congenital Diaphragmatic Hernia (Figs. 9–10 and 9–11). Posterolateral defect of the diaphragm is the only relatively common congenital abnormality of the diaphragm. It occurs about once in 2000 births and leads to *herniation of abdominal contents* into the thoracic cavity. This abnormality results *from defective formation and/or*

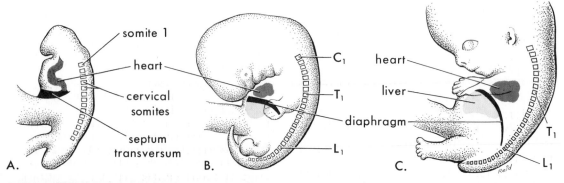

A.   B.   C.

Figure 9–9 Diagrams illustrating positional changes of the developing diaphragm. *A,* Stage 11 of development (about 24 days). The septum transversum (primordium of the diaphragm, in part) is at the level of the third, fourth, and fifth cervical segments. *B,* Stage 17 of development (about 41 days). *C,* Stage 21 of development (about 52 days).

vertebra   defect

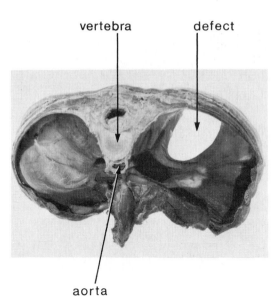

aorta

Figure 9–10   Photograph of a transverse section through the thoracic region of a newborn infant, viewed from above. Note the large left posterolateral defect of the diaphragm. *Half actual size.*

fusion of the pleuroperitoneal membrane, which normally separates the pleural and peritoneal cavities. The defect, usually unilateral, consists of a large opening (often called the *foramen of Bochdalek*) in the posterior or the posterolateral region of the diaphragm (Fig. 9–10). *The defect occurs five times more often on the left side than on the right* (Verhagen, 1967); this is likely related to the earlier closure of the right pleuroperitoneal opening. Normally the pleuroperitoneal membranes fuse with other diaphragmatic components by the end of the sixth week (Fig. 9–7C). If a pleuroperitoneal membrane is unfused when the intestines return to the abdomen from the umbilical cord during the tenth week (see Fig. 12–11), the intestines pass into the thorax. Often, the stomach, spleen, and most of the intestines herniate into the thorax (Fig. 9–11). In rare instances, the liver and kidneys also pass into the thoracic cavity and *displace the lungs and heart.* Sometimes, the viscera can move freely through the defect; they may therefore be in the thoracic cavity, especially when the infant is lying down, or in the abdominal cavity when the infant is upright.

If the abdominal viscera are in the thoracic cavity at birth, the initiation of respiration may be impaired. Because the abdominal organs are most often in the left side of the thorax, the heart and mediastinum are usually displaced to the right. The lungs are often hypoplastic and greatly reduced in size. The growth retardation of the lungs results from lack of room in the thorax for them to develop normally. The affected lung is usually aerated and often achieves its normal size after reduction of the herniated viscera and repair of the defect in the diaphragm.

Eventration of the Diaphragm.   In this rare condition, half the diaphragm has defective musculature and balloons into the thoracic cavity as an aponeurotic sheet. As a result, there is an upward displacement of abdominal contents into an outpouching of the diaphragm. This malformation results mainly from failure of muscular tissue to extend into the pleuroperitoneal membrane on the affected side.

Eventration is not a herniation, but is an upward displacement into a sac-like structure. The clinical manifestations of an eventration may simulate a congenital diaphragmatic hernia (Vaughan et al., 1979).

Congenital Epigastric Hernia.   These uncommon hernias occur in the midline between the xiphoid process and the umbilicus. They are similar to umbilical hernias except for their location. Epigastric hernias result from failure of the lateral body folds to fuse completely during transverse folding in the fourth week (Fig. 9–2C and F).

Congenital Hiatal Hernia.   Rarely, there may be partial congenital herniation of the stomach through an excessively large esophageal hiatus. Although hiatal hernia is an acquired lesion usually occurring during adult life, a congenitally enlarged esophageal hiatus may be the predisposing factor.

Retrosternal Hernia.   Large herniations may occur rarely through the *sternocostal hiatus* (foramen of Morgagni) for the superior epigastric vessels in the retrosternal area. There may be herniation of intestine into the pericardial sac (Vaughan et al., 1979), or, conversely, the heart may pass into the peritoneal cavity in the epigastric region. This large defect is commonly associated with defects in the umbilical region.

Fatty herniations through the sternocostal hiatus are often observed by radiologists and pathologists in the retrosternal areal, but they are of no clinical significance.

Congenital Pericardial Defects.   Defective formation and/or fusion of the pleuropericardial membranes, which normally separate the pericardial and pleural cavities, is a rare abnormality; this defect results in a congenital defect of the pericardium, usually on the left side. The pericardial cavity communicates with the pleural cavity, and in very rare cases, a portion of the atrium herniates into the pleural cavity at each heartbeat (Gray and Skandalakis, 1972).

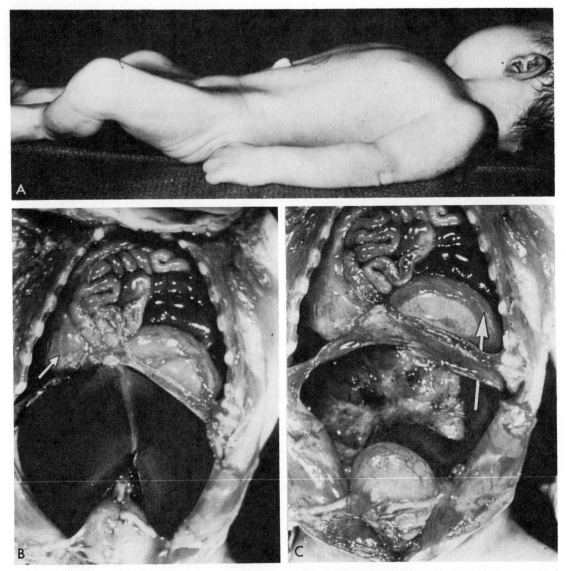

Figure 9–11   A, Photograph of an infant with a large left posterolateral diaphragmatic defect similar to that shown in Figure 9–10. Note the relatively flat abdomen resulting from herniation of the abdominal contents into the thorax through the defect. B, The thoracic and abdominal cavities opened at autopsy to show the intestines and other viscera in the thoracic cavity. The arrow indicates the displaced heart. C, The liver has been removed to show that only distal parts of the large intestine have remained in the abdominal cavity. The arrow passes through the diaphragmatic defect. (Courtesy of Dr. Jan Hoogstraten, Children's Centre, Winnipeg, Canada.)

SUMMARY

The *intraembryonic coelom* begins to develop near the end of the third week. By the beginning of the fourth week, it appears as a continuous *horseshoe-shaped cavity* in the cardiogenic and lateral mesoderm. The curve, or bend, of the "horseshoe" represents the future *pericardial cavity,* and the lateral parts, or limbs, represent the future *pleural and peritoneal cavities.*

During folding of the embryonic disc in the fourth week, the lateral parts of the coelom are brought together on the ventral aspect of the embryo, where they merge in the region of the future peritoneal cavity. As the peritoneal portions of the intraembryonic coelom come together, the splanchnic mesoderm encloses the primitive gut and suspends it from the dorsal body wall by a double-layered membrane known as the *dorsal mesentery.*

Until the seventh week, the pericardial

cavity communicates with the peritoneal cavity through paired *pericardioperitoneal canals*. During the fifth and sixth weeks, partitions, or membranes, form near the cranial and caudal ends of these canals. The cranial *pleuropericardial membranes* separate the pericardial cavity from the pleural cavities, and the caudal *pleuroperitoneal membranes* separate the pleural cavities from the peritoneal cavity.

The diaphragm develops from four main structures: (1) the septum transversum, (2) the pleuroperitoneal membranes, (3) the dorsal mesentery of the esophagus, and (4) the body wall.

Posterolateral defect of the diaphragm is the common type of congenital diaphragmatic defect, and it is usually associated with herniation of abdominal viscera into the thoracic cavity (congenital diaphragmatic hernia). The defect occurs five times more often on the left side than on the right and results from failure of the pleuroperitoneal membrane on the affected side to fuse with the other diaphragmatic components and separate the pleural and peritoneal cavities.

SUGGESTIONS FOR FURTHER READING

Gray, S. W., and Skandalakis, J. E.: The diaphragm: *in* S. W. Gray and J. E. Skandalakis (Eds.): The Embryological Basis For the Treatment of Congenital Defects. Philadelphia, W. B. Saunders Co., 1972, pp. 359–383. A well-organized account of the development of the diaphragm and a critical review of the embryogenesis of congenital malformations of the diaphragm. There is also a discussion of the diagnosis of these malformations and the principles underlying their correction.

Hamilton, W. J., and Mossman, H. W.: *Human Embryology. Prenatal Development of Form and Function.* 4th ed. Cambridge, W. Heffner and Sons, Ltd., 1972, pp. 78–80; 291–293; 365–374. Excellent accounts of the development of the body cavities and the diaphragm.

Schaffer, A. J., and Avery, M. E.: Disorders of the diaphragm: *in* A. J. Schaffer and M. E. Avery (Eds.): *Diseases of the Newborn.* 4th ed. Philadelphia, W. B. Saunders Co., 1977, pp. 183–192; 330–336. Contains short descriptions of congenital diaphragmatic defects and their differential diagnosis.

CLINICALLY ORIENTED PROBLEMS

1. A newborn infant suffered from severe *respiratory distress.* The abdomen was unusually small and intestinal peristaltic movements were heard over the left side of the chest. What congenital malformation would you suspect? Explain the signs described above. How would the diagnosis likely be established?

2. What congenital malformation could result in herniation of intestine into the pericardial cavity? What is the embryological basis of this abnormality?

3. How common is *posterolateral defect of the diaphragm?* How do you think a newborn infant in whom this diagnosis is suspected should be positioned (i.e., tilted up or down)? Why would this positional treatment be given?

4. A baby was born with a hernia in the midline, between the xiphoid process and the umbilicus. Name this type of hernia. Is it common? What is the embryological basis of this congenital malformation?

The answers to these questions are given at the back of the book.

REFERENCES

Andersen, D. M.: Effect of diet during pregnancy on the incidence of congenital hereditary diaphragmatic hernia in the rat. *Am. J. Pathol. 25*:163, 1949.

Areechon, W., and Reid, L.: Hypoplasia of the lung with congenital diaphragmatic hernia. *Br. Med. J. 1*:230, 1963.

Arey, L. B.: *Development Anatomy: A Textbook and Laboratory Manual of Embryology.* Revised 7th ed. Philadelphia, W. B. Saunders Co., 1974, pp. 272–294.

Avery, M. E., Fletcher, B. D., and Williams, R. G.: *The Lung and Its Disorders in the Newborn Infant.* 4th ed. Philadelphia, W. B. Saunders Co., 1981, 182–184.

Bremer, J. L.: The diaphragm and diaphragmatic hernia. *Arch. Pathol. 36*:539, 1943.

Butler, N., and Claireaux, A. E.: Congenital diaphragmatic hernia as a cause of perinatal mortality. *Lancet 1*:659, 1962.

Corliss, C. E.: *Patten's Human Embryology. Elements of Clinical Development.* New York, McGraw-Hill Book Co., 1976, pp. 307–324.

Ellis, K., Leeds, N. E., and Himmelstein, A.: Congenital deficiencies in the parietal pericardium. A review with two new cases including successful diagnosis by plain roentgenography. *Am. J. Roentgenol. 82*:125, 1959.

Fosberg, R. G., Jakubiak, J. W., and Delaney, T. B.: Congenital partial absence of the pericardium. *Ann. Thorac. Surg. 5*:171, 1968.

Gross, R. E.: *The Surgery of Infancy and Childhood.* Philadelphia, W. B. Saunders Co., 1953.

Laxdal, O. E., McDougall, H., and Mellen, G. W.: Congenital eventration of the diaphragm. *N. Engl. J. Med. 250*:401, 1954.

McNamara, J. J., Eraklis, A. J., and Gross, R. E.: Congenital posterolateral diaphragmatic hernia in the newborn. *J. Thorac. Cardiovasc. Surg. 55*:55, 1968.

Scothorne, R. J.: Early development: *in* R. Passmore and J. S. Robson (Eds.): *A Companion to Medical Studies.* Vol. I: *Anatomy, Biochemistry, Physiology and Related Subjects.* Oxford, Blackwell Scientific Publications, 1968, pp. 18.4–18.13.

Swenson, O.: *Pediatric Surgery.* New York, Appleton-Century-Crofts, Inc., 1958, pp. 196–201.

Tarnay, T. J.: Diaphragmatic hernia. *Ann. Thorac. Surg. 5*:66, 1968.

Vaughan, V. C. McKay, R. J., and Behrman, R. E.: *Nelson Textbook of Pediatrics.* 11th ed. Philadelphia, W. B. Saunders Co., 1979.

Verhagen, A. R.: Respiratory system; *in* A. Rubin (Ed.): *Handbook of Congenital Malformations.* Philadelphia, W. B. Saunders Co., 1967, pp. 157–169.

Warkany, J., Roth, C. B., and Wilson, J. G.: Multiple congenital malformations: A consideration of etiologic factors. *Pediatrics 1*:462, 1948.

Wells, L. J.: Development of the human diaphragm and pleural sacs. *Contr. Embryol. Carneg. Instn. 35*:107, 1954.

THE BRANCHIAL APPARATUS AND THE HEAD AND NECK

The branchial apparatus consisting of (1) *branchial arches,* (2) *pharyngeal pouches,* (3) *branchial grooves,* and (4) *branchial membranes* (Fig. 10–1), contributes greatly to the formation of the head and neck. The cranial region of an early human embryo during the fourth week somewhat resembles a fish embryo of a comparable stage. By the end of the embryonic period, these ancestral structures have either become rearranged and adapted to new functions or disappeared. *Most congenital malformations of the head and neck originate during transformation of the branchial apparatus into adult derivatives.*

Studying the development of the human branchial apparatus may be confusing if the function of the branchial apparatus in lower forms of life is not understood first. In fish and larval amphibians, the branchial apparatus forms a system of gills for exchanging oxygen and carbon dioxide between the blood and the water. (The adjective "branchial" is from the Greek *branchia,* meaning "gill.") The branchial arches support the gills.

A branchial apparatus develops in human embryos, but no gills form. Because gills do not form in the human embryo, some authors prefer to use the term *pharyngeal arch* instead of *branchial arch,* but the term chosen in the *Nomina Embryologica* is the one used in this book.

THE BRANCHIAL ARCHES

Branchial arches begin to develop early in the fourth week as *neural crest cells* (see Fig. 18–8) migrate into the future head and neck region. The first branchial arch, the primordium of the jaws, appears as a slight surface elevation lateral to the developing pharynx (Fig. 10–1). Soon this and the other branchial arches appear as obliquely disposed, rounded ridges on each side of the future head and neck regions (Fig. 10–1*B* to *C*).

By the end of the fourth week, four well-defined pairs of branchial arches are visible externally (Fig. 10–1*D*). The fifth and sixth arches are small and cannot be seen on the surface of the embryo.

The arches are separated from each other by prominent clefts called *branchial grooves* (Fig. 10–1*B* to *D*) and are numbered in a craniocaudal sequence.

The first branchial arch, often called the *mandibular arch,* develops two prominences (elevations): (1) The larger *mandibular prominence* (process) forms the lower jaw, or mandible, and (2) the smaller *maxillary prominence* (process) gives rise to the maxilla, the zygomatic bone, and the squamous part of the temporal bone (Fig. 10–3).

The second branchial arch, often called the *hyoid arch,* contributes to the hyoid bone and adjacent regions of the neck (Fig. 10–3*G*). The arches caudal to the hyoid arch are referred to by number only. Structures derived from the branchial arches are summarized in Table 10–1.

The branchial arches support the lateral walls of the cranial part of the foregut, or *primitive pharynx.* The mouth initially appears as a slight depression of the surface ectoderm, called the *stomodeum,* or *primitive mouth* (Fig. 10–1*D* to *G*). At first, the stomodeum is separated from the primitive embryonic pharynx by a bilaminar membrane, the *oropharyngeal membrane* (buccopharyngeal membrane). It is composed of ectoderm externally and endoderm internally. This membrane ruptures at about 24 days, bringing the primitive gut, or digestive tract, into communication with the amniotic cavity.

Branchial Arch Components (Fig. 10–2).
Each arch consists at first of mesenchyme derived from the lateral mesoderm

179

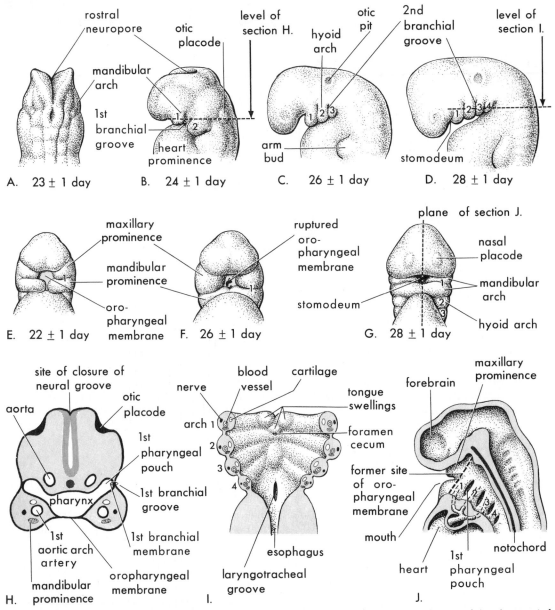

Figure 10–1   Drawings illustrating the human branchial apparatus during stages 10 to 13 of development. *A*, Dorsal view of the cranial part of an early embryo. *B* to *D*, Lateral views, showing later development of the branchial arches. *E* to *G*, Facial views, illustrating the relationship of the first branchial arch to the stomodeum. *H*, Transverse section through the cranial region of an embryo. *I*, Horizontal section through the cranial region of an embryo, illustrating the branchial arch components and the floor of the primitive pharynx. *J*, Sagittal section of the cranial region of an embryo, illustrating the openings of the pharyngeal pouches in the lateral wall of the primitive pharynx.

(see Fig. 4–5*G*), and is covered externally by ectoderm and internally by endoderm. Soon *neural crest cells* (see Fig. 18–8) migrate into the branchial arches and surround the central core of mesenchymal cells. It is the migration of the neural crest cells into the arches and their proliferative activity that produce the discrete swellings demarcating each of the branchial arches. These neural crest cells are unique in that, despite their ectodermal origin, they make a major contribution to the mesenchyme in the head (Noden, 1980). *Mesenchyme* (embryonic connective tissue) gives rise to skeletal, connective, muscular,

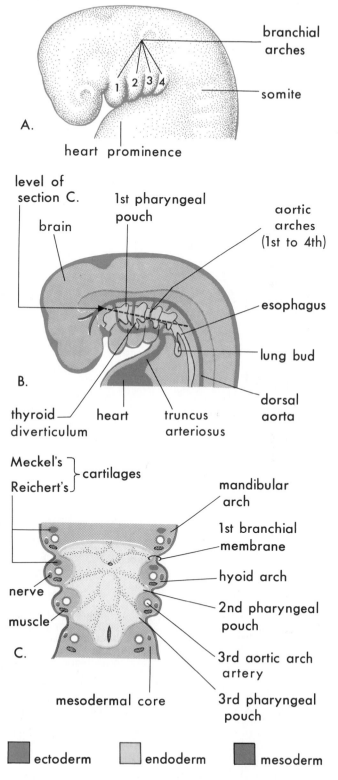

Figure 10–2 *A,* Drawing of the head and neck region of an embryo during stage 13 of development (about 28 days), illustrating the branchial apparatus. *B.,* Schematic drawing showing the pharyngeal pouches and the aortic arches or branchial arch arteries. *C,* Horizontal section through the embryo, showing the floor of the primitive pharynx and illustrating the germ layer of origin of the branchial arch components.

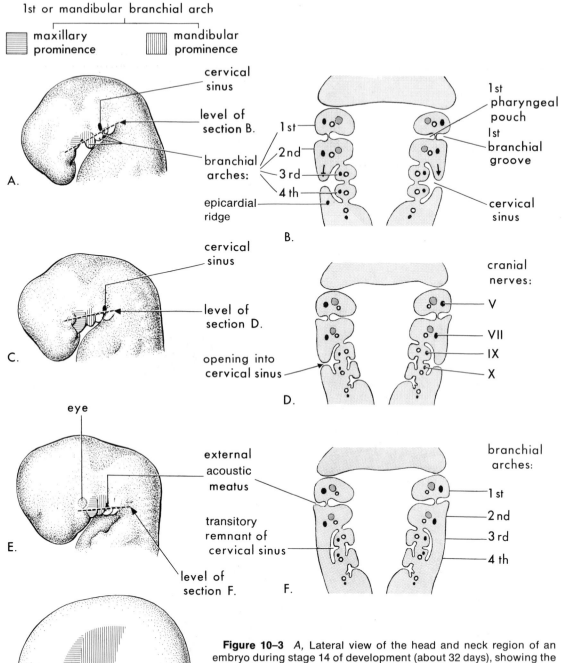

Figure 10–3 *A,* Lateral view of the head and neck region of an embryo during stage 14 of development (about 32 days), showing the branchial arches and the cervical sinus. *B,* Diagrammatic horizontal section through the embryo, illustrating the growth of the second branchial arch over the third and fourth arches. *C,* An embryo during stage 15 of development (about 33 days). *D,* Horizontal section through the embryo, illustrating closure of the cervical sinus. *E,* An embryo during stage 17 of development (about 41 days). *F,* Horizontal section through the embryo, showing the transitory cystic remnant of the cervical sinus. *G,* 20-week fetus, illustrating the areas of the face derived from the first branchial arch.

and ensheathing cells of the peripheral nervous system. Mesenchyme derived from neural crest cells is often called *ectomesenchyme,* or mesectoderm, to differentiate it from mesenchyme derived from mesoderm.

The mesenchyme in each branchial arch gives rise to muscles (e.g., muscles of mastication, Table 10–1), cartilage, and bone. *The neural crest cells give rise to specific skeletal structures;* for example, those cells in the first arch form the cartilage of the first arch (Meckel's cartilage), the maxilla, the mandible, various ligaments, and two of the three ossicles of the middle ear (Fig. 10–4).

A typical branchial arch contains an artery, a cartilaginous bar, a muscular component, and a nerve (Fig. 10–2). The nerves grow into the arches from the brain (Fig. 10–6) and are derived from neuroectoderm.

FATE OF THE BRANCHIAL ARCHES

The branchial arches contribute extensively to the formation of the face, the neck, the nasal cavities, the mouth, the larynx, and the pharynx. *The first branchial arch is involved with development of the face* and is discussed with this region (Fig. 10–20). Small elevations, or hillocks, develop at the dorsal ends of the first and second arches surrounding the first branchial groove (Fig. 10–20E_1);

these hillocks gradually fuse to form the *auricle, or pinna, of the external ear.*

During the fifth week, the *second branchial arch* overgrows the third and fourth arches, forming an ectodermal depression known as the *cervical sinus* (Fig. 10–3A to D). Gradually, the second to fourth branchial grooves and the cervical sinus are obliterated. This occurs as the second branchial arch merges with the *epicardial ridge,* giving the neck a smooth contour (Fig. 10–3F and G). The branchial arches caudal to the first one make little contribution to the skin of the head and the neck.

Derivatives of the Branchial Arch Arteries. The transformation of the aortic arches into the adult arterial pattern is described with the circulatory system in Chapter 14 (see Fig. 14–32).

In fishes, the branchial arch arteries supply blood to the capillary network of the gills. In the human embryo, the blood in these arteries passes right through the branchial arches and enters the dorsal aortae (Fig. 10–2B).

Derivatives of the Branchial Arch Cartilages (Fig. 10–4; Table 10–1). The dorsal end of the *first arch cartilage* (Meckel's cartilage) is closely related to the developing ear and becomes ossified to form two middle ear bones, the *malleus* and the *incus.* The intermediate portion of the cartilage re-

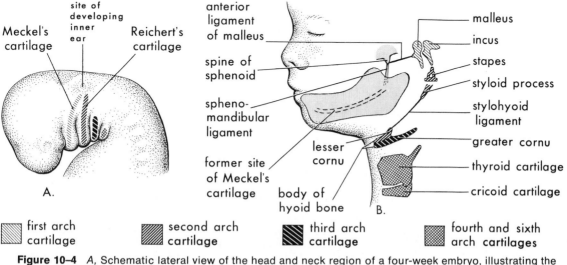

Figure 10–4 *A,* Schematic lateral view of the head and neck region of a four-week embryo, illustrating the location of the branchial arch cartilages. *B,* Similar view of a 24-week fetus, illustrating the adult derivatives of the branchial arch cartilages. Note that the mandible is formed by membranous ossification of the mesenchymal tissue surrounding Meckel's cartilage. This cartilage acts as a template, or guide, but does not contribute directly to the formation of the mandible.

TABLE 10–1 STRUCTURES DERIVED FROM BRANCHIAL ARCH COMPONENTS*
AND INNERVATION OF THE BRANCHIAL ARCHES

Arch	Nerve	Muscles	Skeletal Structures	Ligaments
First-Mandibular	Trigeminal† (V)	Muscles of mastication‡ Mylohyoid and anterior belly of digastric Tensor tympani Tensor veli palatini	Malleus Incus	Anterior ligament of malleus Sphenomandibular ligament
Second-Hyoid	Facial (VII)	Muscles of facial expression§ Stapedius Stylohyoid Posterior belly of digastric	Stapes Styloid process Lesser cornu of hyoid Upper part of body of the hyoid bone	Stylohyoid ligament
Third	Glossopharyngeal (IX)	Stylopharyngeus	Greater cornu of hyoid Lower part of body of the hyoid bone	
Fourth and Sixth‖	Superior laryngeal branch of vagus (X). Recurrent laryngeal branch of vagus (X)	Cricothyroid Levator veli palatini Constrictors of pharynx Intrinsic muscles of larynx	Thyroid cartilage Cricoid cartilage Arytenoid cartilage Corniculate cartilage Cuneiform cartilage	

*The derivatives of the aortic arches are described in Chapter 14 (see Fig. 14–32).
†The ophthalmic division does not supply any branchial components.
‡Temporalis, masseter, medial and lateral pterygoids.
§Buccinator, auricularis, frontalis, platysma, orbicularis oris and oculi.
‖The fifth branchial arch is sometimes absent. When present, it is rudimentary like the sixth arch and usually has no recognizable cartilage bar. The cartilaginous components of the fourth and sixth arches fuse to form the cartilages listed.

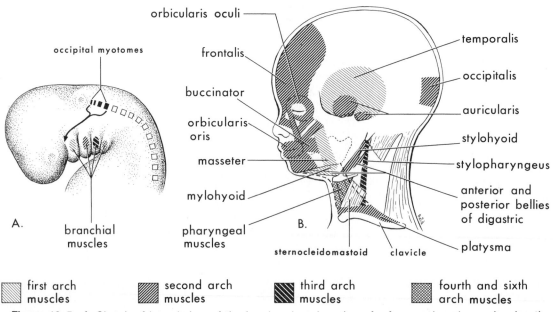

Figure 10–5 A, Sketch of lateral view of the head and neck region of a four-week embryo, showing the branchial muscles. The arrow shows the pathway taken by myoblasts from the occipital myotomes to form the tongue musculature. B, Sketch of the head and neck of a 20-week fetus dissected to show the muscles derived from the branchial arches. Parts of the platysma and sternocleiodomastoid muscles have been removed to show the deeper muscles. Note that myoblasts from the second branchial arch migrate from the neck region to the head and give rise to the muscles of facial expression. Thus, these muscles are supplied by the facial nerve, the nerve of the second branchial arch.

gresses, and its perichondrium forms the *anterior ligament of the malleus* and the *sphenomandibular ligament*. The ventral portion of the first arch cartilage largely disappears as the mandible develops around it by intramembranous ossification (Fig. 10–4B).

The dorsal end of the *second arch cartilage* (Reichert's cartilage) is also closely related to the developing ear and ossifies to form the *stapes* of the middle ear and the *styloid process* of the temporal bone. The portion of cartilage between the styloid process and the hyoid bone regresses, and its perichondrium forms the *stylohyoid ligament*. The ventral end of the second arch cartilage ossifies to form the lesser cornu and superior part of the body of the *hyoid bone* (Fig. 10–4B).

The *third arch cartilage* is located in the ventral portion of the arch and ossifies to form the greater cornu and inferior part of the body of the hyoid bone.

The *fourth and sixth arch cartilages* fuse to form the *laryngeal cartilages* (Table 10–1), except for the epiglottis. The cartilage of the epiglottis develops from mesenchyme in the *hypobranchial eminence* (Fig. 10–18A), a derivative of the third and fourth branchial arches.

Derivatives of the Branchial Arch Muscles. The muscular components of the branchial arches form various striated muscles in the head and the neck (Table 10–1). The development and migration of these muscles are illustrated in Figure 10–5.

Derivatives of the Branchial Arch Nerves (Fig. 10–6). Each branchial arch is supplied by its own cranial nerve. The nerves supplying the branchial muscles, listed in Table 10–1, are classified as *branchial efferent nerves*. Because mesenchyme from the branchial arches contributes to the dermis and mucous membranes of the head and the neck, these areas are supplied with sensory, or *branchial afferent,* fibers.

The facial skin is supplied by the fifth cranial *(trigeminal)* nerve; however, only the caudal two branches *(maxillary and mandibular)* supply derivatives of the first branchial arch (Fig. 10–6B). These branches also innervate the teeth and the mucous membranes of the nasal cavities, the palate, the mouth, and the tongue (Fig. 10–6C).

The seventh cranial *(facial)* nerve, the ninth cranial *(glossopharyngeal)* nerve, and the tenth cranial *(vagus)* nerve supply the

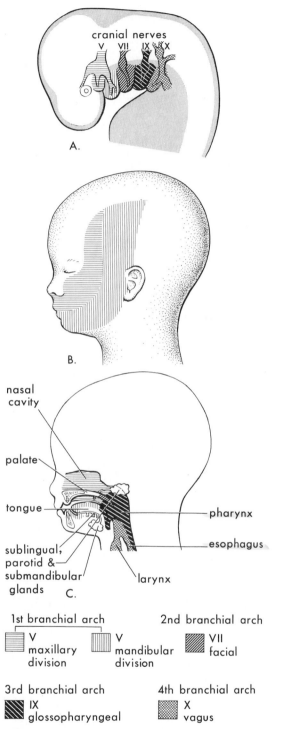

Figure 10–6 *A,* Lateral view of the head and neck region of a 4-week embryo, showing the nerves supplying the branchial arches. *B,* Sketch of a 20-week fetal head, showing the superficial distribution of the two caudal branches of the first arch nerve (V). *C,* Sagittal section of the fetal head, showing the deep distribution of sensory fibers of the branchial nerves to the teeth and the mucosa of the tongue, pharynx, nasal cavity, palate, and larynx.

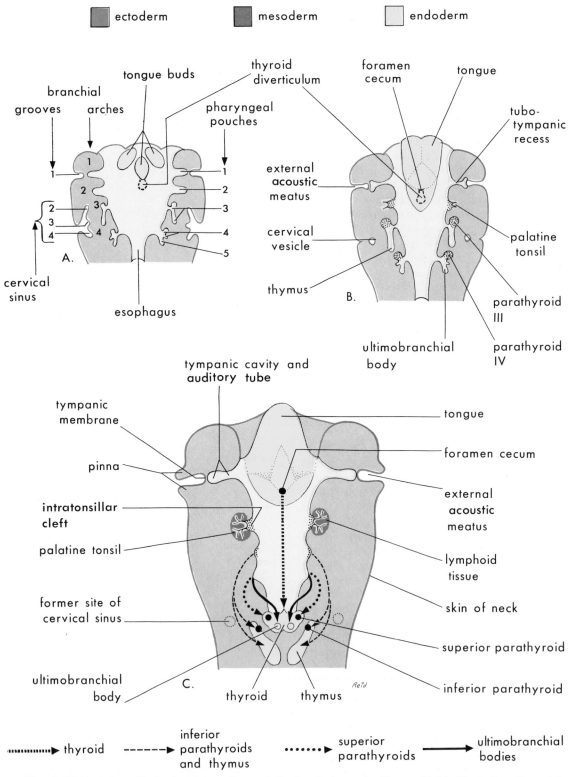

Figure 10–7 Schematic horizontal sections at the level shown in Figure 10–3A, illustrating the adult derivatives of the pharyngeal pouches. *A,* Five weeks. *B,* Six weeks. *C,* Seven weeks.

second (hyoid), third, and caudal (fourth to sixth) branchial arches, respectively. The fourth arch is supplied by the superior laryngeal branch of the vagus, and the sixth by the recurrent laryngeal branch. The nerves of the second to sixth arches have little cutaneous distribution, but they innervate the mucous membranes of the tongue, the pharynx, and the larynx, as illustrated in Fig. 10–6C.

THE PHARYNGEAL POUCHES

The primitive pharynx widens cranially and narrows caudally as it joins the esophagus. The endoderm of the pharynx lines the inner aspects of the branchial arches and passes into balloon-like diverticula called *pharyngeal pouches* (Figs. 10–1H to J and 10–2B and C). Pairs of pouches develop in a craniocaudal sequence between the branchial arches, e.g., the first pouch lies between the first and second branchial arches.

There are four well-defined pairs of pouches; the fifth pair is absent or rudimentary. The endoderm of the pouches contacts the ectoderm of the branchial grooves, and together they form thin double-layered *branchial membranes* that separate the pharyngeal pouches and the branchial grooves (Fig. 10–1H).

DERIVATIVES OF THE PHARYNGEAL POUCHES

The First Pharyngeal Pouch (Figs. 10–7 and 10–8). This pouch expands into an elongate *tubotympanic recess* and envelops the middle ear bones. The expanded distal portion of this recess contacts the first branchial groove, where it later contributes to the formation of the *tympanic membrane.* The tubotympanic recess gives rise to the *tympanic cavity* and the *mastoid antrum,* which develop later as a dorsal expansion of the tympanic cavity. The connection of the tubotympanic recess with the pharynx gradually elongates to form the *auditory tube* (pharyngotympanic tube, or Eustachian tube).

The Second Pharyngeal Pouch (Figs. 10–7 and 10–8). Although it is largely obliterated as the palatine tonsil develops, part of this pouch remains as the *intratonsillar cleft.* The endoderm of this pouch proliferates and forms buds that grow into

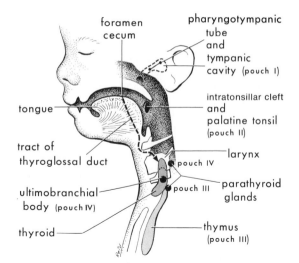

Figure 10–8 Schematic sagittal section of a 20-week fetal head showing the adult derivatives of the pharyngeal pouches and descent of the thyroid gland.

the surrounding mesenchyme. The central parts of these buds break down, forming the tonsillar crypts. The pouch endoderm forms the surface epithelium and the lining of the *crypts of the palatine tonsil.* At about 20 weeks, the mesenchyme surrounding the crypts differentiates into lymphoid tissue and soon becomes organized into *lymphatic nodules.*

The Third Pharyngeal Pouch (Figs. 10–7 and 10–8). This pouch expands into a solid dorsal bulbar portion and a hollow ventral elongate portion. Its connection with the pharynx reduces to a narrow duct that soon degenerates. By the sixth week, the epithelium of each dorsal bulbar portion begins to differentiate into an *inferior parathyroid gland* (often called *parathyroid III* after its pouch of origin). The epithelium of the elongate ventral portions of the two pouches proliferates, obliterating the cavities of the pouches. These primordia of the thymus migrate medially, where they meet and fuse to form the *thymus.* The primordia of the thymus and the parathyroid glands lose their connections with the pharynx and then migrate caudally. Later, the parathyroid glands separate from the thymus and come to lie on the dorsal surface of the thyroid gland, which by this stage has descended from the foramen cecum of the tongue (Fig. 10–7C).

Histogenesis of the Thymus. The thymus develops from tubes of epithelial cells that are derived from the third pair of endodermal pharyngeal pouches and from the mesenchyme into which these tubes grow. The epithelial tubes soon become solid cords that proliferate and give rise to side branches. Each side branch becomes the core of a lobule of the thymus. Some cells of the epithelial cords become arranged around a central point, forming small groups of cells called *Hassall's corpuscles* (Ham and Cormack, 1979). Other cells of the epithelial cords tend to spread apart, but they retain connections with each other to form an *epithelial reticulum*. The mesenchyme between the epithelial cords forms incomplete, thin partitions (septa), and *lymphocytes* soon appear and fill the interstices between the epithelial cells. The lymphocytes are derived from *hematopoietic stem cells* (colony-forming units) [CFU's], or a special type of immediate progeny).

Growth and development of the thymus is not complete at birth. It is relatively large during the perinatal period and may extend superiorly through the superior aperture of the thorax into the root of the neck.

During late childhood, as puberty is reached, the thymus begins to diminish in relative size (undergoes involution). By adulthood, it is often scarcely recognizable.

The Fourth Pharyngeal Pouch (Figs. 10–7 and 10–8). This pouch also expands into a dorsal bulbar portion and a ventral elongate portion, and its connection with the pharynx also becomes reduced to a narrow duct that soon degenerates. By the sixth week, each dorsal portion develops into a *superior parathyroid gland* (often called *parathyroid IV* after its pouch of origin), which comes to lie on the dorsal surface of the thyroid gland.

As described, the parathyroid glands derived from the third pouches descend with the thymus, and, as a result, they are carried to a more inferior position than the parathyroid glands derived from the fourth pharyngeal pouches (Figs. 10–7 and 10–8). This explains why the parathyroid glands derived from the third pair of pouches are located inferior to those derived from the fourth pouches.

Histogenesis of the Parathyroid Glands. The epithelium of the dorsal parts of the third and fourth pouches proliferates during the fifth week and forms small nodules on the dorsal aspect of each pouch. Vascular mesenchyme soon grows into these nodules, forming a capillary network. The *chief* or *principal cells* differentiate during the embryonic period and are believed to become functionally active in regulating fetal calcium metabolism. The *oxyphil cells* do not differentiate until about five to seven years after birth.

The ventral elongate portion of each fourth pouch develops into an *ultimobranchial body*, which received its name because it is the last of the series of structures derived from the pharyngeal pouches. The ultimobranchial body fuses with the thyroid gland and subsequently disseminates to give rise to the *parafollicular cells*, or *C cells*, of the thyroid gland. They are called *C cells* to indicate that they produce *calcitonin*, a hormone involved in the regulation of the normal calcium level in body fluids (Copp et al., 1962). The C cells appear to differentiate from neural crest cells that migrate into the caudal branchial arches (Pearse and Polak, 1971).

The Fifth Pharyngeal Pouch. This *rudimentary structure*, when it develops, becomes part of the fourth pharyngeal pouch which develops into an ultimobranchial body.

THE BRANCHIAL GROOVES

The future neck region of the human embryo exhibits four branchial grooves on each side during the fourth and fifth weeks (Figs. 10–1 and 10–3). These dorsoventral grooves *separate the branchial arches externally*. Only one pair of branchial grooves contributes to adult structures. The first branchial groove persists as the epithelium of the *external acoustic meatus* (Fig. 10–7C). The other branchial grooves come to lie in a depression called the *cervical sinus* and are obliterated with it as the neck develops (Figs. 10–3 and 10–7).

THE BRANCHIAL MEMBRANES

Four branchial membranes appear on each side of the future neck region of the human embryo during the fourth week (Figs. 10–1H and 10–2C). *They form where the epithelia of a branchial groove and a pharyngeal pouch approach each other,* but they are temporary structures in the human embryo. The endoderm of the pharyngeal pouches and the ectoderm of the branchial grooves are soon separated by mesoderm. Only one pair of branchial membranes contributes to the formation of adult structures. The first branchial membrane, along with the intervening layer of mesoderm, gives rise to the *tympanic membrane*, or eardrum (Fig. 10–7C).

BRANCHIAL ANOMALIES

Congenital malformations of the head and of the neck originate mainly during transfor-

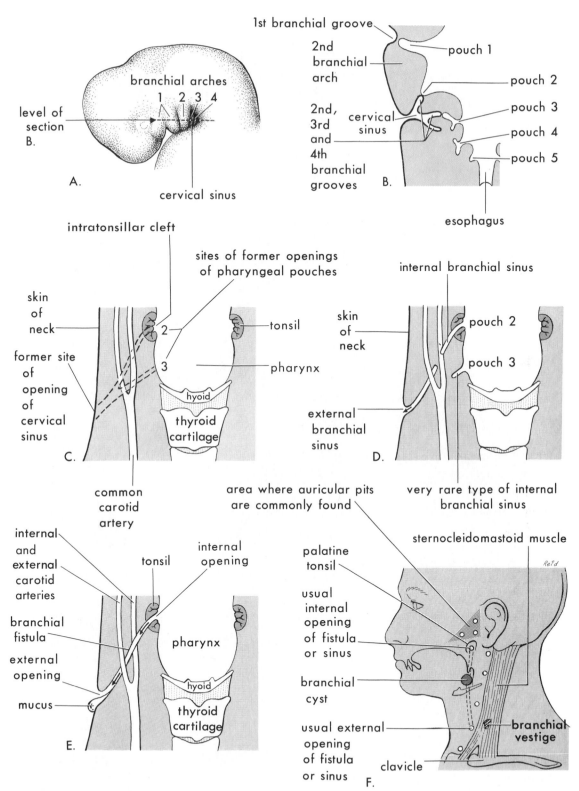

Figure 10–9 *A,* Drawing of the head and neck region of a five-week embryo, showing the cervical sinus normally present at this stage of development. *B,* Horizontal section through the embryo, illustrating the relationship of the cervical sinus to the branchial arches and pharyngeal pouches. *C,* Diagrammatic sketch of the adult neck region, indicating the former sites of openings of the cervical sinus and the pharyngeal pouches. The broken lines indicate possible courses of branchial fistulas. *D,* Similar sketch showing the embryological basis of various types of branchial sinus. *E,* Drawing of a branchial fistula resulting from persistence of parts of the second branchial groove and the second pharyngeal pouch. *F,* Sketch showing possible sites of branchial cysts and openings of branchial sinuses and fistulas. A branchial vestige is also illustrated.

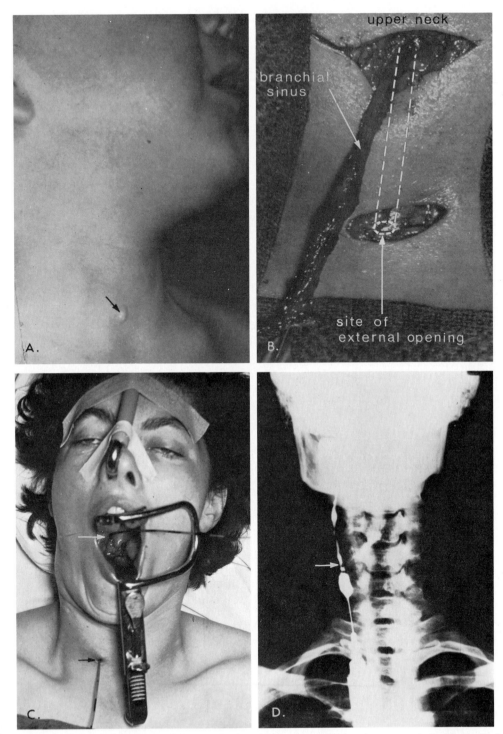

Figure 10–10 *A,* Photograph of a child's neck showing mucus dripping from an external branchial sinus (arrow). *B,* Photographs of a branchial sinus taken during excision. The external opening in the skin of the neck and the original course of the sinus in the subcutaneous tissue are indicated by broken lines. (From Swenson, O.: *Pediatric Surgery,* 1958. Courtesy of Appleton-Century-Crofts, Inc.) *C,* Photograph illustrating a branchial fistula in an adult female. The catheter enters the internal opening in the intratonsillar cleft (arrow), passes through the fistula, and leaves by the opening on the surface of the neck (arrow). *D,* Radiograph showing the course of the fistula (arrow) through the neck. (Courtesy of Dr. D. A. Kernahan, The Children's Memorial Hospital, Chicago.)

mation of the branchial apparatus into adult structures. Most of these abnormalities represent *remnants of the branchial apparatus* that normally disappear as these structures develop. Most of *these malformations are uncommon.*

Congenital Auricular Sinuses and Cysts. Small blind pits, or cysts, in the skin are commonly found in a triangular area anterior to the ear (Fig. 10–9*F*), but may occur in other sites around the auricle or in the lobule. Although some pits and cysts are remnants of the first branchial groove, others may represent ectodermal folds sequestered during formation of the external ear (Gray and Skandalakis, 1972).

Branchial Sinus, or Lateral Cervical Sinus. Branchial sinuses are uncommon, and almost all that open externally on the side of the neck result from failure of the second branchial groove and the cervical sinus to obliterate (Figs. 10–9 and 10–10*B*). A blind pit or channel then remains, which typically opens on the line of the anterior border of the sternocleidomastoid muscle in the inferior third of the neck (Figs. 10–9*D* and *F* and 10–10*B*). Often there is an intermittent discharge of mucus from the opening (Fig. 10–10*A*).

External branchial sinuses are commonly detected during infancy owing to the discharge of material from their orifices on the neck. Branchial sinuses are bilateral in about 10 per cent of cases and are commonly associated with auricular sinuses in the lobule of the auricle.

Internal branchial sinuses opening into the pharynx are very rare. Because they usually open into the intratonsillar cleft or near the palatopharyngeal arch (Fig. 10–9*D* and *F*), almost all these sinuses result from persistence of part of the second pharyngeal pouch.

Branchial Fistula (Fig. 10–10C). An abnormal tract (canal) opening both on the side of the neck and in the pharynx is called a *branchial fistula* (Fig. 10–9*E* and *F*). It is usually the result of persistence of parts of the second branchial groove and second pharyngeal pouch. The fistula ascends from its cervical opening through the subcutaneous tissue, the platysma muscle, and the deep fascia to reach the carotid sheath. It then passes between the internal and external carotid arteries and usually opens in the intratonsillar cleft (Figs. 10–8, 10–9*E*, and 10–10*C* and *D*). In older patients, there may occasionally be a disagreeable taste in the mouth, owing to the discharge of material into the pharynx from the fistula.

Branchial Cyst, or Lateral Cervical Cyst. The third and fourth branchial arches are normally buried in the *cervical sinus* (Fig. 10–9). Remnants of parts of the cervical sinus and/or the

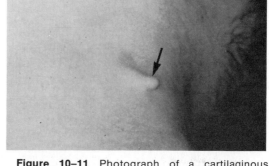

Figure 10–11 Photograph of a cartilaginous branchial vestige under the skin of a child's neck. (From Swenson, O.: *Pediatric Surgery,* 1958. Courtesy of Appleton-Century-Crofts, Inc.)

second branchial groove may persist and form spherical or elongate cysts (Fig. 10–9*B* and *F*). Although they may be associated with branchial sinuses and drain through them, these cysts often lie free in the neck just inferior to the angle of the mandible. They may, however, develop anywhere along the anterior border of the sternocleidomastoid muscle.

These cysts often do not become apparent until late childhood or early adulthood, when they produce a slowly enlarging, painless swelling in the neck. The cysts enlarge owing to the accumulation of fluid and cellular debris derived from desquamation of their epithelial linings.

Branchial Vestige. Normally, the branchial cartilages disappear, except for those parts that form ligaments or bones (Fig. 10–4*B* and Table 10–1). Very rarely, cartilaginous or bony remnants of branchial arch cartilages may appear under the skin on the side of the neck (Fig. 10–11). These are usually found anterior to the inferior third of the sternocleidomastoid muscle (Fig. 10–9*F*).

The First Arch Syndrome (Fig. 10–12). Maldevelopment of the components of the first branchial arch results in various congenital malformations of the eyes, the ears, the mandible, and the palate that together constitute the first arch syndrome. This set of symptoms is believed to be caused by *insufficient migration of cranial neural crest cells* into the first branchial arch during the fourth week. There are two main manifestations of the first arch syndrome.

In *Treacher Collins syndrome* (mandibulofacial dysostosis), which is caused by an autosomal dominant gene, there is malar hypoplasia with downslanting palpebral fis-

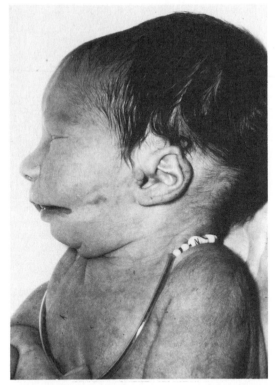

Figure 10–12 Photograph of an infant with the first arch syndrome, a pattern of malformations resulting from insufficient migration of neural crest cells into the first branchial arch. Note the following: deformed auricle of the external ear, preauricular appendage, defect in cheek between the ear and the mouth, hypoplasia of the mandible, and large mouth. (Courtesy of Dr. T. V. N. Persaud, Professor of Anatomy, University of Manitoba, Winnipeg, Canada.)

sures, defects of the lower eyelid, deformed external ear, and, sometimes, abnormalities of the middle and inner ears (Hayden and Arnold, 1968, and Smith, 1976).

In *Pierre Robin syndrome,* striking hypoplasia of the mandible, cleft palate, and defects of the eye and the ear are found (Goodman and Gorlin, 1970). *In the Robin morphogenetic complex, the initiating defect was the small mandible (micrognathia),* which resulted in posterior displacement of the tongue and obstruction to the full closure of the palatine processes (Fig. 10–23) and a U-shaped bilateral cleft palate (Fig. 10–29).

Congenital Thymic Aplasia and Absence of the Parathyroid Glands (DiGeorge Syndrome). Infants with this disease are born without thymus and parathyroid glands, but ectopic glandular tissues have been found at autopsy in some cases (Rosen, 1965). The disease is characterized by congenital hypoparathyroidism, increased susceptibility to infections, malformations of the mouth (shortened philtrum of lip or fish-mouth deformity), low-set notched ears, nasal clefts, thyroid hypoplasia, and cardiac abnormalities (defects of the aortic arch and heart).

This syndrome results from failure of the third and fourth pharyngeal pouches to differentiate into the thymus and the parathyroid glands. The facial abnormalities result from abnormal development of the first arch components during formation of the face and the ears. There is no known genetic cause for these abnormalities; the syndrome may result from a teratogen acting during

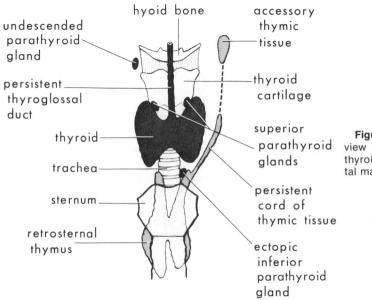

Figure 10–13 Drawing of an anterior view of the thyroid, thymus, and parathyroid glands showing various congenital malformations.

the fourth to sixth weeks, when the branchial arch components are transforming into adult derivatives. *Thymic transplantation* has been successful in treatment of DiGeorge syndrome (Vaughan et al., 1979).

Accessory Thymic Tissue (Fig. 10–13). An isolated portion of thymic tissue may persist in the neck, often in close association with one of the inferior parathyroid glands.

Variations in the shape of the thymus sometimes occur. It may exhibit slender cords, or prolongations, into the neck on each side, anterolatreal to the trachea. These processes may be connected to the inferior parathyroid glands by fibrous strands.

Ectopic and Supernumerary Parathyroid Glands (Fig. 10–13). The parathyroid glands are highly variable in number and location. They may be found anywhere near or within the thyroid or thymus glands. Rarely, an inferior parathyroid gland may fail to descend and will remain near the bifurcation of the common carotid artery. In other cases, it may accompany the thymus into the thorax. The variability of the position of the parathyroid glands may create a problem during thyroid and parathyroid surgery.

In rare cases, there are more than four parathyroid glands. Supernumerary parathyroid glands probably form by division of the primordia of the original glands. Absence of a parathyroid gland results from failure of development of one of the primordia or from atrophy of the gland early in development.

DEVELOPMENT OF THE THYROID GLAND

The thyroid gland is the *first endocrine gland to appear in embryonic development*. It begins to develop during stage 11 of development (about 24 days) from a median endodermal thickening in the floor of the primitive pharynx (Fig. 10–14), just caudal to the future site of the *median tongue bud* (Fig. 10–18*A*). This thickening soon forms a downgrowth known as the *thyroid diverticulum* (Figs. 10–2*B*, 10–7*A*, and 10–14*A*). As the embryo elongates and the tongue grows, the developing thyroid descends in the substance of the neck, passing ventral to the developing hyoid bone and laryngeal cartilages. It is connected to the tongue by a narrow canal, the *thyroglossal duct* (Fig. 10–14*B* and *C*); its opening in the tongue is called the *foramen cecum* (Figs. 10–18 and 10–19).

The thyroid diverticulum grows rapidly and divides into two lobes. The right and left

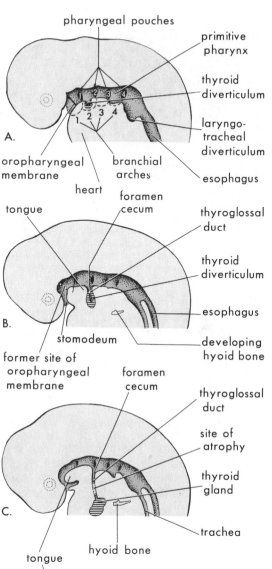

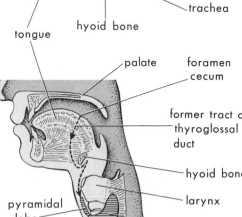

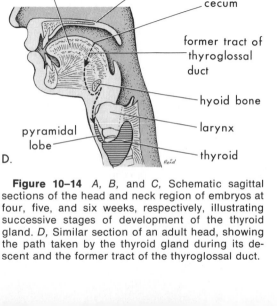

Figure 10–14 *A, B,* and *C,* Schematic sagittal sections of the head and neck region of embryos at four, five, and six weeks, respectively, illustrating successive stages of development of the thyroid gland. *D,* Similar section of an adult head, showing the path taken by the thyroid gland during its descent and the former tract of the thyroglossal duct.

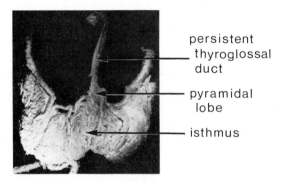

Figure 10–15 Photograph of the anterior surface of an adult thyroid gland, showing persistence of the thyroglossal duct. The pyramidal lobe, ascending from the superior border of the isthmus, represents a persistent portion of the inferior end of the thyroglossal duct. This conical lobe is seldom median, lying more often on the left, as in this specimen.

lobes are connected by an *isthmus* that lies anterior to the second and third tracheal rings. By seven weeks, the thyroid gland has usually reached its final site in the inferior part of the neck. By this time, the thyroglossal duct has normally disappeared.

A *pyramidal lobe* (Fig. 10–14D), which extends superiorly from the isthmus, is present in about 50 per cent of people. This lobe may be attached to the hyoid bone by fibrous or muscular tissue. The pyramidal lobe represents a persistent portion of the inferior end of the thyroglossal duct (Figs. 10–13D, 10–14, and 10–15). The original opening of the thyroglossal duct persists as a vestigial

pit, the foramen cecum of the tongue (Figs. 10–18 and 10–19).

Histogenesis of the Thyroid Gland. At first, the thyroid primordium consists of a solid mass of endodermal cells. It is later broken up into a network of epithelial cords, or plates, by invasion of the surrounding vascular mesenchyme. By the tenth week, these cords have divided into small cellular groups. Soon, a lumen forms in each of these clusters, and the cells become arranged in a single layer around the lumen. During the eleventh week, colloid begins to appear in these structures, called *thyroid follicles*, and thereafter *thyroxine* can be demonstrated. For a detailed account of the histogenesis of the thyroid gland, see Shepard (1975).

CONGENITAL MALFORMATIONS OF THE THYROID GLAND

Thyroglossal Cysts and Sinuses. Cysts may form anywhere along the course followed by the thyroglossal duct during descent of the thyroid gland from the tongue (Fig. 10–16). Normally, the thyroglossal duct atrophies and disappears, but remnants of it may persist and give rise to cysts in the tongue or in the midline of the neck, usually just inferior to the hyoid bone (Fig. 10–16B). The swelling usually develops as a painless, progressively enlarging, and movable mass. In some cases, an opening through the skin exists as a result of perforation following infection of the cyst. This forms a *thyroglos-*

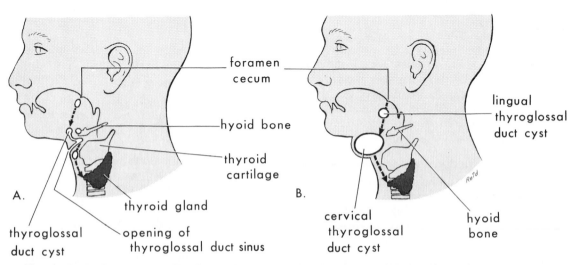

Figure 10–16 A, Diagrammatic sketch of the head, showing the possible locations of thyroglossal duct cysts. A thyroglossal duct sinus is also illustrated. The broken line indicates the course taken by the thyroglossal duct during descent of the thyroid gland from the foramen cecum to its final position anterior to the trachea. B, A similar sketch illustrating lingual and cervical thyroglossal duct cysts. Most cysts are located near the hyoid bone.

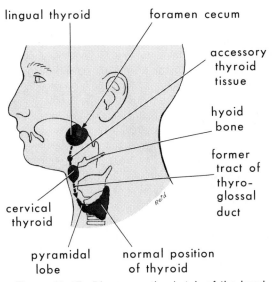

Figure 10–17 Diagrammatic sketch of the head, showing the usual sites of ectopic thyroid tissue. The broken line indicates the path followed by the thyroid gland during its descent and the former tract of the thyroglossal duct.

sal duct sinus that usually opens in the midline of the neck anterior to the laryngeal cartilages (Fig. 10–16A).

Ectopic Thyroid Gland and Accessory Thyroid Tissue. Rarely, the thyroid fails to descend, resulting in a *lingual thyroid* (Fig. 10–17). Very rarely, incomplete descent results in the thyroid gland appearing high in the neck at or just inferior to the hyoid bone. *Accessory thyroid tissue* may be functional, but it is often of insufficient size to maintain normal function if the thyroid gland is removed. Accessory thyroid tissue may result from pieces of the gland that become separated from the main gland, but it usually originates from remnants of the thyroglossal duct. It may be found anywhere from the tongue to the usual site of the thyroid gland.

DEVELOPMENT OF THE TONGUE

Around the end of the fourth week, a median, somewhat triangular elevation appears in the floor of the pharynx just rostral to the foramen cecum. This elevation, the *median tongue bud* (tuberculum impar), gives the first indication of tongue development (Fig. 10–18A). Soon, two oval *distal tongue buds* (lateral lingual swellings) develop on each side of the median tongue bud. These elevations result from proliferation of mesenchyme in the ventromedial parts of the first pair of branchial arches. The

distal tongue buds rapidly increase in size, merge with each other, and overgrow the median tongue bud.

The merged distal tongue buds form the *anterior two thirds*, or *oral part, of the*

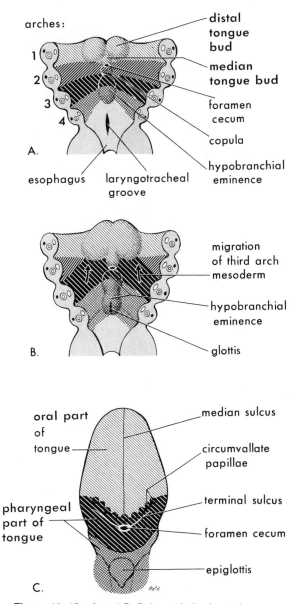

Figure 10–18 A and B, Schematic horizontal sections through the pharynx at the level shown in Figure 10–3A, showing successive stages in the development of the tongue during the fourth and fifth weeks. C, Adult tongue showing the branchial arch derivation of the nerve supply of the mucosa.

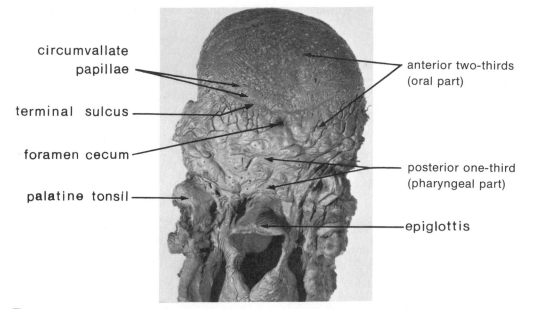

circumvallate
papillae

terminal sulcus

foramen cecum

palatine tonsil

anterior two-thirds
(oral part)

posterior one-third
(pharyngeal part)

epiglottis

Figure 10–19   Photograph of the dorsum of an adult tongue. The foramen cecum indicates the site of origin of the thyroid diverticulum and thyroglossal duct in the embryo. The sulcus terminalis demarcates the developmentally different pharyngeal and oral parts of the tongue.

tongue (Fig. 10–18C). The plane of fusion of the distal tongue buds is indicated superficially by the *median sulcus* of the tongue (Fig. 10–18C) and internally by the fibrous *median septum*. The median tongue bud forms no recognizable portion of the adult tongue.

The *posterior third*, or *pharyngeal part, of the tongue* is initially indicated by two elevations that develop caudal to the foramen cecum (Fig. 10–18A): (1) the *copula* (L., bond, tie), formed by fusion of the ventromedial parts of the second branchial arches; and (2) the large *hypobranchial eminence,* which develops caudal to the copula from mesoderm in the ventromedial parts of the third and fourth branchial arches.

As the tongue develops, the copula is gradually overgrown by the hypobranchial eminence and disappears (Fig. 10–18B and C). As a result, the posterior third of the tongue develops from the cranial part of the hypobranchial eminence. The line of fusion of the anterior and posterior parts of the tongue is roughly indicated by the V-shaped groove called the *terminal sulcus* (Figs. 10–18C and 10–19).

Branchial arch mesenchyme forms the connective tissue and the lymphatic and blood vessels of the tongue, and probably some of its muscle fibers. Most of the tongue

musculature, however, is derived from myoblasts that migrate from the *myotomes of the occipital somites* (Fig. 10–5A). These myoblasts (primitive muscle cells) migrate into the tongue, where they differentiate into the muscles. The hypoglossal nerve (CN XII) accompanies the myoblasts during their migration and innervates the *tongue musculature* when it develops.

Papillae and Taste Buds.   The *papillae* of the tongue appear during stage 22 (about 54 days). The vallate and foliate papillae appear first in close relationship to the terminal branches of the glossopharyngeal nerve. The fungiform papillae appear later, near the terminations of the chorda tympani branch of the facial nerve. All the papillae soon develop *taste buds*. The filiform papillae develop during the early fetal period. Responses can be induced in the face by bitter-tasting substances at 26 to 28 weeks because reflex pathways between taste buds and facial muscles are established by this stage.

The Nerve Supply of the Tongue.   Development of the tongue from the branchial arches explains its nerve supply. The sensory nerve supply to the mucosa of almost the entire *anterior two thirds of the tongue* (oral part) is from the lingual branch of the mandibular division of the trigeminal nerve (Fig. 10–18C), *the nerve of the first branchial arch,* which forms the median and distal tongue buds. Although the facial nerve is the nerve of the second branchial arch, its chorda tympani branch supplies the taste buds in the anterior two thirds of the tongue, except for the vallate papillae. Because the second arch component, the copula, is overgrown by the third arch, the facial nerve does not supply any of the mucosa of the tongue (Fig. 10–18*B* and *C*). The vallate papillae in the anterior two thirds of the tongue are innervated by the glossopharyngeal nerve of the third branchial arch (Fig. 10–18*C*). The reason usually given for this is that the mucosa of the posterior third of the tongue is pulled slightly forward as the tongue develops.

The *posterior third of the tongue* (pharyngeal part) is innervated mainly by the glossopharyngeal nerve of the third branchial arch. The superior laryngeal branch of the vagus nerve of the fourth arch supplies a small area of the tongue anterior to the epiglottis (Fig. 10–18*C*). All muscles of the tongue are supplied by the *hypoglossal nerve*, except for the palatoglossus, which is supplied by the vagus.

CONGENITAL MALFORMATIONS OF THE TONGUE

Congenital Cysts and Fistulae (Fig. 10–16). Cysts within the tongue substance, just superior to the hyoid bone, are usually derived from *remnants of the thyroglossal duct*. They may enlarge and produce symptoms of pharyngeal discomfort and/or *dysphagia* (difficulty in swallowing).

Fistulae in the tongue are also derived from persistence of the thyroglossal duct, and they open through the *foramen cecum* into the mouth.

Ankyloglossia (Tongue-Tie). The frenulum normally connects the inferior surface of the anterior part of the tongue to the floor of the mouth. In tongue-tie, the frenulum extends to near the tip of the tongue and interferes with its free protrusion. This abnormality occurs in about one in 300 North American infants (Witkop et al., 1967). Usually, the frenulum stretches with time so that surgical correction of the malformation is rarely necessary (Vaughan et al., 1979).

Macroglossia. An excessively large tongue is not common and results from generalized hypertrophy of the tongue. These cases usually result from lymphangioma or muscular hypertrophy (Schaffer and Avery, 1977).

Microglossia. An abnormally small tongue is rare and is usually associated with micrognathia (underdeveloped mandible with recession of the chin).

Cleft Tongue. Rarely, incomplete fusion of the distal tongue buds posteriorly may result in a median groove, or *cleft of the tongue;* usually, the cleft does not extend to the tip.

Bifid Tongue. Complete failure of fusion of the distal tongue buds results in a bifid tongue, a common malformation in South American infants (Witkop et al., 1967).

The Salivary Glands (Fig. 10–6*C*). These glands begin as solid proliferation of cells from the epithelium of the mouth during the sixth and seventh weeks.

The *parotid gland* develops from buds that arise from the ectodermal lining of the stomodeum, or primitive mouth. These buds branch to form solid cords with rounded ends; later, these cords develop lumina to form ducts, and the rounded ends of the cords differentiate into acini. The capsule and connective tissue develop from the surrounding mesenchyme.

The *submandibular glands* develop from the endoderm in the floor of the mouth. A solid cellular process grows backward lateral to the tongue; it later branches and differentiates as described for the parotid gland. Lateral to the tongue, a linear groove forms that later closes over to form the submandibular duct.

The *sublingual glands* appear slightly later than the other glands and develop as multiple buds of the endoderm in the paralingual sulcus (Fig. 10–6*C*).

DEVELOPMENT OF THE FACE

The *five facial primordia* appear around the stomodeum or primitive mouth early in the fourth week (Figs. 10–1*E* and 10–20*A*).

1. The frontonasal prominence (elevation), formed by the proliferation of mesenchyme ventral to the forebrain, constitutes the cranial boundary of the stomodeum.

2. The paired *maxillary prominences* of the first branchial arch form the lateral boundaries, or sides, of the stomodeum.

3. The paired *mandibular prominences* of this same arch constitute the caudal boundary of the stomodeum.

The mesoderm of the five facial primordia is continuous from one prominence to the other. There are no internal divisions corresponding to the grooves demarcating the prominences externally.

The development of the face occurs mainly between the fifth and eighth weeks (Fig. 10–20*A* to *G*). The facial proportions develop during the fetal period (Fig. 10–20*H* and *I*). The lower jaw, or mandible, is the first part of the face to form. This results from merging of the medial ends of the two mandibular prominences during the fourth week.

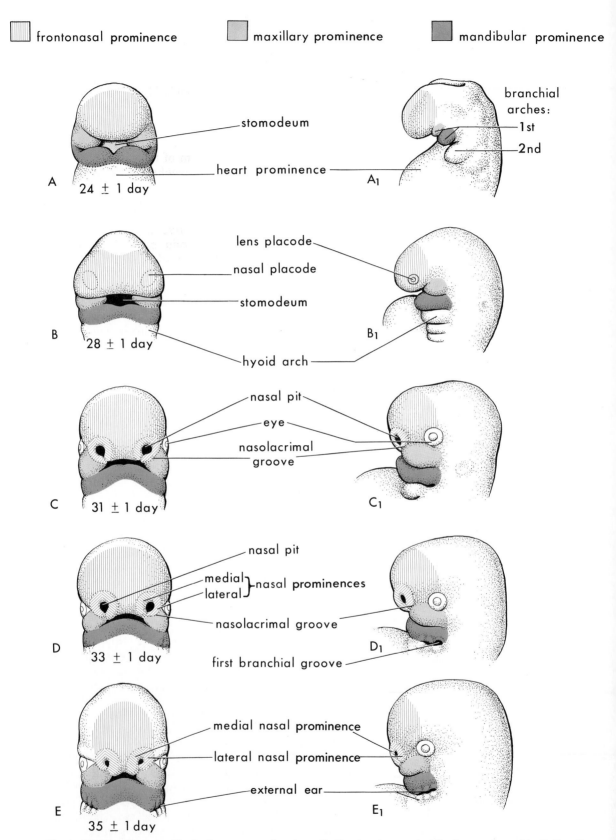

frontonasal prominence maxillary prominence mandibular prominence

Figure 10–20 Diagrams illustrating progressive stages in the development of the human face. *A* to *E*, Fourth and fifth weeks.

Illustration continued on opposite page

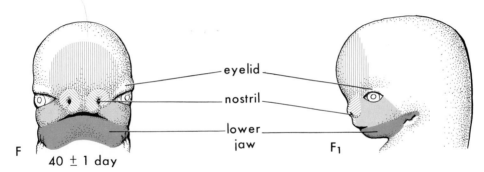

eyelid

nostril

lower
jaw

F
40 ± 1 day

F₁

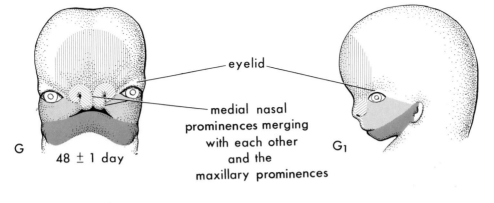

eyelid

medial nasal
prominences merging
with each other
and the
maxillary prominences

G
48 ± 1 day

G₁

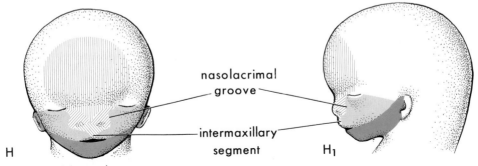

nasolacrimal
groove

intermaxillary
segment

H
10 weeks

H₁

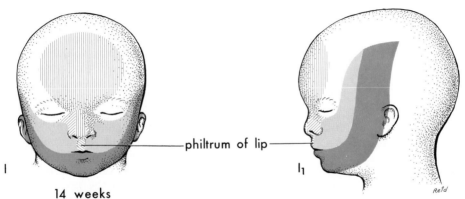

philtrum of lip

I

I₁

Reid

14 weeks

Figure 10–20 *Continued.   F to I,* Sixth to fourteenth weeks.

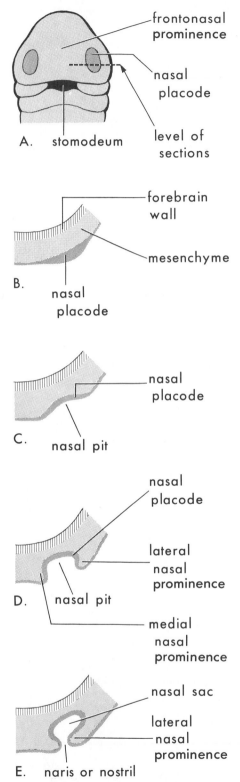

A. stomodeum

frontonasal prominence

nasal placode

level of sections

B. nasal placode

forebrain wall

mesenchyme

C. nasal pit

nasal placode

D. nasal pit

nasal placode

lateral nasal prominence

medial nasal prominence

E. naris or nostril

nasal sac

lateral nasal prominence

Figure 10–21 Progressive stages in the development of a nasal sac (future nasal cavity) during the fourth week, shown in transverse sections through the left side of the developing nose.

By the end of the fourth week, bilateral oval-shaped thickenings of the surface ectoderm, called *nasal placodes,* develop on each side of the lower part of the frontonasal prominence (Figs. 10–20 and 10–21B). Mesenchyme proliferates at the margins of these placodes, producing horseshoe-shaped elevations, the sides of which are called the *medial and lateral nasal prominences* (elevations). The nasal placodes now lie in depressions called *nasal pits* (Fig. 10–20C and 10–21C and D).

The maxillary prominences enlarge and grow rapidly toward each other and the medial nasal prominences (Fig. 10–20D and E). Each lateral nasal prominence is separated from the maxillary prominences by a cleft, or furrow, called the *nasolacrimal groove* (Fig. 10–20C and D). By the end of the fifth week, the eyes are slightly forward on the face, and the external ear has begun to develop. By this time, each maxillary prominence has merged with the lateral nasal prominence along the line of the *nasolacrimal groove.* This establishes a continuity between the side of the nose, formed by the lateral nasal prominence, and the upper cheek region, formed by the maxillary prominence.

The Nasolacrimal Duct and Lacrimal Sac. Each duct develops from a linear thickening of ectoderm that forms in the floor of the nasolacrimal groove. The thickening gives rise to a solid epithelial cord that separates from the ectoderm and sinks into the mesenchyme. Later, this cord becomes canalized to form the nasolacrimal duct; its cranial end expands to form the lacrimal sac. Eventually, this duct drains into the inferior meatus in the lateral wall of the nasal cavity. Not uncommonly, the caudal part of this duct fails to canalize, resulting in a congenital malformation known as *atresia of the nasolacrimal duct.* Development of the lacrimal glands is described in Chapter 19.

During the sixth and seventh weeks, the medial nasal prominences merge with each other and the maxillary prominences (Fig. 10–20F and G). When the medial nasal prominences merge with each other, they form an *intermaxillary segment* (Figs. 10–20H and 10–22D). This segment gives rise to (1) the middle portion, or *philtrum,* of the upper lip, (2) the *premaxillary part of the maxilla* (Figs. 10–22F and 10–24B) and its associated gingiva (gum), and (3) the *primary palate.*

The lateral parts of the upper lip, most of the maxilla, and the secondary palate form from the maxillary prominences (Figs. 10–

20*H* and *I* and 10–23). These prominences merge laterally with the mandibular prominences.

The primitive lips and cheeks are invaded by second branchial arch mesenchyme, which gives rise to the facial muscles (Fig. 10–5 and Table 10–1). These *muscles of facial expression* are supplied by the facial nerve (CN VII), the nerve of the second branchial arch (Table 10–1). The mesenchyme of the first pair of branchial arches gives rise to the *muscles of mastication* and a few others, all of which are innervated by the trigeminal nerves (CN V), which supply the first pair of branchial arches.

The *frontonasal prominence* forms the forehead and the dorsum and the apex of the nose. The sides (alae) of the nose are derived from the lateral nasal prominences (Fig. 10–20*H* and *I*). The *fleshy nasal septum* and the *philtrum of the upper lip* are formed by the medial nasal prominences. The *maxillary prominences* form the upper cheek regions and most of the upper lip (Fig. 10–20*I*). The *mandibular prominences* give rise to the lower lip, the chin, and the lower cheek regions. In addition to the fleshy derivatives just described, various bony derivatives are formed from the facial prominences (e.g., the frontonasal prominences give rise to the frontal and nasal bones).

Until the end of the sixth week, the primitive jaws are solid masses of tissue. The lips and *gingivae* (gums) begin to develop when a linear thickening of the ectoderm, the *labiogingival lamina* (Fig. 10–23*B*), grows into the underlying mesenchyme. Gradually, this lamina largely degerates, leaving a *labiogingival groove*, or *lip sulcus*, between the lips and the gingivae (Fig. 10–23*H*). A small area of the labiogingival lamina persists in the midline, forming the *frenulum*, which attaches each lip to the gingiva.

Final development of the face occurs slowly and results mainly from changes in the proportion and relative position of the facial components. During the early fetal period, the nose is flattened and the mandible is underdeveloped (Fig. 10–20*H*); they obtain their characteristic form when facial development is complete (Fig. 10–20*I*). The brain enlarges, creating a prominent forehead; the eyes move medially, and the external ears rise. The smallness of the face at birth results from (1) the rudimentary upper and lower jaws, (2) the unerupted teeth, and (3) the small size of the nasal cavities and maxillary air sinuses.

DEVELOPMENT OF THE NASAL CAVITIES

As the face develops, the *nasal placodes* becomes depressed, forming *nasal pits* (Fig. 10–21). Growth of the surrounding mesenchyme, forming the medial and lateral *nasal prominences,* causes deepening of the nasal pits and formation of primitive *nasal sacs* (Fig. 10–21*E*). Each nasal sac grows dorsocaudally, ventral to the developing brain (Fig. 10–25*A*). At first, these sacs are separated from the oral cavity by the *oronasal membrane* (Fig. 10–25*A* and *B*), but this membrane soon ruptures, bringing the nasal and oral cavities into communication (Fig. 10–25*C*). The regions of continuity are the *primitive choanae*, which lie posterior to the primary palate.

After the *secondary palate* develops, the choanae are located at the junction of the nasal cavity and the pharynx (Fig. 10–25*D*). When the lateral palatine processes fuse with each other and the nasal septum, the oral and nasal cavities are again separated (Fig. 10–23*G*). This fusion also results in separation of the nasal cavities from each other.

While these changes are occurring, the *superior, middle,* and *inferior conchae* develop as elevations on the lateral wall of each nasal cavity (Figs. 10–23*E* and 10–25*D*). In addition, the ectodermal epithelium in the roof of each nasal cavity becomes specialized as the *olfactory epithelium* (Fig. 10–26). Some cells differentiate into *olfactory cells,* which give origin to fibers that grow into the *olfactory bulbs* of the brain (Figs. 10–23*E* and 10–25*D*).

The Paranasal Air Sinuses. Some paranasal sinuses develop during late fetal life; the remainder develop after birth. They form as diverticula of the walls of the nasal cavities.

The *maxillary sinuses* are about 3 to 4 mm in diameter in the newborn infant, and there are only a few small anterior and posterior ethmoidal air cells. *No frontal or sphenoidal sinuses are present at birth.* The maxillary sinuses grow slowly until puberty and are not fully developed until all the permanent teeth have erupted in early adulthood.

The *ethmoidal sinuses* are small before the

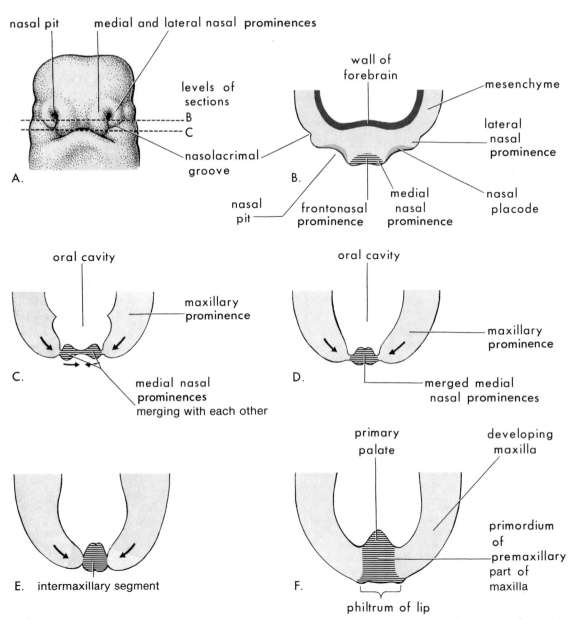

Figure 10–22 Diagrams illustrating development of the upper jaw and lip. *A,* Facial view of a five-week embryo. *B* and *C,* Sketches of horizontal sections at the levels shown in *A.* The arrows in *C* indicate subsequent growth of the maxillary and medial nasal prominences toward each other. *D* to *F,* Similar sections of older embryos illustrating the merging of the medial nasal prominences with each other and their merging with the maxillary prominences to form the upper lip. (Modified from Patten, 1961.)

age of two years, and they do not begin to grow rapidly until six to eight years of age. Around the age of two years, the two most anterior ethmoidal sinuses grow into the frontal bone, forming a frontal sinus on each side. Usually, the *frontal sinuses* are visible in radiographs by the seventh year.

The septum between the right and left frontal sinuses is rarely in the median plane. The two most posterior ethmoidal sinuses grow into the sphenoid bone at about the age of two years, forming two *sphenoid sinuses.* In adults, the sphenoid sinuses vary greatly in size.

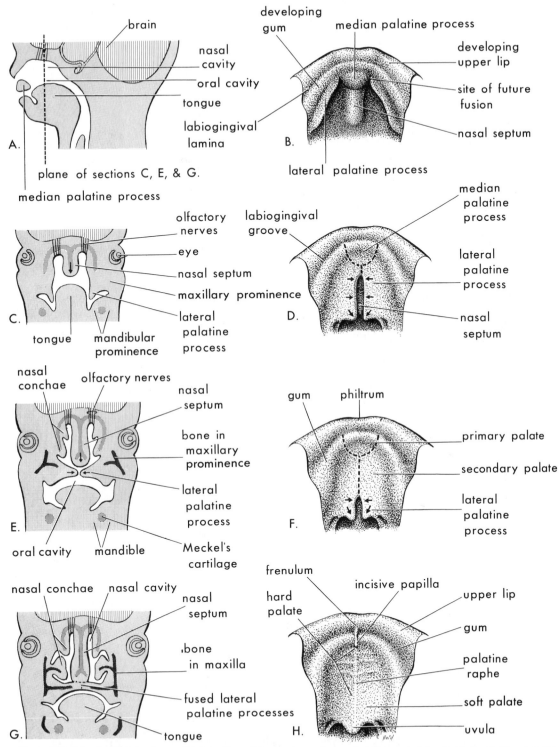

Figure 10–23 *A*, Sketch of a sagittal section of the embryonic head at the end of the sixth week showing the median palatine process, or primary palate. *B, D, F,* and *H,* Drawings of the roof of the mouth from the sixth to twelfth weeks illustrating development of the palate. The broken lines in *D* and *F* indicate sites of fusion of the palatine processes; the arrows indicate medial and posterior growth of the lateral palatine processes. *C, E,* and *G,* Drawings of frontal sections of the head illustrating fusion of the lateral palatine processes with each other and the nasal septum, and separation of the nasal and oral cavities.

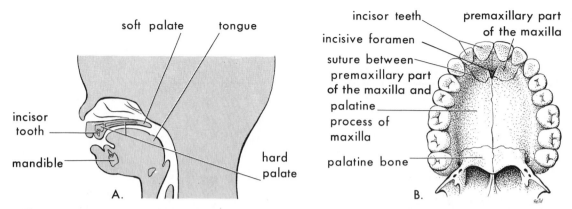

Figure 10–24 *A,* Drawing of a sagittal section of a 20-week fetal head, illustrating the parts of the palate. *B,* The bony palate and alveolar arch of a young adult. The suture between the premaxillary part of the maxilla (sometimes called the incisive bone) and the fused palatine processses of the maxillae is usually visible only in skulls of young persons. It is not usually visible in the hard palates of most dried skulls because they are often from older adults.

The Vomeronasal Organs (Fig. 10–26). During the late embryonic period, the epithelium invaginates on each side of the nasal septum, just superior to the primitive palate, forming a pair of diverticula known as the *vomeronasal organs* (of Jacobson). A vomeronasal cartilage develops ventral to each diverticulum. The vomeronasal organs are 4 to 8 mm long during the sixth fetal month, when they are fully developed. At this stage, they are lined by neurosensory epithelium similar to the olfactory epithelium. A vomeronasal nerve projects to a small accessory olfactory bulb. During late fetal life, these organs begin to regress and usually disappear completely along with their nerves and accessory bulbs. The *vomeronasal cartilages* are usually the only adult

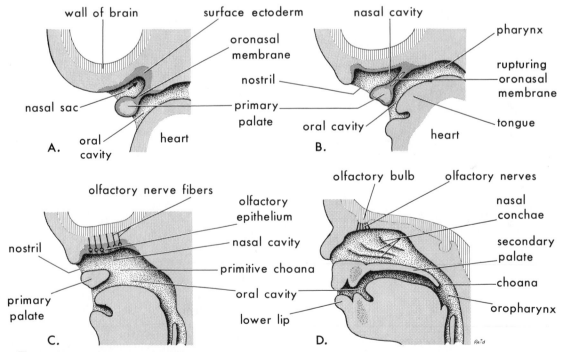

Figure 10–25 Drawings of sagittal sections of the head showing development of the nasal cavities. The nasal septum has been removed. *A,* Five weeks. *B,* Six weeks, showing breakdown of the oronasal membrane. *C,* Seven weeks, showing the nasal cavity communicating with the oral cavity and development of the olfactory epithelium. *D,* 12 weeks, showing the palate and the lateral wall of the nasal cavity. Most of the definitive palate is formed by the secondary palate.

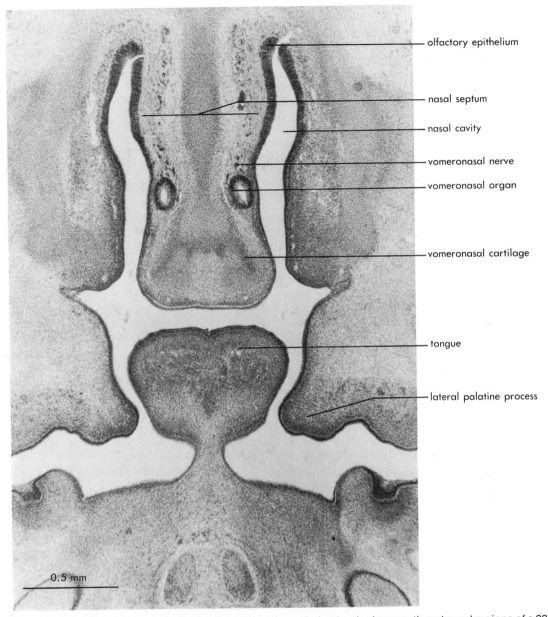

olfactory epithelium

nasal septum

nasal cavity

vomeronasal nerve

vomeronasal organ

vomeronasal cartilage

tongue

lateral palatine process

0.5 mm

Figure 10–26 Photomicrograph of a frontal section through the developing mouth and nasal regions of a 22 mm human embryo of about 54 days (×c.50). (Courtesy of Dr. Kunwar Bhatnagar, Associate Professor of Anatomy, University of Louisville, Louisville, Ky.)

remnants. These narrow strips of cartilage are located between the inferior edge of the cartilage of the nasal septum and the vomer.

Remnants of one or both vomeronasal organs may persist as cysts that may present wide orifices opening in the nasal vestibule on each side of the nasal septum. The remnants of these organs

usually remain undetected and asymptomatic, but in some cases they have been linked with various pathological conditions (Gabriele, 1967). *Atavistic remnants* in man, the vomeronasal organs are well developed in mammals and are considered to be olfactory chemoreceptor organs that aid the sense of smell, reproduction, and feeding.

DEVELOPMENT OF THE PALATE

The palate develops from two primordia: the *primary palate* and the *secondary palate*. Although palatogenesis begins toward the end of the fifth week, fusion of the palate's parts is not complete until about the twelfth week.

The Primary Palate (Figs. 10–22, 10–23, 10–24, and 10–25). The *primary palate,* or *median palatine process,* develops at the end of the fifth week from the innermost part of the intermaxillary segment of the maxilla. This segment, formed by merging of the medial nasal prominences, forms a wedge-shaped mass of mesoderm between the internal surfaces of the maxillary prominences of the developing maxillae.

The primary palate becomes the premaxillary part of the maxilla (Fig. 10–24B), which contains the incisor teeth. The primary palate gives rise to only a very small part of the adult hard palate (i.e., the small part anterior to the incisive foramen).

The Secondary Palate (Figs. 10–23, 10–24, 10–25, and 10–26). The secondary palate is the primordium of the hard and soft palates, extending from the region of the incisive foramen posteriorly. The secondary palate develops from two horizontal mesodermal projections that extend from the internal aspects of the maxillary prominences, called the *lateral palatine processes*. These shelf-like structures, often called *palatine shelves,* initially project downward on each side of the tongue (Fig. 10–23C). As the jaws and the neck develop, the tongue moves downward. As *palatogenesis* proceeds, the lateral palatine processes elongate and begin to move to a horizontal position superior to the tongue during the seventh week. The lateral palatine processes approach each other and fuse in the midline (Fig. 10–23E). They also fuse with the primary palate and the *nasal septum.*

The nasal septum develops as a downgrowth from the merged medial nasal prominences (Fig. 10–23D to H). The fusion between the nasal septum and the palatine processes begins anteriorly during the ninth week and is completed posteriorly in the region of the *uvula* by the twelfth week. Bone gradually develops in the primary palate, forming the *premaxillary part of the maxilla,* which carries the incisor teeth (Fig. 10–24B). Concurrently, bone extends from the maxillae and palatine bones into the lateral palatine processes to form the *hard palate* (Figs. 10–23G and 10–24B). The posterior portions of the lateral palatine processes do not become ossified but extend beyond the nasal septum and fuse to form the *soft palate and uvula* (Fig. 10–23D, F, and H). The *palatine raphe* permanently indicates the line of fusion of the lateral palatine processes (Fig. 10–23H).

A small *nasopalatine canal* persists in the midline of the palate between the premaxillary part of the maxilla and the palatine processes of the maxillae. Although eventually this canal is almost obliterated, it is represented in the adult hard palate by the *incisive foramen* (Fig. 10–24B). An irregular suture runs from the incisive foramen to the alveolar process between the lateral incisor and canine teeth on each side. *The incisive foramen serves as a landmark between the primary and secondary palates.*

CLEFT LIP AND CLEFT PALATE

Clefts of the upper lip and the palate are common and are especially conspicuous because they cause abnormalities of facial appearance and speech. There are two major groups of cleft lip and palate:

1. *Clefts involving the upper lip and the anterior part of the maxilla,* with or without involvement of parts of the remaining hard and soft regions of the palate; and

2. *Clefts involving the hard and soft regions of the palate.*

The term *complete cleft* indicates the maximum degree of clefting of any particular type; for example, a *complete cleft of the posterior palate* is a malformation in which the cleft extends through the soft palate and anteriorly to the incisive foramen.

The *incisive foramen* (Fig. 10–24B) serves as a landmark for distinguishing anterior from posterior cleft malformations. *Anterior cleft malformations* include cleft lip, with or without cleft of the alveolar part of the maxilla. A complete anterior cleft malformation is one in which the cleft extends through the lip and the alveolar part of the maxilla to the incisive foramen, separating the primary and secondary palates. *Posterior cleft malformations* include clefts of the secondary, or posterior, palate that extend through the soft palate and the hard palate to the incisive foramen, separating the secondary palate from the primary palate.

Anterior and posterior cleft malformations are embryologically distinct. *Anterior cleft*

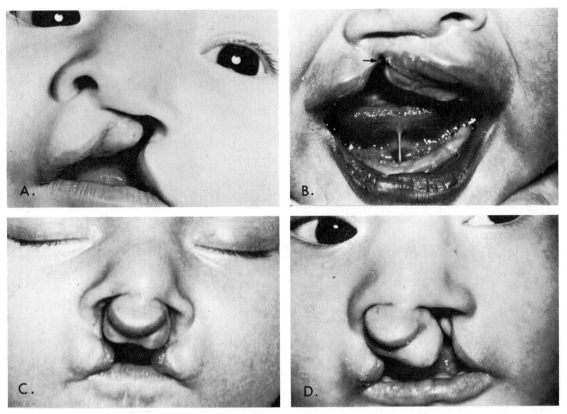

Figure 10–27 Photographs illustrating clefts of the lip. This malformation used to be referred to as "hare lip"; this is an inappropriate term because the hare's lip is divided in the midline. *A* and *B*, Unilateral cleft lip. The cleft in *B* is incomplete; the arrow indicates a band of tissue (Simonart's band) connecting the parts of the lip. *C* and *D*, Bilateral cleft lip. (Courtesy of Dr. D. A. Kernahan, The Children's Memorial Hospital, Chicago.)

malformations are caused by defective development of the primary palate and result from a deficiency of mesenchyme in the maxillary prominence(s) and the intermaxillary segment. *Posterior cleft malformations are caused by defective development of the secondary palate* and result from growth distortions of the lateral palatine processes, which prevent their fusion.

Cleft Lip (Figs. 10–27 and 10–28). Clefts involving the upper lip, with or without cleft palate, occur about once in 1000 births, but their frequency varies widely among ethnic groups (Thompson and Thompson, 1980). About 60 to 80 per cent of affected infants are males. The clefts vary from small notches of the vermilion border of the lip to larger divisions that extend into the floor of the nostril and through the alveolar part of the maxilla. Cleft lip can be either unilateral or bilateral.

Unilateral cleft lip results from failure of the maxillary prominence on the affected side to unite with the merged medial nasal prominences (Fig. 10–28*C* to *H*). This is the consequence of failure of the mesenchymal masses to merge and the mesenchyme to proliferate and push out the overlying epithelium (Fig. 10–28*D*). The result is a *persistent labial groove* (Fig. 10–28*C*). In addition, the epithelium in the labial groove becomes stretched, and then breakdown of tissues in the floor of the persistent groove leads to division of the lip into medial and lateral parts (Figs. 10–27*A* and 10–28*G* and *H*). Sometimes, a bridge of tissue called *Simonart's band* joins the parts of the incomplete cleft lip (Fig. 10–27*B*).

Bilateral cleft lip results from failure of the mesenchymal masses of the maxillary prominences to meet and unite with the merged medial nasal prominences. The epithelium in both labial grooves becomes stretched and breaks down. In bilateral cases, the defects may be similar or dissimilar, with varying degrees of defect on each side. In complete

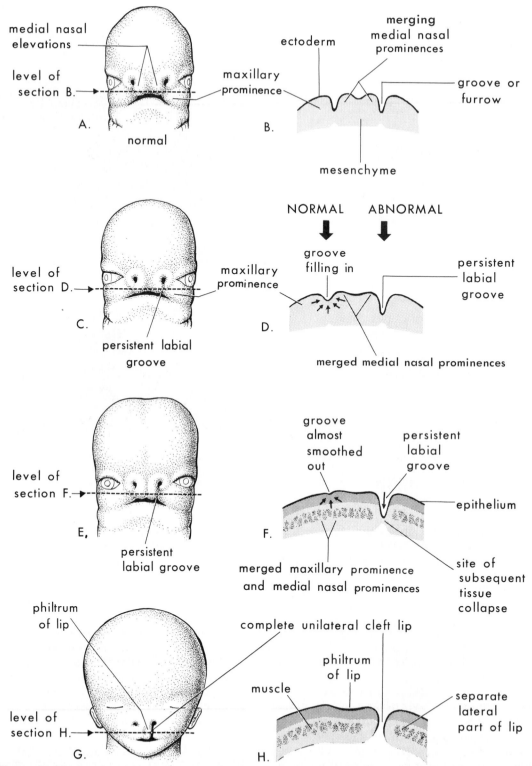

Figure 10–28 Drawings illustrating the embryological basis of complete unilateral cleft lip. *A,* Five-week embryo. *B,* Horizontal section through the head, illustrating the grooves between the maxillary prominences and the merging medial nasal prominences. *C,* Six-week embryo, showing a persistent labial groove on the left side. *D,* Horizontal section through the head, showing the groove gradually filling in on the right side because of proliferation of the mesenchyme (arrows). *E,* Seven-week embryo. *F,* Horizontal section through the head, showing that the epithelium on the right has almost been pushed out of the groove between the maxillary prominence and medial nasal prominence. *G,* 10-week fetus with a complete unilateral cleft lip. *H,* Horizontal section through the head after stretching of the epithelium and breakdown of the tissues in the floor of the persistent labial groove on the left side, forming a complete unilateral cleft lip.

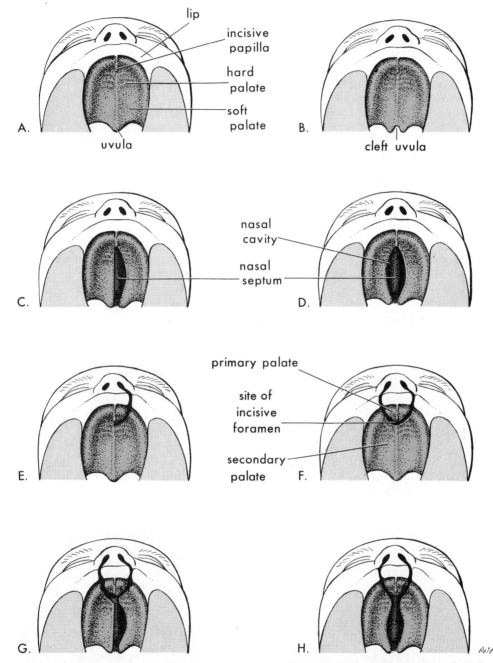

Figure 10–29 Drawings of various types of cleft lip and cleft palate. *A*, Normal lip and palate. *B*, Cleft uvula. *C*, Unilateral cleft of the posterior or secondary palate. *D*, Bilateral cleft of the posterior palate. *E*, Complete unilateral cleft of the lip and alveolar process with a unilateral cleft of the anterior or primary palate. *F*, Complete bilateral cleft of the lip and alveolar process with bilateral cleft of the anterior palate. *G*, Complete bilateral cleft of the lip and alveolar process with bilateral cleft of the anterior palate and unilateral cleft of the posterior palate. *H*, Complete bilateral cleft of the lip and alveolar process with complete bilateral cleft of the anterior and posterior palate. Although not illustrated here, cleft lip may occur without cleft palate (see Fig. 10–27A).

bilateral cleft of the upper lip and alveolar processes, the intermaxillary segment hangs free and projects anteriorly (Figs. 10–27C and 10–30B). Such defects are especially deforming because of the loss of continuity of the orbicularis oris muscle, which closes the mouth and purses the lips as in whistling.

Median Cleft Lip (Fig. 10–31A). This extremely rare defect of the upper lip is caused by mesodermal deficiency, which results in partial or complete failure of the medial nasal prominences to merge and form the intermaxillary segment. A midline cleft of the upper lip is a characteristic feature of the Mohr syndrome, which is transmitted as an autosomal recessive trait (Goodman and Gorlin, 1970). Median cleft of the lower lip (Fig. 10–31B) is also very rare and is caused by failure of the mesenchymal masses of the mandibular prominences to meet and merge completely.

Cleft Palate (Figs. 10–29 and 10–30). Cleft palate, with or without cleft lip,

occurs about once in 2500 births and is more common in females than in males. The cleft may involve only the uvula, giving it a fishtail appearance (Fig. 10–29B), or it may extend through the soft and hard palates (Figs. 10–29C and D and 10–30C and D). In severe cases associated with cleft lip, the cleft in the palate extends through the alveolar process and lip on both sides (Figs. 10–29G and H and 10–30B).

The embryological basis of cleft palate is failure of the mesenchymal masses of the lateral palatine processes to meet and fuse with each other, with the nasal septum, and/or with the posterior margin of the median palatine process, or primary palate (Fig. 10–29). Such clefts may be unilateral or bilateral and are classified into three groups:

1. *Clefts of the anterior, or primary, palate,* i.e., clefts anterior to the incisive fora-

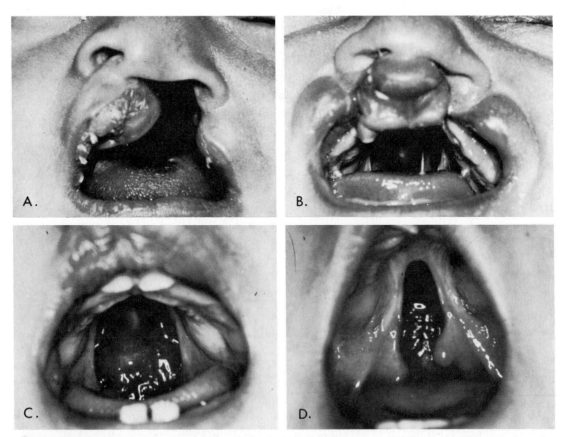

Figure 10–30 Photographs illustrating congenital malformations of the lip and palate. *A,* Complete unilateral cleft of the lip and alveolar process. *B,* Complete bilateral cleft of the lip and alveolar process with bilateral cleft of the anterior palate. *C* and *D,* Bilateral cleft of the posterior or secondary palate; the lip is normal (Courtesy of Dr. Harry Medovy, Children's Centre, Winnipeg, Canada.)

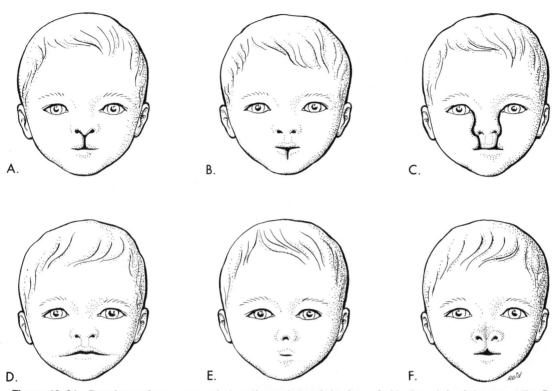

Figure 10–31 Drawings of rare congenital malformations of the face. *A*, Median cleft of the upper lip. *B*, Median cleft of the lower lip. *C*, Bilateral oblique facial clefts with complete bilateral cleft lip. *D*, Macrostomia or lateral facial cleft. *E*, Single nostril and microstomia; these malformations are not usually associated. *F*, Bifid nose and incomplete median cleft lip.

men resulting from failure of the mesenchymal masses of the lateral palatine processes to meet and fuse with the mesenchyme of the primary palate (Fig. 10–29*E* and *F*).

2. *Clefts of the anterior and posterior palate*, i.e., clefts involving both the primary and secondary palates, resulting from the failure of the mesenchymal masses of the lateral palatine processes to meet and fuse with the mesenchyme of the primary palate, with each other, and with the nasal septum (Fig. 10–29*G* and *H*).

3. *Clefts of the posterior, or secondary, palate*, i.e., clefts posterior to the incisive foramen, resulting from failure of the mesenchymal masses of the lateral palatine processes to meet and fuse with each other and with the nasal septum (Fig. 10–29*B* to *D*).

Causes of Cleft Lip and Cleft Palate. *The great majority of cases of cleft lip and cleft palate are determined by multiple factors,* genetic and possibly also nongenetic, each causing only a minor developmental defect. This is called *multifactorial inheritance.* These factors seem to operate by influencing the amount of neural crest mesenchyme that migrates into the embryonic facial primordia (Ross and Johnston, 1972). If this amount is insufficient, clefting of the lip and/or palate occurs. Some clefts of the lip and/or palate appear as part of syndromes determined by single mutant genes (Fraser, 1980, and Thompson and Thompson, 1980). Other cases are parts of chromosomal syndromes, especially trisomy 13 (see Fig. 8–6). A few cases of cleft lip and/or palate appear to have been caused by teratogenic agents (e.g., anticonvulsant drugs; Hanson, 1980).

Studies of twins indicate that genetic factors are of more importance in cleft lip, with or without cleft palate, than in cleft palate alone. A sibling of a child with a cleft palate has an elevated risk of having a cleft palate, but no increased risk of having a cleft lip. Clefts of the lip and alveolar process that continue through the palate are usually transmitted through a male sex-linked gene (Matthews, 1971). When neither parent is affected, the *recurrence risk* in subsequent siblings (brother or sister) is about 4 per cent. For further examples of recurrence risks, see Thompson and Thompson (1980).

The fact that the palatine processes fuse about a week later in females (Burdi, 1969) may explain why isolated cleft palate is more common in females than in males (e.g., among Japanese, they occur in 0.63 per 1000 live births, and, of those, 34 per cent are in males and 66 per cent in females; Witkop et al., 1967).

Facial Clefts. Various types of facial cleft may occur, but they are all *extremely rare.* Severe clefts are usually associated with gross malformations of the head. In *median cleft of the mandible* (Fig. 10–31B), there is a deep cleft resulting from failure of the mesenchymal masses of the mandibular prominences of the first branchial arch to merge completely with each other. *Oblique facial clefts* (orbitofacial fissures) are often bilateral and extend from the upper lip to the medial margin of the orbit (Fig. 10–31C). When this occurs, the nasolacrimal ducts are open grooves (persistent nasolacrimal grooves). Oblique facial clefts associated with cleft lip result from failure of the mesenchymal masses of the maxillary prominences to merge with the lateral and medial nasal prominences. *Lateral, or transverse, facial clefts* run from the mouth toward the ear; bilateral clefts result in a very large mouth, a condition called *macrostomia* (Fig. 10–31D). This abnormality results from failure of the lateral mesenchymal masses of the maxillary and mandibular prominences to merge. In severe cases, the cheek is cleft almost to the ear.

Other Facial Malformations. *Congenital microstomia* (small mouth) results from excessive merging of the mesenchymal masses of the maxillary and mandibular prominences of the first arch (Fig. 10–31E). In severe cases, the abnormality may be associated with underdevelopment (hypoplasia) of the mandible. *Absence of the nose occurs* when no nasal placodes form. A *single nostril* results when only one nasal placode forms (Fig. 10–31E). *Bifid nose* results when the medial nasal prominences do not merge completely; the nostrils are widely separated and the nasal bridge is bifid (Fig. 10–31F). In mild forms a small groove is present in the tip of the nose.

SUMMARY

During the fourth and fifth weeks, the primitive pharynx is bounded laterally by distinctive bar-like *branchial arches.* Each arch consists of a core of mesenchymal tissue covered externally by *surface ectoderm* and internally by *endoderm.* The mesenchymal core also contains numerous *cranial neural crest cells* that migrate into the branchial arches. Each branchial arch also contains an artery, a cartilage bar, a nerve, and a muscular component. Externally, between the arches, are *branchial grooves.* Internally, between the arches are extensions of the pharynx called *pharyngeal pouches.* The ectoderm of each branchial groove contacts the endoderm of each pharyngeal pouch, thus forming *branchial membranes.* The pharyngeal pouches and the branchial arches, grooves, and membranes make up the *branchial apparatus,* which contributes greatly to the formation of the head and neck.

Development of the tongue, the face, the lips, the jaws, the palate, the pharynx, and the neck largely involves transformation of the branchial apparatus into adult structures. The derivatives of the various branchial arch components are summarized in Table 10–1, and the adult derivatives of the pharyngeal pouches are illustrated in Figure 10–7. The branchial grooves disappear except for the first, which persists as the *external acoustic meatus.* The branchial membranes also disappear, except for the first, which becomes the *tympanic membrane.* The pharyngeal pouches give rise to the tympanic cavity and the mastoid antrum, the auditory tube, the palatine tonsil, the thymus, and the parathyroid glands.

The *thyroid gland* develops from a downgrowth from the floor of the pharynx in the region where the tongue develops. The parafollicular cells in the thyroid gland are derived from the *ultimobranchial bodies,* formed mainly by the fourth pair of pharyngeal pouches.

Most congenital malformations of the head and the neck originate during transformation of the branchial apparatus into adult structures. *Branchial cysts, sinuses,* or *fistulas* may develop from parts of the second branchial groove, the cervical sinus, or the second pharyngeal pouch that fail to obliterate. An *ectopic thyroid gland* results when the thyroid gland fails to descend completely from its site of origin in the tongue. The thyroglossal duct may persist, or remnants of it may give rise to *thyroglossal duct cysts;* these cysts, if infected, may form *thyroglossal duct sinuses* that open in the midline of the neck.

Because of the complicated development

of the face and palate, congenital malformations resulting from an arrest of development and/or a failure of fusion of the prominences and processes involved in the development of the face and palate are not uncommon.

Cleft lip is the most common congenital abnormality of the face. Although cleft lip is frequently associated with cleft palate, cleft lip and cleft palate are etiologically distinct malformations that involve different developmental processes occurring at different times. *Cleft lip* results from failure of the mesenchymal masses of the medial nasal and the maxillary prominences to merge, whereas *cleft palate* results from failure of the mesenchymal masses of the palatine processes to fuse.

The great majority of cases of cleft lip, with or without cleft palate, are caused by a combination of genetic and environmental factors *(multifactorial inheritance)*. These factors appear to act by influencing the amount of *neural crest mesenchyme* that develops in the maxillary prominences of the first branchial arch. If this amount is insufficient, clefting of the lip and/or palate occurs.

SUGGESTIONS FOR ADDITIONAL READING

Bernstein, L.: Congenital malformations of the mouth and face; in C. F. Ferguson and E. L. Kendig, Jr. (Eds.): *Disorders of the Respiratory Tract in Children. Vol. II. Pediatric Otolaryngology.* 2nd ed. Philadelphia, W. B. Saunders Co., 1972, pp. 1024–1036. A good account of congenital malformations involving the mouth and face. Clinical photographs and treatments are described.

Jaffee, B. F.: The branchial arches; normal development and abnormalities; in C. F. Ferguson and E. L. Kendig, Jr. (Eds.): *Disorders of the Respiratory Tract in Children. Vol. II. Pediatric Otolaryngology.* 2nd ed. Philadelphia, W. B. Saunders Co., 1972, pp. 1118–1125. Clinical presentations of branchial sinuses and cysts, of thyroglossal duct sinuses and cysts, and of other malformations of the pharynx. Good clinical photographs are included.

Pratt, R. M., and Christiansen, R. L. (Eds.): *Current Research Trends in Prenatal Craniofacial Development.* New York, Elsevier North-Holland, 1980. This book represents the proceedings of a conference held at the National Institute of Health in Bethesda, Maryland. There are 24 papers on recent and productive areas of prenatal craniofacial development, normal and abnormal.

Ross, R. B., and Johnston, M. C.: *Cleft Lip and Palate.* Baltimore, The Williams & Wilkins Co., 1972. A broad-based general text giving a comprehensive analysis of basic cleft lip and cleft palate problems, as well as an analysis of clinical procedures.

CLINICALLY ORIENTED PROBLEMS

1. A two-year-old boy had had an intermittent discharge of mucoid material from a small opening in the side of his neck, but the discharge had stopped a week ago. There was extensive redness and swelling in the inferior third of the neck, anterior to the sternocleidomastoid muscle. What is the probable embryological basis of the intermittent discharge?
2. During a partial, or subtotal, thyroidectomy, a surgeon could locate only one inferior parathyroid gland. Where might the other one be located? What is the embryological basis for this condition?
3. A young woman consulted her doctor about a midline swelling in her neck, just inferior to her hyoid bone. What kind of a cyst might be present? Discuss its embryological basis.
4. A male infant was born with a unilateral cleft lip extending into the floor of the nose and through the alveolar part of the maxilla. Neither parent had cleft lip or cleft palate. Is this malformation more common in males? What is the chance that the next child will have a cleft lip?
5. An epileptic mother who was treated with anticonvulsant drugs during pregnancy gave birth to a child with cleft lip and cleft palate. Is there any evidence indicating that these drugs increase the incidence of cleft lip and cleft palate? Discuss the respective etiologies of these two malformations.

The answers to these questions are given at the back of the book.

REFERENCES

Albers, G. D.: Branchial anomalies. *J.A.M.A.* *183*:399, 1963.

Anast, C. S.: Calcitonin; *in* L. I. Gardner (Ed.): *Endocrine and Genetic Diseases of Childhood and Adolescence.* 2nd ed. Philadelphia, W. B. Saunders Co., 1975, pp. 425–456.

Burdi, A. R.: Sexual differences in closure of the human palatal shelves. *Cleft Palate J.* *6*:1, 1969.

Carter, C. D.: Incidence and aetiology; *in* A. P. Norman (Ed.): *Congenital Abnormalities in Infancy.* 2nd ed. Oxford, Blackwell Scientific Publications, 1971.

Copp, D. H., Cameron, E. C., Cheney, B. A., Davidson, A. G. F., and Henze, K. G.: Evidence for calcitonin — a new hormone from the parathyroid that lowers blood calcium. *Endocrinology 70*:638, 1962.

Crelin, E. S.: *Development of the Upper Respiratory System.* New Jersey, Clinical Symposia, Vol. 28, No. 3, 1976.

Curtis, E., Fraser, F. C., and Warburton, D.: Congenital cleft lip and palate. Risk figures for counseling. *Am. J. Dis. Child. 102*:853, 1961.

Diewert, V. M.: The role of craniofacial growth in palatal shelf elevation; *in* R. M. Pratt and R. Christiansen (Eds.): *Current Research Trends in Prenatal Craniofacial Development.* New York, Elsevier North-Holland, 1980, pp. 165–186.

Fogh-Anderson, P.: Inheritance patterns for cleft lip and cleft palate; *in* S. Pruzansky (Ed.): *Congenital Anomalies of the Face and Associated Structures.* Springfield, Illinois, Charles C Thomas, Publisher, 1961, pp. 123–133.

Fraser, F. C.: Genetic counseling in some common pediatric diseases. *Pediatr. Clin. North Am. 5*:475, 1958.

Fraser, F. C.: The multifactorial/threshold concept — uses and misuses. *Teratology 14*:267, 1976.

Fraser, F. C.: The genetics of cleft lip and palate: yet another look; *in* R. M. Pratt and R. L. Christiansen (Eds.): *Current Research Trends in Prenatal Craniofacial Development.* New York, Elsevier North-Holland, 1980, pp. 357–366.

Gabriele, O. F.: Persistent vomeronasal organ. *Am. J. Roentgenol. Radium Ther. Nucl. Med. 99*:697, 1967.

Goodman, R. M., and Gorlin, R. J.: *The Face in Genetic Disorders.* St. Louis, The C. V. Mosby Co., 1970.

Gorlin, R. J., Pindborg, J. J., and Cohen, M. M.: *Syndromes of the Head and Neck.* 2nd ed. New York, McGraw-Hill Book Co., 1975.

Goss, A. N.: Human palatal development *in vitro. Cleft Palate J. 12*:210, 1975.

Gray, S. W., and Skandalakis, J.: *Embryology for Surgeons. The Embryological Basis for the Treatment of Congenital Defects.* Philadelphia, W. B. Saunders Co., 1972, pp. 15–61.

Ham, A. W., and Cormack, D. H.: *Histology.* 8th ed. Philadelphia, J. B. Lippincott Co., 1979, p. 336.

Hanson, J. W.: Patterns of abnormal human craniofacial development; *in* R. M. Pratt and R. L. Christiansen (Eds.): *Current Research Trends in Prenatal Craniofacial Development.* New York, Elsevier North-Holland, 1980, pp. 345–366.

Hayden, G. D., and Arnold, G. G.: The ear; *in* E. L. Kendig, Jr. and V. Chernick (Eds.): *Disorders of the Respiratory Tract in Children.* 3rd ed. Philadelphia, W. B. Saunders Co., 1977.

Karmody, C. S.: Autosomal dominant first and second arch syndrome; *in* D. Bergsma (Ed.): *Malformation Syndromes.* New York, International Medical Book Corp., Vol. 10, 1974, p. 31.

Kernahan, D. A., and Stark, R. B.: A new classification for cleft lip and cleft palate. *Plast. Reconstr. Surg. 22*:435, 1958.

Lavelle, C. L. B.: An analysis of foetal craniofacial growth. *Ann. Hum. Biol. 1*:269, 1974.

MacCollum, D. W., and Rubin, A.: Cleft lip and cleft palate; *in* A. Rubin (Ed.): *Handbook of Congenital Malformations.* Philadelphia, W. B. Saunders Co., 1967, pp. 114–117.

MacCollum, D. W., and Witkop, C. J., Jr.: Lateral and oblique facial clefts and Robin syndrome; *in* A. Rubin (Ed.): *Handbook of Congenital Malformations.* Philadelphia, W. B. Saunders Co., 1967, pp. 112–114.

McKenzie, J.: The first arch syndrome. *Dev. Med. Child. Neurol. 8*:55, 1966.

Marshall, S. F., and Becker, W. F.: Thyroglossal cysts and sinuses. *Ann. Surg. 129*:642, 1949.

Martins, A. G.: Lateral cervical sinus and pre-auricular sinuses. *Br. Med. J. 5*:255, 1961.

Mathews, D. N.: Hare lip and cleft palate; *in* J. C. Mustardé (Ed.): *Plastic Surgery in Infancy and Childhood.* Edinburgh, E. and S. Livingstone, Ltd., 1971, pp. 1–36.

Meller, S. M.: Morphological alterations in the prefusion human palatal epithelium; *in* R. M. Pratt and R. L. Christiansen (Eds.): *Current Research Trends in Prenatal Craniofacial Development.* New York, Elsevier North-Holland, 1980, pp. 221–234.

Melsen, B.: Palatal growth studies on human autopsy material. *Am. J. Orthod. 68*:42, 1975.

Moore, M. A. S., and Owen, J. J. T.: Experimental studies on the development of the thymus. *J. Exp. Med. 126*:715, 1967.

Moseley, J. M., Mathews, E. W., Breed, R. H., Galante, E., Tse, A., and MacIntyre, I.: The ultimobranchial origin of calcitonin. *Lancet 1*:108, 1968.

Mustardé, J. C.: *Plastic Surgery in Infancy and Childhood.* Edinburgh, E. and S. Livingstone, Ltd., 1971.

Noden, D. M.: The migration and cytodifferentiation of cranial neural crest cells; *in* R. M. Pratt and R. L. Christiansen (Eds.): *Current Research Trends in Prenatal Craniofacial Development.* New York, Elsevier North-Holland, 1980, pp. 3–25.

Norman, A. P.: *Congenital Abnormalities in Infancy.* 2nd ed. Oxford, Blackwell Scientific Publications, 1971.

Patten, B. M.: Normal development of the facial region; *in* S. Pruzansky (Ed.): *Congenital Anomalies of the Face and Associated Structures.* Springfield, Illinois, Charles C Thomas, Publisher, 1961, pp. 11–45.

Patten, B. M.: *Human Embryology.* 3rd ed. New York, Blakiston Division. McGraw-Hill Book Co., 1968, pp. 427–448.

Pearse, A. G. E., and Polak, J. M.: The neural crest origin of the endocrine polypeptide cells of the APUD series. *Gut 12*:783, 1971.

Remnick, H.: *Embryology of the Face and Oral Cavity.* Rutherford, N. J., Fairleigh Dickinson University Press, 1970.

Rosen, F. S.: The thymus gland and the immune deficiency syndromes; *in* M. Samter, D. W. Talmage, B. Rose, W. B. Sherman and J. H. Vaughan (Eds.): *Immunological Diseases,* Vol. I. 2nd ed. Boston, Little, Brown and Co., 1965, pp. 506–507.

Ross, R. B., and Johnston, M. C.: *Cleft Lip and Palate.* Baltimore, The Williams & Wilkins Co., 1972.

Sanel, F. T.: Ultrastructure of differentiating cells during thymus histogenesis. A light and electron microscopic study of epithelial and lymphoid cell differentiation during thymus histogenesis in C 57 black mice. *Z. Zellforsch. 83*:8, 1967.

Schaffer, A. J., and Avery, M. E.: *Diseases of the Newborn.* 4th ed. Philadelphia, W. B. Saunders Co., 1977.

Shepard, T. H.: Development of the thyroid gland; *in* L. I. Gardner (Ed.): *Endocrine and Genetic Diseases of Childhood and Adolescence.* 2nd ed. Philadelphia, W. B. Saunders Co., 1975, pp. 220–225.

Small, A.: The surgical removal of branchial sinus. *Lancet 2*:891, 1960.

Smith, D. W.: *Recognizable Patterns of Human Malformations: Genetic, Embryologic and Clinical Aspects.* 2nd ed. Philadelphia, W. B. Saunders Co., 1976.

Smith, D. W.: Patterns of malformation; *in* V. C. Vaughan, R. J. McKay, and R. E. Behrman (Eds.): *Nelson Textbook of Pediatrics.* 11th ed. Philadelphia, W. B. Saunders Co., 1979, pp. 2035–2051.

Sperber, G. H.: Development of the dentition (odontogenesis); *in* Sperber, G. H.: *Craniofacial Embryology.* 2nd ed. Bristol, John Wright and Sons Ltd. (distributed by Year Book Medical Publishers, Inc., Chicago), 1976.

Stark, R. B.: Embryology, pathogenesis and classification of cleft lip and cleft palate; *in* S. Pruzansky (Ed.): *Congenital Anomalies of the Face and Associated Structures.* Springfield, Illinois, Charles C Thomas, Publisher, 1961, pp. 66–84.

Swenson, O.: *Pediatric Surgery.* New York, Appleton-Century-Crofts, Inc., 1958.

Thompson, J. S., and Thompson, M. W.: *Genetics in Medicine.* 3rd ed. Philadelphia, W. B. Saunders Co., 1980.

Vaughan, V. C., McKay, R. J., and Behrman, R. E. (Eds.): *Nelson Textbook of Pediatrics.* 11th ed. Philadelphia, W. B. Saunders Co., 1979, p. 594.

Villee, D. B.: *Human Endocrinology. A Developmental Approach.* Philadelphia, W. B. Saunders Co., 1975, p. 54.

Walker, D. G.: *Malformations of the Face.* Edinburgh, E. and S. Livingstone, Ltd., 1961.

Warbrick, J. G.: The early development of the nasal cavity and upper lip in the human embryo. *J. Anat. 94*:351, 1960.

Wilson, C. P.: Lateral cysts and fistulas of the neck of developmental origin. *Ann. R. Coll. Surg. Engl. 17*:1, 1955.

Witkop, C. J., MacCollum, D. W., and Rubin, A.: Cleft lip and cleft palate; *in* A. Rubin (Ed.): *Handbook of Congenital Malformations.* Philadelphia, W. B. Saunders Co., 1967, pp. 103–156.

11

THE RESPIRATORY SYSTEM

The lower respiratory system begins to form during stage 12 of development (26 to 27 days) and is first indicated by a median *laryngotracheal groove* in the caudal end of the ventral wall of the pharynx, the *primitive pharyngeal floor* (Fig. 11–1). This groove produces a ridge on the external surface of the primitive pharynx.

The endoderm lining the laryngotracheal groove gives rise to the epithelium and glands of the larynx, the trachea, and the bronchi and to the pulmonary lining epithelium. The connective tissue, the cartilage, and the smooth muscle of these structures develop from splanchnic mesenchyme surrounding the foregut (Fig. 11–4).

The laryngotracheal groove deepens and the external ridge expands to form a *laryngotracheal diverticulum* ventral to the primitive pharynx (Fig. 11–2A). As this diverticulum grows ventrocaudally from the pharyngeal floor, it is invested with splanchnic mesenchyme.

The laryngotracheal diverticulum soon becomes separated from the foregut by longitudinal ridges called *tracheoesophageal*

folds (Fig. 11–2D). These folds soon fuse to form a partition, known as the *tracheoesophageal septum* (Fig. 11–2E), which divides the foregut into a ventral portion, the *laryngotracheal tube* (primordium of the larynx, the trachea, and the lungs), and a dorsal portion, the *esophagus* (Fig. 11–2F).

The laryngotracheal tube opens into the pharynx. Its opening becomes the *laryngeal aditus,* the entrance to the vestibule of the larynx. The laryngotracheal tube and the surrounding splanchnic mesenchyme give rise to the larynx, the trachea, the bronchi, and the lungs (Fig. 11–3, 11–4, and 11–5).

Development of most of the *upper respiratory system* (the nose, the nasopharynx, and the oropharynx) is described in Chapter 10.

DEVELOPMENT OF THE LARYNX

The larynx develops from the endodermal lining of the cranial end of the laryngotracheal tube and from the surrounding mesenchyme derived from the fourth and sixth pairs of branchial arches (Fig. 11–3A). This mesenchyme proliferates rapidly at the cra-

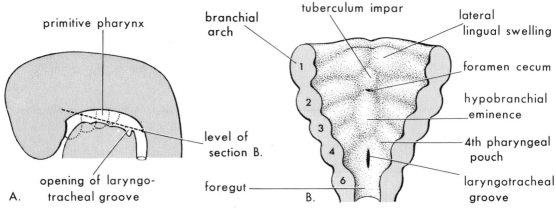

Figure 11–1 *A,* Diagrammatic sagittal section of the cranial half of an embryo during stage 12 of development (about 26 days) showing the laryngotracheal groove. *B,* Horizontal section at the level shown in *A,* illustrating the floor of the primitive pharynx and the location of the laryngotracheal groove. The opening into the laryngotracheal groove represents the future inlet of the larynx.

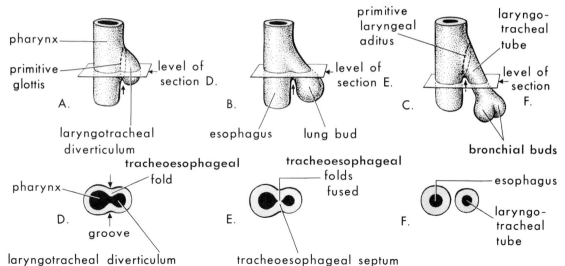

Figure 11–2 Successive stages of development of the tracheoesophageal septum during the fourth week. *A, B,* and *C,* Lateral views of the caudal part of the primitive pharynx illustrating partitioning of the foregut into the esophagus and the laryngotracheal tube. *D, E,* and *F,* Transverse sections illustrating development of the tracheoesophageal septum and separation of the foregut into the laryngotracheal tube and the esophagus.

nial end of the laryngotracheal tube, producing paired *arytenoid swellings* (Fig. 11–3*B*). These swellings convert the slit-like aperture, or *primitive glottis,* into a T-shaped opening, and they reduce the developing laryngeal lumen to a narrow slit. The laryngeal epithelium proliferates rapidly, resulting

in a temporary occlusion of the laryngeal lumen. Recanalization occurs by the tenth week. During this process, the *laryngeal ventricles* form. These recesses are bounded by folds of mucous membranes that become the *vocal folds* and the *vestibular folds*. Laryngeal cartilages develop within the ary-

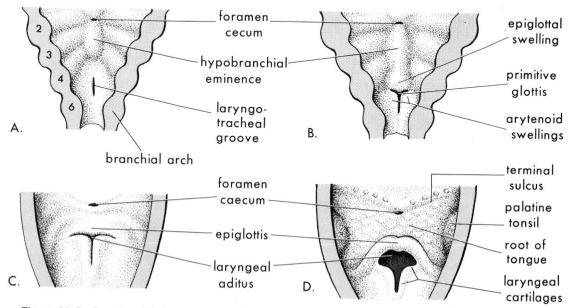

Figure 11–3 Drawings illustrating successive stages of development of the larynx. *A,* 4 weeks. *B,* 5 weeks, *C,* 6 weeks, *D,* 10 weeks. The internal lining of larynx is of endodermal origin. The cartilages and muscles of the larynx arise from the mesenchyme of the fourth and sixth pair of branchial arches. Note that the laryngeal inlet, or aditus, changes in shape from a slit-like opening into a T-shaped inlet as the mesenchyme proliferates.

tenoid swellings from the cartilages of the branchial arches (see Table 10–1). The epiglottis develops from the caudal part of the *hypobranchial eminence,* a derivative of the third and fourth branchial arches (Fig. 11–3B to D).

The *laryngeal muscles* develop from muscle elements in the fourth and sixth branchial arches and are therefore innervated by the laryngeal branches of the vagus nerves that supply those arches (see Table 10–1). Persons desiring a detailed description of the development of the larynx should consult Hast (1976) and/or Crelin (1976).

Laryngeal Web. This rare malformation results from incomplete recanalization of the larynx during the tenth week. A membranous web forms at the level of the vocal folds, partially obstructing the airways.

DEVELOPMENT OF THE TRACHEA

The laryngotracheal tube distal to the larynx gives rise to the trachea and the lung buds. The endodermal lining of the segment of the laryngotracheal tube between the de-

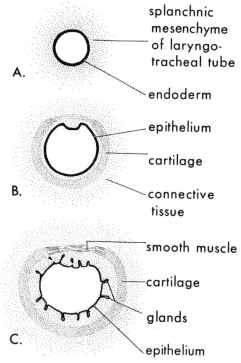

Figure 11–4 Drawings of transverse sections through the laryngotracheal tube illustrating progressive stages of development of the trachea. *A,* 4 weeks. *B,* 10 weeks. *C,* 11 weeks.

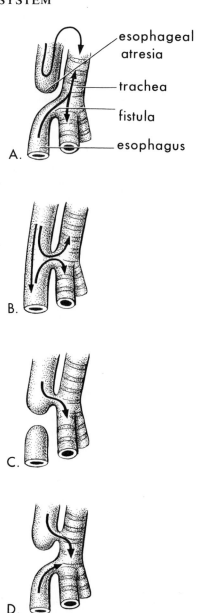

Figure 11–5 Sketches illustrating the four main varieties of tracheoesophageal fistula. Possible direction(s) of flow of contents is indicated by arrows. Esophageal atresia, as illustrated in *A,* occurs in about 90 per cent of cases. The abdomen rapidly becomes distended as the intestines fill with air. In *C,* air cannot enter the lower esophagus and the stomach.

veloping larynx and the lung buds gives rise to the epithelium and the glands of the trachea. The cartilage, the connective tissue, and the muscle are derived from the surrounding splanchnic mesenchyme (Fig. 11–4).

Tracheosophageal Fistula (Fig. 11–5).

A communication, or fistula, connecting the trachea and the esophagus occurs about once in every 2500 births; most infants affected are males. Tracheoesophageal fistula is usually associated with *esophageal atresia;* in all cases, there is an abnormal communication between the trachea and the esophagus.

Tracheoesophageal fistula results from incomplete division of the foregut into respiratory and digestive portions during the fourth and fifth weeks. Incomplete fusion of the tracheoesophageal folds results in a defective tracheoesophageal septum, leaving a communication between the trachea and the esophagus. There are four main varieties of tracheoesophageal fistula. The most common abnormality is for the upper portion of the esophagus to end blindly (esophageal atresia) and for the lower portion to join the trachea near its bifurcation (Fig. 11–5A). Other varieties of this malformation are illustrated in Fig. 11–5B to D.

Infants with esophageal atresia and tracheoesophageal fistula (Fig. 11–5A) cough and choke on swallowing owing to excessive amounts of saliva accumulating in the mouth and the upper respiratory tract. When the infant swallows milk and saliva, they rapidly fill the esophageal pouch and are regurgitated. They then pass into the trachea, resulting in gagging, coughing, and *respiratory distress.*

Gastric contents may also reflux through the fistula into the trachea and the lungs from the stomach. This may result in pneumonia or pneumonitis (inflammation of the lungs).

An excess of amniotic fluid (*polyhydramnios*) may be associated with esophageal atresia and tracheoesophageal fistula (Fig. 11–5A), because amniotic fluid may not pass to the stomach and intestines for absorption and subsequent placental transfer to the mother's blood for disposal.

Tracheal Stenosis and Atresia.   Narrowing, or stenosis, and closure, or atresia, of the trachea are rare malformations and are usually associated with one of the varieties of tracheoesophageal fistula (Fig. 11–5). Stenoses and atresias probably result from unequal partitioning of the foregut into the esophagus and the trachea.

Tracheal Diverticulum.   This rare deformity consists of a blind, bronchus-like projection from the trachea. Such a diverticulum may terminate in normal-appearing lung tissue, forming a so-called *tracheal lobe.*

DEVELOPMENT OF THE BRONCHI AND THE LUNGS

A *lung bud* develops at the caudal end of the laryngotracheal tube (Fig. 11–6A) and soon divides into two knob-like *bronchial buds* (Fig. 11–6B). These endodermal buds, together with the surrounding splanchnic mesenchyme, differentiate into the bronchi and their ramifications in the lungs.

Early in the fifth week, each bronchial bud enlarges to form a primitive *primary bronchus* (Fig. 11–6E). The embryonic right bronchus is slightly larger than the left one and is oriented more vertically; this embryonic relationship persists, and consequently a foreign body is more liable to enter the right primary bronchus than the left.

During the fifth week, the endodermal lung buds grow laterally into the medial walls of the *pericardioperitoneal canals,* or *primitive pleural cavities* (Fig. 11–7A). Concurrently, the primary bronchi subdivide into *secondary bronchi* (Fig. 11–6F). On the right, the superior secondary bronchus will supply the superior lobe of the lung, whereas the inferior secondary bronchus soon subdivides into two bronchi, one to the middle lobe of the right lung and the other to the inferior lobe (Fig. 11–6G). On the left, the two secondary bronchi supply the superior and inferior lobes of the lung.

Each secondary bronchus subsequently undergoes progressive dichotomous branching; that is, each branch bifurcates repeatedly into branches. Tertiary (segmental) bronchi, 10 in the right lung and 8 or 9 in the left, begin to form by the seventh week (Crelin, 1975). As this occurs, the surrounding mesenchymal tissue divides (Fig. 11–6G). Each tertiary (segmental) bronchus with its surrounding mass of mesenchyme will form a *bronchopulmonary segment.* By 24 weeks, about 17 orders of branches have formed and the respiratory bronchioles are present (Fig. 11–8A). An additional seven orders of airways develop after birth.

As the bronchi develop, cartilaginous rings, or plates, develop from the surrounding mesenchyme. This splanchnic mesenchyme also gives rise to the bronchial smooth musculature and connective tissue and to the pulmonary connective tissue and capillaries.

As the lungs develop, they acquire a layer of *visceral pleura* from the splanchnic mesenchyme (Fig. 11–7B). With expansion, the

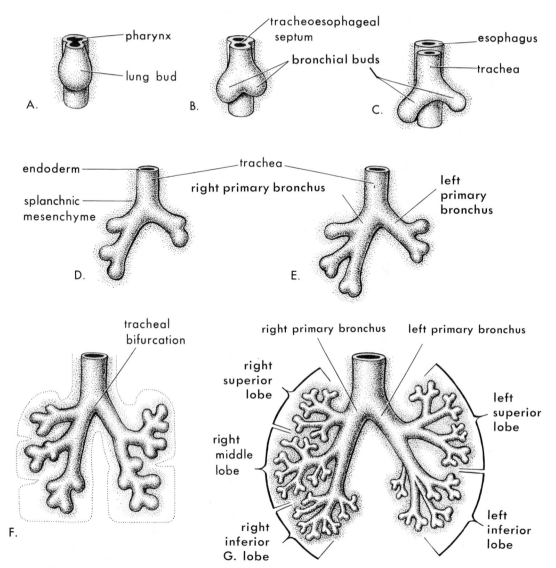

Figure 11-6 Drawings of ventral views illustrating successive stages in the development of the bronchi and the lungs. *A* to *C*, 4 weeks. *D* and *E*, 5 weeks, *F*, 6 weeks. *G*, 8 weeks.

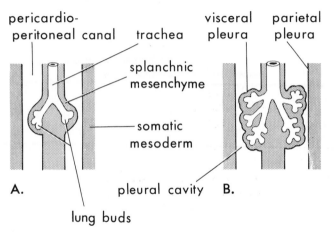

Figure 11-7 Diagrams illustrating the growth of the developing lungs into the splanchnic mesenchyme of the medial walls of the pericardioperitoneal canals (primitive pleural cavities), and the development of the layers of the pleura. *A*, 5 weeks. *B*, 6 weeks.

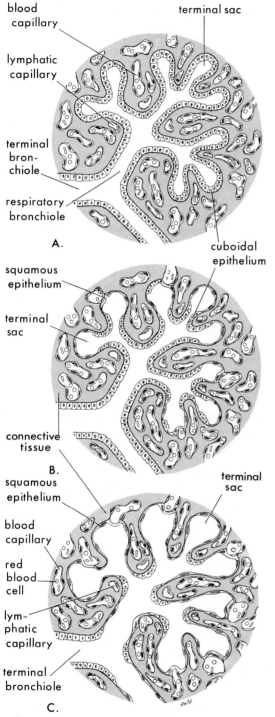

A.

B.

C.

Figure 11–8 Diagrammatic sketches of sections illustrating progressive stages of lung development. *A,* Late canalicular period (about 24 weeks). *B,* Early terminal sac period (about 26 weeks). *C,* Newborn infant. Early alveolar period. Note the thin alveolocapillary membrane. Note also that some of the capillaries have begun to bulge into the terminal sacs (future alveoli).

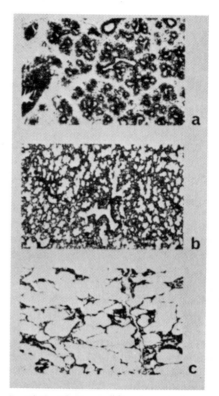

Figure 11–9 Sections of lungs at various stages of development, showing the changing appearance of the lung tissue. *A,,* Pseudoglandular period: 5 to 17 weeks. *B,* Canalicular period: 16 to 25 weeks. *C,* Terminal sac period: 24 weeks to birth (From Reid, L.: *The Pathology of Emphysema,* 1967. Courtesy of Lloyd-Luke [Medical Books] Ltd.)

lungs and the pleural cavities grow caudally into the mesenchyme of the body wall and soon come to lie close to the heart (see Fig. 9–5). The thoracic body wall becomes lined by a layer of *parietal pleura,* derived from the somatic mesoderm (Fig. 11–7B).

Lung development is divided into four stages:

1. The Pseudoglandular Period (5 to 17 weeks). During this period, the developing lung somewhat resembles a gland (Fig. 11–9A) because the bronchial divisions are differentiating into the air-conducting system. By 17 weeks, all major elements of the lung have formed *except* those involved with gas exchange. Respiration is not possible during this period.

2. The Canalicular Period (16 to 25 weeks). This period overlaps the pseudoglandular period because cranial segments of the lungs mature faster than caudal ones. During the canalicular period (Figs. 11–8A and 9B), the lumina of the bronchi and bronchioles become much larger, and the lung tissue becomes highly vascular. By 24 weeks, each terminal bronchiole has given rise to two or more *respiratory bronchioles* (Fig. 11–8A). Respiration is possible toward the end of this period because some thin-walled saccules, called *terminal sacs* (primitive alveoli), have developed at the ends of the respiratory bronchioles, and these regions are well vascularized.

3. The Terminal Sac Period (24 weeks to birth). During this period, many more terminal sacs develop (Fig. 11–8C and 11–9C), and the epithelium of the terminal sacs becomes very thin. Capillaries begin to bulge into the sacs (Fig. 11–8B and C). The terminal sacs are lined by continuous flattened epithelial cells of endodermal origin, known as *type 1 alveolar epithelial cells*. The capillary network proliferates rapidly in the mesenchyme around the developing alveoli, and there is concurrent active development of lymphatic capillaries (Fig. 11–8A and B).

By 25 to 28 weeks, the fetus usually weighs about 1000 gm and sufficient terminal sacs are present to permit survival of a prematurely born infant. Before this, the lungs are usually incapable of providing adequate gas exchange, partly because the alveolar surface area is insufficient and the vascularity underdeveloped. *It is not the presence of a thin alveolar epithelium so much as the development of an adequate pulmonary vasculature that is critical to the survival of premature infants.*

4. The Alveolar Period (late fetal period to about eight years). The epithelial lining of the terminal sacs attenuates to an extremely thin squamous epithelial layer. The type I alveolar cells become so thin that the underlying capillaries bulge into the space of each terminal sac (Fig. 11–8C). By the late fetal period, the lungs are capable of respiration because the alveolocapillary (respiratory) membrane is sufficiently thin to allow gas exchange. Although the lungs do not begin to perform this vital function until birth, they must be well developed so that they are capable of functioning at birth.

At the beginning of the alveolar period, each respiratory bronchiole terminates in a cluster of thin-walled terminal sacs, separated from each other by loose connective tissue (Fig. 11–8B and C). These terminal sacs represent the future alveolar ducts. Alveolar ducts probably do not exist before birth. Characteristic mature alveoli do not form for some time after birth. Before birth, the immature alveoli appear as small bulges from the walls of the terminal sacs and the respiratory bronchioles.

At birth, the primitive alveoli enlarge slightly as the lungs expand, but the increase in size of the lungs until the age of 3 years results mainly from an increase in the number of respiratory bronchioles and primitive alveoli rather than from an increase in the size of the alveoli (Crelin, 1975). From the third to the eighth year or so, the number of immature alveoli continues to increase. Unlike mature alveoli, immature alveoli have the potential for forming additional primitive alveoli. As primitive alveoli increase in size, they become mature alveoli.

One eighth to one sixth of the adult number of alveoli are present in a newborn infant. On chest radiographs, therefore, the lungs of newborn infants are denser than adult lungs.

It is well established that respiratory movements occur before birth, exerting sufficient force to cause aspiration of some amniotic fluid into the lungs. The developing lungs at birth are about half inflated with liquid derived from the lungs, the amniotic cavity, and the tracheal glands. Aeration of the lungs at birth is not so much the inflation of an empty, collapsed organ, but rather the rapid replacement of intra-alveolar fluid by air (Emery, 1969). This fluid in the lungs is cleared by three routes: (1) through the mouth and nose by pressure on the thorax during delivery; (2) into the pulmonary capillaries; and (3) into the lymphatics and the pulmonary arteries and veins.

In the fetus near term, the pulmonary lymphatic vessels are relatively larger and more numerous than in the adult (Crelin, 1975). Lymph flow is high during the first few hours after birth, and then it diminishes.

Of medicolegal significance is the fact that the lungs of a stillborn infant are firm and will sink when placed in water at autopsy because they contain fluid and not air.

Pulmonary Surfactant. At 23 to 24 weeks, the *type II alveolar epithelial cells* (cuboidal cells of terminal sacs, Fig. 11–8B) begin to secrete *surfactant*, a substance capable of lowering the surface tension at the air-alveolar interface, thereby maintaining patency of the alveoli and preventing *atelectasis* (collapse of the lung).

By 28 to 32 weeks, the amount of surfactant is sufficient to prevent alveolar collapse if a baby is born prematurely and begins breathing. When air expands the primitive alveoli during the first breath, surfactant is rapidly expelled into the alveolar spaces, preventing the development of an air-water interface of high surface tension (Crelin, 1975). As a result, the alveoli do not collapse, but retain a certain amount of air.

Respiratory Distress Syndrome. Infants born prematurely, with weights up to 1 to 1.5 kg, are most susceptible to the *respiratory distress syndrome* (Page et al., 1981). Shortly after birth, the infant develops rapid and labored breathing.

A deficiency of pulmonary surfactant appears to be a major cause of *hyaline membrane disease,* a common cause of death in the perinatal period (Hislop and Reid, 1974). The lungs are underinflated and the alveoli contain a fluid of high protein content that resembles a hyaline (glassy) membrane.

Hyaline membrane disease is a major cause of the respiratory distress syndrome in newborn infants (Villee et al., 1973). This membrane is believed to be derived from a combination of substances in the circulation and injured pulmonary epithelium. Page et al. (1981) suggested that prolonged intrauterine asphyxia may produce irreversible changes in the alveolar cells, making them incapable of producing surfactant. However, there are probably several causes for absence or deficiency of surfactant in premature and full-term infants.

All the factors controlling surfactant production have not been identified, but thyroxine is known to be a potent stimulator of surfactant production (Crelin, 1975). *The amount of surfactant increases during the terminal stages of pregnancy, particularly during the last two weeks before birth.*

CONGENITAL MALFORMATIONS OF THE LUNGS

Lobe of the Azygos Vein. Abnormal fissures and lobes of the lungs are common and usually insignificant. A lobe of the azygos vein appears in the right lung in about 1 per cent of people. It develops when the apical bronchus grows superiorly medial to the arch of the azygos vein instead of lateral to it. As a result, the azygos vein comes to lie at the bottom of a deep fissure in the superior lobe of the right lung.

Congenital Cysts of the Lungs. These cysts are thought to be formed by the dilation of the terminal, or larger, bronchi (Salzberg, 1977). If several cysts are present, the lungs have a honeycomb appearance in radiographs.

Agenesis of the Lungs. Absence of the lung(s) results from failure of the lung bud(s) to develop. Agenesis of one lung is commoner than bilateral agenesis, yet both conditions are rare. *Unilateral pulmonary agenesis* is compatible with life. The heart and other mediastinal structures are shifted to the affected side, and the existing lung is hyperexpanded.

Hypoplasia of the Lungs. In infants with *posterolateral diaphragmatic hernia* (see Fig. 9–11), the lungs are usually unable to develop normally because they are compressed by the abnormally positioned abdominal viscera.

Accessory Lung. An accessory lung is a rare congenital abnormality. It is almost always located at the base of the left lung. It does not communicate with the *tracheobronchial tree,* and its blood supply is usually systemic rather than pulmonary in origin.

SUMMARY

The lower respiratory system begins to develop around the middle of the fourth week from a median longitudinal *laryngotracheal groove* in the floor of the primitive pharynx. This groove deepens to produce a *laryngotracheal diverticulum,* which soon becomes separated from the foregut by the *tracheoesophageal septum* to form the esophagus and the *laryngotracheal tube.* The endodermal lining of this tube gives rise to the epithelium of the lower respiratory organs and the tracheobronchial glands. The splanchnic mesenchyme surrounding this tube forms the connective tissue, the cartilage, the muscle, and the blood and lymphatic vessels of these organs.

Branchial arch mesenchyme contributes to the formation of the epiglottis and the connective tissue of the larynx. The laryngeal muscles and the cartilage skeleton of the larynx are derived from mesenchyme in the caudal branchial arches.

Distally, the laryngotracheal tube divides into two knob-like *bronchial buds* by the early part of the fifth week. Each bud soon enlarges to form a *primary bronchus* and then each of these gives rise to two new bronchial buds, which develop into *secondary bronchi.* The right inferior secondary bronchus soon divides into two bronchi. The secondary bronchi supply the lobes of the developing lungs. Branching continues until about 17 orders of branches have formed. Additional airways are formed after birth, until about 24 orders of branches are formed.

Lung development is divided into four stages: (1) the *pseudoglandular period*, 5 to 17 weeks, when the bronchi and the terminal bronchioles form; (2) the *canalicular period*, 16 to 25 weeks, when the lumina of the bronchi and terminal bronchioles enlarge, the respiratory bronchioles and the alveolar ducts develop, and the lung tissue becomes highly vascular; (3) the *terminal sac period*, 24 weeks to birth, when the alveolar ducts give rise to terminal sacs (primitive alveoli). The terminal sacs are initially lined with cuboidal epithelium that begins to attenuate to squamous epithelium at about 26 weeks. By this time, the capillary network has proliferated close to the alveolar epithelium, and the lungs are usually sufficiently well developed to permit survival of the fetus if born prematurely; and (4) the *alveolar period*, the final stage of lung development, which occurs during the late fetal period to about eight years of age.

The respiratory system develops so that it is capable of immediate function at birth. To be capable of respiration, the lungs must acquire an alveolocapillary membrane that is sufficiently thin, and an adequate amount of *surfactant* must be present. A deficiency of surfactant appears to be responsible for the failure of primitive alveoli to remain open, resulting in *hyaline membrane disease* and respiratory distress.

Growth of the lungs after birth results mainly from an increase in the number of respiratory bronchioles and alveoli. *New alveoli form for at least eight years after birth.*

Major congenital malformations of the lower respiratory system are rare, except for tracheoesophageal fistula, which is usually associated with esophageal atresia. These malformations result from faulty partitioning of the foregut into the esophagus and the trachea during the fourth and fifth weeks.

SUGGESTIONS FOR ADDITIONAL READING

Avery, M. E., Fletcher, B. D., and Williams R.: *The Lung and Its Disorders in the Newborn Infant.* 4th ed. Philadelphia, W. B. Saunders Co., 1981. A scholarly presentation of the pathological and pathophysiological bases of lung disorders, including a thorough discussion of lung development, aeration of the lung at birth, and disorders of respiration in the neonatal period.

Crelin, E. S.: *Development of the Lower Respiratory System.* New Jersey, Clinical Symposia, Vol. 27, No. 4, 1975. An up-to-date review of the structure, function, and development of the lower respiratory system. The article is beautifully illustrated by Dr. Netter.

Hast, H. M.: Developmental anatomy of the larynx; in R. Hinchcliffe, and D. Harrison (Eds.): *Scientific Foundations of Otolaryngology.* London, W. Heinemann Medical Books Ltd., 1976, pp. 529–535. This chapter presents a comprehensive, well-illustrated account of the development of the larynx.

Hislop, A., and Reid, L.: Growth and development of the respiratory system — anatomical development; in J. A. Davis and J. Dobbing (Eds.): *Scientific Foundations of Paediatrics.* Philadelphia, W. B. Saunders Co., 1974, pp. 214–253. A comprehensive account of the growth and development of the lungs, both normal and abnormal, from the embryonic period until childhood and adolescence.

Villee, C. A., Villee, D. B., and Zuckerman, J.: *Respiratory Distress Syndrome.* New York, Academic Press, Inc., 1973. This book is based on the proceedings of a conference on fetal lung development and function in relation to respiratory problems of the newborn infant.

CLINICALLY ORIENTED PROBLEMS

1. Choking and continuous coughing were observed in a newborn infant. There was an excessive amount of mucous secretion and saliva in the mouth of the infant, who also experienced difficulty in breathing. The pediatrician was unable to pass a catheter through the esophagus into the stomach. What congenital malformations would be suspected? What kind of an examination do you think would be used to confirm the diagnosis?

2. A premature infant developed rapid, shallow respirations shortly after birth, and a diagnosis of *respiratory distress syndrome* (RDS) was made. How do you think the infant might attempt to overcome his or her inadequate

exchange of oxygen and carbon dioxide? What disease commonly causes RDS? A deficiency of what substance is associated with RDS?

3. What is the most common type of tracheoesophageal fistula? What is its embryological basis?

4. A newborn infant with esophageal atresia experienced respiratory distress with cyanosis shortly after birth. X-ray films demonstrated air in the infant's stomach. How did it get there? What other problem might result in an infant with this fairly common type of malformation?

The answers to these questions are given at the back of the book.

REFERENCES

Bertalanffy, F. D.: Respiratory tissue: structure, histophysiology, cytodynamics. Part I. Review and basic cytomorphology. *Int. Rev. Cytol. 16*:233, 1964.

Bertalanffy, F. D., and Leblond, C. P.: Structure of respiratory tissue. *Lancet 2*:1365, 1955.

Boyden, E. A.: The pattern of the terminal airspaces in a premature infant of 30–32 weeks that lived nineteen and a quarter hours. *Am. J. Anat. 126*:31, 1969.

Boyden, E. A.: Development of the human lung; in J. Brennemann (Ed.): *Practice of Pediatrics,* Vol. IV. Hagerstown, Harper & Row, Publishers, 1972.

Bucher, U., and Reid, L.: Development of the intrasegmental bronchial tree: the pattern of branching and development of cartilage at various stages of intrauterine life. *Thorax 16*:207, 1961.

Chernick, V., and Avery, M. E.: The functional basis of respiratory pathology; in E. L. Kendig, Jr., and V. Chernick (Eds.): *Disorders of the Respiratory Tract in Children.* Vol. 1. Pulmonary Disorders. 3rd ed. Philadelphia, W. B. Saunders Co., 1977; pp. 3–61.

Conen, P. E., and Balis, J. U.: Electron microscopy in study of lung development; in J. Emery (Ed.): *The Anatomy of the Developing Lung.* London, William Heinemann, Ltd., 1969, pp. 18–48.

Crelim, E. S.: *Development of the Upper Respiratory System.* New Jersey, Clinical Symposia, Vol. 28, No. 3, 1976.

Davis, J. A.: The first breath and development of lung tissue; in E. E. Philipp, J. Barnes, and M. Newton (Eds.): *Scientific Foundations of Obstetrics and Gynecology.* London, William Heinemann, Ltd., 1970, pp. 401–406.

Davis, M. E., and Potter, E. L.: Intrauterine respiration of the human fetus. *J.A.M.A. 131*:1194, 1946.

Dawes, G. S.: Fetal blood gas homeostasis; in G. E. W. Wolstenholme and M. O'Connor (Eds.): *Foetal Autonomy.* London, J. & A. Churchill Ltd., 1969, pp. 162–172.

Emery, J.: *The Anatomy of the Developing Lung.* London, William Heinemann, Ltd., 1969.

Godfrey, S.: Growth and development of the respiratory system — functional development; in J. A. Davis and J. Dobbing (Eds.): *Scientific Foundations of Paediatrics.* Philadelphia, W. B. Saunders Co., 1974, pp. 254–270.

Gray, S. W., and Skandalakis, J. E.: *Embryology for Surgeons. The Embryological Basis for the Treatment of Congenital Defects.* Philadelphia, W. B. Saunders Co., 1972, pp. 283–322.

Hollinger, P. H., Johnson, K. C., and Schiller, F.: Congenital anomalies of the larynx. *Ann. Otol. 63*:581, 1954.

Landing, B. H.: Anomalies of the respiratory tract. *Pediatr. Clin. North Am.,* Feb., 1957, pp. 73–102.

Lind, J., Tahti, E., and Hirvensalo, M.: Roentgenologic studies of the size of the lungs of the newborn baby before and after aeration. *Ann. Paediatr. Fenn. 12*:20, 1966.

Low, F. N., and Sampaio, M. M.: The pulmonary alveolar epithelium as an endodermal derivative. *Anat. Rec. 127*:51, 1957.

Norman, A. P.: *Congenital Abnormalities in Infancy.* 2nd ed. Oxford, Blackwell Scientific Publications, 1971.

Oliver, R. E.: Fetal lung liquids. *Fed. Proc. 36*:2669, 1977.

O'Rahilly, R., and Boyden, E. .: The timing and sequence of events in the development of the human respiratory system during the embryonic period proper. *Z. Anat. Entwicklungsgesch. 141*:237, 1973.

O'Rahilly, R., and Tucker, J. A.: The early development of the larynx in staged human embryos. Part 1. Embryos of the first five weeks (to stage 15). *Ann. Otol. Rhinol. Laryngol. 82*(Suppl. 7):1, 1973.

Page, E. W., Villee, C. A., and Villee, D. B.: *Human Reproduction: Essentials of Reproductive and Perinatal Medicine.* 3rd ed. Philadelphia, W. B. Saunders Co., 1981.

Pattle, R. E.: The development of the foetal lung; in G. E. W. Wolstenholme and M. O'Connor (Eds.): *Foetal Autonomy.* London, J. & A. Churchill, Ltd., 1969, pp. 132–142.

Potter, E. L.: *Pathology of the Fetus and Infant.* 2nd ed. Chicago, Year Book Medical Publishers Inc., 1961.

Reid, D. E., Ryan, K. J., and Benirschke, K.: *Principles and Management of Human Reproduction.* Philadelphia, W. B. Saunders Co., 1972, pp. 794–802.

Salzberg, A. M., and Ehrlich, F. E.: Congenital malformations of the lower respiratory tract; in E. L. Kendig, Jr. and V. Chernick (Eds.): *Disorders of the Respiratory Tract in Children.* Vol. I. Pulmonary Disorders. 3rd ed. Philadelphia, W. B. Saunders Co., 1977, pp. 213–252.

Smith, E. I.: The early development of the trachea and oesophagus in relation to atresia of the oesophagus and tracheo-oesophageal fistula. *Contr. Embryol. Carneg. Instn. 245*:36, 1957.

Swenson, O.: *Pediatric Surgery.* New York, Appleton-Century-Crofts, Inc., 1958.

Thomas, L. B., and Boyden, E. A.: Agenesis of the right lung. *Surgery 31*:429, 1952.

Towers, B.: Amniotic fluid and the foetal lung. *Nature 183*:1140, 1959.

Vaughan, V. C., McKay, R. J., and Behrman, R. E.: *Nelson Textbook of Pediatrics.* 11th ed. Philadelphia, W. B. Saunders Co., 1979.

Waterson, D. J., Carter, R. E., and Aberdeen, E.: Oesophageal atresia: tracheo-oesophageal fistula, a study of survival in 218 infants. *Lancet 1*:819, 1962.

Wells, L. J., and Boyden, E. A.: The development of the bronchopulmonary segments in human embryos of horizons XVII and XIX. *Am. J. Anat. 95*:163, 1954.

12

THE DIGESTIVE SYSTEM

Development of the oral cavity, the tongue, the salivary glands, and the pharynx is described in Chapter 10.

The *primitive gut* forms during the fourth week, as the head, tail, and lateral folds incorporate the dorsal part of the *yolk sac* into the embryo (see Fig. 5–1). The endoderm of the primitive gut gives rise to most of the epithelium and glands of the digestive tract; the epithelium at the cranial and caudal extremities of the tract is derived from ectoderm of the *stomodeum* (primitive mouth) and the *proctodeum,* respectively (Figs. 12–1 and 12–21). The muscular and fibrous elements of the digestive tract, and the visceral peritoneum, are derived from the splanchnic

mesenchyme surrounding the endodermal lining of the primitive gut.

For descriptive purposes, the primitive gut is divided into three parts: the *foregut,* the *midgut,* and the *hindgut* (Fig. 12–1).

THE FOREGUT

The derivatives of the foregut are as follows:

1. The *pharynx* and its derivatives (discussed in Chapter 10)
2. The *lower respiratory tract* (described in Chapter 11)
3. The *esophagus*
4. The *stomach*
5. The *duodenum* proximal to the opening of the bile duct
6. The *liver* and *pancreas*
7. The *biliary apparatus*

All these foregut derivatives, except the pharynx, the respiratory tract, and most of the esophagus, are supplied by the *celiac artery* (Fig. 12–1), the artery that supplies the foregut.

DEVELOPMENT OF THE ESOPHAGUS

The partitioning of the trachea from the esophagus by the *tracheoesophageal septum* is described in Chapter 11 and is illustrated in Figure 11–2. Initially, the esophagus is very short (Fig. 12–1), but it elongates rapidly, reaching its final relative length by about seven weeks. Elongation of the esophagus results mainly from cranial body growth, i.e., from "ascent" of the pharynx rather than from "descent" of the stomach (Gray and Skandalakis, 1972).

The epithelium of the esophagus and the esophageal mucous glands are derived from endoderm. The epithelium of the esophagus proliferates and almost obliterates the lumen, but recanalization of the esophagus normally occurs by the end of the embryonic period.

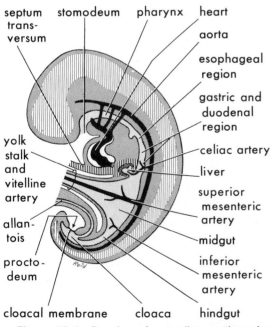

septum transversum stomodeum pharynx heart
aorta
esophageal region
gastric and duodenal region
yolk stalk and vitelline artery
celiac artery
liver
superior mesenteric artery
allantois
midgut
proctodeum
inferior mesenteric artery
cloacal membrane cloaca hindgut

Figure 12–1 Drawing of a median section of a four-week embryo showing the early digestive system and its blood supply. The primitive gut is a long tube extending the length of the embryo. It is formed by incorporation of the dorsal part of the yolk sac into the embryo (see Fig. 5–1). Its blood vessels are derived from the vessels that supplied the yolk sac.

The striated muscle in the upper esophagus, forming the muscularis externa, is derived from mesenchyme in the caudal branchial arches. The smooth muscle of the esophagus develops from the surrounding splanchnic mesenchyme. Both types of muscle are innervated by components of the *vagus nerves* that supply the caudal branchial arches (see Chapter 10).

Esophageal Atresia (see Fig. 11–5A). Esophageal atresia usually occurs with *tracheoesophageal fistula*. It may occur as a separate malformation, but this is exceedingly rare. Atresia probably results from deviation of the *tracheoesophageal septum* (see Fig. 11–2) in a posterior direction, but atresia could also result from failure of esophageal recanalization during the embryonic period.

When there is esophageal atresia, amniotic fluid cannot pass to the intestines for absorption and subsequent transfer to the placenta for disposal. This results in *polyhydramnios*, the accumulation of an excessive amount of amniotic fluid.

Newborn infants with esophageal atresia usually appear healthy, and their first one or two swallows are normal. Suddenly, fluid returns through the nose and mouth, and *respiratory distress* occurs. Confirmation of the presence of esophageal atresia can be made by demonstrating radiographically that a radiopaque catheter cannot pass into the stomach.

Esophageal Stenosis. Congenital stenosis, or narrowing of the esophagus, can occur in any place, but it is usually present in the distal third as either a web or a long segment of esophagus with only a thread-like lumen. Esophageal stenosis results from incomplete recanalization of the esophagus during the eighth week of development.

Short Esophagus. The esophagus may be abnormally short, and a portion of the stomach may be displaced superiorly through the *esophageal hiatus* into the thorax.

DEVELOPMENT OF THE STOMACH

The stomach first appears as a fusiform dilatation of the caudal part of the foregut (Figs. 12–1 and 12–2A). This primordium soon enlarges and broadens ventrodorsally (Fig. 12–2B). During the following weeks, the dorsal border grows faster than the ventral border, producing the *greater curvature* (Fig. 12–2C).

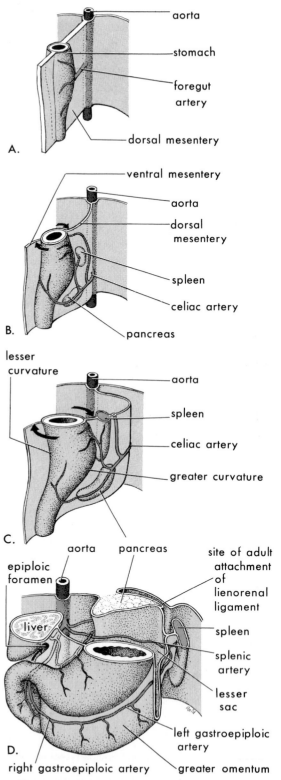

Figure 12–2 Drawings illustrating development and rotation of the stomach, and formation of the omental bursa (lesser sac) and the greater omentum. *A*, about 28 days. *B*, about 35 days. *C*, about 40 days. *D*, about 48 days.

Rotation of the Stomach. As the stomach acquires its adult shape, it slowly rotates 90 degrees in a clockwise direction around its longitudinal axis. The effects of this rotation on the stomach are as follows (Fig. 12–2C and D):

1. The ventral border (lesser curvature) moves to the right.

2. The dorsal border (greater curvature) moves to the left.

3. The original left side becomes the ventral surface.

4. The original right side becomes the dorsal surface.

Before rotation, the cranial and caudal ends of the stomach are located in the midline (Fig. 12–2A). During rotation and further growth of the stomach, the cranial region (future *cardiac part*) moves to the left and slightly inferiorly, and the caudal region (future *pyloric part*) moves to the right and superiorly. As a result, the stomach assumes its final position, with its long axis almost transverse to the long axis of the body (Fig. 12–2D).

The rotation and growth of the stomach explain why the left vagus nerve supplies the anterior wall of the adult stomach and the right vagus innervates the posterior wall.

Mesenteries of the Stomach. The stomach is suspended from the dorsal wall of the abdominal cavity by the *dorsal mesentery*, or *dorsal mesogastrium* (Fig. 12–2A). The dorsal mesogastrium is originally in the median plane, but it is carried to the left during rotation of the stomach and formation of the *omental bursa*, or *lesser sac* of peritoneum. (Fig. 12–2 A to C). A *ventral mesentery*, or *ventral mesogastrium* (Fig. 12–2 B) exists only in the region of the inferior end of the esophagus, the stomach, and the superior part of the duodenum. It attaches the stomach and duodenum to the developing liver and the ventral abdominal wall (Fig. 12–2D).

The Omental Bursa, or Lesser Sac of Peritoneum. Isolated clefts develop between the mesenchymal cells in the thick dorsal mesogastrium (Fig. 12–3A and B) and coalesce to form a single cavity, which is the primordium of the *omental bursa*, or lesser sac (Fig. 12–3C and D). Rotation of the stomach is thought to pull the dorsal mesogastrium to the left, thereby helping to form this large *compartment*, or *recess*, of the peritoneal cavity. It expands transversely and cranially and comes to lie between the stomach and the posterior abdominal wall. It is called a *bursa* (L., purse) because it facilitates movement of the stomach.

The superior part of the cranial extension of the lesser sac is cut off as the diaphragm develops, forming a closed space, or sac, known as the *infracardiac bursa*. If it persists, it usually lies medial to the basal part of the right lung. The inferior portion of the superior part of the cranial extension of the lesser sac persists as the *superior recess of the lesser sac*.

As the stomach enlarges, the lesser sac also expands and acquires an *inferior recess* between the layers of the elongate dorsal mesogastrium, the *greater omentum* (L., "fat skin"). This four-layered membrane overhangs the developing intestines. Later, the inferior recess almost disappears as the layers of the greater omentum fuse (Fig. 12–13F). The lesser sac communicates with the peritoneal cavity, or *greater peritoneal sac*, through a small opening, the *epiploic foramen* (Figs. 12–2D and 12–3H). In the adult, this foramen is located posterior to the free edge of the lesser omentum at the upper border of the superior part of the duodenum.

Congenital Hypertrophic Pyloric Stenosis. Malformations of the stomach are very rare, except for pyloric stenosis. It affects 1 in every 150 male and 1 in every 750 female infants. In infants with this abnormality, there is a marked thickening of the *pylorus*, the distal sphincteric region of the stomach. The circular and, to a lesser degree, the longitudinal muscle in the pyloric region are hypertrophied. This results in *severe narrowing* (stenosis) of the pyloric canal and an obstruction to the passage of food. Although the cause of congenital pyloric stenosis is unknown, the high incidence of the condition in both infants of monozygotic twins, suggests the involvement of genetic factors.

DEVELOPMENT OF THE DUODENUM

Early in the fourth week, the duodenum develops from the caudal part of the foregut and the cranial part of the midgut. The junction of the two parts in the adult is just distal to the origin of the bile duct. These parts of the foregut and the midgut grow rapidly and form a C-shaped loop that projects ventrally (Fig. 12–4B to D). The junction of the foregut and the midgut is at the apex of this embryonic duodenal loop (Fig. 12–4D). As the stomach rotates (Fig. 12–2),

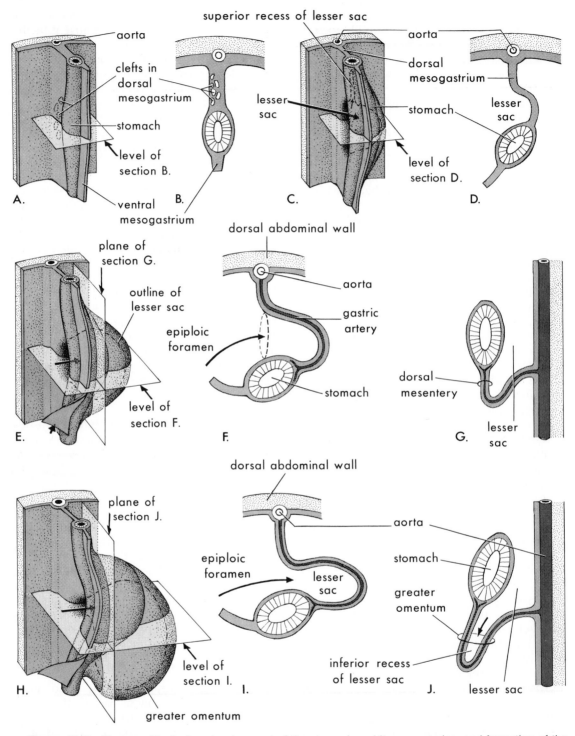

Figure 12–3 Diagrams illustrating development of the stomach and its mesenteries, and formation of the omental bursa, or lesser sac of the peritoneum. *A*, 5 weeks. *B*, Transverse section showing clefts in the dorsal mesogastrium. *C*, Later stage after coalescence of the clefts to form the lesser sac. *D*, Transverse section showing the initial appearance of the lesser sac. *E*, The dorsal mesentery has elongated and the lesser sac has enlarged. *F* and *G*, Transverse and longitudinal sections respectively showing elongation of the dorsal mesogastrium and expansion of the lesser sac. *H*, 6 weeks, showing the greater omentum and expansion of the lesser sac. *I* and *J*, Transverse and longitudinal sections respectively showing the inferior recess of the lesser sac and the epiploic foramen.

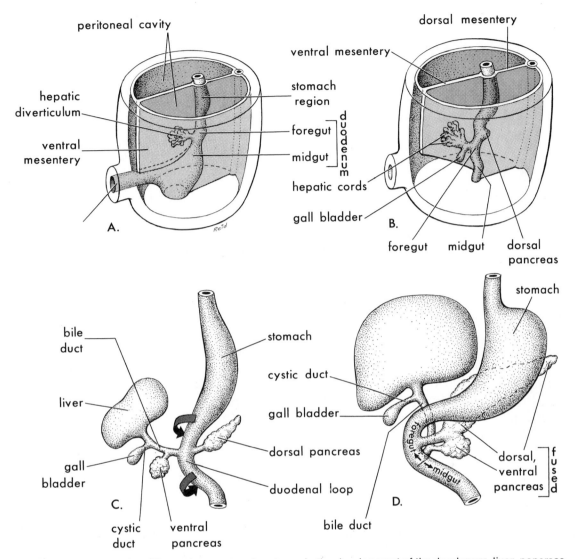

Figure 12-4 Drawings illustrating progressive stages in the development of the duodenum, liver, pancreas, and extrahepatic biliary apparatus. *A,* 4 weeks. *B* and *C,* 5 weeks. *D,* 6 weeks. The pancreas develops from dorsal and ventral buds that fuse to form the definitive pancreas. Note that, as the result of the positional changes of the duodenum, the entrance of the bile duct into the duodenum gradually shifts from its initial position to a posterior one. This explains why the bile duct in the adult passes posterior to the duodenum and the head of the pancreas.

the developing duodenal loop rotates to the right where it comes to lie retroperitoneally (Fig. 12–13 *F*). Because of its derivation from both the foregut and the midgut, the duodenum is supplied by branches of the celiac and superior mesenteric arteries.

During the fifth and sixth weeks, the lumen of the duodenum becomes reduced and may be temporarily obliterated by epithelial cells, but it normally recanalizes by the end of the embryonic period.

Most of the ventral mesentery of the duo-denum disappears. The free border of the part that remains forms the anterior border of the epiploic foramen (Fig. 12–11*E*).

Duodenal Stenosis (Fig. 12–5A). Narrowing of the duodenal lumen usually results from incomplete recanalization of the duodenum. It may also be caused by pressure from an *anular pancreas* (Fig. 12–9) or peritoneal bands. Most stenoses involve the horizontal (third) and/or ascending (fourth) parts of the duodenum. Hence the vomitus usually contains bile.

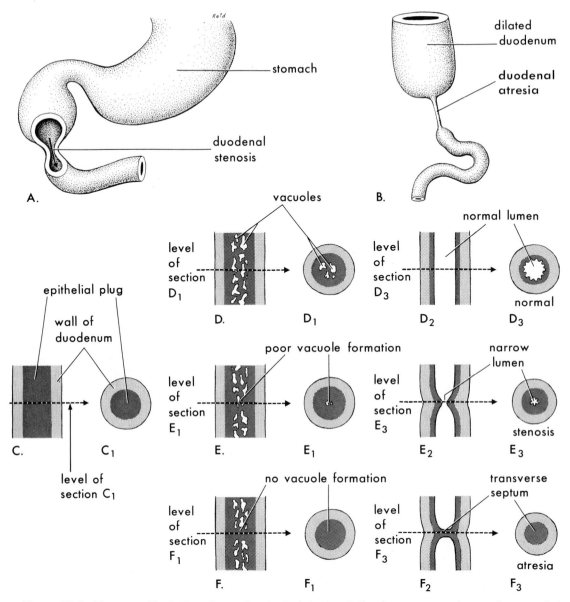

Figure 12-5 Diagrams illustrating the embryological basis of the two common types of congenital intestinal obstruction. *A*, Duodenal stenosis. *B*, Duodenal atresia. *C* to *F*, Diagrammatic longitudinal and transverse sections of the duodenum showing (1) normal recanalization (*D* to *D₃*), (2) stenosis (*E* to *E₃*) and (3) atresia (*F* to *F₃*). Most duodenal atresias occur in the descending (second) and horizontal (third) parts of the duodenum.

Duodenal Atresia (Fig. 12–5*B*). Blockage of the lumen of the duodenum is not common, except in premature infants and in those with Down syndrome (see Fig. 8–4). During the solid stage of duodenal development occurring in some embryos, the lumen is completely filled with epithelial cells. If reformation of the lumen fails to occur by a process of vacuolization (Fig. 12–5*D*), a short segment of the duodenum is occluded.

Most atresias involve the descending (second) and horizontal (third) parts of the duodenum, and are located distal to the opening of the bile duct.

In infants with duodenal atresia, vomiting begins within a few hours of birth, before ingestion of any fluids. *The vomitus almost always contains bile.* Often there is distention of the epigastrium, resulting from an overfilled stomach and upper duodenum.

Other severe congenital malformations are associated with duodenal atresia, e.g., Down syndrome, anular pancreas (Fig. 12–9), cardiovascular abnormalities, and anorectal malformations. *Polyhydramnios* usually occurs because the duodenal atresia prevents normal absorption of amniotic fluid from the intestine.

Diverticula of the Duodenum. These herniations are common and usually occur singly. They can be shown radiologically during barium studies but are often not observed during autopsies. The origin of these diverticula is uncertain, but they are believed to form where a duct or blood vessel pierces the gut wall.

DEVELOPMENT OF THE LIVER AND BILIARY APPARATUS

The liver, the gallbladder, and the biliary duct system arise as a ventral bud from the most caudal part of the foregut early in the fourth week (Fig. 12–4*A*). This *hepatic diverticulum* extends into the *septum transversum* as rapidly proliferating cell strands. It rapidly enlarges and divides into two parts as it *grows between the layers of the ventral mesentery* (Fig. 12–4*B*).

The *large cranial part* is the primordium of the liver. The proliferating endodermal cells give rise to interlacing cords of liver cells and to the epithelial lining of the *intrahepatic portion of the biliary apparatus* (Fig. 12–4*B*). The *hepatic cords* anastomose around preexisting endothelium-lined spaces, which are the primordia of the *hepatic sinusoids.* The fibrous and *hemopoietic tissue* and the *Kupffer cells* of the liver are derived from the splanchnic mesenchyme of the septum transversum.

The liver grows rapidly and soon fills most of the abdominal cavity (see Fig. 5–18*C* and *D*). Initially, the right and left lobes are about the same size, but the right lobe becomes much larger. The caudate and quadrate lobes develop as subdivisions of the right lobe.

Hemopoiesis begins during the sixth week, giving the liver a bright reddish appearance. This hemopoietic activity is mainly responsible for the relatively large size of the liver during the second month. *By nine weeks, the liver accounts for about 10 per cent of the total weight of the fetus.* This hemopoietic activity subsides during the terminal stages of pregnancy. By term, the weight of the liver is only 5 per cent of the total body weight. *Bile formation by the hepatic cells begins during the twelfth week.*

The *small caudal part* of the hepatic diverticulum expands to form the *gallbladder;* its stalk becomes the *cystic duct* (Fig. 12–4*C*). Initially, the extrahepatic biliary apparatus is occluded with endodermal cells, but it is later recanalized. The stalk connecting the hepatic and cystic ducts to the duodenum becomes the *bile duct.* Initially, this duct attaches to the ventral aspect of the duodenal loop, but, as the duodenum grows and rotates, the entrance of the bile duct is carried around to the dorsal aspect of the duodenum (Fig. 12–4*C* and *D*). *Bile pigments (bilirubin) begin to form during the 13- to 16-week period* and enter the duodenum, giving its contents (*meconium*) a dark green color.

The *ventral mesentery* (Fig. 12–6) is a thin, double-layered membrane that gives rise to three structures: (1) the *lesser omentum,* which passes from the liver to the ventral border of the stomach (*hepatogastric ligament*) and from the liver to the duodenum (*hepatoduodenal ligament*) (Fig. 12–7); (2) the *falciform ligament* (Fig. 12–6), which extends from the liver to the anterior (ventral) abdominal wall. The *umbilical vein* passes in the inferior free border of the falciform ligament on its way from the umbilical cord to the liver (Fig. 12–7). The ventral mesentery also gives rise to (3) *the visceral peritoneum of the liver* (Fig. 12–6*D*). The liver is covered by visceral peritoneum, except for an area that is in direct contact with the diaphragm; this is called the *bare area of the liver* (Fig. 12–7).

Variations of the Liver and Biliary Apparatus. Variations of liver lobulation are not uncommon, but congenital malformations are rare. Variations of the hepatic ducts, the bile duct, and the cystic duct are common. *Accessory hepatic ducts are common,* and an awareness of their possible presence is of surgical importance. In some cases, the cystic duct opens into an accessory hepatic duct rather than into the common hepatic duct.

Extrahepatic Biliary Atresia. This is the most serious malformation of the extrahepatic biliary system and occurs once in about every 20,000 births (Stowens, 1959). The occlusion likely results from persistence of the solid stage of duct development, possibly owing to the presence of some noxious agent during embryonic development, but it could

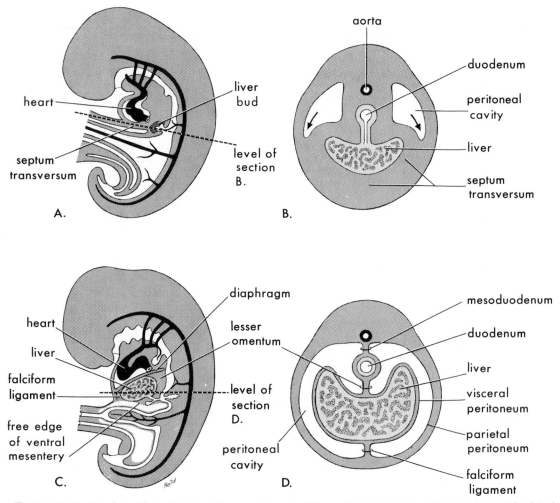

Figure 12–6 Drawings illustrating how the caudal part of the septum transversum becomes stretched and membranous, thereby forming the ventral mesentery. *A,* Sagittal section of a four-week embryo. *B,* Transverse section through the embryo showing expansion of the peritoneal cavity (arrows). *C,* Sagittal section of a five-week embryo. *D,* Transverse section through the embryo after formation of the dorsal and ventral mesenteries. Note that the liver is joined to the ventral abdominal wall and to the stomach by the falciform ligament and lesser omentum, respectively.

result from liver infection during late fetal development (Myers et al., 1956).

DEVELOPMENT OF THE PANCREAS

The pancreas develops from *dorsal and ventral pancreatic buds* of endodermal cells that arise from the caudal part of the foregut that is developing into the proximal part of the duodenum (Figs. 12–4, 12–6, and 12–8). The larger dorsal bud appears first and grows rapidly into the dorsal mesentery (Figs. 12–7 and 12–10). The ventral bud develops near the entry of the bile duct into the duodenum. When the duodenum rotates to the right (clockwise) and becomes C-shaped, the ven-

tral pancreatic bud is carried dorsally with the bile duct and comes to lie in the mesoduodenum, inferior and posterior to the dorsal pancreatic bud (Fig. 12–8C). It then fuses with the dorsal bud (Fig. 12–8D and G.)

The ventral bud forms the *uncinate process* and the inferior part of the *head* of the pancreas. *Most of the pancreas is derived from the dorsal bud.* As the pancreatic buds fuse, the ducts anastomose. The *main pancreatic duct* forms from the duct of the ventral bud and the distal part of the duct of the dorsal bud. The proximal part of the duct of the dorsal bud often persists as an *acces-*

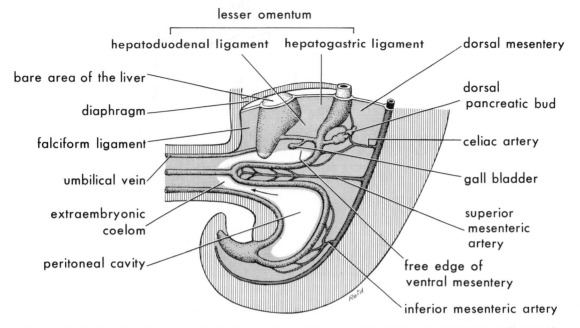

lesser omentum

hepatoduodenal ligament   hepatogastric ligament   dorsal mesentery

bare area of the liver

diaphragm

falciform ligament

umbilical vein

extraembryonic coelom

peritoneal cavity

dorsal pancreatic bud

celiac artery

gall bladder

superior mesenteric artery

free edge of ventral mesentery

inferior mesenteric artery

Figure 12–7 Sketch of the caudal half of an embryo at the end of the fifth week, showing the liver and its associated ligaments, viewed from the left. The arrow indicates the communication of the peritoneal cavity with the extraembryonic coelom. Owing to the rapid growth of the liver and the midgut loop, the abdominal cavity temporarily becomes too small to contain the developing intestines. Consequently, they enter the extraembryonic coelom in the umbilical cord (see also Figure 12–12).

sory pancreatic duct that opens at the summit of the *minor duodenal papilla,* located about 2 cm cranial to the main duct (Fig. 12–8G). The two ducts often communicate with each other. In about 9 per cent of people the pancreatic duct systems fail to fuse and the original double ducts persist.

Histogenesis of the Pancreas. The pancreatic parenchyma is derived from endoderm, which forms a network of tubules. Early in the fetal period, *acini* begin to develop from cell clusters around the end of these tubules, or primitive ducts. The *islets of Langerhans* develop from groups of cells that separate from the tubules and soon come to lie between the acini. *Insulin secretion begins at about 20 weeks.* The connective tissue covering and the septa of the pancreas develop from the surrounding splanchnic mesenchyme.

Malformations of the Pancreas. Heterotopic, or accessory, pancreatic tissue is most often located in the wall of the stomach or duodenum, or in a Meckel's diverticulum (Figs. 12–17 and 12–18).

Anular pancreas is a rare malformation that warrants description because it may cause duodenal obstruction (Fig. 12–9C). The anular portion of the pancreas consists of a thin flat band of pancreatic tissue surrounding the second part of the duodenum. *An anular pancreas may cause*

obstruction of the duodenum shortly after birth. In other cases, the obstruction develops later in life if inflammation or malignant disease develops in this ring-like pancreas. *Males are affected much more frequently than females.* Althoughother explanations have been offered, this abnormality probably results from the growth of a bifid ventral pancreatic bud around the duodenum (Fig. 12–9). The portions of the ventral bud then fuse with the dorsal bud, forming a pancreatic ring (Fig. 12–9C).

DEVELOPMENT OF THE SPLEEN

Development of the spleen is described here, because this organ is derived from a mass of mesenchymal cells between the layers of the dorsal mesogastrium. The spleen, a large, vascular, lymphatic organ, acquires its characteristic shape early in the fetal period (Figs. 12–2, 12–8, and 12–10).

As the stomach rotates, the left surface of the mesogastrium fuses with the peritoneum over the left kidney. This fusion explains the dorsal attachment of the *lienorenal ligament,* and why the adult *splenic artery,* the largest branch of the celiac trunk, follows a tortuous course posterior to the omental bursa and anterior to the left kidney (Fig. 12–10C).

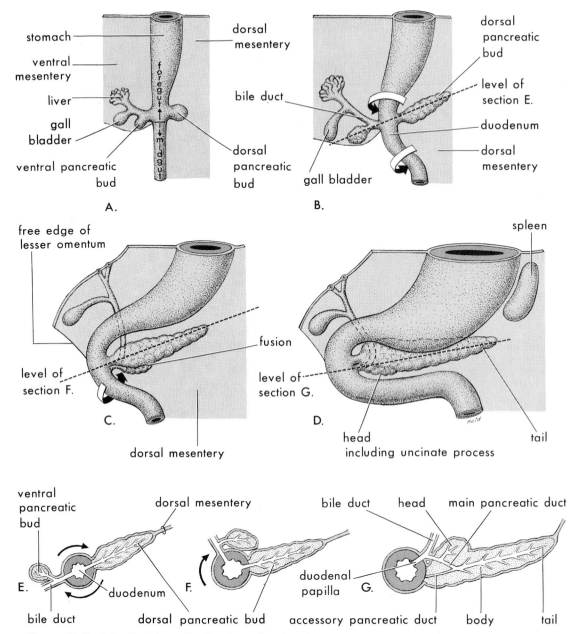

Figure 12–8 *A* to *D,* Schematic drawings showing the successive stages in the development of the pancreas from the fifth to the eighth weeks. *E* to *G,* Diagrammatic transverse sections through the duodenum and the developing pancreas. Growth and rotation (arrows) of the duodenum bring the ventral pancreatic bud towards the dorsal bud, and they subsequently fuse. Note that the bile duct initially attaches to the ventral aspect of the duodenum and is carried around to the dorsal aspect as the duodenum rotates. The main pancreatic duct is formed by the union of the distal part of the dorsal pancreatic duct and the entire ventral pancreatic duct. The proximal part of the dorsal pancreatic duct usually obliterates, but it may persist as an accessory pancreatic duct.

Histogenesis of the Spleen. The mesenchymal cells differentiate and form the capsule and connective tissue framework and parenchyma of the spleen. *The organ functions as a hematopoietic center until late feta! life,* but the spleen *retains its potentiality for blood cell formation* even in adult life.

Accessory Spleens. One or more of these organs can occur in an individual, and they most commonly appear near the hilum of the spleen.

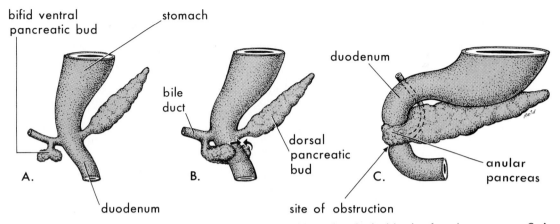

Figure 12–9 A and B, Drawings illustrating the probable embryological basis of anular pancreas. C, An anular pancreas encircling the duodenum. Anular pancreas may cause no compression of the duodenum, but it sometimes produces complete obstruction (*atresia*) or partial obstruction (*stenosis*). Thus, anular pancreas may cause no symptoms, or it may simulate duodenal stenosis (Fig. 12–5A) or duodenal atresia (Fig. 12–5B). In most cases, the anular pancreas encircles the second part of the duodenum, distal to the hepatopancreatic ampulla.

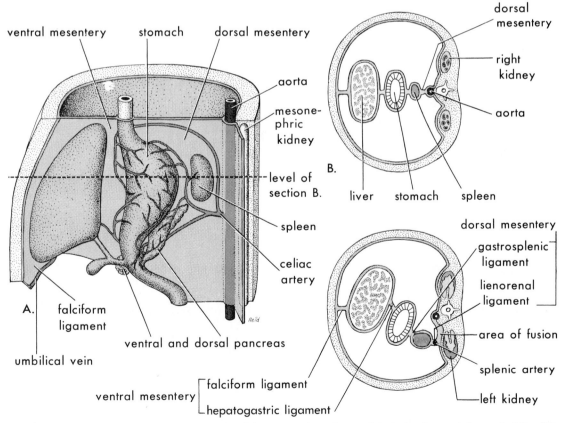

Figure 12–10 A, Drawing of the left side of the stomach and associated structures at the end of the fifth week. Note that the pancreas, the spleen, and the celiac artery are between the layers of the dorsal mesogastrium. B, Transverse section through the liver, the stomach, and the spleen at the level shown in A illustrating their relationship to the dorsal and ventral mesenteries. C, Transverse section through a fetus showing fusion of the dorsal mesogastrium with the peritoneum of the posterior abdominal wall.

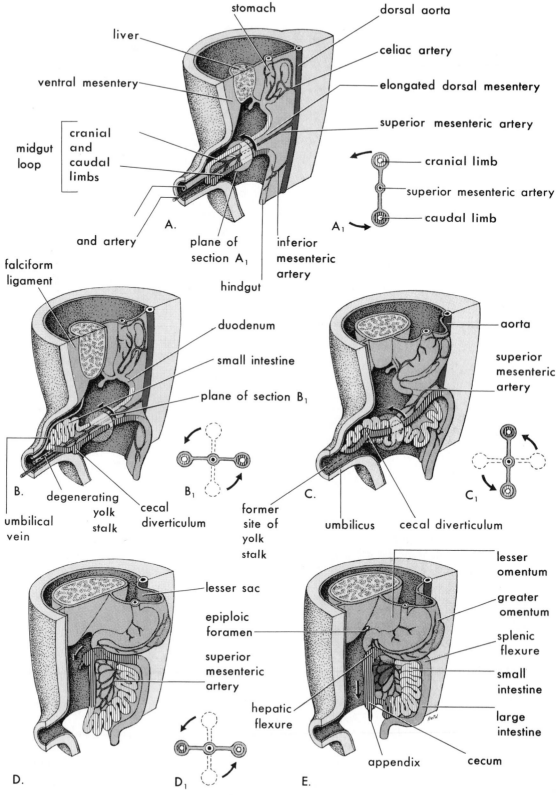

Figure 12–11 *See legend on opposite page*

An accessory spleen may be embedded partly or wholly in the tail of the pancreas, or within the gastrolienal ligament. *Accessory spleens occur in about 10 per cent of people;* usually they are about 1 cm in diameter.

THE MIDGUT

The *derivatives of the midgut* are as follows:

1. The *small intestines,* including the part of the duodenum distal to the opening of the bile duct
2. The *cecum* and *appendix*
3. The *ascending colon*
4. The right half to two thirds, or *proximal part, of the transverse colon.*

All of the aforementioned midgut derivatives are supplied by the *superior mesenteric artery* (Figs. 12–1 and 12–11).

Initially, the midgut is suspended from the dorsal abdominal wall by a short mesentery (Fig. 12–1); this mesentery elongates rapidly (Fig. 12–11A). At first, the midgut communicates widely with the yolk sac (Fig. 12–1), but this connection becomes reduced to a narrow *yolk stalk,* or vitelline duct (Fig. 12–11A). Elongation of the midgut occurs faster than does elongation of the embryo's body and development of the abdominal cavity. As a result, a series of intestinal movements occur.

ROTATION AND FIXATION OF THE MIDGUT

Herniation of the Midgut (Figs. 12–11A and 12–12). As the midgut elongates, it forms a ventral U-shaped umbilical loop of gut, the *midgut loop,* which projects into the umbilical cord (Fig. 12–11A). This so-called *physiological umbilical herniation* occurs at the beginning of the sixth week and is a normal *migration of the midgut into the extraembryonic coelom. (Figs. 9–2E* and 12–11). At this stage the intraembryonic and extraembryonic coeloms communicate at the umbilicus (Fig. 12–7). This umbilical herniation occurs because there is not enough room in the abdomen for the rapidly growing midgut. The shortage of space is caused mainly by the relatively massive liver and kidneys (mesonephric and metanephric).

The midgut loop has a *cranial limb* and a *caudal limb.* The yolk stalk is attached to the apex of the midgut loop at the junction of the two limbs. The cranial limb grows rapidly and forms intestinal loops, but the caudal limb undergoes very little change except for development of the *cecal diverticulum* (Fig. 12–11B). Within the umbilical cord, *the midgut loop rotates 90 degrees counterclockwise* (as viewed from the ventral aspect of the embryo, Fig. 12–11B) around the axis of the *superior mesenteric artery;* this brings the cranial limb of the midgut loop to the right and the caudal limb to the left (Fig. 12–11B). During this rotation, the midgut elongates and coils to form loops of small bowel (jejunum and ileum).

Return of the Midgut (Fig. 12–11C). During the tenth week, the intestines return rapidly to the abdomen; this is often called *reduction of the midgut hernia.* The small intestines (formed from the cranial limb) return first and pass posterior to the superior mesenteric artery. As the intestines return, they undergo a further 180–degree counterclockwise rotation (Fig. 12–11C$_1$ and D$_1$), making a total of 270 degrees. The cecum and appendix now lie in contact with the caudal aspect of the liver (Fig. 12–11D), i.e., in a *subhepatic position. During the terminal months of pregnancy, the cecum grows down into the right iliac fossa* (Fig. 12–11E).

It is not known what causes the intestines to return to the abdomen, but the decrease in

---

Figure 12–11 Drawings showing rotation of the midgut, as seen from the left. A, Around the beginning of the sixth week, showing the umbilical loop of gut, or the midgut loop partially within the umbilical cord. Note the elongated, double-layered dorsal mesentery containing the superior mesenteric artery. A$_1$, Transverse section through the midgut loop illustrates the initial relationship of the limbs of the midgut to the artery. B, Later stage showing the beginning of the midgut rotation. B$_1$ illustrates the 90° counterclockwise rotation that carries the cranial limb to the right. C, About 10 weeks, showing the intestines returning to the abdomen. C$_1$ illustrates a further rotation of 90°. D, About 11 weeks, after return of intestines to the abdomen. D$_1$ shows a further 90°rotation of the gut, with a total of 270°. E, Late fetal period, after descent of the cecum to its normal position and fixation of the gut.

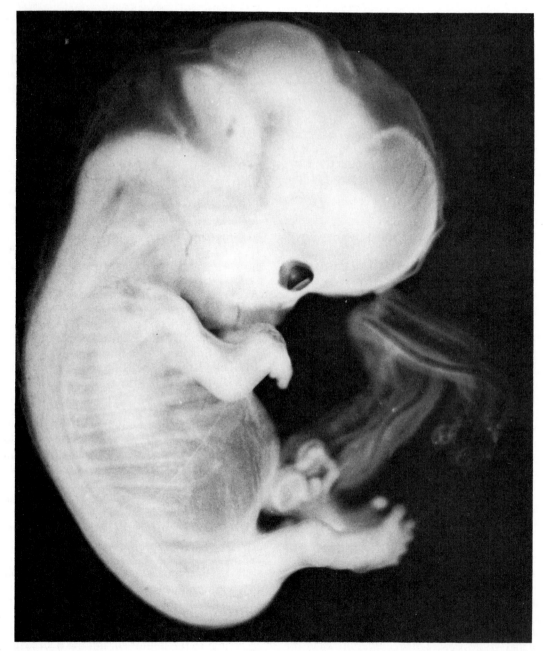

Figure 12–12   Photograph of a 28 mm human embryo during stage 23 of development (about 56 days). Note the herniated midgut loop in the proximal part of the umbilical cord and the umbilical blood vessels. Observe also the cartilaginous ribs, the prominent eye, and the relatively well-developed brain. (Courtesy of Dr. Bruce Fraser, Faculty of Medicine, Memorial University of Newfoundland.)

the relative size of the liver and mesonephric kidneys (see Chapter 13) and the enlargement of the abdominal cavity are important factors.

Fixation of the Intestines (Fig. 12–13).   Lengthening of the proximal part of the colon gives rise to the *ascending colon*

(Fig. 12–11D and E). As the intestines assume their final positions, their mesenteries are pressed against the posterior abdominal wall. The mesentery of the ascending colon fuses with the parietal peritoneum and disappears; consequently, the ascending colon becomes retroperitoneal.

The duodenum, except for the first 2.5 cm (derived from the foregut), also becomes retroperitoneal. The other derivatives of the midgut loop retain their mesenteries (Fig. 12–13). The mesentery of the jejunum and ileum is at first attached to the midline of the posterior abdominal wall (Fig. 12–11A). During rotation of the midgut, this mesentery twists around the origin of the superior mesenteric artery. When the mesentery of the ascending colon disappears, the fan-shaped mesentery of the small intestines acquires a

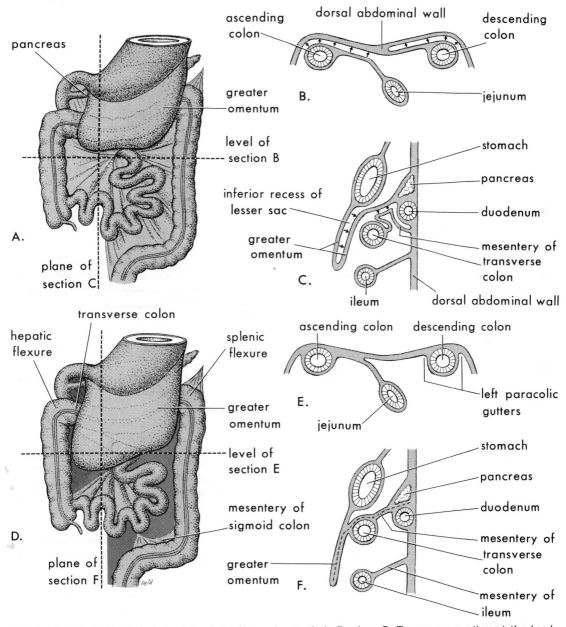

Figure 12–13 A, Ventral view of the intestines prior to their fixation. B, Transverse section at the level shown in A. The arrows indicate areas of subsequent fusion. C, Sagittal section at the plane shown in A showing the greater omentum overhanging the transverse colon. The arrows indicate areas of subsequent fusion. D, Ventral view of the intestines after their fixation. E, Transverse section at the level shown in D after disappearance of the mesentery of the ascending and descending colon. F, Sagittal section at the plane shown in D showing fusion of the greater omentum with the mesentery of the transverse colon and fusion of the layers of the greater omentum.

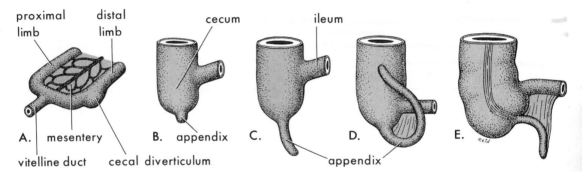

Figure 12–14 Drawings showing successive stages in the development of the cecum and the vermiform appendix. *A,* 6 weeks. *B,* 8 weeks. *C,* 12 weeks. *D,* At birth. Note that the appendix is relatively long and is continuous with the apex of the cecum. *E,* Child. Note that the appendix in the child is relatively short and lies on the medial side of the cecum, but it is still longer than the appendix in the adult. Because the appendix develops as the ascending colon and cecum descend, its final position is retrocolic or retrocecal. In about 64 per cent of people, it is posterior to the cecum (retrocecal). In about 32 per cent of people, it appears as illustrated in *E.*

new line of attachment that passes from the duodenojejunal junction obliquely downward to the ileocecal junction (Fig. 12–13*D*).

Fixation of the Duodenum (Fig. 12–13). Rotation of the stomach and duodenum causes the duodenum and the pancreas to fall to the right, and to become pressed against the posterior abdominal wall. The adjacent layers of peritoneum fuse and subsequently disappear (Fig. 12–13*C* and *F*). Consequently, most of the duodenum and the head of the pancreas become retroperitoneal.

The Cecum and Vermiform Appendix (Figs. 12–11 and 12–14). The primordium of the cecum and appendix, the *cecal diverticulum,* appears in the sixth week. This conical pouch appears on the antimesenteric border of the caudal limb of the midgut loop, just beyond the apex of the loop (Fig. 12–11*B*). The apex of this blind sac does not grow so rapidly and thus the appendix forms.

The appendix is subject to considerable variation in position. As the proximal part of the colon elongates, the cecum and appendix are displaced downward into the right iliac fossa. During this process, the appendix may pass posterior to the cecum (*retrocecal*) or colon (*retrocolic*), or it may descend over the brim of the pelvis (*pelvic,* or descending). *In about 64 per cent of people, the appendix is located retrocecally.*

The appendix increases rapidly in length, so that by birth it is a relatively long, worm-shaped tube (Fig. 12–14*D*). After birth, the wall of the cecum grows unequally, and, consequently, the appendix comes to lie on its medial side (Fig. 12–14*E*).

MALFORMATIONS OF THE MIDGUT

Congenital abnormalities of the intestines are common; most of them result from incomplete rotation and/or failure of fixation of the intestines.

Omphalocele (Fig. 12–15). This condition occurs once in about 6000 births and results from failure of the intestines to return to the abdomen during the second stage of rotation of the midgut loop. The hernia may consist of a single loop of bowel, or it may contain most of the intestines. The covering of the hernial sac is the amniotic epithelium of the umbilical cord.

Faulty closure of the lateral body fold during the fourth week (see Chapter 5) produces a large defect in the anterior abdominal wall and results in most of the abdominal viscera developing outside the embryo in a transparent sac of amnion. This severe type of omphalocele, sometimes called *eventration of the abdominal viscera,* is often associated with exstrophy of the urinary bladder (see Chapter 13).

Normally, after the intestines return from the umbilical cord, the rectus muscles approach each other and the linea alba, closing the circular defect. If a defect remains at this time, the omphalocele is usually small. In both large and small omphaloceles, *the protruding mass is covered by a thin transparent membrane composed of peritoneum and amnion. One third to one half of all infants with omphalocele have other congenital malformations,* e.g., malrotation, Meckel's diverticulum, and cardiovascular defects.

Umbilical Hernia. An umbilical hernia differs from an omphalocele in that *the pro-*

truding mass (omentum, or loop of bowel) is covered by subcutaneous tissue and skin. The hernia usually does not reach its maximum size until the end of the first month after birth. It ranges in size from a marble to a grapefruit. The defect through which the hernia occurs is in the linea alba.

Gastroschisis. This uncommon condition results from a defect of the anterior abdominal wall that permits extrusion of the abdominal contents without involving the umbilical cord. The viscera protrude into the amniotic cavity and are bathed by amniotic fluid. The term gastroschisis, which means "split, or open, stomach," is a misnomer, because it is the anterior abdominal wall that is split, not the stomach. The defect usually occurs on the right side and is more common in males than in females.

Nonrotation (Fig. 12–16A). This relatively common condition, often called "left-sided colon," is generally asymptomatic, but volvulus (twisting) of the intestines may occur. In nonrotation, the midgut loop does not rotate after it enters the abdomen; as a result, the caudal limb of the loop returns to the abdomen first and the small intestine lies on the right side of the abdomen and the entire large intestine on the left. When volvulus occurs, the superior mesenteric artery may be obstructed by the twisting. This results in *infarction* and gangrene of the bowel supplied by it.

Mixed Rotation and Volvulus (Fig. 12–16B). In this condition, the cecum lies inferior to the pylorus and is fixed to the posterior abdominal wall by peritoneal bands that pass over the duodenum. These bands and the frequent presence of volvulus of the intestines usually cause *duodenal obstruction*. This type of malrotation results from failure of the midgut loop to complete the final 90 degrees of rotation; consequently, the terminal part of the ileum returns to the abdomen first.

Reversed Rotation (Fig. 12–16C). Rarely, the intestines rotate in a clockwise rather than a counterclockwise direction. As a result, the duodenum lies anterior to the superior mesenteric artery, and the transverse colon lies posterior to it. The transverse colon may be obstructed by pressure from this artery. *Very rarely,* the small intestine lies on the left side of the abdomen, and

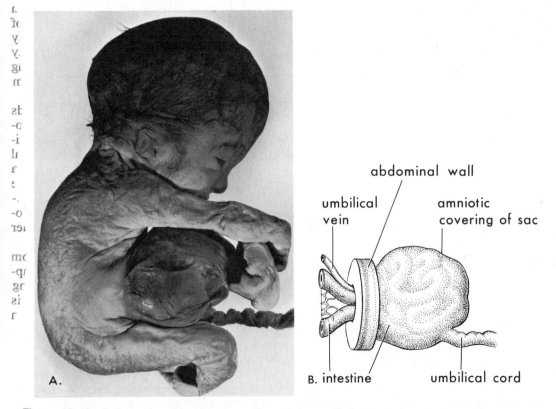

A.

abdominal wall

umbilical vein

amniotic covering of sac

B. intestine

umbilical cord

Figure 12–15 *A,* Large omphalocele in a 28-week fetus. *Half actual size. B,* Drawing illustrating the structure and contents of the hernial sac. The protruding mass of intestines is covered by a thin, transparent membrane composed of peritoneum and amnion. Occasionally, this membrane ruptures prior to or during birth. In this case, the eviscerated intestine lies freely around the gaping defect in the abdominal wall.

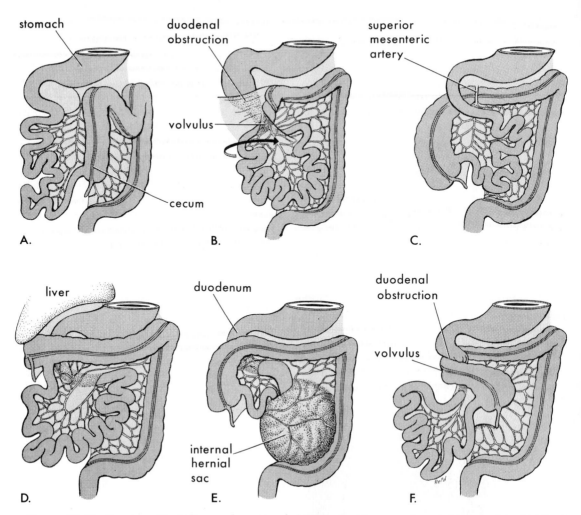

Figure 12–16 Drawings illustrating various abnormalities of midgut rotation. *A*, Nonrotation. *B*, Mixed rotation and volvulus. *C*, Reversed rotation. *D*, Subhepatic cecum and appendix. *E*, Internal hernia. *F*, Midgut volvulus. Malrotation may be present alone and be asymptomatic for life; however, it often becomes symptomatic because volvulus (twisting) occurs owing to the excessive mobility of the bowel.

the large intestine lies on the right side, with the cecum in the center.

Subhepatic Cecum and Appendix (Fig. 12–16D). Failure of the proximal part of the colon to elongate during the third stage of rotation results in the cecum staying up with the liver as the abdomen enlarges. More common in males, this condition occurs in about 6 per cent of fetuses and results in the cecum and the appendix being located in the subcostal region, near the inferior surface of the liver. Some elongation of the colon occurs during childhood; hence, a subhepatic cecum is not so common in adults.

Mobile Cecum. In about 10 per cent of people the cecum has an unusual amount of freedom, so it may even become herniated through the right inguinal canal. This condition results from incomplete fixation of the ascending colon. It is also significant because of the possible variations in position of the appendix and because volvulus of the cecum may occur.

Internal Hernia (Fig. 12–16E). In this condition, the small intestine passes into the mesentery of the midgut loop during return of the midgut to the abdomen, forming a hernia-like sac. This very uncommon condition rarely produces symptoms and is usually detected at autopsy or during an anatomical dissection.

Midgut Volvulus (Fig. 12–16F). In this condition, the small bowel fails to enter the abdominal cavity normally and the mesenteries fail to undergo normal fixation. As a result, twisting of the intestine commonly occurs with incomplete rotation of the midgut loop. The intestine is attached to the posterior abdominal wall at only two points: (1) the duodenum and (2) the proximal colon. The small intestines hang by a narrow stalk, containing the superior mesenteric vessels, and they usually twist around this stalk and become obstructed at or near the duodenojejunal junction. The circulation to the twisted segment is often obstructed, and thus *gangrene* may develop.

Intestinal Stenosis and Atresia (Fig. 12–5). Narrowing, or stenosis, and complete obstruction, or atresia, of the intestinal lumen occur most often in the duodenum and the ileum. The length of the area affected varies. Failure of an adequate number of vacuoles to form during recanalization leaves a transverse diaphragm, producing a so-called *diaphragmatic atresia* (Fig. 12–5F$_2$).

Another possible cause of stenoses and atresias is interruption of the blood supply to a loop of the fetal intestine, the so-called *fetal vascular accident*. A mobile loop of intestine may become twisted, thereby interrupting its blood supply and leading to necrosis of the bowel involved. This segment later becomes a fibrous cord connecting the proximal and distal ends of normal intestine. Similar malformations have been produced experimentally in fetal dogs by ligating the blood vessels to a loop of intestine (Barnard, 1956).

Most jejunoileal atresias are probably caused by infarction of the fetal bowel as the result of impairment of its blood supply. This probably occurs during the tenth week, when the intestines are returning to the abdomen. Malfixation of the gut predisposes it to strangulation and impairment of its blood supply, owing to *volvulus* (twisting of the gut).

Meckel's Diverticulum (Figs. 12–17A to C and 12–18). This ileal diverticulum is one of the most common malformations of

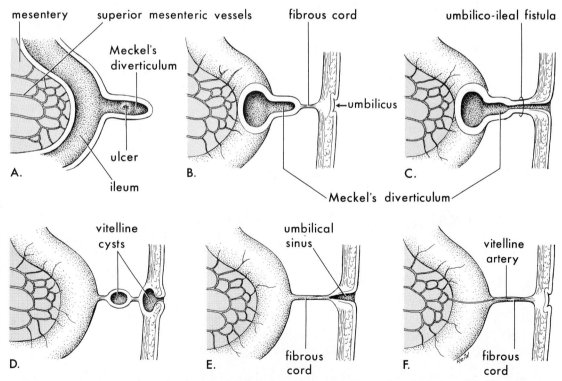

Figure 12–17 Drawings illustrating Meckel's diverticulum and other remnants of the yolk stalk. *A,* Section of the ileum and a Meckel's diverticulum with an ulcer. *B,* An ileal diverticulum, commonly called a Meckel's diverticulum, connected to the umbilicus by a fibrous cord. *C,* Umbilico-ileal fistula resulting from persistence of the entire intra-abdominal portion of the yolk stalk (see also Figure 12–18). *D,* Vitelline cysts at the umbilicus and in a fibrous remnant of the yolk stalk. *E,* Umbilical sinus resulting from the persistence of the yolk stalk near the umbilicus. The sinus is not always connected to the ileum by a fibrous cord as illustrated. *F,* The yolk stalk has persisted as a fibrous cord connecting the ileum with the umbilicus. A persistent vitelline artery extends along the fibrous cord to the umbilicus.

mesentery   ileum

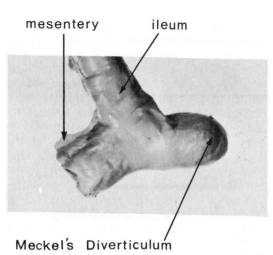

Meckel's Diverticulum

Figure 12–18 Photograph of an ileal diverticulum, which is usually referred to clinically as a Meckel's diverticulum. *Half actual size.* Only a small percentage of these diverticula ever produce symptoms.

the digestive tract; it occurs in 2 to 4 per cent of people. This malformation is three to five times more prevalent in males than in females. A Meckel's diverticulum is of clinical significance because it sometimes becomes inflamed and causes symptoms mimicking appendicitis.

The wall of the diverticulum contains all layers of the ileum and *may contain gastric and pancreatic tissues.* The gastric mucosa often secretes acid, producing ulceration (Fig. 12–17A).

A Meckel's diverticulum represents the remnant of the proximal portion of the *yolk stalk.* Typically, it appears as a finger-like pouch, about 3 to 6 cm long, arising from the antimesenteric border of the ileum (Fig. 12–18) 40 to 50 cm from the ileocecal junction. A Meckel's diverticulum may be connected to the umbilicus by a fibrous cord or a

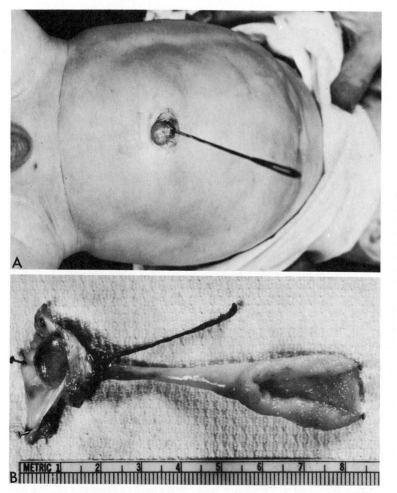

Figure 12–19 *A,* Photograph of the abdomen of an infant with an *umbilico-ileal fistula.* A probe has been inserted into the fistula, extending from the umbilicus to the ileum (a distance of about 5 cm). *B,* The excised fistula showing a granulomatous-looking bulge at the umbilical end and a cone-shaped Meckel's diverticulum at the ileal end. See Figure 12–17C for orientation. (From Schaffer, A. J., and Avery, M. E.: *Diseases of the Newborn.* 4th ed. Philadelphia, W. B. Saunders Co., 1977.)

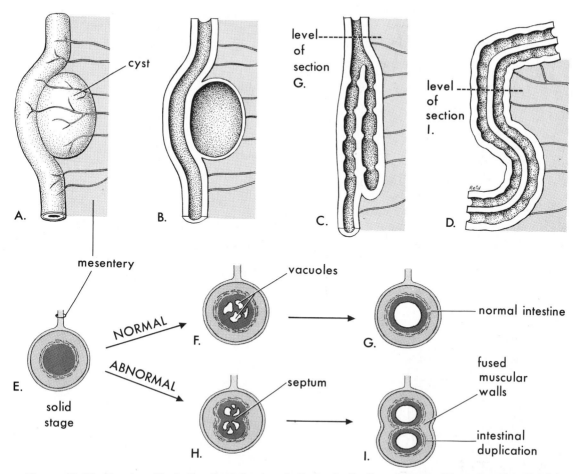

Figure 12–20 Drawings illustrating the following: *A*, Cystic duplication of the small intestine. Note that it is on the mesenteric side and receives branches from the arteries supplying the intestine. *B*, Longitudinal section of the duplication shown in *A*. It does not communicate with the intestine, but its musculature is continuous with the gut wall. *C*, A short tubular duplication of the small intestine. *D*, A long duplication of the small intestine showing a partition consisting of the fused muscular walls. *E*, Transverse section of the intestine during the solid stage. *F*, Normal vacuole formation. *G*, Coalescence of vacuoles to re-form the normal intestinal lumen. *H*, Two groups of vacuoles have formed. *I*, Coalescence of the vacuoles illustrated in *H* results in intestinal duplication.

fistula (Fig. 12–19); other possible remnants of the yolk stalk are illustrated in Figure 12–17*D* to *F*.

Rarely, there is abnormal reconstruction of the endodermal roof of the yolk sac during notochord formation (see Chapter 4); this may result in the endoderm being attached to the notochord. When the dorsal part of the yolk sac is incorporated into the embryo as the primitive gut (see Chapter 5), a cord of endodermal cells passes from the gut to the vertebral column, which has developed around the notochord. This endodermal cord may give rise to a giant diverticulum on the mesenteric side of the gastrointestinal tract. The attachment

of the diverticulum to the developing vertebral column may cause an anterior cleft in the body of the vertebra, or some other vertebral anomaly (Gray and Skandalakis, 1972).

Duplications (Fig. 12–20). Most intestinal duplications may be classified as closed *cystic duplications* (Fig. 12–20*B*) or *tubular duplications*, which communicate with the intestinal lumen (Fig. 12–20*C*). Cystic duplications are more common. Almost all duplications are caused by failure of normal recanalization, which results in the formation of two lumina (Fig. 12–20*H* and *I*). *They lie on the mesenteric side of the intestine.*

THE HINDGUT

The derivatives of the hindgut are as follows:

1. The left one third to one half, or *distal part, of the transverse colon*

2. The *descending colon*
3. The *sigmoid colon*
4. The *rectum*
5. The *superior portion of the anal canal*
6. The *epithelium of the urinary bladder and most of the urethra* (see Chapter 13)

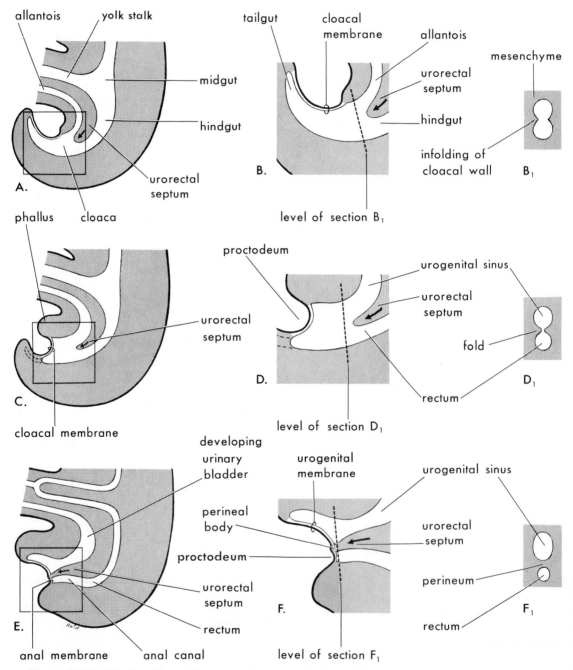

Figure 12–21 Drawings illustrating successive stages in the partitioning of the cloaca into the rectum and urogenital sinus by the urorectal septum. *A, C,* and *E,* Views from the left side at 4, 6, and 7 weeks, respectively. *B, D,* and *F* are enlargements of the cloacal region. *B₁, D₁,* and *F₁* are transverse sections through the cloaca at the levels shown in *B, D,* and *F.* Note that the tailgut (shown in *B*) degenerates and disappears as the rectum forms from the dorsal part of cloaca.

All these hindgut derivatives are supplied by the *inferior mesenteric artery*. The junction between the segment of transverse colon derived from the midgut and that originating from the hindgut is indicated by a change in blood supply from the superior (midgut) to the inferior mesenteric (hindgut) arteries.

The terminal portion of the hindgut called the *cloaca* (Fig. 12–21*A*), is an endoderm-lined cavity that is in contact with the surface ectoderm. The area of contact between the endoderm and the ectoderm is known as the *cloacal membrane* (Fig. 12–21*B*). This membrane is composed of endoderm of the cloaca and ectoderm of the *proctodeum*, or anal pit (Fig. 12–22). The expanded terminal part of the hindgut, the *cloaca* (L., an open drain or canal), receives the *allantois* ventrally and the mesonephric ducts laterally (see Chapter 13 and Fig. 12–21*A*).

Fixation of the Hindgut (Fig. 12–13). The descending colon becomes retroperitoneal when its mesentery fuses with the peritoneum of the left posterior abdominal wall and then disappears. The mesentery of the sigmoid colon is retained but is diminished in length during fixation.

Partitioning of the Cloaca (Fig. 12–21). The cloaca is divided by a coronal sheet or wedge of mesenchyme, the *urorectal septum,* which develops in the angle between the *allantois* and the *hindgut*. As this septum grows caudally toward the cloacal membrane, it develops extensions, like a two-pronged fork, that produce infoldings of the lateral walls of the cloaca (Fig. 12–21*B₁*). These folds grow toward each other and fuse, forming a partition that divides the cloaca in two parts: (1) the *rectum* and *upper anal canal* dorsally, and (2) the *urogenital sinus* ventrally (Fig. 12–21*D* and *F*).

By the end of the sixth week, the urorectal septum has fused with the cloacal membrane, dividing it into a dorsal *anal membrane* and a larger central *urogenital membrane* (Fig. 12–21*E* and *F*). The area of fusion of the urorectal septum with the cloacal membrane becomes the *central perineal tendon,* or *perineal body*. This fibromuscular node is located at the center of the perineum and is the *landmark of the perineum* where several muscles converge to insert into it.

The urorectal septum also divides the *cloacal sphincter* into anterior and posterior parts. The posterior part becomes the *external anal sphincter,* and the anterior part becomes the superficial transversus perinei, the bulbospongiosus, the ischiocavernosis, and the *urogenital diaphragm*. This explains why one nerve, the *pudendal nerve,* supplies all the muscles into which the cloacal sphincter divides.

Mesenchymal proliferations around the anal membrane elevate the surface ectoderm, forming a shallow pit known as the *proctodeum* (anal pit). The anal membrane is now located at the bottom of this pit, or depression (Fig. 12–21*E*). The *anal membrane usually ruptures at the end of the eighth week,* establishing the *anal canal*. This also brings the caudal part of the digestive tract into communication with the amniotic cavity.

The Anal Canal (Fig. 12–22). The superior two thirds (about 25 mm) of this canal is derived from the *hindgut;* the inferior one third (about 13 mm) develops from the *proctodeum*. The junction of the epithelium derived from the ectoderm of the proctodeum and the endoderm of the hindgut is roughly indicated by the *pectinate line* at the level of the anal valves. This line indicates the *approximate* former site of the anal membrane. Inferior to the pectinate line is the *anal pecten,* a bluish-white zone in living persons that measures about 1 cm wide. About 2 cm superior to the anus is the *anocutaneous line* ("white line"), where the anal epithelium

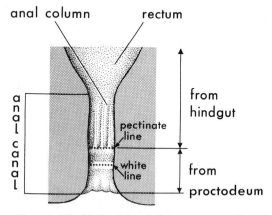

Figure 12–22 Sketch of the rectum and anal canal showing their developmental origins. Note that the superior two thirds of the anal canal is derived from hindgut and is endodermal in origin, whereas the inferior one third of the anal canal is derived from the proctodeum and is ectodermal in origin. Because of their different embryological origins, the superior and inferior parts of the anal canal are supplied by different arteries and nerves, and have different venous and lymphatic drainages.

changes from columnar to stratified squamous cells.

At the *anus,* the epithelium is keratinized and continuous with the skin of the anal region. The other layers of the wall of the anal canal are derived from splanchnic mesenchyme.

Owing to its hindgut origin, the superior part of the anal canal is supplied by the *superior rectal artery,* the continuation of the *inferior mesenteric artery* (hindgut artery), whereas the inferior part of the canal, derived from the proctodeum, is supplied by the *inferior rectal arteries,* branches of the *internal pudendal artery.* The inferior rectal arteries also supply the skin of the anal region.

The venous and lymphatic drainage and the nerve supply also differ because of the different embryological origins of the superior and inferior parts of the anal canal.

MALFORMATIONS OF THE HINDGUT

Congenital Aganglionic Megacolon or Hirschsprung's Disease. Rarely, a portion of the colon is dilated because of the absence of ganglion cells of the myenteric plexus distal to the dilated segment. The dilated colon, or megacolon (Gr. *megas,* big) has normal ganglion cells. The dilation of the colon results from failure of peristalsis in the aganglionic segment, causing failure of onward movement of the intestinal contents. In most cases, only the rectum and the sigmoid colon are involved, but, occasionally, ganglia are absent from more proximal parts of the colon.

Congenital megacolon results from failure of neural crest cells to migrate normally into the wall of the colon. This results in absence of parasympathetic ganglion cells. The cause of failure of the neural crest cells to complete their migration is unknown.

The extent of the aganglionosis varies. The internal anal sphincter is always involved, as well as most of the rectum in many cases.

Imperforate Anus and Related Malformations (Fig. 12–23). Some form of imperforate anus occurs once in about 5000 births; it is more common in males. *Most anorectal malformations result from abnormal development of the urorectal septum,* resulting in incomplete separation of the cloaca into urogenital and anorectal portions (Fig. 12–21).

The following are *low malformations of the anorectal region:*

Anal Agenesis, With or Without Fistula (Fig. 12–23D and E). The anal canal may end blindly and there may be an ectopic opening *(ectopic anus),* or fistula, that commonly opens into the perineum. However, the fistula may open into the vulva in females or the urethra in males.

Anal agenesis with fistula results from incomplete separation of the cloaca by the urorectal septum. It accounts for about 46 per cent of cases.

Anal Stenosis (Fig. 12–23B). The anus is in the normal position, but there is a narrowing of the anal canal. This malformation is probably caused by a slight dorsal deviation of the urorectal septum as it grows caudally to fuse with the cloacal membrane. As a result, the anal membrane (and later the anus) is small. Sometimes, only a small probe can be inserted (the so-called microscopic anus).

Membranous Atresia ("Covered Anus") (Fig. 12–23C). The anus is in the normal position, but a thin layer of tissue separates the anal canal from the exterior. The membrane is thin enough to bulge on straining and appears blue from the presence of meconium behind it. This malformation is very rare and results from failure of the anal membrane to perforate at the end of the eighth week.

The following are *high malformations of the anorectal region:*

Anorectal Agenesis, With or Without Fistula (Fig. 12–23F and G). The rectum ends well above the anal canal; *this is the most common type of anorectal malformation.* Although the rectum may end blindly, there is *usually a fistula* to the urethra in males or the vagina in females. Anorectal agenesis with a fistula results from incomplete separation of the cloaca by the urorectal septum.

Rectal Atresia (Fig. 12–23H and I). The anal canal and the rectum are present, but they are separated by an atretic segment of rectum. The cause of rectal atresia may be abnormal recanalization or defective blood supply, as discussed with atresia of the small intestine.

SUMMARY

The *primitive gut* (foregut, midgut, and hindgut) forms during the fourth week by incorporation of the roof of the yolk sac into the embryo. The endoderm of the primitive

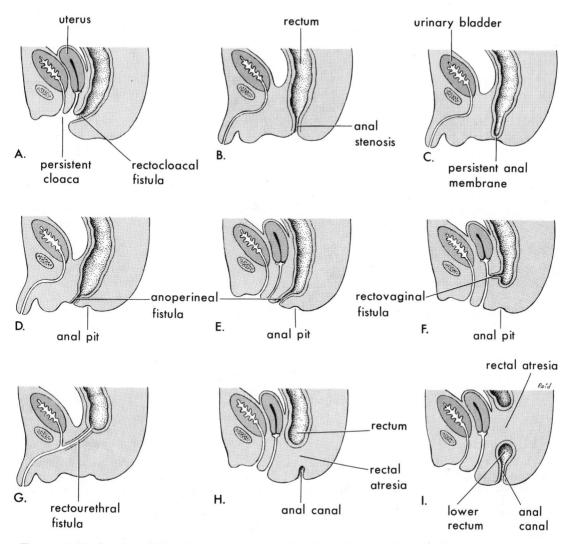

Figure 12–23 Drawings illustrating various anorectal malformations. *A,* Persistent cloaca. Note the common outlet for the intestinal, urinary, and reproductive tracts. This very rare condition usually occurs in females. *B,* Anal stenosis. *C,* Membranous atresia (covered anus). This is a very rare condition. *D* and *E,* Anal agenesis with perineal fistula. *F,* Anorectal agenesis with rectovaginal fistula. *G,* Anorectal agenesis with rectourethral fistula. *H* and *I,* Rectal atresia. Anal agenesis (*D* and *E*) and anorectal agenesis (*F* and *G*) account for over 75 per cent of anorectal malformations.

gut gives rise to the epithelial lining of most of the digestive tract and biliary passages, together with the parenchyma of the liver and pancreas. The epithelium at the cranial and caudal extremities of the digestive tract is derived from the ectoderm of the stomodeum and proctodeum, respectively. The muscular and connective tissue components of these structures are derived from the splanchnic mesenchyme surrounding the primitive gut.

The *foregut* gives rise to the pharynx, the lower respiratory system, the esophagus, the stomach, the duodenum (proximal to the opening of the bile duct), the liver, the pancreas, and the biliary apparatus.

Because the trachea and esophagus have a common origin from the foregut, incomplete partitioning by the tracheoesophageal septum results in stenoses or atresias, with or without fistulas between them.

The *liver bud,* or hepatic diverticulum, is formed from an outgrowth of the endodermal epithelial lining of the foregut. The epithelial liver cords and the primordia of the *biliary system,* which develop from the hepatic diverticulum, grow into the mesenchymal septum transversum. Here they differentiate into the *parenchyma of the liver* and the lining of the ducts of the biliary system between the layers of the ventral mesentery.

Congenital duodenal atresia results from failure of the vacuolization process to occur and to restore the lumen of the duodenum. Obstruction of the duodenum can also be caused by an *anular pancreas*, resulting from the two parts of the developing pancreas surrounding the duodenum.

The *pancreas* is formed by two buds that originate from the endodermal lining of the foregut. When the duodenum rotates to the right, the ventral pancreatic bud moves dorsally and fuses with the dorsal pancreatic bud. In some fetuses, the duct systems of the two buds fail to fuse, and an *accessory pancreatic duct* forms.

The *midgut* gives rise to the duodenum (distal to the bile duct), the jejunum, the ileum, the cecum, the vermiform appendix, the ascending colon, and the right, or proximal, half to two thirds of the transverse colon. The midgut forms a U-shaped intestinal loop that herniates into the umbilical cord during the sixth week because of inadequate room in the abdomen. While in the umbilical cord, the midgut loop rotates counterclockwise through 90 degrees. During the tenth week, the intestines rapidly return to the abdomen and rotate a further 180 degrees during this process.

Omphalocele, malrotations, and abnormalities of fixation result from failure of return or abnormal return of the intestines to the abdomen.

Because the gut is normally occluded at one stage, *stenosis* (narrowing), *atresia* (obstruction), and duplications may result if recanalization fails to occur or occurs abnormally. Various remnants of the yolk stalk may persist; Meckel's *diverticulum* is common, but only a small percentage of these *ileal diverticula* ever become symptomatic.

The *hindgut* gives rise to the left, or distal, one third to one half of the transverse colon, the descending and sigmoid colon, the rectum, and the superior part of the anal canal. The remainder of the anal canal develops from the proctodeum. The caudal part of the hindgut is expanded into the *cloaca,* which is divided by the *urorectal septum* into the urogenital sinus and rectum. The urogenital sinus mainly gives rise to the urinary bladder and the urethra (see Chapter 13). At first, the rectum and the superior part of the anal canal are separated from the exterior by the *anal membrane,* but this normally breaks down by the end of the eighth week.

Most anorectal malformations arise from abnormal partitioning of the cloaca by the urorectal septum into anorectal and urogenital parts. *Deviation of the urorectal septum in a dorsal direction probably causes most of the anorectal abnormalities,* such as rectal atresia and abnormal connections (fistulas) between the rectum and the urethra, the bladder, or the vagina.

SUGGESTIONS FOR ADDITIONAL READING

Crelin, E. S.: Development of Gastrointestinal Tract. New Jersey. *Clin. Symp. 13*:67, 1961. An excellent account of development of the digestive system, superbly illustrated by Dr. Netter. A brief account of clinically important congenital malformations is included.

Estrada, R. L.: *Anomalies of Intestinal Rotation and Fixation.* Springfield, Ill., Charles C Thomas, Publisher, 1968. A very good description of congenital malformations resulting from incomplete rotation of the gut and failure of fixation of the intestines.

Minor, C. L.: Alimentary System; *in* A. Rubin (Ed.): *Handbook of Congenital Malformations.* Philadelphia, W. B. Saunders Co., 1967, pp. 1–14. A concise reference book about malformations of the digestive system. Each malformation is discussed in terms of prevalence, associated anomalies, hereditary factors, treatment, and outlook.

Moore, K. L.: *Clinically Oriented Anatomy.* Baltimore, Williams & Wilkins Company, 1980, Chapter 12. This book contains many clinically oriented comments and patient oriented problems that are related to abnormal human development. It also explains how many anatomical relationships, such as those of omental bursa, make sense when developmental events are considered.

CLINICALLY ORIENTED PROBLEMS

1. A female child was born prematurely at 32 weeks' gestation to a 39-year-old woman whose pregnancy was complicated by *polyhydramnios.* Amniocentesis at 16 weeks showed that the infant had *trisomy 21.* The baby began to vomit within a few hours after birth, and marked dilation of the epigastrium was noted. X-ray films of the abdomen showed gas in the stomach and upper duo-

denum, but no other intestinal gas was observed. *A diagnosis of duodenal atresia was made.* Where does obstruction of the duodenum usually occur? What is the embryological basis of this congenital malformation? What caused distention of the infant's epigastrium? Is duodenal atresia commonly associated with malformations such as *Down syndrome?* What is the basis of *polyhydramnios?*

2. The umbilicus of a newborn infant failed to heal normally. It became swollen and there was persistent discharge from the umbilical stump. After probing, a *sinus tract* was outlined with radiopaque oil during fluoroscopy. The tract was resected on the ninth day after birth and its distal end was found to terminate in a *diverticulum of the ileum.* What is the embryological basis of the sinus tract *(umbilico-ileal fistula)?* What is the name given to this type of ileal diverticulum? Is this condition common?

3. A female infant was born with a small dimple where the anus should have been *(imperforate anus).* Examination of the infant's vagina revealed meconium and an opening *(sinus tract)* in the posterior wall of the vagina. Radiographic examination, using a contrast medium injected through a tiny catheter inserted into the sinus tract, revealed a *fistulous connection with the lower bowel.* With which part of the lower bowel would the fistula probably be connected? Name this malformation. What is the embryological basis of this condition?

4. A newborn infant was born with a light gray, shiny mass measuring the size of an orange and protruding from the umbilical region. *It was covered by a thin transparent membrane.* What is this congenital malformation called? What is the structure of the membrane covering the mass? What would be the composition ot the mass? What is the embryological basis of this protrusion?

5. A newborn infant appeared normal at birth, but vomiting and *abdominal distention* developed after a few hours. The vomitus contained bile, and only a little meconium was passed. X-ray study showed a gas-filled stomach and dilated, gas-filled loops of small bowel, but no air was present in the large intestine. This indicated a congenital *obstruction of the lower small bowel.* What part of the small bowel was probably obstructed? What would the condition be called? Why was a little meconium passed? *What would likely be observed at operation?* What was the probable embryological basis of the condition?

The answers to these questions are given at the back of the book.

REFERENCES

Ackerman, P.: Congenital defects ot the abdominal wall; in A. J. C. Huffstadt (Ed.): *Congenital Malformations.* Amsterdam, Excerpta Medica, 1980, pp. 215–231.

Barnard, C. N.: The genesis of intestinal atresia. *Surg. Forum 7:*393, 1956.

Bremer, J. L.: *Congenital Anomalies of the Viscera.* Cambridge, Massachusetts, Harvard University Press, 1957.

Clift, M. M.: Duplication of the small intestine. *J. Am. Med. Wom. Assoc. 9:*396, 1954.

Dawson, W., and Langman, J.: An anatomical-radiological study on the pancreatic duct pattern in man. *Anat. Rec. 139:*59, 1961.

Dennison, W. M.: Congenital malformations of the rectum and anus. *Glas. Med. J. 36:*283, 1955.

Fallin, L. T.: The development and cytodifferentiation of the islets of Langerhans in human embryos and foetuses. *Acta Anat. 68:*147, 1967.

Fitzgerald, M. J. T., Nolan, J. P., and O'Neill, M. N.: The position of the human caecum in fetal life. *J. Anat. 109:*71, 1971.

Gough, M. H.: Congenital abnormalities of the anus and rectum. *Arch. Dis. Child. 36:*146, 1961.

Gray, S. W., and Skandalakis, J. E.: *Embryology for Surgeons. The Embryological Basis for the Treatment of Congenital Defects.* Philadelphia, W. B. Saunders Co., 1972, pp. 63–281.

Healey, J. E.: *A Symopsis of Clinical Anatomy.* Philadelphia, W. B. Saunders Co., 1969, pp. 141–201.

Houle, M. P., and Hill, P. S.: Congenital absence of the gall bladder. *J. Maine Med. Assoc. 51:*108, 1960.

Kanagasuntheram, R.: Development of the human lesser sac. *J. Anat. 91:*188, 1957.

Kiesemelter, W. B., and Nixon, H. H.: Imperforate anus. I. Its surgical anatomy. *J. Pediatr. Surg. 2*:60, 1967.

Ladd, W. E., and Gross, R. E.: Congenital malformations of the anus and rectum. *Am. J. Surg. 23*:167, 1934.

Martinez, N. S., Morlach, C. G., Dockerty, B., Waugh, J. M., and Weber, H.: Heterotopic pancreatic tissue involving the stomach. *Ann. Surg. 147*:1, 1958.

McKeown, T., MacMahon, B., and Record, R. G.: An investivation of 69 cases of exomphalus. *Am. J. Hum. Genet. 5*:168, 1953.

Miller, V., and Holzel, A.: Growth and development of endodermal structures; *in* J. A. Davis and J. Dobbing (Eds.): *Scientific Foundations of Paediatrics.* Philadelphia, W. B. Saunders Co., 1974, p. 281.

Minor, C. L.: Alimentary system; *in* A. Rubin (Ed.): *Handbook of Congenital Malformations.* Philadelphia, W. B. Saunders Co., 1967, pp. 1–14.

Myers, R. L., Baggenstoss, A. H., Logan, G. B., and Hallenback, G. A.: Congenital atresia of the extrahepatic biliary tract: A clinical and pathological study. *Pediatrics 18*:767, 1956.

Nixon, H. H., and Wilkinson, A. W.: Abnormalities of the gastro-intestinal system; *in* A. P. Norman (Ed.): *Congenital Abnormalities in Infancy.* Oxford, Blackwell Scientific Publications, 1971, pp. 236–255.

Odgers, P. N. B.: Some observations on the development of the ventral pancreas in man. *J. Anat. 65*:1, 1930.

Olson, A. M., and Harrington, S. W.: Esophageal hiatal hernias of the short esophagus type: Etiologic and therapeutic considerations. *J. Thorac. Surg. 17*:189, 1948.

Partridge, J. P., and Gough, M. H.: Congenital abnormalities of the anus and rectum. *Br. J. Surg. 49*:37, 1961.

Rubin, A.: *Handbook of Congenital Malformations.* Philadelphia, W. B. Saunders Co., 1967.

Salebury, A. M., and Collins, R. E.: Congenital pyloric stenosis. *Arch. Surg. 80*:501, 1960.

Schaffer, A. J., and Avery, M. E.: *Diseases of the Newborn,* 4th ed. Philadelphia, W. B. Saunders Co., 1977.

Severn, C. B.: A morphological study of the development of the human liver. I. Development of the hepatic diverticulum *Am. J. Anat. 131*:133, 1971.

Severn, C. B.: A morphological study of the development of the human liver. II. Establishment of liver parenchyma, extrahepatic ducts, and associated venous channels. *Am. J. Anat. 133*:85, 1972.

Solebury, A. M., and Collins, R. E.: Congenital pyloric atresia. *A.M.A. Arch. Surg. 80*:501, 1960.

Stephens, F. D.: *Congenital Malformations of the Rectum, Anus and Genito-Urinary Tracts.* Edinburgh, E. & S. Livingstone, Ltd., 1963.

Stowens, D.: *Pediatric Pathology.* Baltimore, The Williams & Wilkins Co., 1959.

Swenson, O.: *Pediatric Surgery.* New York, Appleton-Century-Crofts, Inc., 1958.

Swenson, O.: Hirschsprung's disease (aganglionic megacolon). *N. Engl. J. Med. 260*:972, 1959.

Thompson, I. M.: On the arteries and ducts in the hepatic pedicle: A study in statistical human anatomy. Berkely, *Univ. Calif. Publ. Anat. 1*:55, 1933.

Weatherhill, D., Forgrave, E. G., and Carpenter, W. S.: Annular pancreas producing duodenal obstruction in the newborn. *A.M.A.J. Dis. Child. 95*:202, 1958.

13

THE UROGENITAL SYSTEM

The Urinary and Genital Systems

Embryologically and anatomically, the urinary (excretory) and genital (reproductive) systems are closely associated. These systems are very closely associated in the adult male, e.g., the urethra conveys both urine and semen. Although the systems are separate in normal adult females, the urethra and the vagina open into a common space between the labia minora, called the *vestibule of the vagina*.

Both the urinary and genital systems develop from the intermediate mesoderm, which extends along the entire length of the dorsal body wall of the embryo (Fig. 13–1*B*). During transverse folding of the embryo, the intermediate mesoderm is carried ventrally and loses its connection with the somites. This longitudinal mass of mesoderm on each side of the primitive aorta in the trunk region is called a *nephrogenic cord* (Figs. 13–1*D*

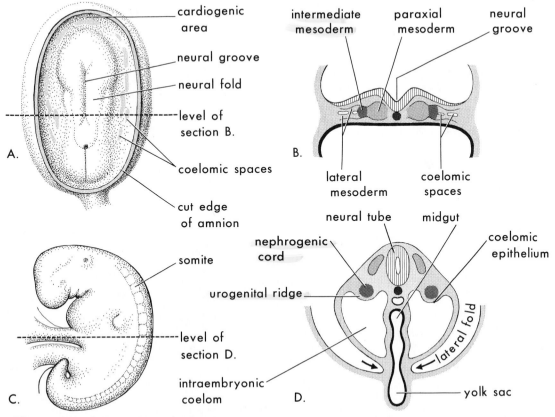

Figure 13–1 *A*, Dorsal view of an embryo during developmental stage 8 (about 18 days). *B*, Transverse section through the embryo showing the original position of the intermediate mesoderm. *C*, Lateral view of an embryo during developmental stage 12 (about 26 days). *D*, Transverse section through the embryo showing the urogenital ridges produced by the nephrogenic cords of the intermediate mesoderm.

and 13–5). On the dorsal wall of the coelomic cavity, these cords produce bilateral longitudinal bulges called the *urogenital ridges,* which give rise to both nephric and genital structures. A clear understanding of genitourinary embryology is necessary because *about 10 per cent of infants are born with some abnormality of the genitourinary system* (Vaughan and Middleton, 1975).

Functionally, the urogenital system can be divided into two parts: the *urinary system* and the *genital system.* Furthermore, development of the urogenital system is easier to understand if the urinary and genital systems are described separately. Development of the urinary system begins first.

THE URINARY SYSTEM

The urinary organs consist of the following structures: (1) the kidneys, which excrete the urine; (2) the ureters, which convey the urine to (3) the urinary bladder, where it is stored temporarily, and (4) the *urethra,* through which the urine is discharged to the exterior.

DEVELOPMENT OF THE KIDNEYS AND URETERS

Three different, slightly overlapping (in time) sets of excretory organs develop in human embryos: the *pronephros,* the *mesonephros,* and the *metanephros.* The first set, or *pronephroi* (plural term), are rudimentary and nonfunctional. They are analogous to the kidneys of some primitive fishes. The second set, or *mesonephroi,* are analogous to the kidneys of fishes and amphibians, and are probably functional for a short time during the early fetal period before they undergo involution. *The third set, or metanephroi, become the permanent kidneys and begin to produce urine when the fetus is 11 to 13 weeks old.*

The Pronephroi ("Forekidneys"). These transitory, nonfunctional structures appear in human embryos early in the fourth week and are represented by a few solid cell clusters, or tubular-arranged structures, in the cervical region (Fig. 13–2A). The pronephric ducts run caudally and open into the cloaca. The pronephroi soon degenerate, but most of their ducts are utilized by the next kidney (Fig. 13–2B).

The Mesonephroi ("Midkidneys"). These large organs appear later in the fourth week, caudal to the rudimentary pronephroi (Fig. 13–2). Each mesonephros may function as an intermim kidney until the permanent kidneys are established. The mesonephroi function in rabbit, cat, and pig embryos, but *there is no proof of mesonephric function in human embryos.* However, their morphological similarity to the mesonephroi in those species suggests that the human mesonephroi may function for a short time in the early fetal period.

The mesenchymal cell clusters in the nephrogenic cords (Fig. 13–4B) develop lumina and become mesonephric vesicles (Fig. 13–4C). Soon, each vesicle grows into an S-shaped *mesonephric tubule* (Fig. 13–4D and E). These tubules grow

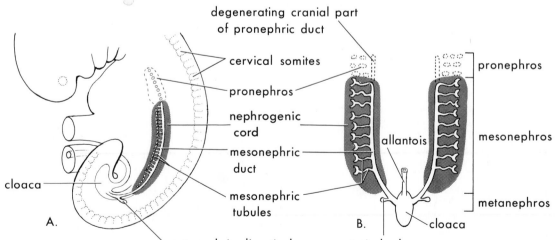

Figure 13–2 Diagrammatic sketches illustrating the three sets of excretory structures present in an embryo during developmental stage 14 (about 32 days). *A,* Lateral view. *B,* Ventral view. For the sake of simplicity, the mesonephric tubules have been pulled out to the sides of the mesonephric ducts. Actually, the tubules lie medial to ducts (see Fig. 13–4F).

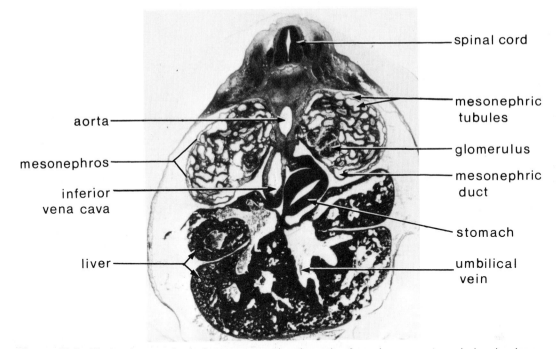

aorta

mesonephros

inferior
vena cava

liver

spinal cord

mesonephric
tubules

glomerulus

mesonephric
duct

stomach

umbilical
vein

Figure 13–3   Photomicrograph of a transverse section through a 6-mm human embryo during developmental stage 14 (about 32 days) showing the mesonephros. The mesonephric tubules exhibit various outlines in this section because they are convoluted and thus are cut simultaneously at several points along their lengths (×12).

laterally until they contact and become continuous with the pronephric duct, now called the *mesonephric duct* (Figs. 13–3 and 13–4*D*). The medial end of each mesonephric tubule expands and becomes invaginated by blood capillaries to form a double-layered cup, the *glomerular (Bowman's) capsule*. The cluster of capillaries that projects into this capsule is called a glomerulus (Figs. 13–3 and 13–4*F*). The capsule and glomerulus together form a *mesonephric (renal) corpuscle*. The intermediate portion of the mesonephric tubule lengthens rapidly and becomes highly convoluted (Fig. 13–4*F*). These tubules develop in a craniocaudal sequence, forming a long, ovoid kidney on each side of the abdominal cavity (Figs. 13–3 and 13–8*A*). As tubules are forming in the lumbar region, those in the thoracic region are degenerating. Thus, no more than 40 mesonephric tubules are present in each kidney at any one time. By the beginning of the fetal period, cranial parts of the mesonephroi have almost completely degenerated and disappeared. The mesonephric ducts and a few tubules persist as genital ducts in males or form vestigial remnants in females (Tables 13–1).

The Metanephroi ("Hindkidneys"). The metanephroi, or *permanent kidneys,* begin to develop early in the fifth week and start to function about six weeks later. Urine formation continues actively throughout fetal life,

and is excreted into the amniotic fluid. Because the placenta adequately eliminates metabolic wastes from the fetal blood, there is no need for the kidneys to function before birth. However, the kidneys must be capable of assuming their excretory and regulatory roles at birth (Page et al., 1981).

The urine mixes with the amniotic fluid, which the fetus drinks. A fetus normally swallows up to several hundred milliliters of amniotic fluid each day. This fluid is absorbed by the intestine; consequently, the fetal kidneys are involved in the regulation of the volume of amniotic fluid. In such abnormal conditions as *renal agenesis* (absence of kidneys) or *urethral obstruction,* the volume of amniotic fluid may be abnormally small, a condition called *oligohydramnios.* Hence, a fetus without kidneys *(bilateral renal agenesis)* continues to develop prenatally, but dies during the postnatal period.

Some cases of *polyhydramnios* (excessive amount of amniotic fluid) are caused by inability of the fetus to absorb amniotic fluid, e.g., due to obstruction of the esophagus or duodenum. The swallowed amniotic fluid is unable to enter the bowel to be absorbed for placental transfer.

The permanent kidney, or metanephros, develops from two different sources: the *metanephric diverticulum,* or ureteric bud,

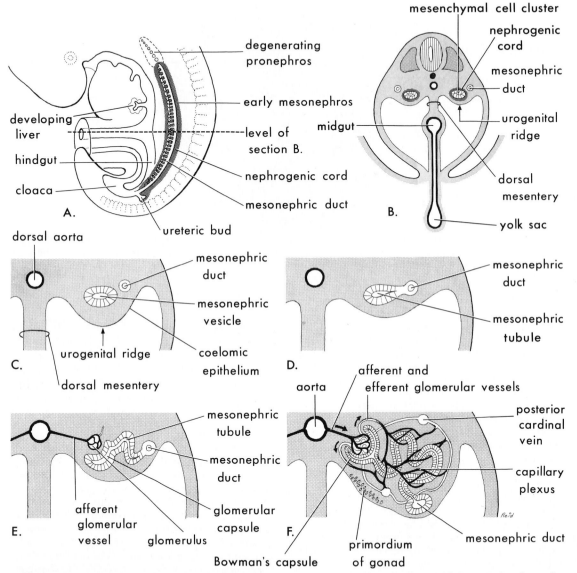

Figure 13–4 *A*, Sketch of a lateral view of a five-week embryo showing the extent of the mesonephros. *B*, Transverse section through the embryo showing the nephrogenic cords from which the mesonephric tubules develop. *C* to *F*, Sketches from transverse sections showing successive stages in the development of a mesonephric tubule.

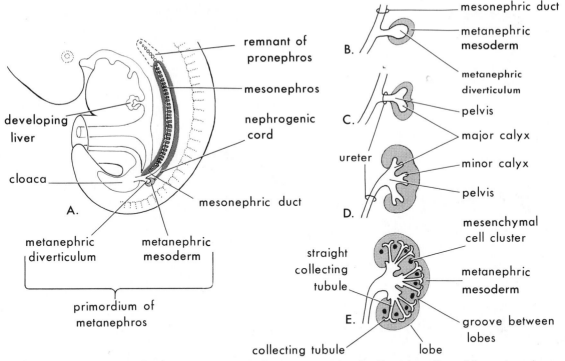

Figure 13–5 *A*, Sketch of a lateral view of a five-week embryo showing the primordium of the metanephros, or permanent kidney. *B* to *E*, Sketches showing successive stages of development of the metanephric diverticulum, or ureteric bud (fifth to eighth weeks) into the ureter, the pelvis, the calyces, and the collecting tubules. The renal lobes illustrated in *E* are also visible in the kidneys of newborn infants (Fig. 13–7).

and the *metanephric mesoderm,* or metanephrogenic blastema (Fig. 13–5*A* and *B*). The metanephric mesoderm is located in the caudal part of the nephrogenic cord. *Both primordia of the metanephros are of mesodermal origin.*

The metanephric diverticulum begins as a dorsal outgrowth, or bud, from the mesonephric duct near its entry into the cloaca (Fig. 13–5*A* and *B*). This *ureteric bud* gives rise to the *ureter,* the *renal pelvis,* the *calyces,* and the *collecting tubules* (Fig. 13–5*C* to *E*). As this diverticulum extends dorsocranially, it penetrates the *metanephric mesoderm,* which forms a cap over its expanded end (Fig. 13–5*B*). The stalk of the metanephric diverticulum becomes the ureter, and its expanded cranial end forms the renal pelvis. The pelvis divides into *major* and *minor calyces* and collecting tubules grow out from the minor calyces (Fig. 13–5*C* to *E*). Each collecting tubule undergoes repeated dichotomous branching, forming successive generations of collecting tubules.

Near the blind end of each arched collecting tubule (Figs. 13–5*E* and 13–6*A*), clusters

of mesenchymal cells in the metanephric mesoderm form small *metanephric vesicles*. These vesicles form as the result of an inductive influence from the arched collecting tubules (Fig. 13–6*B*). The metanephric vesicles soon give rise to renal tubules, or *metanephric tubules* (Fig. 13–6*C*). As these tubules develop, their proximal ends become invaginated by *glomeruli*. The renal corpuscle (glomerulus and Bowman's capsule) and its convoluted tubules form a *nephron* (Fig. 13–6*D*). The distal end of the metanephric tubule (future distal convoluted tubule) contacts an arched collecting tubule, and the two tubules soon become confluent. Hence, a *uriniferous tubule* consists of two embryologically different parts: a *nephron* derived from the metanephric mesoderm and a *collecting tubule* derived from the metanephric diverticulum (Figs. 13–5 and 13–6).

Continued lengthening of the metanephric tubule results in formation of the *proximal convoluted tubule,* the *loop of Henle,* and the *distal convoluted tubule* (Fig. 13–6*D*).

Four orders of collecting tubules are gradually absorbed into the walls of the calyces; conse-

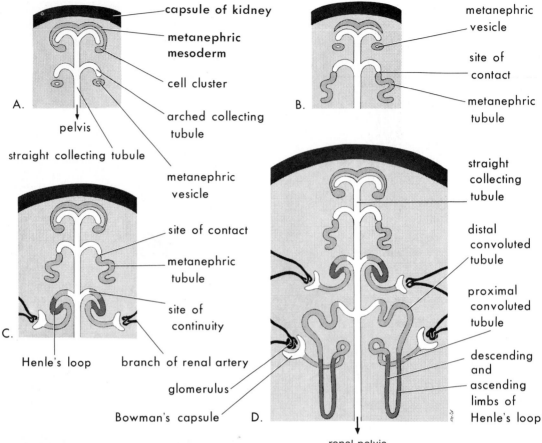

Figure 13-6 Diagrammatic sketches illustrating stages in the development of nephrons. The nephrons become continuous with the collecting tubules to form uriniferous tubules. This process commences around the beginning of the eighth week. It has been estimated that by 11 to 13 weeks, approximately 20 per cent of nephrons are relatively mature. This is when urine formation begins.

The number of nephrons more than doubles from 20 weeks' to 38 weeks' gestation. After examining Figure 13-5 and the above drawings, one should see clearly that the excretory units, or nephrons, are derived from the metanephric mesoderm, and that the collecting system is derived from the metanephric diverticulum.

quently, the associated nephrons lose their connections with collecting tubules. These nephrons normally degenerate and disappear.

The increase in kidney size after birth results mainly from hypertrophy of the nephrons. The outer cortical region of the kidneys of newborn infants contains undifferentiated mesenchyme, from which additional nephrons develop for several months after birth.

The Fetal and Newborn Kidneys (Fig. 13-7). In the fetus and infant, the lobes of the kidney are demarcated on the surface. This lobation usually disappears during infancy as cortical growth of nephrons occurs.

Positional Changes of the Kidney (Figs. 13-8 and 13-9). Initially, the kidneys are in the pelvis, but they gradually come to lie in the abdomen. This migration results mainly from

growth of the embryo's body caudal to the kidneys. In effect, the caudal part of the embryo grows away from the kidneys so that they occupy progressively higher levels. Eventually, they come to lie posterior to the peritoneum on the posterior abdominal wall.

Initially, the hilum of the kidney faces ventrally, but as the kidney ascends it rotates almost 90 degrees so that its hilum is directed anteromedially (Fig. 13-8C and D).

Blood Vessels to the Developing Kidneys (Fig. 13-8). As the kidneys "ascend" from the pelvis to the abdomen, they receive their blood supply from blood vessels close to them. Initially, the renal arteries are branches of the middle sacral and common iliac arteries. As they "ascend" further, the

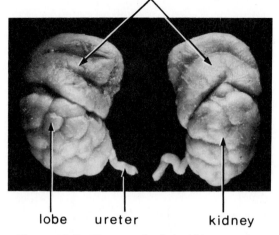

suprarenal or adrenal glands

lobe   ureter   kidney

Figure 13–7 Photograph of the kidneys and suprarenal (adrenal) glands of a 28-week fetus (× 2). The external evidence of the lobes of the kidney normally disappears by the end of the first postnatal year. In rare instances, evidence of the fetal lobes may be visible in adult kidneys, appearing as indentations along the lateral margin of the kidney. Note the large suprarenal, or adrenal, glands. During the first two weeks after birth, the adrenal glands reduce to about half this size (see Fig. 13–19).

kidneys receive their blood supply from arteries that extend laterally from the aorta. When they reach a higher level, the kidneys receive blood from new branches from the aorta at higher levels, and the inferior branches undergo involution and disappear.

The definitive *renal arteries* are not recognizable until the ninth week (Fig. 13–8*D*).

The relatively common variations in the blood supply to the kidneys reflect the manner in which the blood supply continually changes as the kidneys "ascend" from the pelvis to the abdomen.

CONGENITAL MALFORMATIONS OF THE KIDNEYS AND URETERS

Abnormalities of the kidney and ureter occur in 3 to 4 per cent of the population and include variations in blood supply, abnormal positions, and upper urinary tract duplications.

Renal Agenesis (Fig. 13–10A). Unilateral absence of a kidney is relatively common, occurring about once in every 1000 births. *Unilateral renal agenesis* causes no symptoms and is usually not discovered in the neonatal period because the other kidney is able to perform the function of the missing kidney.

Bilateral renal agenesis is rare (about 0.3 per 1000 births), and is incompatible with postnatal life. Most infants afflicted with this malformation die during birth or a few hours later. Because no urine is excreted into the amniotic fluid, *bilateral renal agenesis is associated with oligohydramnios* (deficiency in the amount of amniotic fluid).

Renal agenesis results when the meta-

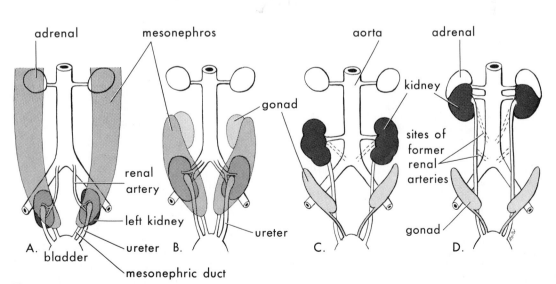

Figure 13–8 Diagrams of ventral views of the abdominopelvic region of embryos and fetuses (sixth to ninth weeks) showing the medial rotation and ascent of the kidneys from the pelvis to the abdomen. This is not a true upward migration, but the result of disproportionate growth of the caudal region of the embryo. Note that as the kidneys "ascend" they are supplied by arteries at successively higher levels, and the hilum is eventually directed anteromedially.

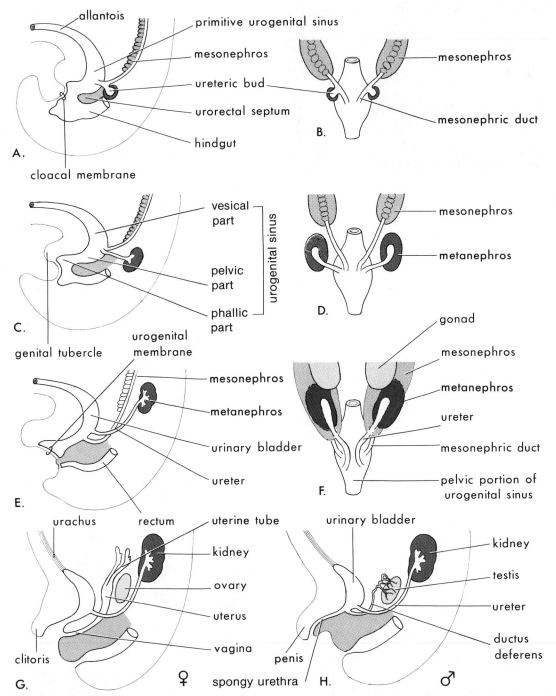

Figure 13–9 Diagrams showing (1) division of the cloaca into the urogenital sinus and the rectum, (2) absorption of the mesonephric ducts, (3) development of the urinary bladder, urethra, and urachus, and (4) changes in the location of the ureters. *A,* Lateral view of the caudal half of a five-week embryo. *B, D,* and *F,* Dorsal views. *C, E, G,* and *H,* Lateral views. The stages shown in *G* and *H* are reached by about 12 weeks.

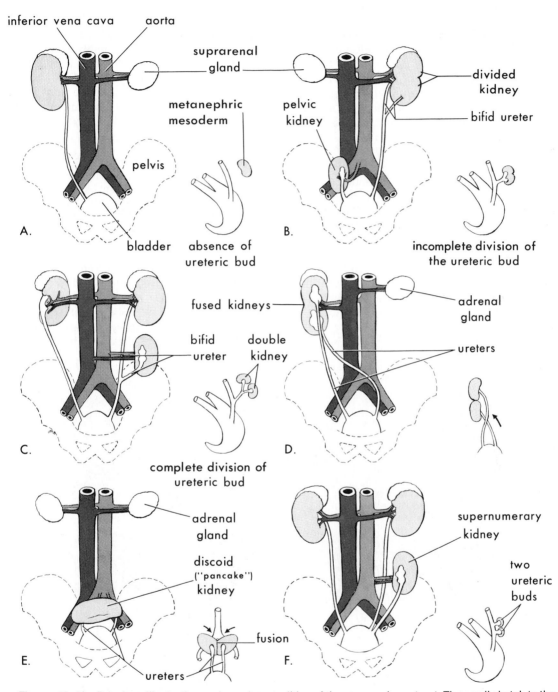

Figure 13–10 Drawings illustrating various abnormalities of the upper urinary tract. The small sketch to the lower right of each drawing illustrates the probable embryological basis of the malformation. *A*, Unilateral renal agenesis. *B*, Right side, pelvic kidney; left side, divided kidney with a bifid ureter. *C*, Right side, malrotation of the kidney; left side, bifid ureter and double kidney. *D*, Crossed renal ectopia. The left kidney crossed to the right side and fused with the right kidney. *E*, "Pancake," or discoid, kidney resulting from fusion of the unascended kidneys. *F*, Supernumerary left kidney resulting from the development of two ureteric buds.

nephric diverticulum fails to develop, or when early degeneration of this bud occurs. Failure of the metanephric diverticulum to penetrate the metanephric mesoderm results in absence of kidney development because no nephrons are induced to develop from the metanephric mesoderm.

Abnormal Rotation of the Kidney (Fig. 13–10C). In this uncommon abnormality, the hilum of the kidney does not face anteromedially. If the hilum faces ventrally, the kidney has retained its original position and has failed to rotate medially (Fig. 13–8). If the hilum faces dorsally, rotation of the kidney has proceeded too far; if it faces laterally, malrotation (i.e., lateral rotation) has occurred. Abnormal rotation is often associated with ectopic kidneys (Fig. 13–10B and F).

Ectopic Kidneys (Figs. 13–10B and E). One or both kidneys may be in an abnormal position. They are lower than usual and malrotated. Most ectopic kidneys are located in the pelvis, but some are low in the abdomen.

Pelvic kidney and other forms of low kidney result from failure of the kidneys to ''ascend.'' Pelvic kidneys may fuse to form a round mass known as a *pancake kidney* (Fig. 13–10E). Ectopic kidneys receive their blood supply from blood vessels near them, and they are often supplied by multiple vessels.

An unusual type of ectopic kidney is *unilateral fused kidney* (Fig. 13–10D). The developing kidneys fuse while in the pelvis, and one kidney ''ascends'' to its normal position, carrying the other one with it across the midline.

Crossed Renal Ectopia (Fig. 13–10D). Occasionally, during its ascent to the abdomen, a kidney may cross to the opposite side. When this occurs, the kidney may fuse with the other kidney, producing a single large kidney. This condition may be distinguished from partial duplication of the kidney by the fact that one ureter descends on one side and the other descends on the other side of the midline to enter the bladder.

Horseshoe Kidney (Fig. 13–11). In 1 in about 600 persons, the kidneys are fused across the midline; usually the inferior poles are fused. The superior poles are fused in less than 10 per cent of cases. The large, U-

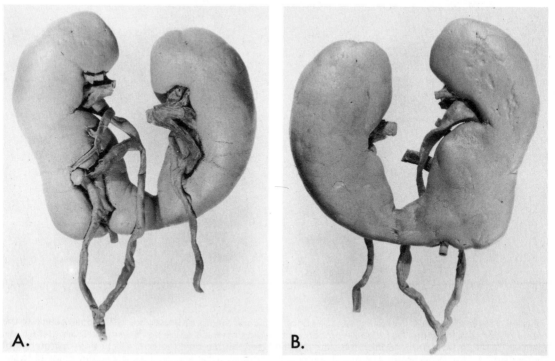

A.   **B.**

Figure 13–11 Photographs of a horseshoe kidney resulting from fusion of the inferior poles of the kidneys. Only rarely do the superior poles of the kidneys fuse, forming a horseshoe kidney that has its opening facing inferiorly. *A*, Anterior view. *B*, Posterior view. *Half actual size.* The larger right kidney has a bifid ureter. Horseshoe kidney is a rather common abnormality (in one of about 600 persons). It is asymptomatic unless urinary outflow is impeded.

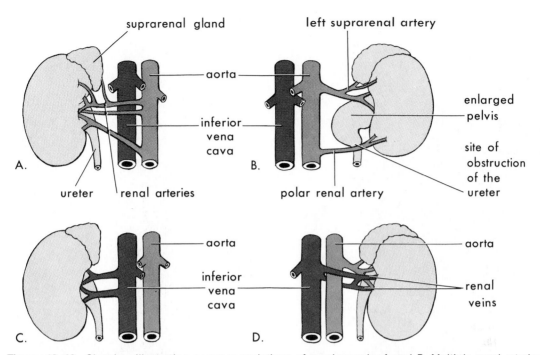

Figure 13–12 Sketches illustrating common variations of renal vessels. *A* and *B*, Multiple renal arteries. Note that some accessory vessels enter the poles of the kidney. *Accessory renal arteries are common*, occurring in about 25 per cent of people. They may arise superior or inferior to the main artery, but they occur more frequently at the former location. The polar renal artery in *B* has obstructed the ureter and produced an enlarged renal pelvis. *C* and *D*, Multiple veins.

shaped kidney usually lies in the hypogastrium at the level of the lower lumbar vertebrae, because normal ascent was prevented by the root of the inferior mesenteric artery. *Horseshoe kidney usually produces no symptoms,* because the collecting system usually develops normally and the ureters usually enter the bladder normally. If urinary outflow is impeded, signs and symptoms of obstruction and/or infection may appear.

Multiple Renal Vessels (Fig. 13–12). Variations in the number of renal arteries and in their position with respect to the renal veins are common. About 25 per cent of kidneys have two or more renal arteries. Supernumerary arteries, usually two or three, are about twice as common as supernumerary veins, and they usually arise at the level of the kidney.

Accessory vessels may arise from the suprarenal artery and pass to the superior pole of the kidney. Polar vessels may also arise from the aorta and pass to the inferior pole of the kidney. Sometimes, an accessory renal artery supplying the inferior pole of the kidney compresses and obstructs the ureter at the uteropelvic junction. Variations in the blood supply of the kidneys are most common in ectopic kidneys. As the kidney moves out of the pelvis, it is supplied by successively higher vessels, and the lower vessels normally degenerate (Fig. 13–8). *Vascular variations result from persistence of embryonic vessels that normally disappear when the definitive renal arteries form.*

Duplications of the Upper Urinary Tract (Figs. 13–10 and 13–13). Duplications of the abdominal part of the ureter and renal pelvis are common, but a supernumerary kidney is rare. These abnormalities result from division of the metanephric diverticulum, or ureteric bud. The extent of ureteral duplication depends on how complete the division of the diverticulum is. Incomplete division of the diverticulum results in a divided kidney with a bifid ureter (Fig. 13–10B). Complete division of the diverticulum results in a supernumerary, or double, kidney with a bifid ureter (Fig. 13–10C) or with separate ureters. A supernumerary kidney with its own ureter probably results from the formation of an extra ureteric bud (Fig. 13–10F).

Ectopic Ureteric Orifices. A ureter that

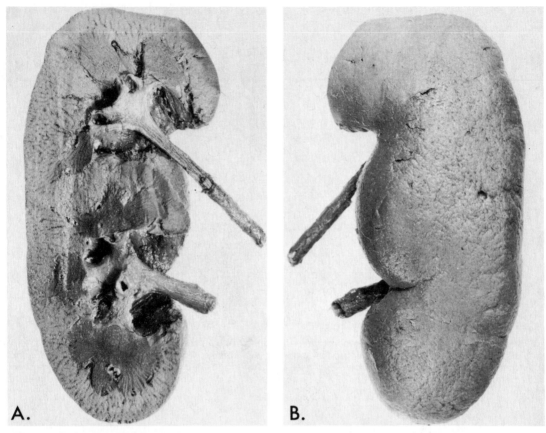

Figure 13–13 Photographs of a kidney with two ureters and two renal pelves. Note that the kidney consists of two parts, each with its own renal pelvis and ureter. This malformation results from incomplete division of the metanephric diverticulum (Fig. 13–9B). A, Longitudinal section through the kidney showing the two renal pelves. B, Anterior surface.

opens anywhere except into the posterosuperior angle of the urinary bladder will have an ectopic ureteric orifice. In males, ectopic ureteric orifices are usually in the prostatic portion of the urethra, but they may be in the prostatic utricle, or the seminal vesicle. In females, ectopic ureteric orifices may be in the urethra, the vagina, or the vestibule of the vagina.

Incontinence is the common complaint when there is an ectopic ureteric orifice, because the urine flowing from the ectopic ureteric orifice is not entering the bladder; instead, it continually dribbles from the urethra in a male, or from the urethra or the vagina in a female.

An abnormal ureteric orifice results when the ureter is not incorporated into the posterior part of the urinary bladder as illustrated in Figure 13–9. Instead, the ureter is carried caudally with the mesonephric duct and is incorporated into the caudal portion of the vesical part of the urogenital sinus. Because this part becomes the prostatic urethra in the male and the urethra in the female, the location of ectopic ureteric orifices in each sex is understandable.

When two ureters form on one side, they usually open into the urinary bladder (Fig. 13–10F). In some cases, the extra ureter is carried caudally and drains into the urethra or the vagina, as discussed previously.

Congenital Bilateral Polycystic Disease of the Kidney. The congenital cystic form of this disease is relatively common; death occurs at or shortly after birth. The kidneys contain multiple small to large cysts, which result in severe renal insufficiency. There are three possible embryological explanations for these malformed kidneys: (1) The cysts result from failure of some metanephric tubules to join with collecting tubules, as shown in Figure 13–6C; (2) the cysts develop from remnants of the first rudimentary nephrons, which normally degenerate; and (3) the collecting tubules develop abnormally. The collecting tubules fail to branch normally and undergo cystic dilation. Abnormal development of the collecting tubules results in failure of a normal number of nephrons to form from the metanephric mesoderm. *Current view favors abnormal development of the collecting tubules as the cause of congenital polycystic kidney.*

DEVELOPMENT OF THE URINARY BLADDER AND URETHRA

Division of the endodermal cloaca by the *urorectal septum* into a dorsal rectum and a ventral urogenital sinus is illustrated in Figure 12–21. For descriptive purposes, the *urogenital sinus* is divided into three parts: a cranial *vesical part* (primitive bladder continuous with the allantois), a middle *pelvic part,* and a caudal *phallic part* that is closed externally by the *urogenital membrane* (Fig. 13–9E).

The Urinary Bladder (Figs. 13–9 and 13–14). The epithelium of the bladder is derived from the endoderm of the vesical part of the urogenital sinus. The lamina propria, the muscle layers, and the serosa (or adventitia) develop from the adjacent splanchnic mesenchyme. Initially, the bladder is continuous with the *allantois* (Fig. 13–9C), but the lumen of this vestigial structure is soon constricted. The allantois then becomes a thick fibrous cord, the *urachus,* attached to the apex of the bladder and the umbilicus (Figs. 13–9 and 13–14). In the adult, the urachus is called the *median umbilical ligament.* As the bladder enlarges, the caudal portions of the mesonephric ducts are incorporated into its dorsal wall (Fig. 13–9). Initially, these ducts contribute to the formation of the mucosa of the *trigone of the bladder*, but the mesodermal epithelium derived from the mesonephric ducts is soon replaced by the endodermal epithelium of the urogenital sinus (Gyllensten, 1949). As the mesonephric ducts are absorbed, the ureters come to open separately into the urinary bladder (Fig. 13–9C to F). Partly because of traction exerted by the kidneys during their cranial migration and partly because the

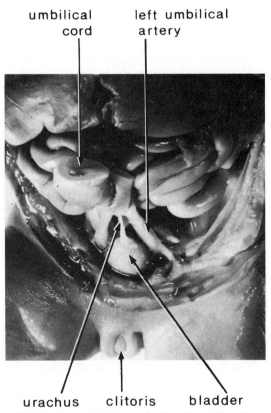

umbilical cord   left umbilical artery

urachus   clitoris   bladder

Figure 13–14 Photograph of a dissection of an 18-week female fetus showing the relations of the urachus to the urinary bladder and the umbilical arteries. Note that the clitoris is still relatively large at this stage.

mesonephric ducts continue to grow downward, the ureters in males open laterally and cranially to the *ejaculatory ducts,* the adult derivatives of the caudal ends of the mesonephric ducts (Fig. 13–24A). The caudal ends of the mesonephric ducts in females subsequently degenerate (Fig. 13–24B).

In infants and children, the urinary bladder, even when empty, is in the abdomen. It begins to enter the pelvis major at about 6 years, but it is not entirely in the pelvis minor until after puberty.

The Male Urethra (Figs. 13–9H and 13–15). The epithelium of the prostatic urethra proximal to the orifices of the ejaculatory ducts is derived from endoderm of the vesical part of the urogenital sinus. The connective tissue and the smooth muscle develop from adjacent splanchnic mesenchyme. The epithelium of the remainder of the prostatic urethra and the membranous urethra is derived from the endoderm of the pelvic part

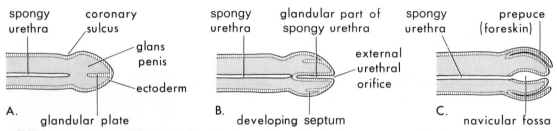

Figure 13–15 Schematic longitudinal sections of the developing penis illustrating development of the prepuce and the glandular portion of the spongy urethra. *A,* 11 weeks. *B,* 12 weeks. *C,* 14 weeks.

of the urogenital sinus. The epithelium of the spongy urethra, except for the glandular part, forms from cells from the phallic part of the urogenital sinus, which grow into the *urethral groove* of the developing penis (Fig. 13–27). The epithelium of the navicular fossa develops by canalization of an ectodermal cord of cells, the *glandular plate,* that extends into the glans from its tip (Fig. 13–15). The connective tissue and smooth mucle develop from the adjacent splanchnic mesenchyme.

The Female Urethra (Fig. 13–9G). The epithelium of the entire urethra is derived from the endodermal vesical part of the urogenital sinus. The connective tissue and smooth muscle develop from the adjacent splanchnic mesenchyme.

CONGENITAL MALFORMATIONS OF THE URACHUS AND URINARY BLADDER

Urachal Malformations (Fig. 13–16). The adult derivative of the urachus is the *median umbilical ligament.* In fetuses, the *urachus* lies between the umbilical arteries and connects the urinary bladder with the umbilicus (Figs. 13–9 and 13–14). A remnant of its lumen usually persists in the inferior part of the urachus, and, in about 50 per cent of cases, the lumen is continuous with the cavity of the bladder (Gray and Skandalakis, 1972). The patent inferior end of the urachus may dilate to form a *urachal sinus* that opens into the bladder (Fig. 13–16*B*). The lumen in the superior part of the urachus may also remain patent and form a urachal sinus that opens at the umbilicus (Fig. 13–16*B*). Very rarely, the entire urachus remains patent and forms a *urachal fistula* that allows urine to escape from its umbilical orifice (Fig. 13–16*C*).

Remnants of the lumen of the urachus may later give rise to *urachal cysts* (Fig. 13–16*A*).

Small cysts are observed in about one third of cadavers examined, but cysts are not detected in living persons unless they become infected and enlarge.

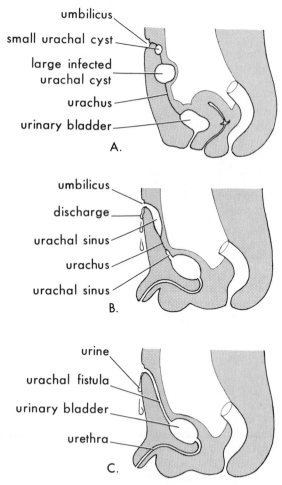

Figure 13–16 Diagrams illustrating malformations of the urachus. *A,* Urachal cysts. The most common site is at the superior end of the urachus, just inferior to the umbilicus. *B,* Two types of urachal sinus; one is continuous with the bladder, and the other opens at the umbilicus. *C,* Patent urachus, or urachal fistula, connecting the bladder and umbilicus.

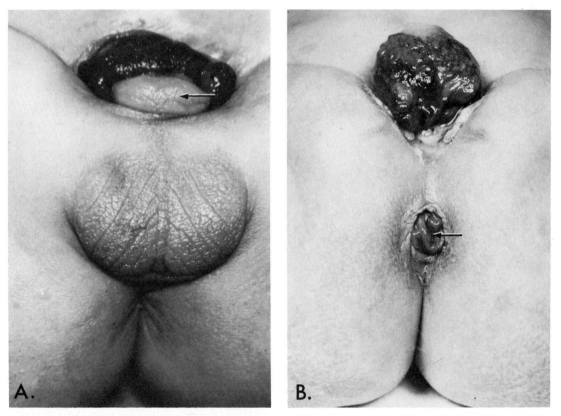

Figure 13–17 Photographs of infants with exstrophy of the bladder. Because of defective closure of the inferior portion of the anterior abdominal wall and the anterior wall of the bladder, the bladder appears as an everted bulging mass inferior to the umbilicus. *A*, Male. Epispadias is also present, and the penis (arrow) is small, flattened, and deeply fissured on the dorsal surface (see also Fig. 13–18*F*). (Courtesy of Dr. Colin C. Ferguson. Children's Centre, Winnipeg, Canada.) *B*, Female. The arrow indicates a slight prolapse of the rectum. (Courtesy of Mr. Innes Williams, Genitourinary Surgeon, The Hospital for Sick Children, Great Ormond Street, London, England.)

Exstrophy of the Bladder (Figs. 13–17 and 13–18). Fortunately, this severe malformation is rare, occurring about once in every 50,000 births. *Exposure and protrusion of the posterior wall of the urinary bladder* characterize this congenital abnormality, which occurs chiefly in males. The trigone of the bladder and the ureteric orifices are exposed, and urine dribbles intermittently on the mucous membrane of the everted bladder. *Epispadias* (Fig. 13–18*F*) and wide separation of the pubic bones are associated with complete exstrophy of the bladder. In some cases, the penis or clitoris is divided and the halves of the scrotum or labia majora are widely separated.

Exstrophy of the bladder is caused by incomplete midline closure of the inferior part of the anterior abdominal wall (Fig. 13–18*F*). The fissure involves not only the anterior abdominal wall but also the anterior wall of the urinary bladder. The defective closure results from failure of mesenchymal cells to migrate between the surface ectoderm and the urogenital sinus during the fourth week (Fig. 13–18*B* and *C*). As a result, no muscle forms in the anterior abdominal wall over the urinary bladder. Later, the thin epidermis and the anterior wall of the bladder rupture, causing a wide communication between the exterior and the mucous membrane of the bladder.

DEVELOPMENT OF THE SUPRARENAL GLANDS

The cortex and medulla of the *suprarenal (adrenal)* glands have different origins (Fig. 13–19): The *cortex* develops from mesoderm, and the *medulla* forms from neuroectoderm (*neural crest cells*).

The cortex is first indicated during the sixth week by an aggregation of mesenchymal cells on each side between the root of the

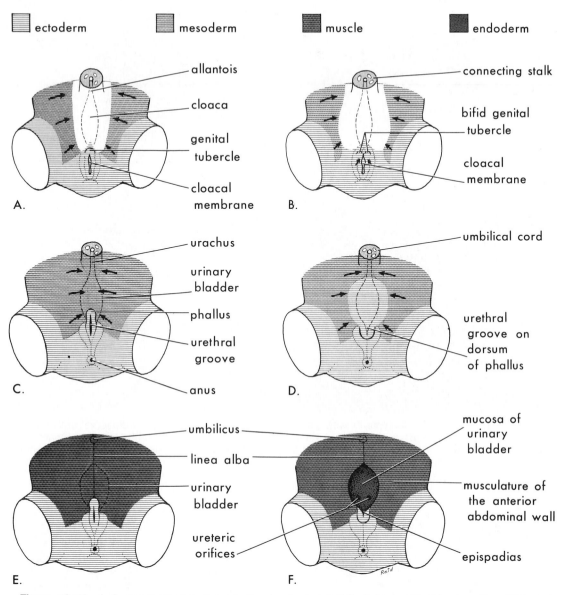

▨ ectoderm   ▨ mesoderm   ▨ muscle   ■ endoderm

A. — allantois, cloaca, genital tubercle, cloacal membrane

B. — connecting stalk, bifid genital tubercle, cloacal membrane

C. — urachus, urinary bladder, phallus, urethral groove, anus

D. — umbilical cord, urethral groove on dorsum of phallus

E. — umbilicus, linea alba, urinary bladder, ureteric orifices

F. — mucosa of urinary bladder, musculature of the anterior abdominal wall, epispadias

Figure 13–18 *A, C,* and *E,* Normal stages in the development of the infraumbilical body wall and the penis during the fourth to eighth weeks. Note that mesoderm and later muscle reinforce the ectoderm of the developing anterior abdominal wall. *B, D,* and *F,* Probable stages in the development of exstrophy of the bladder and epispadias. In *B* and *D,* note that mesoderm fails to extend into the anterior abdominal wall anterior to the urinary bladder. Also note that the genital tubercle is located in a more caudal position than usual, and that the urethral groove has formed on the dorsal (upper) surface of the penis. In *F,* the surface ectoderm and the endodermal anterior wall of the bladder have ruptured, resulting in exposure of the bladder mucosa. Note that the musculature of the anterior abdominal wall is present on each side of the defect. Failure of these muscle layers to meet and fuse in the midline resulted in the rupture of the anterior abdominal wall and bladder. (Based on Patten and Berry, 1952.)

dorsal mesentery and the developing gonad (Fig. 13–20C). The cells that form the *fetal cortex* are derived from the coelomic epithelium lining the posterior abdominal wall, whereas the cells that form the medulla are derived from an adjacent *sympathetic ganglion* (Fig. 13–20C), a derivative of the neural

crest. These cells form a cellular mass on the medial side of the fetal cortex (Fig. 13–19B). As they are encapsulated by the fetal cortex, the neural crest cells differentiate into the *chromaffin cells* of the adrenal medulla.

Concurrently, more mesenchymal cells arise from the coelomic epithelium and en-

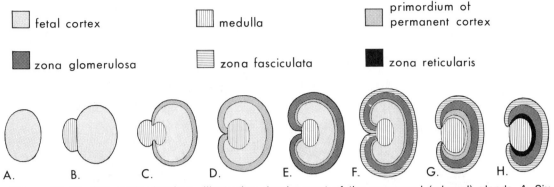

Figure 13–19 Schematic drawings, illustrating development of the suprarenal (adrenal) glands. *A*, Six weeks, showing the mesodermal primordium of the fetal cortex. *B*, Seven weeks, showing the addition of the ectodermal primordium of the suprarenal medulla. *C*, Eight weeks, showing the fetal cortex and the early permanent cortex beginning to encapsulate the medulla. *D* and *E*, Later stages of encapsulation of the medulla by the cortex. *F*, Newborn, showing the fetal cortex and two zones of the permanent cortex. *G*, One year; the fetal cortex has almost disappeared. *H*, Four years, showing the adult pattern of cortical zones. Note that the fetal cortex has disappeared, and that the gland is relatively smaller than it was at birth.

close the fetal cortex. These cells will give rise to the permanent cortex (Fig. 13–19*C*). Differentiation of the characteristic adrenal cortical zones begins during the late fetal period. The zona glomerulosa and the zona fasciculata are present at birth, but the zona reticularis is not recognizable until about the end of the third year (Fig. 13–19*H*).

The suprarenal gland of the human fetus is 10 to 20 times larger than the adult gland, relative to body weight (Gardner, 1975), and is large compared with the kidney (Fig. 13–7). The large adrenals result from the extensive fetal cortex. The adrenal medulla remains relatively small until after birth. The adrenal glands rapidly become smaller postnatally as the fetal cortex regresses. This cortex normally disappears by the end of the first year. The glands lose about one third of their weight during the first two or three weeks after birth and do not regain their original weight until about the end of the second year.

Hyperplasia of the fetal adrenal cortex during the fetal period usually results in female pseudohermaphroditism (Fig. 13–28). The adrenogenital syndrome associated with congenital adrenal hyperplasia manifests itself in various clinical forms that, in most aspects, can be correlated with certain *enzymatic deficiencies of cortisol biosynthesis* (Zurbrügg, 1975). Congenital adrenal hyperplasia is caused by a genetically determined deficiency of adrenal cortical enzymes that are necessary for the synthesis of various steroid hormones (Thompson and Thompson, 1980). The reduced hormone output results in an increased release of ACTH, which causes adrenal hyperplasia and overproduction of androgens by the hyperplastic adrenal glands. In females, this causes masculinization (Fig. 13–29*B*). In males, the excess androgens may cause precocious sexual development.

The concept of a *fetoplacental unit* is that there is mutual transport of steroid intermediates between the fetus and the placenta. Neither the fetus nor the placenta can achieve entire steroid biosynthesis, i.e., they complement each other. (For more information, see Gardner, 1975.)

The adrenal glands are hypoplastic in anencephalic infants owing to failure of normal development of the pituitary gland. This results in an inadequate release of ACTH, which leads to failure of normal development of the fetal adrenal cortex.

THE GENITAL SYSTEM

Although the genetic sex of an embryo is determined at fertilization by the kind of sperm that fertilizes the ovum (see Chapter 2), there is no morphological indication of sex until the seventh week, when the *gonads* (future ovaries or testes) begin to acquire sexual characteristics. The early genital system is similar in both sexes, and this initial period of early genital development is referred to as the *indifferent stage* of the reproductive organs.

DEVELOPMENT OF THE TESTES AND OVARIES

The gonads (testes and ovaries) are derived from three sources: the *coelomic epithelium*, the underlying *mesenchyme*, and the *primordial germ cells*.

The Indifferent Gonads (Figs. 13–20 and 13–21). Gonadal development is first indicated during the fifth week, when a thickened area of coelomic epithelium develops

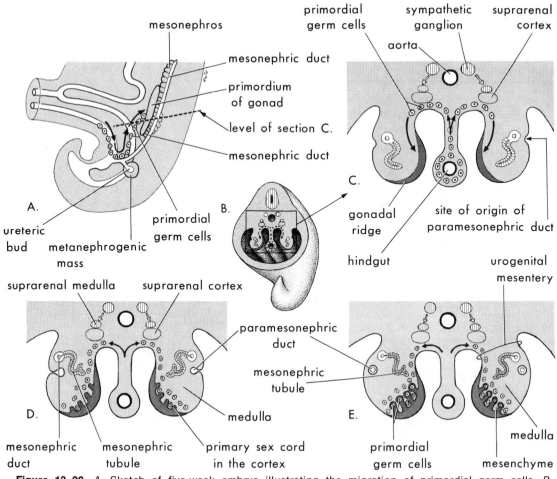

Figure 13–20 *A*, Sketch of five-week embryo illustrating the migration of primordial germ cells. *B*, Three-dimensional sketch of the caudal region of a five-week embryo showing the location and extent of the gonadal ridges on the medial aspect of the urogenital ridges (also see Figure 13–4*F*). *C*, Transverse section showing the primordium of the adrenal glands, the gonadal ridges, and the migration of primordial germ cells. *D*, Transverse section through a six-week embryo showing the primary sex cords and the developing paramesonephric ducts. *E*, Similar section at later stage showing the indifferent gonads and the mesonephric and paramesonephric ducts.

on the medial side of the *mesonephros* (Figs. 13–4*F* and 13–20). Proliferation of the coelomic epithelial cells and of the underlying mesenchyme produces a bulge on the medial side of the mesonephros known as the *gonadal ridge* (Fig. 13–20*C*). Soon, finger-like epithelial cords, called *primary sex cords*, grow into the underlying mesenchyme (Fig. 13–20*D*). The indifferent gonad now consists of an outer *cortex* and an inner *medulla*. In embryos with an XX sex chromosome complex, the cortex normally differentiates into an ovary, and the medulla regresses. In embryos with an XY sex chromosome complex, the medulla normally differentiates into a testis, and the cortex regresses.

The Primordial Germ Cells (Fig. 13–20). Large spherical primitive sex cells,

called *primordial germ cells*, are visible early in the fourth week among the endodermal cells of the wall of the yolk sac near the origin of the allantois. During folding of the embryo (see Chapter 5), part of the yolk sac is incorporated into the embryo, and the primordial germ cells migrate by ameboid movement along the dorsal mesentery of the hindgut to the gonadal ridges. During the sixth week, the primordial germ cells migrate into the underlying mesenchyme and become incorporated in the primary sex cords (Fig. 13–20*E*).

Sex Determination. *Genetic sex* is established at fertilization and depends upon whether an X-bearing sperm or a Y-bearing sperm fertilizes the X-bearing ovum (see Chapter 2). The type of gonads that develop,

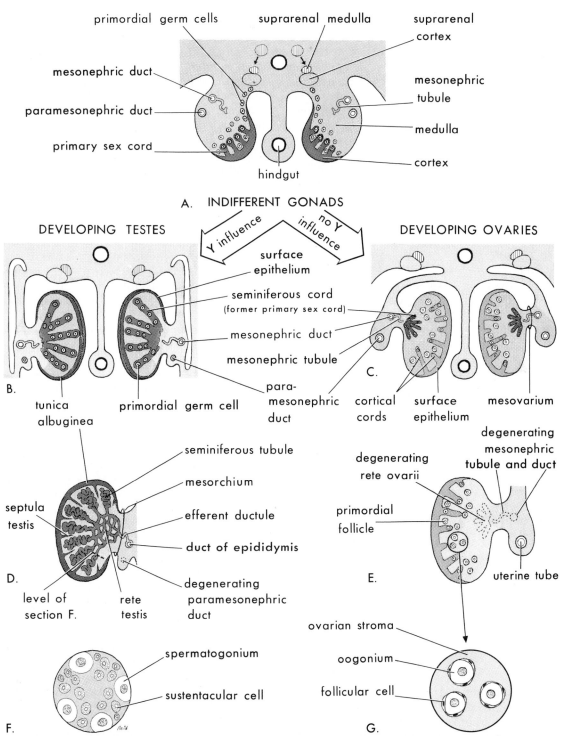

primordial germ cells   suprarenal medulla   suprarenal cortex

mesonephric duct

paramesonephric duct

primary sex cord

mesonephric tubule

medulla

cortex

hindgut

A.   INDIFFERENT GONADS

DEVELOPING TESTES   Y influence   no Y influence   DEVELOPING OVARIES

surface epithelium

seminiferous cord
(former primary sex cord)

mesonephric duct

mesonephric tubule

para-mesonephric duct

B.

tunica albuginea   primordial germ cell

C.

cortical cords   surface epithelium   mesovarium

seminiferous tubule

mesorchium

efferent ductule

duct of epididymis

degenerating paramesonephric duct

septula testis

D.

level of section F.   rete testis

degenerating mesonephric tubule and duct

degenerating rete ovarii

primordial follicle

uterine tube

E.

spermatogonium

sustentacular cell

F.

ovarian stroma

oogonium

follicular cell

G.

Figure 13–21 Schematic sections illustrating the differentiation of the indifferent gonads into testes or ovaries. *A*, Six weeks, showing the indifferent gonads composed of an outer cortex and an inner medulla. *B*, Seven weeks, showing testes developing under the influence of a Y chromosome. Note that the primary sex cords have become seminiferous cords and that they are separated from the surface epithelium by the tunica albuginea. *C*, 12 weeks, showing ovaries beginning to develop in the absence of a Y chromosome influence. Cortical cords have extended from the surface epithelium, displacing the primary sex cords centrally into the mesovarium, where they form the rudimentary rete ovarii. *D*, Testis at 20 weeks, showing the rete testis and the seminiferous tubules derived from the seminiferous cords. An efferent ductule has developed from a mesonephric tubule, and the mesonephric duct has become the duct of the epididymis. *E*, Ovary at 20 weeks, showing the primordial follicles formed from the cortical cords. The rete ovarii derived from the primary sex cords and the mesonephric tubule and duct are regressing. *F*, Section of a seminiferous tubule from a 20-week fetus. Note that no lumen is present at this stage, and that the seminiferous epithelium is composed of two kinds of cells. *G*, Section from the ovarian cortex of a 20-week fetus showing three primordial follicles.

or *gonadal sex*, is determined by the sex chromosome complex (XX or XY). Before the seventh week of embryonic life, the gonads of both sexes are identical in appearance and are referred to as *indifferent gonads* (Fig. 13–21A).

The Y chromosome has a strong testis-determining effect on the medulla of the indifferent gonad. It is the presence of the Y-antigen gene on the Y chromosome that determines testicular differentiation. Under the influence of the Y chromosome, the primary sex cords differentiate into seminiferous tubules (Fig. 13–21B and D).

The absence of a Y chromosome results in formation of an ovary (Fig. 13–21C and E). Thus, the type of sex chromosome complex established at fertilization determines the type of gonad that develops from the indifferent gonad. *The gonads then determine the type of sexual differentiation that occurs in the genital ducts and external genitalia. The androgens produced by the testes determine maleness.* Female sexual differentiation in the fetus does not depend on estrogens; it occurs even if the ovaries are absent.

In embryos with abnormal sex chromosome complexes, the number of X chromosomes appears to be unimportant in sex determination. If a Y chromosome is present, the embryo develops as a male. If there is no Y chromosome, female development occurs. Loss of a sex chromosome (as in XO females) causes *ovarian dysgenesis*, in which the ovaries are represented by gonadal streaks. (For a description of females with XO Turner syndrome, see Chapter 8 and Figure 8–2.)

The loss of an X chromosome does not appear to interfere with the migration of primordial germ cells to the gonadal ridges (Hamerton, 1971) because some germ cells have been observed in the fetal gonads of XO females (Carr et al., 1968). This indicates that two X chromosomes are needed to bring about complete ovarian development.

Development of Testes (Figs. 13–21B, D, and F, and 13–22A and C). In embryos with a Y chromosome, the primary sex cords condense and extend into the medulla of the gonad. Here they branch, and their ends anastomose to form the *rete testis*. The prominent sex cords, called *seminiferous cords* (testicular cords), soon lose their connections with the surface epithelium because of the development of a thick, fibrous capsule called the *tunica albuginea* (Fig. 13–21B and D).

The *development of a thick tunica albugin-*ea is a characteristic and diagnostic feature of testicular development*. Gradually, the enlarging testis separates from the regressing mesonephros and becomes suspended by its own mesentery, the *mesorchium*. The seminiferous cords develop into the *seminiferous tubules,* the *tubuli recti,* and the *rete testis.* The seminiferous tubules become separated by mesenchyme that gives rise to the *interstitial cells* (of Leydig). The interstitial cells produce the male sex hormone *testosterone,* which induces masculine differentiation of the external genitalia. In addition to testosterone, *genital duct inducer and suppressor substances* are produced by the interstitial cells. As described subsequently, these substances induce development of the mesonephric ducts and suppress development of the paramesonephric ducts.

The walls of the seminiferous tubules are composed of two kinds of cells (Fig. 13–21F): supporting, or *sustentacular, cells of Sertoli,* derived from the surface epithelium, and *spermatogonia,* derived from the primordial germ cells. The Sertoli cells constitute most of the seminiferous epithelium in the fetal testis (Figs. 13–21F and 13–22C).

During later development of the testes, the surface epithelium flattens to form the mesothelium on the outer surface of the adult testis. The rete testis becomes continuous with 15 to 20 adjacent, persistent, mesonephric tubules, which become *efferent ductules.* These ductules are connected with the mesonephric duct, which forms the *ductus epididymis* (Figs. 13–21B and D, and 13–24A).

Development of Ovaries (Figs. 13–21C, E, and G, and 13–22B and D). In embryos lacking a Y chromosome (e.g., 46,XX), gonadal development occurs slowly. The ovary is not positively identifiable until about the tenth week. The *primary sex cords* do not become prominent in the gonads of female embryos, but they extend into the medulla and form a rudimentary *rete ovarii.* This structure and the primary sex cords normally degenerate and disappear (Fig. 13–21E). In some females, vestiges of the rete ovarii occur as epithelial strands, or tubules, in the region of the hilum (Copenhaver et al., 1978).

Although it is commonly stated that the sex of gonads can be distinguished during the embryonic period, it must be emphasized that a diagnosis of femaleness can be made only after failure to observe testicular characteristics in a gonad from

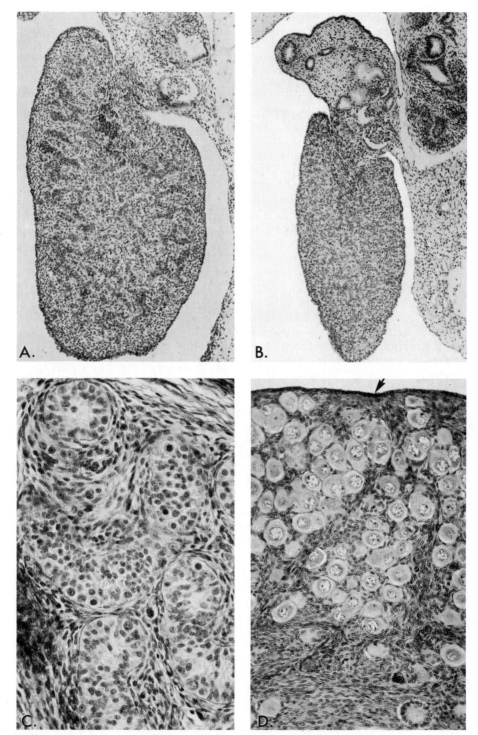

Figure 13–22 Cross sections through the gonads of human embryos and fetuses. *A*, Testis from an embryo of about 43 days, showing the prominent seminiferous or testicular cords (×175). *B*, From an embryo of about the same age, a gonad that may be assumed to be an ovary because of the absence of prominent primary sex cords (×125). *C*, Section of a testis from a male fetus born prematurely at about 21 weeks showing seminiferous tubules composed mostly of sustentacular cells (Sertoli cells). A few large spermatogonia are visible (×475). Compare with Figure 13–21*F*, *D*, Section of an ovary from a 14-day-old female infant showing numerous primordial follicles in the cortex. The arrow indicates the relatively thin surface ("germinal") epithelium (×275). Compare with Figure 13–21*G*. (From van Wagenen, G., and Simpson, M. E.: *Embryology of the Ovary and Testis. Homo sapiens and Macaca mulatta.* 1965. Courtesy of Yale University Press.)

an embryo known to be older than seven weeks (Fig. 13–22*B*). Testes can be recognized as such in embryos around the end of the seventh week. Development of a dense fibrous tunica albuginea in the eighth week is a positive sign of maleness. Ovaries cannot be recognized as such until about sixteen weeks.

During the early fetal period, secondary sex cords called *cortical cords* extend from the superficial epithelium, or *surface epithelium* of the developing ovary into the underlying mesenchyme (Fig. 13–21*C*). This cuboidal surface epithelium is derived from the coelomic epithelium. As the cortical cords increase in size, primordial germ cells are incorporated into them. At about 16 weeks, the cortical cords begin to break up into isolated cell clusters called *primordial follicles,* which consist of an *oogonium* derived from a primordial germ cell, surrounded by a single layer of flattened follicular cells derived from the cortical cords (Fig. 13–21*E* and *G*). Active mitosis of oogonia occurs during fetal life, producing thousands of these primitive germ cells. *No oogonia form postnatally.* Although many oogonia degenerate before birth, the two million or so that remain enlarge to become primary oocytes. When the primary oocyte becomes surrounded by one or more layers of cuboidal cells, or low columnar follicular cells, the structure is called a *primary follicle.* Al-

though a few primary and growing follicles may develop prenatally as a result of stimulation of the fetal ovaries by maternal gonadotropins, most follicles remain quiescent until puberty. The mesenchyme surrounding the follicles forms the ovarian stroma. (For details of follicular development, see Chapter 2.)

After birth, the surface epithelium flattens to a single layer of cuboidal cells that is continuous with the mesothelium of the peritoneum at the hilum of the ovary. The surface epithelium used to be called the germinal epithelium, although there was no convincing evidence that it was the site of germ cell formation. It is now well established that the germ cells differentiate from the primordial germ cells (Figs. 13–20 and 13–21).

The surface epithelium becomes separated from the follicles in the cortex by a thin fibrous capsule, the *tunica albuginea.* As the ovary separates from the regressing mesonephros, it becomes suspended by its own mesentery, the *mesovarium* (Fig. 13–21*C*).

DEVELOPMENT OF THE GENITAL DUCTS

Both male and female embryos have two pairs of genital, or sex, ducts. The male, or mesonephric, ducts play an important part in the development of the male reproductive system, and the female, or paramesonephric,

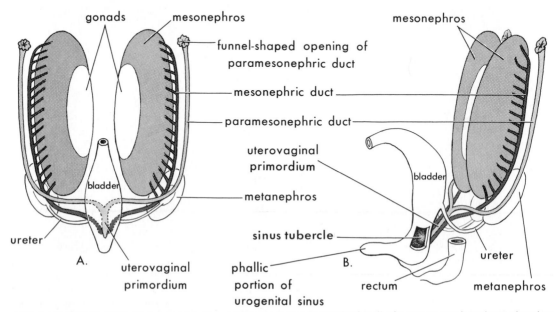

Figure 13–23 *A*, Sketch of a frontal view of the posterior abdominal wall of a seven-week embryo showing the two pairs of genital ducts present during the indifferent stage. *B*, Lateral view of a nine-week fetus showing the sinus (müllerian) tubercle on the posterior wall of the urogenital sinus. It becomes the hymen in females and the seminal colliculus in males.

TABLE 13–1   ADULT DERIVATIVES AND VESTIGIAL REMAINS OF EMBRYONIC UROGENITAL STRUCTURES*

Male	*Embryonic Structure*	*Female*

Testis
    Seminiferous tubules
    Rete testis
	Indifferent Gonad
	Cortex
	Medulla
		Ovary
		    Ovarian follicles
		    Medulla
		    Rete ovarii

Gubernaculum testis	**Gubernaculum**	*Ovarian ligament* *Round ligament of uterus*
Ductuli efferentes	**Mesonephric Tubules**	Epoophoron
Paradidymis		Paroophoron
Appendix of epididymis *Ductus epididymis*	**Mesonephric Duct**	Appendix vesiculosa Duct of epoophoron
Ductus deferens		Duct of Gartner

Ureter, pelvis, calyces and
    collecting tubules
		Ureter, pelvis, calyces and
		    collecting tubules

Ejaculatory duct and seminal *vesicle*		
Appendix of testis	**Paramesonephric Duct**	Hydatid (of Morgagni) *Uterine tube* *Uterus*
		Fibromuscular wall *of vagina*
Urinary bladder *Urethra* (except *navicular fossa*)	**Urogenital Sinus**	*Urinary bladder* *Urethra*
Prostatic utricle		*Vagina*
Prostate gland		*Urethral and paraurethral glands*
Bulbourethral glands		*Greater vestibular glands*
Seminal colliculus	**Sinus** **Tubercle**	Hymen

Penis
    Glans penis
    Corpora cavernosa penis
    Corpus spongiosum penis
Ventral aspect of penis
	Phallus
	
	Urogenital Folds
		Clitoris
		    Glans clitoridis
		    Corpora cavernosa clitoridis
		    Bulb of the vestibule
		Labia minora

Scrotum	**Labioscrotal Swellings**	*Labia majora*

*Functional derivatives are in *italics*.

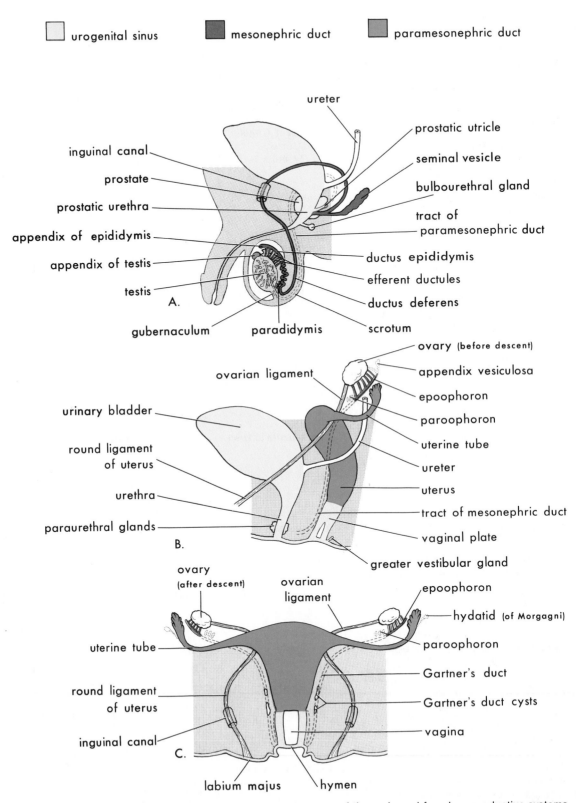

Figure 13–24 Schematic drawings illustrating development of the male and female reproductive systems from the genital ducts and the urogenital sinus. Vestigial structures are also shown. *A,* Reproductive system in a newborn male. *B,* Female reproductive system in a 12-week fetus. *C,* Reproductive system in a newborn female.

ducts play an important part in the development of the female reproductive system.

The Indifferent Stage (Fig. 13–23). When both pairs of genital ducts are present, the genital duct system is in the indifferent stage.

The *mesonephric ducts* drain the mesonephric kidneys and play an essential part in the development of the *male reproductive system* (Fig. 13–24A). They form the epididymides, the ductus deferens, and the ejaculatory ducts when the mesonephroi degenerate. In the female, the mesonephric ducts almost completely disappear; only a few remnants persist (Fig. 13–24 and Table 13–1).

The *paramesonephric ducts* develop on each side from longitudinal invaginations of coelomic epithelium on the lateral aspects of the mesonephroi (Fig. 13–20C). The edges of these invaginations approach each other and fuse to form the paramesonephric ducts (Fig. 13–20D and E). The funnel-shaped cranial ends of the ducts open into the coelomic, or future peritoneal, cavity (Fig. 13–23A). The paramesonephric ducts pass caudally, running parallel to the mesonephric ducts until they reach the caudal region. Here they cross ventral to the mesonephric ducts, come together in the midline, and fuse into a Y-shaped *uterovaginal primordium*, or canal (Fig. 13–23A). This tubular structure projects into the dorsal wall of the urogenital sinus and produces an elevation, called the *sinus tubercle* (Fig. 13–23B). A mesonephric duct enters the urogenital sinus on each side of this tubercle.

Development of Male Genital Ducts and Auxiliary Genital Glands (Figs. 13–24 and 13–25). The fetal testes produce a nonsteroidal *inducer substance* and *androgens*. The inducer substance stimulates development of the mesonephric ducts into the male genital tract, and suppresses development of the paramesonephric ducts. Because the inducer substance also suppresses development, it is often called a *suppressor substance* when it is performing this role.

When the mesonephros degenerates, some mesonephric tubules near the testis persist and are transformed into *efferent ductules*, or *ductuli efferentes* (Fig. 13–24A). These ductules open into the mesonephric duct, which becomes the *ductus epididymis* in this region. Beyond the epididymis, the mesonephric duct acquires a thick investment of smooth muscle and becomes the *ductus deferens*.

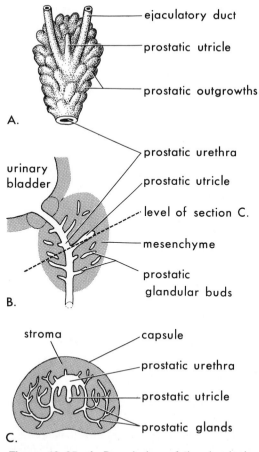

Figure 13–25 *A*, Dorsal view of the developing prostate gland in an 11-week fetus. *B*, Sketch of a sagittal section of the developing prostate gland showing the numerous endodermal outgrowths from the prostatic urethra. The vestigial prostatic utricle is also shown. *C*, Section of the prostate gland at about 16 weeks.

A lateral outgrowth from the caudal end of each mesonephric duct gives rise to a *seminal vesicle*. The part of the mesonephric duct between the duct of this gland and the urethra becomes the *ejaculatory duct*. The remainder of the male genital duct system consists of the urethra (Fig. 13–24A).

The Prostate Gland (Fig. 13–25). Multiple endodermal outgrowths arise from the prostatic portion of the urethra and grow into the surrounding mesenchyme. The glandular epithelium of the prostate differentiates from these endodermal cells, and the associated mesenchyme differentiates into the dense stroma and smooth muscle fibers of the prostate.

The Bulbourethral Glands (Fig. 13–24A). These pea-sized structures, develop from paired endodermal outgrowths from the membranous portion of the urethra. The smooth muscle fibers and

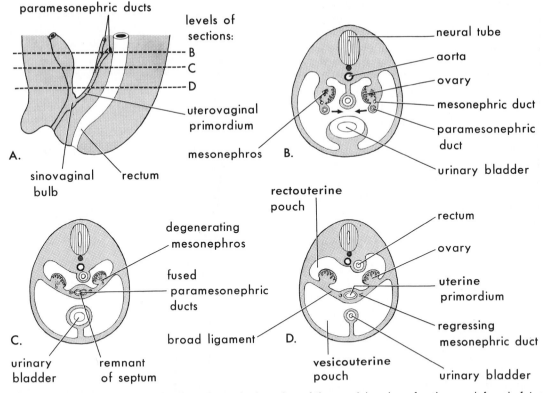

Figure 13–26 *A*, Schematic drawing of a sagittal section of the caudal region of a nine-week female fetus. *B*, Transverse section showing the paramesonephric ducts approaching each other. *C*, Similar section at a lower level illustrating the fused paramesonephric ducts. A remnant of the septum that initially separates them is shown. *D*, Similar section showing the uterovaginal primordium, the broad ligament, and the pouches in the pelvic cavity. Remnants of the mesonephric ducts may persist as ducts of Gartner or give rise to cysts (see Figure 13–24*C*).

the stroma differentiate from the adjacent mesenchyme.

Development of Female Genital Ducts and Auxiliary Genital Glands (Figs. 13–24*B* and *C* and 13–26). In embryos with ovaries, the mesonephric ducts regress and the paramesonephric ducts develop into the female genital tract. Although the testes are essential to the stimulation of male sexual development, *female sexual development in the fetus does not depend on the presence of ovaries.*

The cranial, unfused portions of the paramesonephric ducts develop into the uterine tubes, and the caudal, fused portions form the *uterovaginal primordium,* which gives rise to the epithelium and glands of the uterus and to the fibromuscular vaginal wall. The endometrial stroma and the myometrium are derived from the adjacent mesenchyme.

The uterus is mainly an abdominal organ in the newborn infant, and its cervix is relatively large. During puberty (13 to 15 years) the uterus grows rapidly.

Similar development of the paramesonephric ducts occurs in males if the testes fail to develop or if they are removed before the initiation of differentiation of the genital ducts (Jost, 1970). Removal of the ovaries of female embryos, however, has no effect on *fetal* sexual development (Jost, 1961). This suggests that the testes impose masculinity and repress femininity, and that the *ovaries are not necessary for primary sexual development.*

Fusion of the paramesonephric ducts also brings together two peritoneal folds, forming right and left *broad ligaments* and two peritoneal compartments of the pelvic cavity, the *rectouterine pouch* and the *vesicouterine pouch* (Fig. 13–26*B* to *D*).

Along the sides of the uterus, between the layers of the broad ligament, the mesenchyme proliferates and differentiates into loose connective tissue and smooth muscle, known as the *parametrium.*

Development of the Vagina (Fig. 13–24*B* and *C*).

The vaginal epithelium is derived from the endoderm of the urogenital sinus, and the fibromuscular wall of the vagina develops from the uterovaginal primordium. Contact of the uterovaginal primordium with the urogenital sinus (Fig. 13–23) induces the formation of paired endodermal outgrowths, called *sinovaginal bulbs*. They extend from the urogenital sinus into the caudal end of the uterovaginal primordium. The sinovaginal bulbs soon fuse to form a solid cord, the *vaginal plate* (Fig. 13–24*B*). Later, the central cells of this plate break down, forming the lumen of the vagina. The peripheral cells remain as the vaginal epithelium (Fig. 13–24*C*).

Until late fetal life, the lumen of the vagina is separated from the cavity of the urogenital sinus by a membrane called the *hymen* (Figs. 13–24*C* and 13–27*H*). The hymen usually ruptures during the perinatal period and remains as a thin fold of mucous membrane around the entrance to the vagina.

There has been much controversy about the origin of the vagina. Some authorities consider the superior one third of the vaginal epithelium to be derived from the uterovaginal primordium and the inferior two thirds from the urogenital sinus (Cunha, 1975; O'Rahilly, 1977).

The *prostatic utricle*, a small diverticulum that opens on the seminal colliculus in the prostatic urethra (Fig. 13–24*A*), is probably homologous to the vagina. Formerly, it was considered to represent the remains of the uterovaginal primordium and to be homologous to the uterus.

The *seminal colliculus,* a small elevation in the posterior wall of the prostatic urethra, is the adult derivative of the sinus tubercle and is homologous to the hymen in the female.

Auxiliary Genital Glands. In the female, buds grow out from the urethra into the surrounding mesenchyme and form the *urethral glands* and the *paraurethral glands* (of Skene). These two sets of glands correspond to the prostate gland in the male. Similar outgrowths from the urogenital sinus form the *greater vestibular glands* (of Bartholin), which are homologous with the bulbourethral glands in the male (Table 13–1).

Vestigial Structures Derived from the Genital Ducts (Fig. 13–24 and Table 13–1).

During conversion of the mesonephric and paramesonephric ducts into adult structures, some parts may remain as vestigial structures. These vestiges are rarely seen unless pathological changes develop in them.

Mesonephric Remnants in Males. The blind cranial end of the mesonephric duct may persist as an *appendix of the epididymis;* it is usually attached to the head of the epididymis. Caudal to the efferent ductules (ductuli efferentes), some mesonephric tubules may persist as a small body called the *paradidymis* (Fig. 13–24*A*).

Mesonephric Remnants in Females. The cranial end of the mesonephric duct may persist as a cystic *appendix vesiculosa*. A few blind tubules and a duct, called the *epoophoron,* corresponding to the efferent ductules and duct of the epididymis in the male, may persist in the broad ligament between the ovary and uterine tube. Closer to the uterus, some rudimentary tubules may persist as the *paroophoron.* Parts of the mesonephric duct, corresponding to the portions of the mesonephric duct that form the ductus deferens and the ejaculatory duct, may persist as the duct of Gartner in the broad ligament along the lateral wall of the uterus, or in the wall of the vagina. These remnants may give rise to *Gartner's duct cysts* (Fig. 13–24*C*).

Paramesonephric Remnants in Males. The cranial end of the paramesonephric duct may persist as the vesicular *appendix of the testis* attached to the connective tissue investing the superior pole of the testis (Fig. 13–24*A*).

Paramesonephric Remnants in Females. Part of the cranial end of the paramesonephric duct that does not contribute to the infundibulum of the uterine tube may persist as a vesicular appendage to the tube, called a *hydatid of Morgagni* (Fig. 13–24*C*).

DEVELOPMENT OF THE EXTERNAL GENITALIA

The early development of the external genitalia is similar in both sexes. Distinguishing sexual characteristics begin to appear during the ninth week, but the external genital organs are not fully formed until the twelfth week.

The Indifferent Stage (Fig. 13–27*A* and *B*).

The external genitalia also pass through an undifferentiated state before distinguishing sexual characteristics appear. Early in the fourth week, a *genital tubercle* develops at the cranial end of the cloacal membrane. *Labioscrotal swellings* and *urogenital folds* soon develop on each side of the cloacal membrane. The genital tubercle soon elongates to form a *phallus,* which is as large in females as in males.

When the urorectal septum fuses with the cloacal membrane at the end of the sixth week, it divides the cloacal membrane into a dorsal *anal membrane* and a ventral *urogenital membrane*. These membranes rupture a week or so later, forming the *anus* and the

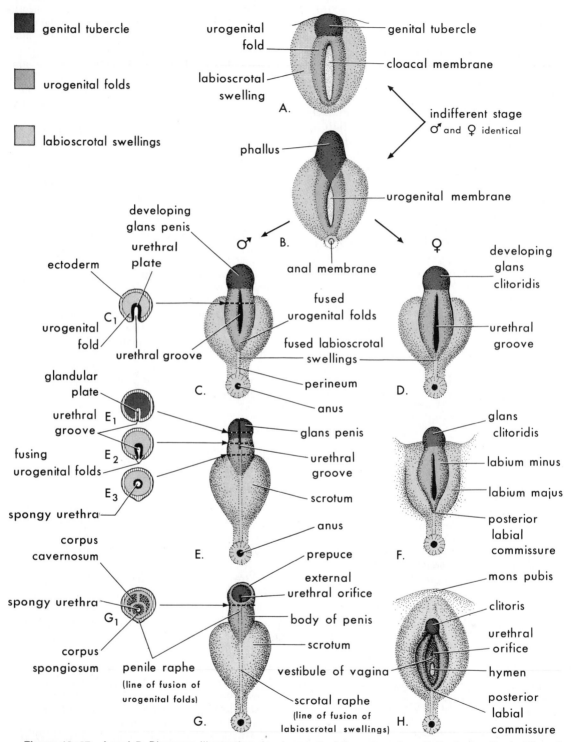

Figure 13–27 *A* and *B,* Diagrams illustrating development of the external genitalia during the indifferent stage (four to seven weeks). *C, E,* and *G,* Stages in the development of male external genitalia at about 9, 11, and 12 weeks, respectively. To the left are schematic transverse sections (*C₁, E₁* to *E₃,* and *G₁*) through the developing penis illustrating formation of the spongy urethra. *D, F,* and *H,* Stages in the development of female external genitalia at 9, 11, and 12 weeks, respectively.

urogenital orifice, respectively. A *urethral groove* that is continuous with the urogenital orifice forms on the ventral surface of the phallus.

Development of Male External Genitalia (Figs. 13–15 and 13–27C, E, and G). Masculinization of the indifferent external genitalia is caused by androgens produced by the testes. As the phallus elongates to form the *penis,* it pulls the *urogenital folds* forward. These folds form the lateral walls of the *urethral groove,* which is on the ventral surface of the penis. This groove is lined by an extension of endoderm, the *urethral plate* (Fig. 13–27C), from the phallic portion of the *urogenital sinus.* The *urogenital folds* fuse with each other along the ventral surface of the penis to form the *spongy urethra.* As a result, the external urethral orifice moves progressively toward the *glans* of the penis.

At the tip of the glans, an ectodermal ingrowth forms a cellular cord called the *glandular plate* (Fig. 13–15A). Subsequent splitting of this plate forms a groove on the ventral surface of the glans that is continuous with the urethral groove in the body of the penis (Fig. 13–27E). Closure of the urethral groove on the glans moves the external urethral orifice to the tip of the glans and joins the two parts of the spongy urethra (Fig. 13–15C).

During the twelfth week, the *prepuce* is developed by a circular ingrowth of the ectoderm at the periphery of the glans penis (Fig. 13–15B). This septum then breaks down except for a small area ventral to the glans that forms the *frenulum.* After breakdown of the septum, the skin and fasciae of the penis are prolonged as a free-fold, or double layer of skin called the prepuce (L., foreskin), which covers the glans penis for a variable extent. For some time, the prepuce is fused to the glans and is usually not retractable at birth. Breakdown of the fused surfaces normally occurs during infancy. The corpora cavernosa penis and the corpus spongiosum penis arise from the mesenchymal tissue in the phallus. The *labioscrotal swellings* grow toward each other and fuse to form the *scrotum* (Fig. 13–27E and G).

Development of Female External Genitalia (Fig. 13–27D, F, and H). In the absence of androgens, feminization of the indifferent external genitalia occurs. The phallus elongates rapidly at first, but its growth grad- ually slows, and it becomes the relatively small *clitoris.* The clitoris develops like the penis, but the urogenital folds do not fuse, except posteriorly, where they fuse to form the *frenulum of the labia minora* (fourchette). The unfused urogenital folds form the *labia minora.* The labioscrotal folds fuse posteriorly to form the *posterior labial commissure* and anteriorly to form the *anterior labial commissure* and a rounded elevation, the *mons pubis.* The labioscrotal folds remain largely unfused and form two large folds of skin called the *labia majora.* The phallic part of the urogenital sinus gives rise to the vestibule of the vagina, into which the urethra, the vagina, and the ducts of the greater vestibular glands open.

CONGENITAL MALFORMATIONS OF THE GENITAL SYSTEM

Because an early embryo has the potential to develop as either a male or a female, errors in sex development may result in various degrees of intermediate sex, a condition known as *intersexuality,* or *hermaphroditism.* A person with ambiguous external genitalia is called an *intersex,* or a hermaphrodite. Intersexual conditions are classified according to the histological appearance of the gonads.

True hermaphrodites have both ovarian and testicular tissue. Some false hermaphrodites, or *pseudohermaphrodites,* have testes and are called male pseudohermaphrodites; others have ovaries and are known as female pseudohermaphrodites. Fortunately, true hermaphroditism is extremely rare and pseudohermaphroditism is relatively uncommon, occurring about once in 25,000 births.

The terms *hermaphrodite* and *intersex* are not applied to persons with conditions such as Klinefelter syndrome or Turner syndrome in which the external genitalia are not ambiguous (see discussion and illustrations in Chapter 8).

Sex chromatin and chromosome studies assist in the clinical evaluation of persons with ambiguous external genitalia (Table 13–2). With the exception of true hermaphrodites, the sex chromatin pattern and karyotype usually correspond to the gonadal sex.

Female Pseudohermaphrodites (Figs. 13–28 and 13–29 and Table 13–2). These persons have chromatin-positive nuclei (see Fig. 8–9B) and a 46,XX chromosome constitution.

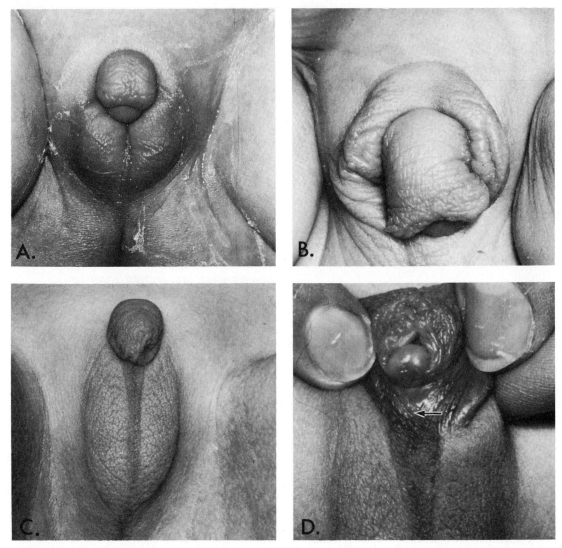

Figure 13–28 Photographs of the external genitalia of female pseudohermaphrodites resulting from congenital virilizing adrenal hyperplasia. The degree of labioscrotal fusion and clitoral hypertrophy depends upon the stage of differentiation at which the fetus is exposed, as well as upon the biological potency of the androgens produced by the hyperplastic adrenal glands. *A,* External genitalia of a newborn female, exhibiting enlargement of the clitoris and fusion of the labia majora. *B,* External genitalia of a female infant showing considerable enlargement of the clitoris. The labia majora are rugose as in a scrotum. *C* and *D,* External genitalia of a six-year-old girl showing an enlarged clitoris and fused labia majora forming a scrotum-like structure. In *D,* note the glans clitoridis and the opening of the urogenital sinus (arrow). See also Figure 13–29B.

The most common single cause of female pseudohermaphroditism is the adrenogenital syndrome, resulting from congenital virilizing adrenal hyperplasia. There is no ovarian abnormality, but the excessive adrenal production of androgens causes masculinization of the external genitalia during the fetal period, varying from enlargement of the clitoris to almost masculine genitalia. Commonly, there is clitoral hypertrophy, partial fusion of the labia majora, and a persistent urogenital sinus (Fig. 13–29). In *extremely* rare cases, fusion of the labioscrotal and urogenital folds is so complete that the urethra traverses the clitoris. Infants with such advanced masculinization are raised as males.

Infants with the adrenogenital syndrome (hyperfunction of the adrenal cortices associated with *ambiguous external genitalia*)

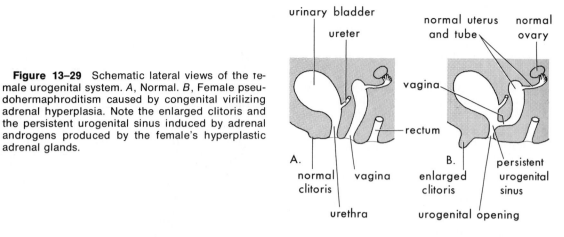

Figure 13–29 Schematic lateral views of the female urogenital system. *A*, Normal. *B*, Female pseudohermaphroditism caused by congenital virilizing adrenal hyperplasia. Note the enlarged clitoris and the persistent urogenital sinus induced by adrenal androgens produced by the female's hyperplastic adrenal glands.

TABLE 13–2   DIFFERENTIAL DIAGNOSIS OF PERSONS WITH AMBIGUOUS EXTERNAL GENITALIA*

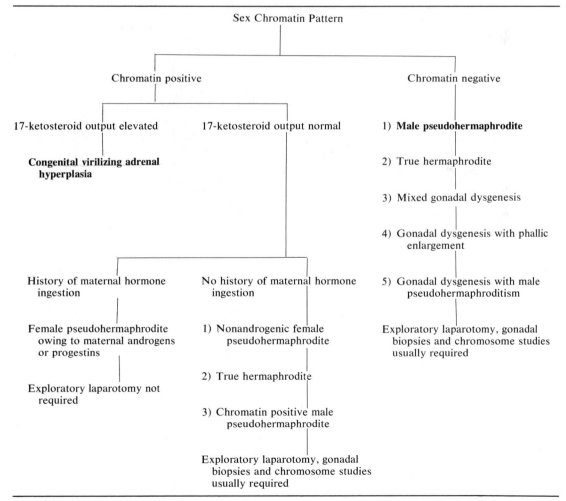

*Based on Federman (1967), Hamerton (1971) and Gray and Skandalakis (1972). For full discussions of the differential diagnosis of patients with ambiguous and abnormal genitalia, see Schlegel and Gardner (1975), Page et al. (1981), and Villee (1975).

are the most frequently encountered group of intersexes, accounting for about one half of all such cases.

Prompt recognition and treatment of the associated adrenal imbalance are most important (Schlegel and Gardner, 1975). Several distinct genetic and clinical forms of this condition are known, all inherited as autosomal recessive traits (Thompson and Thompson, 1980). Each is characterized by a block in a specific step in cortisol biosynthesis, resulting in an increased secretion of ACTH and hyperplasia of the fetal suprarenal glands. This results in masculinization of female fetuses. If the condition is not treated, the virilization continues into adulthood.

Female pseudohermaphrodites who do not have congenital virilizing adrenal hyperplasia are very rare. The administration of certain progestins to a mother during pregnancy may cause similar abnormalities of the fetal external genitalia (see Table 8–3 and Fig. 8–16). In rare instances, a masculinizing tumor such as *arrhenoblastoma* in the mother has caused female pseudohermaphroditism in her infant.

Male Pseudohermaphrodites (Table 13–2). These intersexes usually have *chromatin-negative nuclei* (see Fig. 8–9A) and a 46,XY chromosome constitution. The external and internal genitalia are variable, owing to varying degrees of development of the phallus and the paramesonephric ducts, caused by inadequate production of testosterone and inducer/suppressor substances by the fetal testes. Male pseudohermaphroditism would also result if the fetal hormones were produced after the period of tissue sensitivity of the sexual structures had passed.

True Hermaphrodites (Table 13–2). Persons with this *extremely rare condition* usually have chromatin-positive nuclei and a 46,XX chromosome constitution. Some have chromatin-negative nuclei and a 46,XY chromosome constitution, and others are mosaics (persons with more than one cell population, e.g., XX/XY).

True hermaphrodites have both testicular and ovarian tissue, either as two separate organs or as a single ovotestis. These tissues are not usually functional, but they are histologically identifiable. Oogenesis and spermatogenesis may occur in the same patient (Greene et al., 1954). The physical appearance may be male or female, but the external genitalia are ambiguous. This condition results from an error in sex determination, and *ovotestes* form if both the medulla and the cortex of the indifferent gonads develop.

CONDITIONS RELATED TO INTERSEXUALITY

Testicular Feminization (Fig. 13–30). *Persons with this rare condition appear as normal females despite the presence of testes and XY sex chromosomes.* The external genitalia are female, but the vagina ends blindly in a pouch, and the uterus and uterine tubes are absent or rudimentary.

At puberty, there is normal development of breasts and female appearance, but menstruation does not occur and pubic hair may be scanty or absent. The *psychosexual orientation of these persons is entirely female.*

The testes are usually intra-abdominal or inguinal, but they may descend into the labia majora.

The testes produce normal levels of testosterone. It is firmly established that the absence of masculinization results from a resistance to the action of testosterone at the peripheral cellular level in the labioscrotal and urogenital folds. Current evidence suggests that the defect is in the androgen receptor mechanism. The inducer and suppressor substances produced by the fetal testes normally cause male duct development (e.g., ductus epididymis) and suppress female duct development (e.g., absence of the uterus and uterine tubes).

Embryologically, these females represent an extreme form of male pseudohermaphroditism, but they are not intersexes in the usual sense because they have normal feminine external genitalia. *These patients are always reared as females.* Usually, the testes are removed as soon as they are discovered to avoid adverse effects on the psychosexual orientation of the individual.

The condition is genetically determined and appears to be transmitted as a sex-linked recessive gene. Testicular feminization occurs in about 1 in 50,000 females. This error in sex development is often detected when these females seek medical advice about primary *amenorrhea* (failure of menstruation to begin at puberty), sterility, or inguinal hernia, or when they undergo sex tests (buccal smears for sex chromatin determination) before participating in international athletic events.

Mixed Gonadal Dysgenesis. Persons with this very rare condition usually have chromatin-negative nuclei, a testis on one side, and an undifferentiated gonad on the other. The internal genitalia are female, but male derivatives of the mesonephric ducts are sometimes present. The

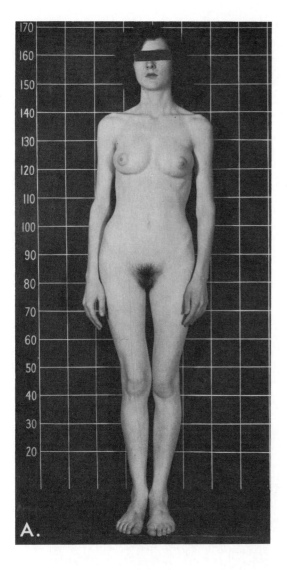

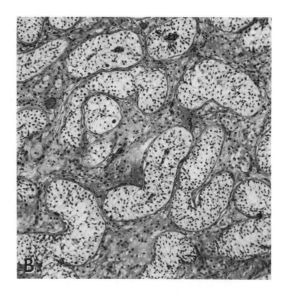

Figure 13–30 *A*, Photograph of a 17-year-old female with the testicular feminization syndrome. The external genitalia are female, but the patient is 46,XY and has testes. *B*, Photomicrograph of a section through a testis removed from the inguinal region of this girl showing seminiferous tubules lined by Sertoli cells. There are no germ cells and the interstitial cells are hypoplastic. (From Jones, H. W., and Scott, W. W.: *Hermaphroditism, Genital Anomalies and Related Endocrine Disorders.* 1958. Courtesy of the Williams & Wilkins Co., Baltimore.)

external genitalia range from normal female, through intermediate states, to normal male. At puberty, neither breast development nor menstruation occurs, but varying degrees of virilization are common. For a full discussion of gonadal dysgenesis and its variants, see Hamerton (1971) and Moore (1967).

Hypospadias (Fig. 13–31A to C). Once in about every 300 males, the external urethral orifice is on the ventral surface of the penis instead of at the tip of the glans. Usually, the penis is underdeveloped and curved downward, or ventrally, a condition known as *chordee.*

There are four types of hypospadias: *glandular, penile, penoscrotal,* and *perineal.* The glandular and penile types constitute about 80 per cent of cases. Hypospadias is the

result of an inadequate production of androgens by the fetal testes, which results in failure of fusion of the urogenital folds and incomplete formation of the spongy urethra. Differences in the timing and the degree of hormonal failure account for the variety of hypospadias.

In perineal hypospadias (not illustrated), the labioscrotal folds also fail to fuse. The external urethral orifice is located between the unfused halves of the scrotum. Because the genitalia in this rare type are ambiguous, persons with perineal hypospadias and cryptorchidism (undescended testes) are sometimes diagnosed as male pseudohermaphrodites.

Epispadias (Fig. 13–31D). Once in about 30,000 male infants, the urethra opens on the dorsal surface of the penis. Although epispadias

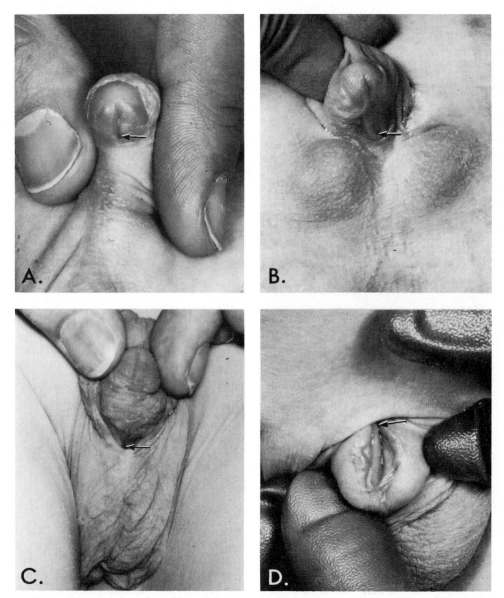

Figure 13–31 Photographs of penile malformations. *A*, Glandular hypospadias. *This constitutes the simplest and most common form of hypospadias.* The external urethral orifice is indicated by the arrow. There is a shallow pit at the usual site of the orifice. Note that the prepuce does not encircle the glans and that there is a moderate degree of chordee, causing the penis to curve ventrally. (From Jolly, H.: *Diseases of Children.* 2nd ed. 1968. Courtesy of Blackwell Scientific Publications.) *B*, Penile hypospadias. The penis is short and curved (chordee). The external urethral orifice (arrow) is near the penoscrotal junction. *C*, Penoscrotal hypospadias. The external urethral orifice (arrow) is located at the penoscrotal junction. *D*, Epispadias. The external urethral orifice (arrow) is on the dorsal (upper) surface of the penis near its origin. (Courtesy of Mr. Innes Williams, Genitourinary Surgeon, The Hospital for Sick Children. Great Ormond Street, London, England.)

may occur as a separate entity, it is often associated with exstrophy of the bladder (Fig. 13–18*F*). The defect may involve the glans or the entire penis.

The embryological basis of epispadias is unclear, but it appears that the genital tubercle develops more caudally than in normal embryos (Fig. 13–18*B*), and when the urogenital membrane ruptures, the urogenital sinus opens on the dorsal surface of the penis, where the urethral groove subsequently develops.

Epispadias also occurs in females, but it is very rare. It consists of a fissure in the superior wall of the urethra that opens on the dorsal surface of the clitoris.

Agenesis, or Absence, of the Penis. This extremely rare condition results from failure of the genital tubercle to develop. The urethra usually opens into the perineum near the anus. Usually the scrotum is normal and the testes descend into it.

Bifid Penis and Double Penis. These abnormalities are very rare. Bifid penis results from failure of fusion of the two parts of the genital tubercle, and is often associated with exstrophy of the bladder (Fig. 13–17). Double penis results from the formation of two genital tubercles. These malformations are often associated with urinary tract abnormalities and imperforate anus.

Micropenis. The penis is so small that it is almost hidden by the suprapubic pad of fat. This condition results from a functional hormonal deficiency of the fetal testes and is commonly associated with hypopituitarism.

Retroscrotal Penis, or Transposition of the Penis and Scrotum. The penis is located posterior to the scrotum. This condition appears to result from failure of the labioscrotal folds to shift caudally as they fuse to form the scrotum. Another possibility is that the labioscrotal folds develop anterior to the genital tubercle and then fuse to form the scrotum.

UTEROVAGINAL MALFORMATIONS

Various types of uterine duplication and vaginal malformation result from the follow-ing: (1) improper fusion of the paramesonephric ducts, which causes malformation of the uterovaginal primordium (Figs. 12–23, 13–24, and 13–32); (2) incomplete development of one paramesonephric duct; (3) failure of parts of one or both ducts to develop; and (4) incomplete canalization of the vaginal plate (Fig. 13–24). Double uterus (*uterus didelphys*), resulting from failure of fusion of the inferior parts of the paramesonephric ducts, may be associated with a double or a single vagina (Fig. 13–32*A* and *B*). If the doubling involves only the superior portion of the body of the uterus, the condition is called *bicornuate uterus* (Fig. 13–32*C* and *D*). If one paramesonephric duct is retarded in its growth and does not fuse with the other duct, a *bicornuate uterus with a rudimentary horn* develops (Fig. 13–32*D*). This rudimentary horn may not communicate with the cavity of the uterus. In this case, the endometrium of the rudimentary horn may undergo cyclic menstrual bleeding, leading to distention of the horn because the blood cannot pass to the vagina. In some cases, the uterus appears normal externally but is divided internally by a thin septum (Fig. 13–32*E*). Very

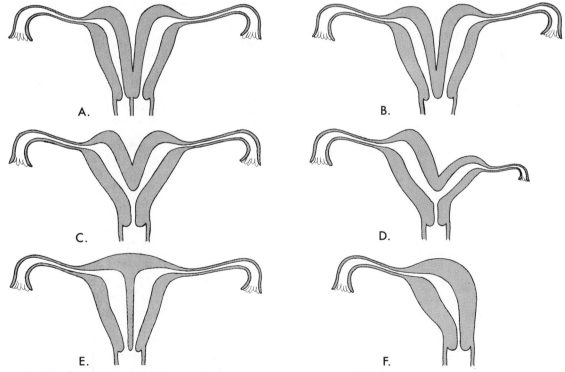

Figure 13–32 Drawings illustrating various types of congenital uterine abnormalities. *A*, Double uterus (uterus didelphys) and double vagina. *B*, Double uterus with single vagina. *C*, Bicornuate uterus. *D*, Bicornuate uterus with a rudimentary left horn. *E*, Septate uterus. *F*, Unicornuate uterus.

rarely, one paramesonephric duct degenerates or fails to form. This results in a *unicornuate uterus* with one uterine tube (Fig. 13–32*F*). *Absence of the uterus* results when the paramesonephric ducts degenerate (as they normally do in male fetuses). When these ducts fail to develop, failure of induction of the vagina usually occurs also because the sinovaginal bulbs are induced to form by the contact of the uterovaginal primordium with the urogenital sinus.

Once in about every 4000 females a condition known as *absence of the vagina* occurs. This results from failure of the sinovaginal bulbs of the urogenital sinus to develop and form the vaginal plate. When the vagina is absent, the uterus is usually absent also, for the reason mentioned previously. Complete absence of the vagina usually creates serious psychological problems (Page et al., 1981). Fortunately, an artificial vagina can be constructed.

Failure of canalization of the vaginal plate results in *vaginal atresia*. The most minor type results from failure of the hymen to rupture, a condition known as *imperforate hymen*. Membranous occlusion at higher levels also occurs and may be mistaken for imperforate hymen. In either case, if there is an excessive secretion of watery fluid, a condition known as *hydrometrocolpos* (distention of the uterus and vagina by fluid other than blood or pus) develops.

DEVELOPMENT OF THE INGUINAL CANALS

The inguinal canals form pathways for the testes to descend through the anterior abdominal wall into the scrotum. Inguinal canals develop in female embryos even though the ovaries do not enter the inguinal canals, except under extremely rare intersexual conditions.

As the mesonephros degenerates, a ligament called the *gubernaculum* descends on each side of the abdomen from the inferior poles of the gonads (Fig. 13–33*A*). Each gubernaculum passes obliquely through the developing anterior abdominal wall (sites of the future inguinal canals) and attaches to a *labioscrotal swelling* (future half of scrotum or a labium majus). Later, an evagination of peritoneum, the *processus vaginalis*, develops on each side ventral to the gubernaculum and herniates through the lower abdominal wall along the path formed by the gubernacu-

lum (Fig. 13–33*B*). Each processus vaginalis carries extensions of the layers of the abdominal wall before it that form the walls of the inguinal canal. In males, they also become the coverings of the spermatic cord and testis (Fig. 13–33*E*). The opening produced in the transversalis fascia by the processus vaginalis becomes the *deep inguinal ring*, and the opening in the external oblique aponeurosis forms the *superficial inguinal ring*.

Descent of the Testes (Fig. 13–33). By about 28 weeks, the testes have descended from the posterior abdominal wall to the deep inguinal rings. This change occurs as the pelvis enlarges and the trunk of the embryo elongates. Because the gubernaculum does not grow as rapidly as the body wall, the testis descends. Little is known about the cause(s) of testicular descent through the inguinal canals and into the scrotum, but it is generally believed that the process is controlled by hormones (gonadotropins and androgens).

The exact role of the gubernaculum in this process is uncertain. Initially, it forms a path through the anterior abdominal wall for the processus vaginalis to follow during formation of the inguinal canal. The gubernaculum also anchors the testis to the scrotum and appears to aid the descent of the testis. It is generally believed, however, that the gubernaculum does not pull the testis into the scrotum. Passage of the testis through the inguinal canal may be aided by the increase in intra-abdominal pressure that results from the growth of the abdominal viscera. There is agreement that gonadotropic and androgenic hormones regulate testicular descent; this belief is supported by clinical evidence that the administration of these hormones induces descent in many cases of cryptorchidism or undescended testis (Brunet et al., 1958). At one time, injections of human chorionic gonadotropin (hCG) were often used in the treatment of cryptorchidism (undescended testes); however, most people believe that hormones will cause the descent of only those testes that would descend anyway in the course of time (Villee, 1975).

Descent of the testis through the inguinal canal usually begins during the twenty-eighth week and takes two or three days. The testis moves posterior to the peritoneum and the processus vaginalis. About four weeks later (about 32 weeks), the testis enters the scrotum. After the testis passes into the scrotum, the inguinal canal contracts around the spermatic cord. In full-term newborn boys, over 97 per cent have both testes well down in the

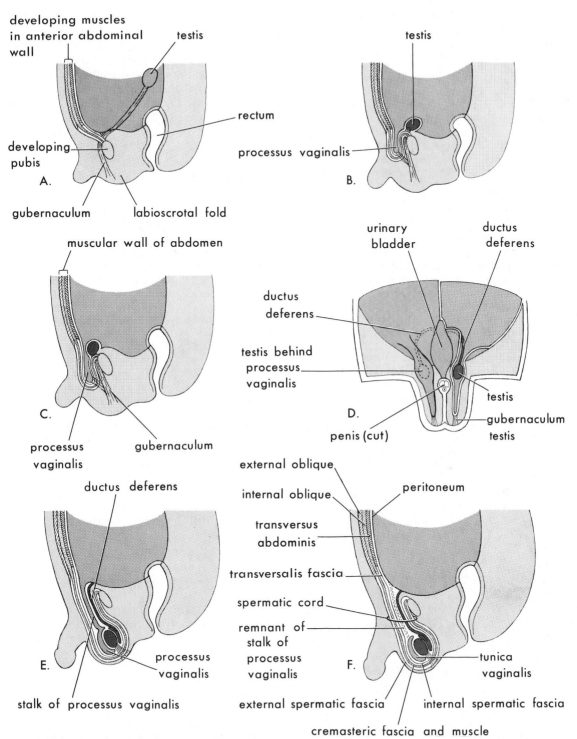

developing muscles in anterior abdominal wall

testis

rectum

developing pubis

A.

gubernaculum

labioscrotal fold

testis

processus vaginalis

B.

muscular wall of abdomen

C.

processus vaginalis

gubernaculum

urinary bladder

ductus deferens

ductus deferens

testis behind processus vaginalis

testis

D.

gubernaculum testis

penis (cut)

ductus deferens

external oblique

internal oblique

peritoneum

transversus abdominis

transversalis fascia

spermatic cord

remnant of stalk of processus vaginalis

E.

processus vaginalis

stalk of processus vaginalis

F.

tunica vaginalis

external spermatic fascia

internal spermatic fascia

cremasteric fascia and muscle

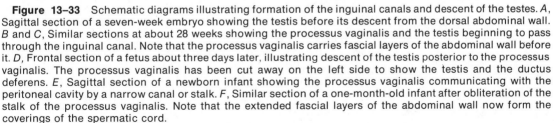

Figure 13–33 Schematic diagrams illustrating formation of the inguinal canals and descent of the testes. *A,* Sagittal section of a seven-week embryo showing the testis before its descent from the dorsal abdominal wall. *B* and *C,* Similar sections at about 28 weeks showing the processus vaginalis and the testis beginning to pass through the inguinal canal. Note that the processus vaginalis carries fascial layers of the abdominal wall before it. *D,* Frontal section of a fetus about three days later, illustrating descent of the testis posterior to the processus vaginalis. The processus vaginalis has been cut away on the left side to show the testis and the ductus deferens. *E,* Sagittal section of a newborn infant showing the processus vaginalis communicating with the peritoneal cavity by a narrow canal or stalk. *F,* Similar section of a one-month-old infant after obliteration of the stalk of the processus vaginalis. Note that the extended fascial layers of the abdominal wall now form the coverings of the spermatic cord.

scrotum (Scorer, 1974). During the first three months after birth, some testes, undescended at birth, descend into the scrotum.

The mode of descent of the testis explains why the ductus deferens crosses anterior to the ureter (Fig. 13–24A) and also explains the course of the testicular vessels. These vessels form when the testis is on the posterior abdominal wall; as the testis descends, it carries its ductus deferens and vessels with it. As the testis and its associated structures descend, they become ensheathed by the fascial extensions of the abdominal wall (Fig. 13–33F). The extension of the transversalis fascia becomes the *internal spermatic fascia;* the internal oblique muscle gives rise to the *cremasteric fascia and muscle,* and the external oblique aponeurosis forms the *external spermatic fascia.* Within the scrotum, the testis projects into the distal end of the processus vaginalis. During the perinatal period, the connecting stalk of the processus vaginalis normally obliterates, isolating the *tunica vaginalis* as a peritoneal sac related to the testis (Fig. 13–33F).

Descent of the Ovary. The ovary also descends from the posterior abdominal wall to a point just inferior to the pelvic brim. It reaches this position by about the twelfth week, and then the gubernaculum attaches to the uterus near the site of entry of the uterine tube. The cranial part of the gubernaculum becomes the *ovarian ligament;* the caudal part forms the *round ligament of the uterus* (Fig. 13–24C). The round ligament of the uterus passes through the inguinal canal and terminates in the labium majus. The relatively small processus vaginalis in the female, often called the *canal of Nuck,* usually obliterates and disappears long before birth.

Cryptorchidism, or Undescended Testis (Fig. 13–34A). This condition occurs in about 30 per cent of premature males and in about 3 per cent of full term males. Cryptorchidism may be unilateral or bilateral. In most cases, the testes descend into the scrotum by the end of the first year. If both testes remain within or just outside

the abdominal cavity, they fail to mature histologically, and sterility is apparently certain. *An undescended testis is unable to produce mature sperms,* presumably because of the higher temperature in the abdominal cavity or the inguinal canal. A cryptorchid testis may be located in the abdominal cavity or anywhere along the usual path of descent of the testis, but usually it lies in the inguinal canal. The cause of most cases of cryptorchidism is unknown, but failure of normal androgen production by the fetal testes appears to be a factor.

The high incidence of cryptorchidism in intersexuality suggests that hormonal imbalance inhibits testicular descent, but failure of androgen production is not demonstrable in most cases of cryptorchidism. Anatomical conditions (e.g., absence or smallness of the superficial inguinal ring) and pathological factors appear to be involved in some cases. For a discussion of testicular maldescent, its causes and treatment, see Mack (1969) and Scorer (1974).

Ectopic Testis (Fig. 13–34B). Occasionally, the testis, after traversing the inguinal canal, deviates from the usual path of descent and becomes lodged in various abnormal locations: interstitial (external to the aponeurosis of the external oblique muscle); in the thigh (femoral triangle); dorsal to the penis; or on the opposite side (crossed ectopia). All types of ectopic testis are rare, but interstitial ectopia occurs most frequently (Warkany, 1971). Ectopic testis may result when a portion of the gubernaculum passes to an abnormal location and the testis later follows it.

Congenital Inguinal Hernia (Fig. 13–35). If the canal connecting the tunica vaginalis and the peritoneal cavity fails to close during early infancy, a condition known as *persistent processus vaginalis* exists, and loops of intestine may herniate through it into the scrotum or labium majus (Fig. 13–35B). Congenital inguinal hernia is much

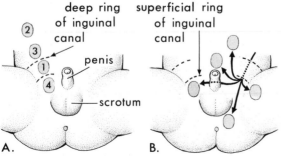

Figure 13–34 Diagrams showing the sites of cryptorchid and ectopic testes. *A,* Positions of cryptorchid (undescended) testes, numbered in order of frequency. *B,* Positions of ectopic testes; about 5 per cent of undescended testes are ectopic.

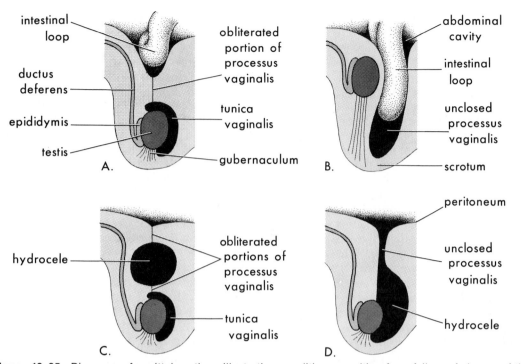

Figure 13–35 Diagrams of sagittal sections illustrating conditions resulting from failure of closure of the processus vaginalis. *A*, Incomplete congenital inguinal hernia resulting from persistence of the proximal part of the processus vaginalis. *B*, Complete congenital inguinal hernia into the scrotum resulting from persistence of the entire processus vaginalis. Cryptorchidism, a commonly associated malformation, is also illustrated. *C*, Large cyst that arose from an unobliterated portion of the processus vaginalis. This condition is called a hydrocele of the spermatic cord. *D*, Hydrocele of the testis and spermatic cord resulting from peritoneal fluid passing into an unclosed processus vaginalis.

more common in males and is often associated with cryptorchidism on the affected side.

Usually, the processus vaginalis in females disappears completely. A diverticulum into which the intestine may herniate may remain, or a remnant of it may remain and give rise to a cyst in the inguinal region or the labium majus.

Hydrocele (Fig. 13–35). Occasionally, the abdominal end of the processus vaginalis remains open but is too small to permit herniation. Peritoneal fluid passes into the processus vaginalis and forms a *hydrocele of the testis* and *spermatic cord* (Fig. 13–35D). If the middle portion of the canal of the processus vaginalis remains open, fluid may accumulate and give rise to a *hydrocele of the spermatic cord* (Fig. 13–35C).

SUMMARY

The urogenital system develops from the intermediate mesoderm, the coelomic epithelium, and the endoderm of the urogenital sinus.

Three successive sets of kidneys develop: (1) the nonfunctional *pronephroi*; (2) the *mesonephroi,* which may serve as temporary excretory organs, and (3) the functional *metanephroi,* or permanent kidneys.

The metanephros develops from two sources: (1) the *metanephric diverticulum,* or ureteric bud, which gives rise to the ureter, the renal pelvis, the calyces, and the collecting tubules; and (2) the *metanephric mesoderm,* which gives rise to the nephrons. At first, the kidneys are located in the pelvis, but they gradually "ascend" to the abdomen. This is not a true upward migration but results from the disproportionate growth of the lumbar and sacral regions.

The urinary bladder develops from the urogenital sinus and the surrounding splanchnic mesenchyme. The entire female urethra and almost all of the male urethra have a similar origin.

Developmental abnormalities of the kidney and excretory passages are relatively common. Incomplete division of the metanephric diverticulum results in double ureter and supernumerary kidney. Failure of the kidney to "ascend" from its embryonic position in the pelvis results in ectopic kidney and abnormal rotation. Various congenital cystic conditions may result from failure of nephrons derived from the metanephric mesoderm to connect with collecting tubules derived from the metanephric diverticulum.

The genital, or reproductive, system develops in close association with the urinary, or excretory, system. Genetic sex is established at fertilization, but the gonads do not begin to attain sexual characteristics until the seventh week, and the external genitalia do not acquire distinct masculine or feminine characteristics until the twelfth week.

The *primordial germ cells* are first recognized on the yolk sac and migrate to the developing gonads.

The reproductive organs in both sexes develop from primordia that appear identical at first. During this *indifferent stage,* an embryo has the potential to develop into either a male or a female. *Gonadal sex* is controlled by the Y chromosome, which exerts a positive testis-determining action on the *indifferent gonad*. It is the presence of the Y-antigen gene on the Y chromosome that determines testicular differentiation. In the presence of a Y chromosome, testes develop and produce an *inducer substance* that *stimulates* development of the mesonephric ducts into the male genital ducts. *Androgens* from the fetal testes stimulate development of the indifferent external genitalia into the penis and scrotum. The *suppressor substance* inhibits development of the paramesonephric ducts.

In the absence of a Y chromosome and in the presence of two X chromosomes, ovaries develop, the mesonephric ducts regress, the paramesonephric ducts develop into the uterus and uterine tubes, the vagina develops from the uterovaginal primordium and the urogenital sinus, and the indifferent external genitalia develop into the clitoris and labia.

Persons with *true hermaphroditism*, an extremely rare condition, have both ovarian and testicular tissue and variable genitalia.

Errors in sexual differentiation may cause *pseudohermaphroditism*. In the male, this results from failure of the fetal testes to produce adequate amounts of masculinizing hormones, or from production of the hormones after the tissue sensitivity of the sexual structures has passed. *In the female, pseudohermaphroditism usually results from virilizing adrenal hyperplasia,* a disorder of the fetal adrenal glands that causes excessive production of androgens and development of the external genitalia in a male direction.

Most abnormalities of the female genital tract result from incomplete fusion of the paramesonephric ducts.

Cryptorchidism and ectopic testes result from abnormalities of testicular descent. *Congenital inguinal hernia* and hydrocele result from abnormalities in the obliteration of the *processus vaginalis*.

Failure of the urogenital folds to fuse normally results in various types of *hypospadias*. Usually the urethral orifice is at the base of the glans penis.

SUGGESTIONS FOR ADDITIONAL READING

Houston, I. B., and Oetliker, O.: The growth and development of the kidneys; *in* J. A. Davis and J. Dobbing (Eds.): *Scientific Foundations of Paediatrics*. Philadelphia, W. B. Saunders Co., 1974, pp. 297–306. A brief description of the anatomic development of the kidney is followed by an extensive account of its functional development and the changes that occur at birth.

Page, E. W., Villee, C. A., and Villee, D. B.: *Human Reproduction. Essentials of Reproductive and Perinatal Medicine*. 3rd ed. Philadelphia, W. B. Saunders Co., 1981. See, in particular, Chapter 2, "Development and Anatomy of the Reproductive Tracts," which gives a concise, well-illustrated account of the embryological development of the urogenital system, with emphasis on aspects that explain the numerous congenital malformations of the male and female reproductive organs encountered in clinical practice.

Scorer, C. G.: The descent of the testis; *in* J. A. Davis and J. Dobbing (Eds.): *Scientific Foundations of Paediatrics*. Philadelphia, W. B. Saunders Co., 1974, pp. 464–468. An authoritative account of the descent of the testis and of the management of the undescended testis.

Visser, H. K. A.: Sexual differentiation in the fetus and newborn; *in* J. A. Davis and J. Dobbing (Eds.): *Scientific Foundations of Paediatrics*. Philadelphia, W. B. Saunders Co., 1974, pp. 455–468. An up-to-date review of sexual differentiation, in which the following are discussed: the role of the Y chromosome in the development of the fetal testis; the role of the fetal testis in the development of genital ducts and external genital organs; the role of androgenic hormones in the sexual differentiation of the central nervous system; imprinting of psychosexual differentiation of the brain; abnormalities of gonadal differentiation; female pseudohermaphroditism: and male pseudohermaphroditism.

CLINICALLY ORIENTED PROBLEMS

1. A 3-year-old girl was still in diapers because she was continually wet with urine. The pediatrician thought she saw urine coming from the infant's vagina. An *intravenous urogram* showed two renal pelves and two ureters on the right side. One ureter was clearly observed to enter the bladder, but the termination of the other one was not clearly seen.

   A *pediatric urologist*, who had been consulted, examined the girl under general anesthesia and observed a small opening in the posterior wall of the vagina. He passed a tiny catheter into it, and, upon injecting a radiopaque solution, he was able to demonstrate that the opening in the vagina was the orifice of the second ureter.

   What is the embryological basis of the two renal pelves and ureters? Describe the embryological basis of the *ectopic ureteric orifice.* What is the anatomical basis of the continual dribbling of urine into the vagina?

2. A seriously injured young man suffered a cardiac arrest. After *cardiopulmonary resuscitation (CPR),* his heart began to beat again, but spontaneous respirations did not occur. Artificial respiration was instituted, but there was *no electroencephalographic (EEG) evidence of brain activity.* After two days, the man's family agreed that there was no hope of his recovery, and they asked to donate his kidneys for transplantation.

   The radiologist carried out *femoral artery catheterization and aortography* (radiographic visualization of the aorta and its branches). This technique showed a single large renal artery on the right, but two renal arteries on the left, one medium in size and the other small. Only the right kidney was used for transplantation because it is more difficult to implant small arteries than large ones. Grafting of the small accessory renal artery into the aorta would be difficult because of its size, and part of the kidney would die if one of the arteries was not successfully grafted.

   Are accessory, or supernumerary, renal arteries common? What is the embryological basis of the two left renal arteries? In what other circumstance might a supernumerary renal artery be of clinical significance?

3. A 32-year-old woman with a short history of *cramping lower abdominal pain* and tenderness underwent a *laparotomy* (incision of the abdominal wall) because of a suspected ectopic pregnancy. The operation revealed a pregnancy in a *rudimentary right uterine horn.* The gravid uterine horn was totally removed. Is this type of uterine malformation common? What is the embryological basis of the rudimentary uterine horn?

4. During the physical examination of a newborn male infant, it was observed that the urethra opened on the ventral surface of the penis at the junction of the glans and body (shaft). The glans was curved toward the undersurface of the penis. Give the medical terms for the malformations described. What is the embryological basis of the abnormal urethral orifice? Is this malformation common? Discuss its etiology.

5. A 10-year-old boy suffered *pain in his left groin* while attempting to lift a heavy box. Later, he noticed a lump in his left groin. When he told his mother about the lump, she arranged an appointment with the family doctor. After a physical examination, a diagnosis of *indirect inguinal hernia* was made. Explain the embryological basis of this type of inguinal hernia. Based on your embryological knowledge, list the layers of the spermatic cord that would cover the hernial sac.

The answers to these questions are given at the back of the book.

REFERENCES

Antell, L.: Hydrocolpos in infancy and childhood. *Pediatrics 10*:306, 1952.

Ashley, D. J. B.: *Human Intersex*. Edinburgh, E. & S. Livingstone, Ltd., 1962.

Austin, C. R.: Sex chromatin in embryonic and fetal tissues; *in* K. L. Moore (Ed.): *The Sex Chromatin*. Philadelphia, W. B. Saunders Co., 1966, pp. 241–254.

Backhouse, K. M., and Butler, H.: The development of the coverings of the testis cord. *J. Anat. 92*:645, 1966.

Bartrina, J.: Hypospadias; *in* M. N. Rashad and W. R. M. Morton (Eds.): *Selected Topics on Genital Anomalies and Related Subjects*. Springfield, Ill., Charles C Thomas, Publisher, 1969, pp. 744–765.

Baxter, T. J.: Cysts arising in the renal tubules. *Arch. Dis. Child. 40*:464, 1965.

Brunet, J., DeMowbray, R. R., and Bishop, P. M. F.: Management of undescended testis. *Br. Med. J. 1*:1367, 1958.

Bulmer, D.: The development of the human vagina. *J. Anat. 91*:490, 1957.

Carr, D. H., Haggar, R. A., and Hart, A. G.: Germ cells in the ovaries of XO female infants. *Am. J. Clin. Pathol. 49*:521, 1968.

Copenhaver, W. M., Kelly, D. E., and Wood, R. L.: *Bailey's Textbook of Histology*. 17th ed. Baltimore, Williams & Wilkins Company, 1978.

Cunha, G. R.: The dual origin of vaginal epithelium. *Am. J. Anat. 143*:387, 1975.

Davidson, W. M., and Ross, G. I. M.: Bilateral absence of kidneys and related congenital anomalies. *J. Pathol. Bacteriol. 68*:459, 1954.

Dewhurst, C. J.: Foetal sex and development of genitalia; *in* E. E. Philipp, J. Barnes, and M. Newton (Eds.): *Scientific Foundations of Obstetrics and Gynecology*. London, William Heinemann, Ltd., 1970, pp. 173–181.

Federman, D. D.: *Abnormal Sexual Development: A Genetic and Endocrine Approach to Differential Diagnosis*. Philadelphia, W. B. Saunders Co., 1967.

Franchi, L. L.: The ovary; *in* E. E. Philipp, J. Barnes, and M. Newton (Eds.): *Scientific Foundations of Obstetrics and Gynecology*. London, William Heinemann, Ltd., 1970, pp. 107–131.

Fukuda, T.: Ultrastructure of primoridal germ cells in human embryos. *Virchows Arch. B. Cell Pathol. 20*:85, 1975.

Gardner, L. I.: Development of the normal fetal and neonatal adrenal; *in* L. I. Gardner (Ed.): *Endocrine and Genetic Diseases of Childhood and Adolescence*. 2nd ed., Philadelphia, W. B. Saunders Co., 1975, pp. 460–476.

Glenister, T. W. A.: A correlation of the normal and abnormal development of the penile urethra and of the infra-umbilical abdominal wall. *Br. J. Urol. 30*:117, 1958.

Gray, S. W., and Skandalakis, J. E.: *Embryology for Surgeons: The Embryological Basis for the Treatment of Congenital Defects*. Philadelphia, W. B. Saunders Co., 1972, pp. 443–693.

Greene, R., Matthews, D., Hughesdon, P. E., and Howard, A.: A case of true hermaphroditism. *Br. J. Surg. 40*:263, 1954.

Grobstein, C.: Some transmission characteristics of the tubule inducing influence on mouse metanephrogenic mesenchyme. *Exp. Cell Res. 13*:575, 1957.

Grumbach, M. M., and Barr, M. L.: Cytologic tests of chromosomal sex in relation to sexual anomalies in man. *Recent Prog. Horm. Res. 14*:255, 1958.

Gyllensten, L.: Contributions to embryology of the urinary bladder: development of definitive relations between openings of the Wolffian ducts and ureters. *Acta Anat. 7*:305, 1949.

Hamerton, J. L.: *Human Cytogenetics*. Vol. II. *Clinical Cytogenetics*. New York, Academic Press, Inc., 1971, pp. 169–195.

Hollinshead, W. H.: *Anatomy for Surgeons*. Vol. II. *The Thorax, Abdomen and Pelvis*. New York, Harper & Row, Publishers, 1956.

Jirásek, J. E.: *Development of the Genital System and Male Pseudohermaphroditism*. Baltimore, Johns Hopkins Press, 1971.

Jones, H. H. and Scott, W. W.: *Hermaphroditism, Genital Anomalies and Related Endocrine Disorders*. Baltimore, The Williams & Wilkins Co., 1958.

Jost, A.: The role of fetal hormones in prenatal development. *Harvey Lect. 55*:201, 1961.

Jost, A.: Development of sexual characteristics. *Sci. J. 6*:67, 1970.

Leeson, T. S.: The fine structure of the mesonephros of the 17-day rabbit embryo. *Exp. Cell Res. 12*:670, 1957.

Lennox, B.: The sex chromatin in hermaphroditism; *in* K. L. Moore (Ed.): *The Sex Chromatin*. Philadelphia, W. B. Saunders Co., 1966, pp. 387–402.

Mack, W. S.: Testicular maldescent; *in* M. N. Rashad and W. R. M. Morton (Eds.): *Selected Topics on Genital Anomalies and Related Subjects*. Springfield, Ill., Charles C Thomas, Publisher, 1969, pp. 729–743.

McCrory, W. W.: *Developmental Nephrology*. Cambridge, Harvard University Press, 1972.

Mittwoch, U.: How does the Y chromosome affect gonadal differentiation? *Philos. Trans. R. Soc. Lond. [Biol. Sci.] 259*:113, 1970.

Monie, I. W.: Double ureter in two human embryos. *Anat. Rec. 103*:195, 1949.

Monie, I. W., and Sigurdson, L. A.: A proposed classification for uterine and vaginal anomalies. *Amer. J. Obstet. Gynecol. 59*:696, 1950.

Moore, K. L.: Sex, intersex and the chromatin test. *Mod. Med. Can. 15*:71, 1960.

Moore, K. L.: Sex determination, sexual differentiation and intersex development. *Can. Med. Assoc. J. 97*:292, 1967.

Morton, W. R. M.: Development of the urogenital systems; *in* M. N. Rashad and W. R. M. Morton (Eds.): *Selected Topics on Genital Anomalies and Related Subjects*. Springfield, Ill., Charles C Thomas, Publisher, 1969, pp. 14–35.

Oliver, J.: *Nephrons and Kidneys: A Quantitative Study of Developmental and Evolutionary Mammalian Renal Architectonics*. New York, Harper & Row, Publishers, 1968.

O'Rahilly, R.: The development of the vagina in the human: *in* R. J. Blandau and D. Bergsma (Eds.): *Morphogenesis and Malformations of the Genital Systems*. Original Article Series, New York, Alan, R. Liss, 1977.

Osathanondh, V., and Potter, E. L.: Pathogenesis of polycystic kidneys. *Arch. Pathol. 77*:459, 1964.

Osathanondh, V., and Potter, E. L.: Development of

the human kidney as shown by microdissection. IV, V. *Arch. Pathol. 82*:391, 1966

Patten, B. M. and Barry, A.: The genesis of exstrophy of the bladder and epispadias. *Am. J. Anat. 90*:35, 1952.

Pearson, P. L., Borrow, M., and Vosa, C. G.: Technique for identifying Y chromosomes in human interphase nuclei. *Nature 226*:78, 1970.

Pinkerton, J. H. M., McKay, D. G., Adams, E. C., and Hertig, A. H.: Development of the human ovary — a study using histochemical techniques. *Obstet. Gynecol. 18*:152, 1961.

Polani, P.: Hormonal and clinical aspects of hermaphroditism and the testicular feminizing syndrome in man. *Philos. Trans. R. Soc. Lond. [Biol. Sci.] 259*:187, 1970.

Potter, E. L.: *Pathology of the Fetus and Infant.* 2nd ed. Chicago, Year Book Medical Publishers, Inc., 1961.

Rickham, P. P.: The incidence and treatment of ectopia vesicae. *Proc. R. Soc. Med. 54*:389, 1961.

Rubin, A. (Ed.): *Handbook of Congenital Malformations.* Philadelphia, W. B. Saunders Co., 1967, pp. 296–352.

Saxen, L.: Embryonic induction. *Clin. Obstet. Gynecol. 18*:149, 1975.

Schlegel, R. J., and Gardner, L. I.: Ambiguous and abnormal genitalia in infants: differential diagnosis and clinical management; *in* L. I. Gardner (Ed.): *Endocrine and Genetic Diseases of Childhood and Adolescence.* 2nd ed. Philadelphia, W. B. Saunders Co., 1975, pp. 571–609.

Scorer, C. G., and Farrington, G. H.: *Congenital Deformities of the Testis and Epididymis.* London, Butterworth & Co. (Publishers), Ltd., 1971.

Scorer, C. G.: The descent of the testis. *Arch. Dis. Child. 39*:605, 1964.

Simpson, J. L.: *Disorders of Sexual Differentiation: Etiology and Clinical Delineation.* New York, Academic Press, 1976.

Smith, E. C., and Orkin, L. A.: A clinical and statistical study of 471 congenital anomalies of the kidney and ureter. *J. Urol. 53*:11, 1945.

Stempfel, R. S., Jr.: Abnormalities of sexual differentiation; *in* L. I. Gardner (Ed.): *Endocrine and Genetic Diseases of Childhood and Adolescence.* 2nd ed. Philadelphia, W. B. Saunders Co., 1975, pp. 551–571.

Sucheston, M. E., and Cannon, M. S.: Development of zonular patterns in the human adrenal gland. *J. Morphol. 126*:477, 1968.

Thompson, J. S., and Thompson, M. W.: *Medical Gentics.* 3rd ed., Philadelphia, W. B. Saunders Co., 1980, pp. 166–180.

Torrey, T. W.: The early development of the human nephros. *Contr. Embryol. Carneg. Instn. 35*:175, 1954.

van Wagenen, G., and Simpson, M. E.: *Embryology of the Ovary and Testis – Homo. Sapiens and Macaca Mulatta.* New York, Yale University Prèss, 1965.

Vaughan, E. D., Jr., and Middleton, G. W.: Pertinent genitourinary embryology. Review for the practicing urologist. *Urology. 6*:139, 1975.

Villee, D. B.: The adrenal gland of the newborn: congenital adrenocortical hyperplasia; *in* D. B. Villee: *Human Endocrinology, A Developmental Approach.* Philadelphia, W. B. Saunders Co., 1975, pp. 145–174.

Warkany, J.: *Congenital Malformations: Notes and Comments.* Chicago, Year Book Medical Publishers, Inc., 1971, pp. 1037–1146.

Willis, R. A.: *The Borderland of Embryology and Pathology.* 2nd ed. London, Butterworth & Co. (Publishers), Ltd., 1958.

Wislocki, G. B.: Observations on descent of testes in macaque and in chimpanzee. *Anat. Rec. 57*:133, 1933.

Witschi, E.: Migration of the germ cells of human embryos from the yolk sac to the primitive gonadal folds. *Contr. Embryol. Carneg. Instn. 32*:67, 1948.

Zurbrügg, R. P.: Congenital adrenal hyperplasia; *in* L. I. Gardner (Ed.): *Endocrine and Genetic Diseases in Childhood and Adolescence.* 2nd ed. Philadelphia, W. B. Saunders Co., 1975, pp. 476–500.

14

THE CIRCULATORY SYSTEM

The Cardiovascular and Lymphatic Systems

THE CARDIOVASCULAR SYSTEM

The cardiovascular system is the first system to function in the embryo; blood begins to circulate by the end of the third week. This precocious development is necessary because the rapidly growing embryo needs an efficient method of acquiring nutrients and disposing of waste products. This precocity is correlated with the absence of a significant amount of nutritive yolk in the ovum and yolk sac.

The earliest blood vessels are derived from *angioblastic tissue,* which arises from the mesenchyme covering the yolk sac, within the connecting stalk, and in the wall of the chorionic sac (see Fig. 4–9). The process of blood vessel development, called *angiogenesis,* is described in Chapter 4.

The primitive vessels cannot be distinguished structurally as arteries or veins but are named according to their future fates and relationship to the heart. The present chapter deals mainly with normal and abnormal development of the heart and great vessels.

DEVELOPMENT OF THE VEINS

Three systems of paired veins drain into the heart of 4-week-old embryos (Figs. 14–1 and 14–2): (1) The *vitelline veins* return blood from the yolk sac; (2) the *umbilical veins* bring oxygenated blood from the chorion (part of early placenta); and (3) the *cardinal veins* return blood from the body of the embryo.

The *vitelline veins* follow the yolk stalk into the embryo and ascend on each side of the foregut. After passing through the septum transversum, they enter the *sinus venosus* of the heart (Fig. 14–1). As the endodermal liver primordium grows into the septum transversum (see Chapter 12), the *hepatic cords* anastomose around pre-existing endothelium-lined spaces. These spaces, the primordia of the *hepatic sinusoids,* later become linked to the vitelline veins. The *hepatic veins* form from the remains of the right vitelline vein in the region of the developing liver. The *portal vein* develops from an anastomotic network formed around the duodenum by the vitelline veins (Fig. 14–3B).

The *umbilical veins* are transformed as follows: (1) The right umbilical vein and the part of the left umbilical vein between the liver and the sinus venosus degenerate (Fig. 14–3B); and (2) the persistent part of the left umbilical vein carries all the blood from the placenta to the fetus. Concurrently a large channel, or shunt, called the *ductus venosus,* develops within the liver and connects the umbilical vein with the inferior vena cava. *The ductus venosus acts as a bypass through the liver, enabling some blood from the placenta to pass almost directly to the heart* (Fig. 14–38). After birth the umbilical vein and the ductus venosus are largely obliterated and subsequently become the *ligamentum teres* and ligamentum venosum, respectively (Fig. 14–39).

During the early stages of development, the *cardinal veins* (colored blue in Figure 14–2) constitute the main venous drainage system of the embryo. The anterior and posterior cardinal veins drain the cranial and caudal parts of the embryo, respectively. They empty into the *sinus venosus* of the primitive heart through a short *common cardinal vein* on each side (Figs. 14–1 to 14–3).

During the eighth week the *anterior cardinal veins* become connected by an oblique

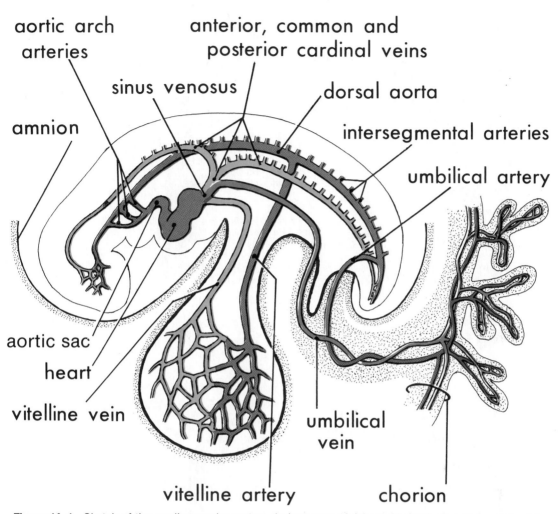

aortic arch arteries

anterior, common and posterior cardinal veins

sinus venosus

dorsal aorta

amnion

intersegmental arteries

umbilical artery

aortic sac

heart

vitelline vein

umbilical vein

vitelline artery

chorion

Figure 14–1 Sketch of the cardiovascular system during stage 12 (about 26 days), showing vessels of the left side only. The umbilical vein is shown in red because it is carrying oxygenated blood and nutrients from the chorion (embryonic part of the placenta) to the embryo. The umbilical arteries are colored medium red to indicate that they are carrying poorly oxygenated umbilical blood and waste products to the chorion.

anastomosis (Fig. 14–3B). This communication shunts blood from the left to the right anterior cardinal vein and becomes the *left brachiocephalic vein* when the caudal part of the left anterior cardinal vein degenerates (Fig. 14–3C). The right anterior cardinal vein and the right common cardinal vein become the superior vena cava (Figs. 14–2 and 14–3).

The *posterior cardinal veins* develop primarily as the vessels of the mesonephric kidneys (see Chapter 13) and largely disappear with these organs. The only adult derivatives of the posterior cardinal veins are the *root of the azygos vein* and the *common iliac veins* (Fig. 14–2D).

The subcardinal veins and the supracardinal veins gradually replace and supplement

the posterior cardinal veins (Fig. 14–2). The *subcardinal veins* appear first (colored red in Figure 14–2). They are connected with each other via the *subcardinal anastomosis* and with the posterior cardinal veins through the mesonephric sinusoids. The subcardinal veins form the stem of the *left renal vein,* the *suprarenal veins,* the *gonadal veins* (e.g., ovarian), and a segment of the *inferior vena cava* (Fig. 14–2D).

The *supracardinal veins* (colored yellow in Figure 14–2) are the last set of vessels to develop. They become disrupted in the region of the kidneys (Fig. 14–2C). Cranial to this they become united by an anastomosis that is represented in the adult by the *azygos* and *hemiazygos veins* (Figs. 14–2D and 14–

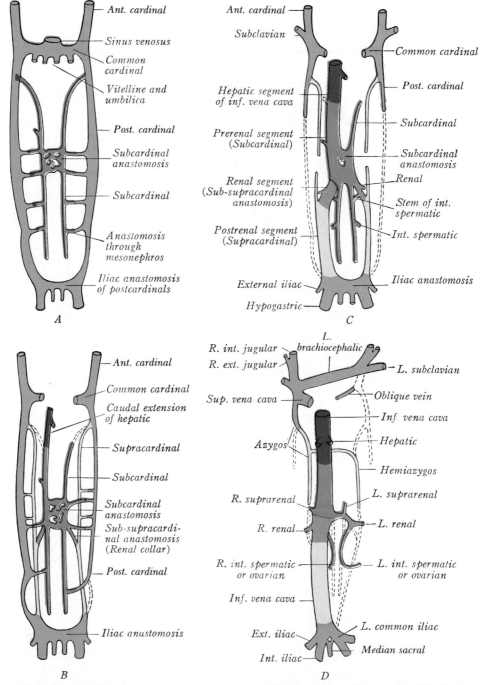

Figure 14–2 Drawings illustrating transformation of the primitive veins of the trunk of the human embryo, shown by diagrams in ventral view. Initially, three systems of veins are present (colored blue): the umbilical veins from the chorion (Fig. 14–1), the vitelline veins from the yolk sac, and the cardinal veins from the body of the embryo. Next, the subcardinal veins (colored red) appear, and finally the supracardinal veins (colored yellow) develop. The transformations leading to the adult venous plan occur during the sixth to eighth weeks. *A,* 6 weeks. *B,* 7 weeks. *C,* 8 weeks. *D,* adult. (From Arey, L. B.: Developmental Anatomy. Revised 7th ed. Philadelphia, W. B. Saunders Co., 1974.)

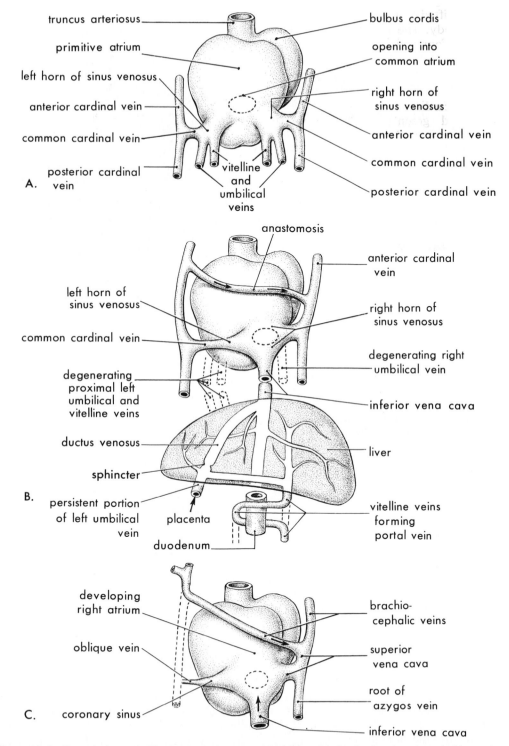

Figure 14–3 Dorsal views: *A,* The heart during stage 11 (about 24 days), showing the primitive atrium and sinus venosus. *B,* Seven weeks, showing the enlarged right horn of the sinus venosus and the venous circulation through the liver. The organs are not drawn to scale. *C,* Eight weeks, indicating the adult derivatives of the cardinal veins.

3). Caudal to the kidneys, the left supracardinal vein degenerates, but the right supracardinal vein becomes the inferior part of the inferior vena cava.

Development of the inferior vena cava (Fig. 14–2) results from a series of changes in the primitive veins of the trunk that occur as blood, returning from the caudal part of the embryo, is shifted from the left to the right side of the body. The inferior vena cava is composed of four main segments: (1) a *hepatic segment* (colored purple) derived from the hepatic vein (proximal part of right vitelline vein) and hepatic sinusoids; (2) a *prerenal segment* (colored red) derived from the right subcardinal vein; (3) a *renal segment* (colored green) derived from the subcardinal-supracardinal anastomosis; and (4) a *postrenal segment* (colored yellow) derived from the right supracardinal vein.

ABNORMALITIES OF THE VENAE CAVAE

Double Superior Vena Cava (Fig. 14–4). Persistence of the left anterior cardinal vein results in the presence of a left superior vena cava; hence there are two superior venae cavae. The anastomosis that usually forms the left brachiocephalic vein (Figs. 14–2 and 14–3) is small or absent, as in Figure 14–4. The abnormal left superior vena cava, derived from the left anterior cardinal and common cardinal veins, opens into the right atrium via the coronary sinus.

Left Superior Vena Cava. Uncommonly, the left anterior cardinal vein and the left common cardinal vein form a left superior vena cava, and the right anterior cardinal vein and the common cardinal vein, which usually form the superior vena cava, degenerate. Blood from the right side is carried by the brachiocephalic vein to the left superior vena cava, which opens into the coronary sinus.

Absence of the Hepatic Portion of the Inferior Vena Cava. Occasionally segments of the inferior vena cava fail to form, or they do not join in a normal manner. As a result, blood from the inferior parts of the body drains into the right atrium through the azygos and hemiazygos veins and the superior vena cava. The hepatic veins open into the right atrium separately.

Double Inferior Vena Cava. In these unusual cases, the inferior vena cava inferior

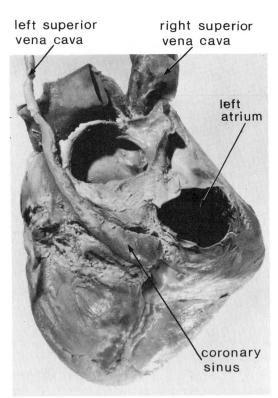

left superior vena cava   right superior vena cava

left atrium

coronary sinus

Figure 14–4 Photograph of the posterior aspect of an adult heart with double superior vena cava; the small left superior vena cava opens into the coronary sinus. Parts of the walls of the atria have been removed.

to the renal veins is represented by two more or less symmetrical vessels. This condition probably results from failure of an anastomosis to develop between the primitive veins of the trunk (Fig. 14–2B). As a result, the inferior part of the left supracardinal vein fails to disappear, as illustrated in Figure 14–2D, and persists as a second inferior vena cava.

DEVELOPMENT OF THE PRIMITIVE HEART AND ARTERIES

The *primitive arterial system* is illustrated in Figures 14–1, 14–9, and 14–31. As the *branchial arches* form during the fourth and fifth weeks, they are penetrated by arteries arising from the *aortic sac*, which are referred to as *aortic arches*. On each side of the embryo, the aortic arches terminate in a single *dorsal aorta* (Figs. 14–31 and 14–32). Just caudal to the eleventh somite, the dorsal aortae fuse to form the *dorsal aorta*.

Thirty or so branches of the dorsal aorta, called *intersegmental arteries* (Fig. 14–1), pass between and carry blood to the somites and their derivatives. Most *somatic arteries* are derived from the intersegmental arteries or their modifications. In the neck, the dorsal intersegmental arteries join together to form a longitudinal artery on each side, called the *vertebral artery*. Most of the original connections of the intersegmental arteries to the dorsal aorta disappear, except the more caudal ones, which remain as the origins of the vertebral arteries from the subclavian arteries. In the thorax, the dorsal intersegmental arteries remain as the *intercostal arteries*. The dorsal intersegmental arteries become the *lumbar arteries* in the lumbar region, and the fifth pair of lumbar intersegmental arteries become the *common iliac arteries*. In the sacral region the intersegmental arteries form the lateral sacral arteries. The dorsal aorta becomes the *median sacral artery,* which anastomoses with the lateral sacral arteries.

The unpaired, ventral midline branches of the dorsal aorta pass to the yolk sac, the allantois, and the chorion (Fig. 14–1). The vitelline arteries pass to the yolk sac and later to the primitive gut, which forms from the yolk sac (see Chapter 12). Three of the vitelline arteries remain as the *celiac artery* to the foregut, the *superior mesenteric artery*

to the midgut, and the *inferior mesenteric artery* to the hindgut (see Fig. 12–1).

The paired umbilical arteries pass through the connecting stalk (later the *umbilical cord*) and become continuous with the *chorionic vessels* in the embryonic part of the placenta *(chorion)*. The umbilical arteries carry deoxygenated blood to the placenta (Figs. 14–1 and 14–38). The proximal parts of the umbilical arteries become the *internal iliac* and *superior vesical arteries*, whereas the distal parts obliterate after birth and become the *medial umbilical ligaments* (Fig. 14–39).

The major changes leading to the definitive arterial system, especially the *transformation of the aortic arches* (Fig. 14–32), are described later.

Heart development is first indicated in embryos of 18 or 19 days. In the *cardiogenic area* the splanchnic mesenchymal cells ventral to the pericardial coelom aggregate and arrange themselves side-by-side as two longitudinal cellular strands called *cardiogenic cords* (Fig. 14–5). These cellular cords become canalized to form two thin-walled endothelial tubes called *endocardial heart tubes* (Figs. 14–6 and 14–7B). As the lateral folds develop, the heart tubes gradually approach each other and fuse to form a single

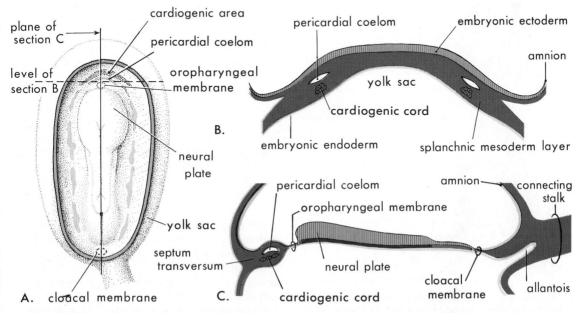

Figure 14–5 *A,* Dorsal view of an embryo during developmental stage 8 (about 18 days). *B,* Transverse section through the embryo demonstrating the cardiogenic cords and their relationship to the pericardial coelom. *C,* Longitudinal section through the embryo illustrating the relationship of the cardiogenic cords to the oropharyngeal (buccopharyngeal) membrane, the pericardial coelom (cavity), and the septum transversum.

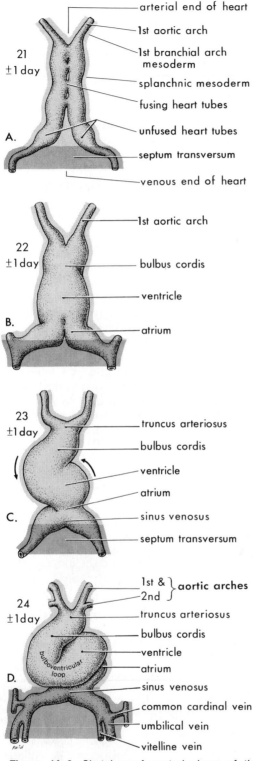

Figure 14–6 Sketches of ventral views of the developing heart during stages 9 to 11 (about 20 to 25 days), showing fusion of the heart tubes, which forms a single heart tube. Bending of the heart tube, which forms a bulboventricular loop, is also illustrated.

heart tube. This fusion begins at the cranial end of the tubes (Fig. 14–6A) and extends caudally until a single tube is formed (Fig. 14–6C).

As the heart tubes fuse, the mesenchyme around them thickens to form a *myoepicardial mantle* (Fig. 14–7C and D). At this stage the developing heart is a simple tube, separated from another tube (the myoepicardial mantle) by gelatinous connective tissue called *cardiac jelly*. The inner endocardial tube is destined to become the internal endothelial lining of the heart, called the *endocardium*. The myoepicardial mantle gives rise to the *myocardium* (muscular wall) and the *epicardium* or visceral pericardium (Fig. 14–7F).

With development of the head fold (see Figs. 5–1 and 5–2), the heart and pericardial cavity come to lie ventral to the foregut and caudal to the oropharyngeal membrane (Fig. 14–8).

Concurrently, the tubular heart elongates and develops alternate dilatations and constrictions. The primordia of the *bulbus cordis, ventricle,* and *atrium* appear first (Fig. 14–6B), but the *truncus arteriosus* and the *sinus venosus* are soon recognizable (Fig. 14–6C). The truncus arteriosus is continuous caudally with the bulbus cordis and cranially it enlarges slightly to form the *aortic sac,* from which the *aortic arches* arise (Figs. 14–1 and 14–9). The sinus venosus is a large venous sinus which receives the *umbilical, vitelline,* and *common cardinal veins* from the chorion (primitive placenta), yolk sac, and embryo, respectively (Figs. 14–1 and 14–6D).

The arterial and venous ends of the heart tube are fixed by the branchial arches and the septum transversum, respectively (Fig. 14–6A). Because the bulbus cordis and ventricle grow faster than the other regions, the heart tube bends upon itself, forming a U-shaped *bulboventricular loop* (Figs. 14–6D and 14–7E); later an *S-shaped heart* forms (Fig. 14–9C).

As the primitive heart bends, the atrium and the sinus venosus come to lie dorsal to the bulbus cordis, truncus arteriosus, and ventricle (Fig. 14–9). By this stage, the sinus venosus has developed lateral expansions called *right* and *left horns* (Fig. 14–9B).

Formation of the Pericardial Cavity. As the heart tube elongates and bends, it gradually sinks into the dorsal wall of the pericardial cavity (Figs. 14–7C and D and

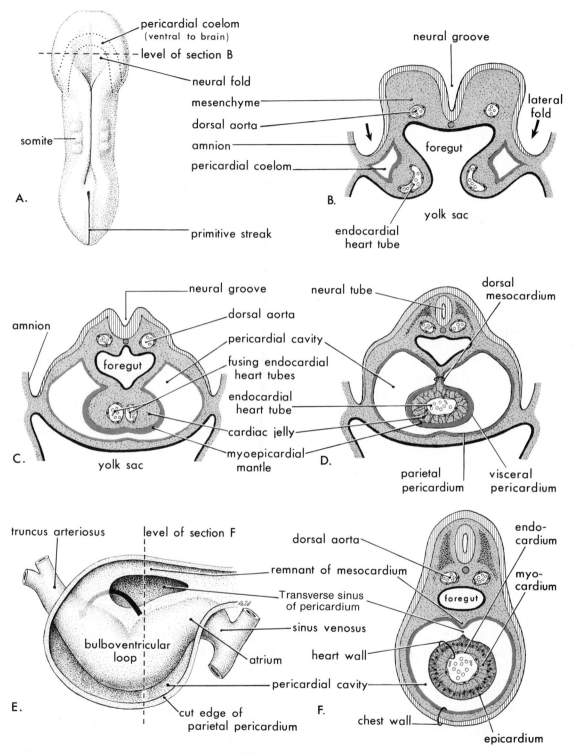

Figure 14-7 *A,* Dorsal view of an embryo during stage 9 (about 20 days). *B,* Transverse section through the heart region, showing the heart tubes and the lateral folds (arrows). *C,* Transverse section through a slightly older embryo, showing the formation of the pericardial cavity and the heart tubes about to fuse. *D,* Similar section during stage 10 (about 22 days), showing the single heart tube suspended by the dorsal mesocardium. *E,* Schematic drawing of the heart during stage 13 (about 28 days), showing degeneration of the dorsal mesocardium and formation of the transverse sinus of the pericardium. *F,* Transverse section through this embryo after disappearance of the dorsal mesocardium, showing the layers of the heart wall.

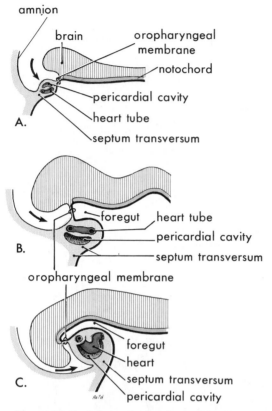

Figure 14–8 Schematic drawings of longitudinal sections through the cranial half of human embryos during the fourth week, showing the effect of the head fold (arrow) on the position of the heart and other structures. As the head fold develops, the heart tube and the pericardial cavity come to lie ventral to the foregut and caudal to the oropharyngeal membrane.

14–8*C*). Initially the heart is suspended from the dorsal wall of this cavity by a mesentery, the *dorsal mesocardium* (Fig. 14–7*D*). The central part of this mesentery soon degenerates, forming a communication, the *transverse pericardial sinus,* between the right and left sides of the pericardial cavity (Fig. 14–7*E* and *F*).

Formation of the Heart Wall (Fig. 14–7). As the heart tubes fuse, the splanchnic mesenchyme around them proliferates and forms a thick layer of cells, the *myoepicardial mantle*. These cells differentiate into (1) *myoblasts,* which form the thick myocardium, and (2) *mesothelial cells,* which form the outermost layer, called the *epicardium* or visceral pericardium. The myoepicardial mantle is separated from the endothelium lining the heart tube by *cardiac jelly,* a gelatinous connective tissue that forms the

subendocardial tissue. The heart wall now consists of three layers: (1) the outer *epicardium;* (2) the middle *myocardium;* and (3) the inner *endocardium* (Fig. 14–7*F*).

Circulation Through the Primitive Heart. Contractions of the heart begin by day 22; these originate in the muscle, i.e., they are of myogenic origin. The muscle layers of the atrium and ventricle are continuous, and contractions occur in peristalsis-like waves that begin in the sinus venosus. At first the circulation through the heart and the embryo is of an ebb-and-flow type, but by the end of the fourth week, coordinated contractions of the heart result in a unidirectional flow.

Blood returns to the sinus venosus from (1) the embryo, via the *common cardinal veins;* (2) the developing placenta, via the *umbilical veins;* and (3) the yolk sac, via the *vitelline veins* (Fig. 14–1). Blood from the sinus venosus enters the atrium via the *sinoatrial orifice* (Fig. 14–20*B*). Its flow is controlled by the *sinoatrial valves* (Fig. 14–9*A*), which fuse cranially to form a marked projection, the *septum spurium* (Fig. 14–20*A*) in the roof of the right side of the atrium. The blood then passes through the atrioventricular canal into the ventricle. When the ventricle contracts, blood is pumped through the bulbus cordis and truncus arteriosus into the aortic sac, from which it passes to the *aortic arches* of the branchial arches (Fig. 14–9*C*). The blood then passes to the dorsal aortae for distribution to the embryo, yolk sac and placenta (Fig. 14–1).

PARTITIONING OF THE ATRIOVENTRICULAR CANAL, THE ATRIA, AND THE VENTRICLES

Partitioning of the atrioventricular canal, the atrium, and the ventricle begins around the middle of the fourth week and is essentially complete by the end of the fifth week. Although they are described separately, it must be emphasized that these processes occur concurrently.

Partitioning of the Atrioventricular Canal (Figs. 14–9, 14–10, and 14–11). During the fourth week, bulges form on the dorsal and ventral walls of the atrioventricular canal. These bulges, called *atrioventricular endocardial cushions,* are at first filled with cardiac jelly but are later invaded by mesenchymal cells. During the fifth week,

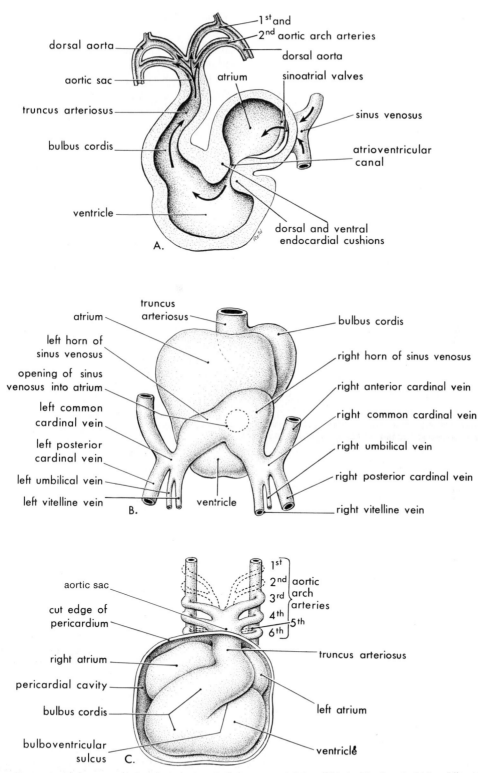

Figure 14–9 *A*, Diagrammatic sagittal section during stage 11 (about 24 days), showing blood flow through the primitive heart. *B*, Dorsal view of the heart during stage 12 (about 26 days), illustrating the horns of the sinus venosus and the dorsal location of the primitive atrium. *C*, Ventral view of the heart and aortic arches during stage 15 (about 35 days). The ventral wall of the pericardial sac has been removed to show the heart in the pericardial cavity. Note that the heart now has the general external appearance of an adult heart.

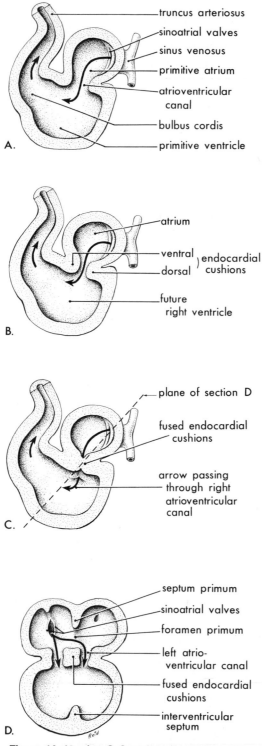

truncus arteriosus

sinoatrial valves

sinus venosus

primitive atrium

atrioventricular canal

bulbus cordis

A.   primitive ventricle

atrium

ventral ⎫ endocardial
dorsal   ⎭ cushions

future right ventricle

B.

plane of section D

fused endocardial cushions

arrow passing through right atrioventricular canal

C.

septum primum

sinoatrial valves

foramen primum

left atrio-ventricular canal

fused endocardial cushions

interventricular septum

D.

Figure 14–10 *A to C*, Sketches of sagittal sections of the heart during the fourth and fifth weeks, illustrating division of the atrioventricular canal. *D*, Frontal section of the heart at the plane shown in *C*. The interatrial and interventricular septa have also started to develop.

these actively growing atrioventricular endocardial cushions approach each other and fuse, forming the *septum of the atrioventricular canal*. It divides the atrioventricular canal into *right and left atrioventricular canals* (Fig. 14–10D).

Partitioning of the Primitive Atrium (Figs. 14–10, 14–11, and 14–12). The primitive atrium is divided into right and left atria by the formation and subsequent fusion of two septa, the septum primum and the septum secundum. The septum primum also fuses with the septum of the atrioventricular canal (fused endocardial cushions).

The *septum primum,* a thin, crescent-shaped membrane, grows ventrally from the dorsocranial wall, or roof, of the primitive atrium. As this curtain-like septum grows, a large opening, the *foramen primum,* exists between its caudal free edge and the endocardial cushions. The foramen primum becomes progressively smaller and is obliterated when the septum primum fuses with the fused endocardial cushions *(atrioventricular septum).*

Before the foramen primum is obliterated, perforations appear in the dorsal part of the septum primum and coalesce to form another opening, the *foramen secundum* (Figs. 14–11 and 14–12). Concurrently, the free edge of the septum primum fuses with the left side of the fused endocardial cushions, thus obliterating the foramen primum (Fig. 14–11D).

Toward the end of the fifth week, another crescentic membrane, the *septum secundum,* grows from the ventrocranial wall of the atrium, immediately to the right of the septum primum (Fig. 14–12D). As this thick, incomplete septum grows, it gradually covers the foramen secundum in the septum primum. The septum secundum forms an incomplete partition and leaves an oval opening, called the *foramen ovale* (Fig. 14–$12E_1$). The cranial part of the septum primum, which is attached to the roof of the left atrium, gradually disappears Fig. 14–$12G_1$); the remaining part of the septum primum, attached to the fused endocardial cushions, becomes the *valve of the foramen ovale* (Fig. 14–$12H_1$).

Before birth, the foramen ovale allows most of the blood entering the right atrium from the inferior vena cava to pass into the left atrium (Fig. 14–38). *After birth,* the foramen ovale normally closes, and the intera-

Text continued on page 313

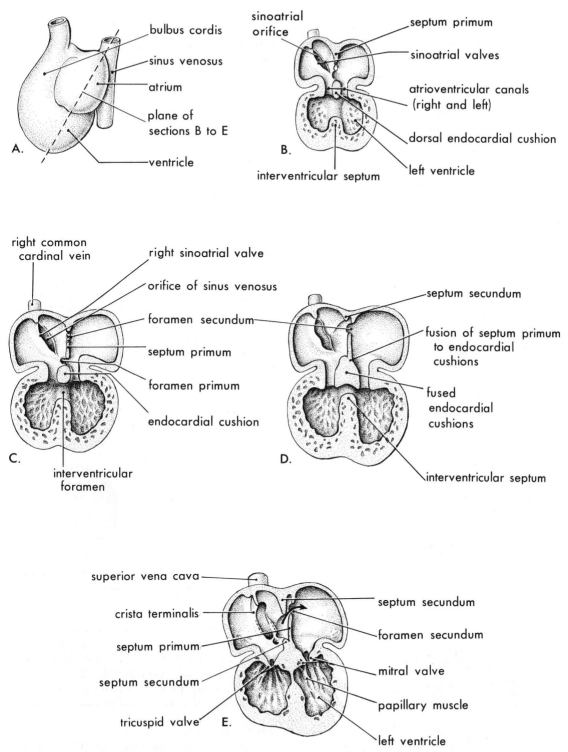

Figure 14–11 Drawings of the developing heart, showing partitioning of the atrioventricular canal, the atrium and the ventricle. *A,* Sketch showing the plane of frontal sections *B* to *E. B,* During stage 13 (about 28 days), showing the early appearance of the septum primum, the interventricular septum, and the dorsal endocardial cushion. *C,* Section of the heart during stage 14 (about 32 days), showing perforations in the dorsal part of the septum. *D,* Section of the heart during stage 15 (about 35 days), showing the foramen secundum. *E,* About eight weeks, showing the heart after partitioning into four chambers. (Adapted from various sources, especially Patten, 1968.)

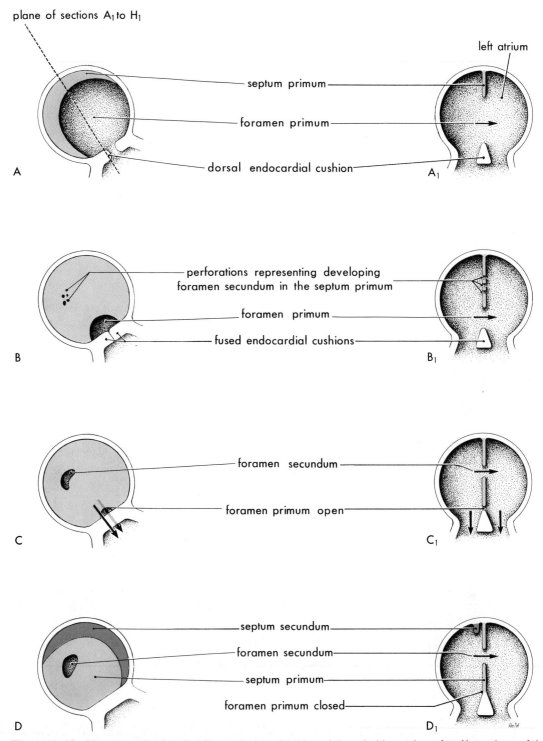

Figure 14–12 Diagrammatic sketches illustrating partitioning of the primitive atrium. *A* to *H* are views of the developing interatrial septum as viewed from the right side. *A₁* to *H₁* are frontal sections of the developing interatrial septum at the plane shown in *A*.

Illustration continued on opposite page

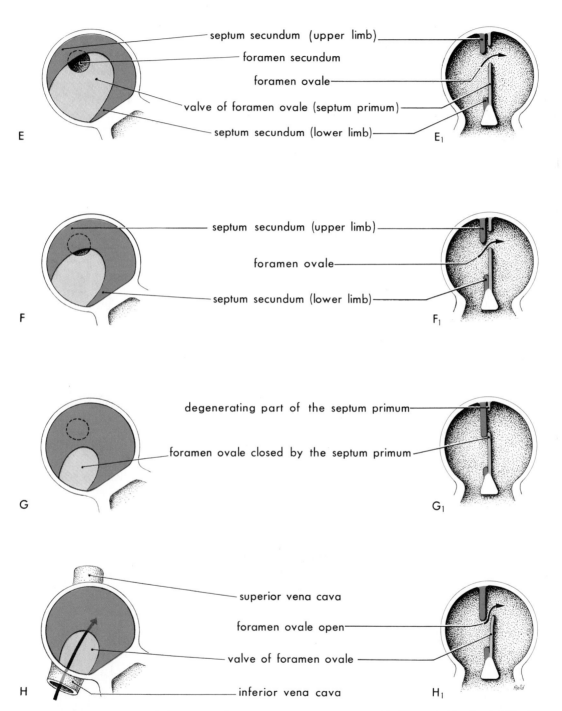

Figure 14–12 *Continued.* The valve-like nature of the foramen ovale is illustrated in G₁ and H₁. When pressure in the right atrium exceeds that in the left atrium, blood passes from the right to the left side of the heart. When the pressures are equal, the septum primum closes the foramen ovale.

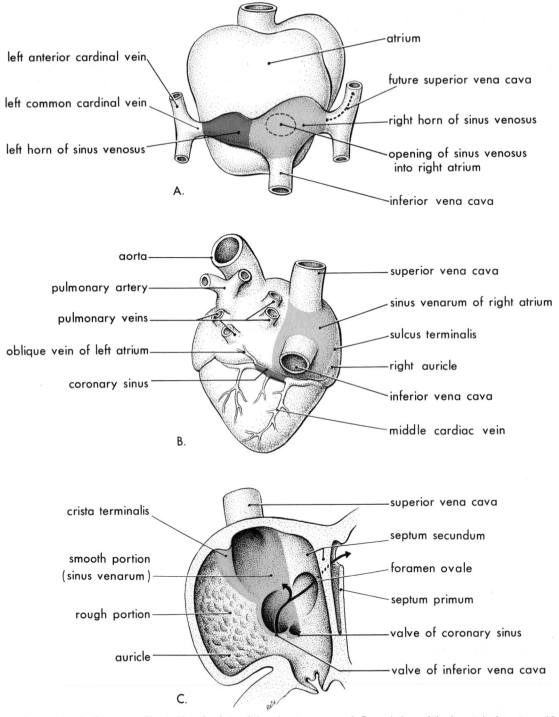

Figure 14–13 Diagrams illustrating the fate of the sinus venosus. *A*, Dorsal view of the heart during stage 12 (about 26 days), showing the sinus venosus. The umbilical and vitelline veins are not shown. *B*, Dorsal view at eight weeks after incorporation of the right horn of the sinus venosus into the right atrium. The left horn of the sinus venosus has become the coronary sinus. *C*, Internal view of the fetal atrium, showing (1) the smooth part (sinus venarum) of the wall of the right atrium derived from the right horn of the sinus venosus and (2) the crista terminalis and the valves of the inferior vena cava and coronary sinus derived from the right sinoatrial valve.

trial septum becomes a complete partition when the septum primum fuses with the septum secundum (Fig. 14–39).

Changes in the Sinus Venosus and Associated Veins (Figs. 14–3 and 14–13). Initially the sinus venosus opens into the center of the primitive atrium and the right and left horns of the sinus are about the same size. Progressive enlargement of the right horn results from two left-to-right shunts of blood, so that by the end of the fourth week, the right horn is noticeably larger than the left (Fig. 14–3C). As this occurs, the sinoatrial orifice moves to the right and opens in the part of the primitive atrium that will become the adult right atrium (Fig. 14–20B).

The *first left-to-right shunt of blood* results from transformation of the vitelline and umbilical veins (discussed previously).

The *second left-to-right shunt of blood* occurs when the anterior cardinal veins become connected by an oblique anastomosis (Fig. 14–3B and C). This communication shunts blood from the left to the right anterior cardinal vein and eventually becomes the left *brachiocephalic vein* (Fig. 14–2D). The right anterior cardinal and the right common cardinal veins become the *superior vena cava*.

The *results of these two venous shunts* are that (1) the left horn of the sinus venosus decreases in size and importance, and (2) the right horn enlarges and receives all the blood from the head and neck via the superior vena cava, and from the placenta and caudal regions of the body via the inferior vena cava (Figs. 14–3C and 14–38).

Fate of the Sinus Venosus and Formation of the Adult Right Atrium (Fig. 14–13). Initially the sinus venosus is a separate chamber of the heart and opens into the caudal wall of the right atrium (Fig. 14–3A). The *left horn* of the sinus venosus forms the *coronary sinus*, and the *right horn* of the sinus venosus becomes incorporated into the wall of the *right atrium* (Fig. 14–13B and C).

The smooth part of the wall of the right atrium, the *sinus venarum*, into which the great veins open, is derived from the sinus venosus (Fig. 14–13B and C). The remainder of the atrium and the conical muscular pouch called the *auricle*

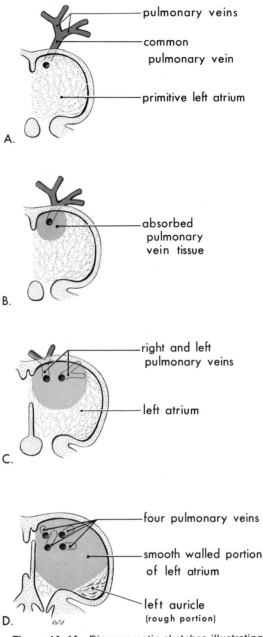

A.

B.

C.

D.

Figure 14–14 Diagrammatic sketches illustrating absorption of the pulmonary veins into the left atrium: *A*, Five weeks, showing the common pulmonary vein opening into the primitive left atrium. *B*, Later stage, showing partial absorption of the common pulmonary vein. *C*, Six weeks, showing the openings of two pulmonary veins into the left atrium resulting from absorption of the common pulmonary vein. *D*, Eight weeks, showing four pulmonary veins with separate atrial orifices. The pulmonary veins and the absorbed pulmonary vein tissue are shown in red.

pulmonary veins

common pulmonary vein

primitive left atrium

absorbed pulmonary vein tissue

right and left pulmonary veins

left atrium

four pulmonary veins

smooth walled portion of left atrium

left auricle (rough portion)

have a rough, trabeculated appearance; they are derived from the primitive right atrium.

The smooth part (sinus venarum) and the rough part (primitive atrium) are demarcated internally by a vertical ridge, the *crista terminalis* (Fig. 14–13C) and externally by a shallow inconspicuous groove, the *sulcus terminalis* (Fig. 14–13B). The crista terminalis represents the cranial part of the right sinoatrial valve (Fig. 14–13C); the caudal part of the right sinoatrial valve forms the valves of the inferior vena cava and the coronary sinus. The left sinoatrial valve fuses with the septum secundum and is incorporated into the interatrial septum.

Formation of the Adult Left Atrium (Fig. 14–14).
Most of the wall of the left atrium is smooth and is derived from the *primitive pulmonary vein*. This vein develops as an outgrowth of the dorsal atrial wall, just to the left of the septum primum. As the atrium expands, the primitive pulmonary vein and its main branches are gradually incorporated into the wall of the left atrium. This results in four pulmonary veins with separate openings into the atrium (Fig. 14–14C and D). The small auricle, derived from the primitive atrium, has a rough, trabeculated appearance.

Anomalous Pulmonary Venous Connections. In *total anomalous pulmonary venous connection*, none of the pulmonary veins connects with the left atrium, but open into the right atrium or into one of the systemic veins or into both. In *partial anomalous pulmonary venous connection*, one or more, but not all, pulmonary veins have similar anomalous connections; the others have normal connections. In both types, the systemic veins involved are variable.

In the early embryo, the lungs are drained by a vascular plexus that communicates with the cardinal veins in many places. Later, the primitive pulmonary vein develops in the sinoatrial region and connects with this plexus. Normally the connections with the cardinal veins then degenerate; if these connections persist, the pulmonary veins may drain into the derivatives of the cardinal veins. Drainage of the pulmonary veins into the right atrium can be explained by a slight shift of the interatrial septum to the left.

Partitioning of the Primitive Ventricle (Figs. 14–11, 14–15, 14–16, 14–17, and 14–18).
Division of the primitive ventricle into right and left ventricles is first indicated by a muscular ridge or fold, the *interventricular septum,* in the floor of the ventricle near its apex. This thick, crescentic membrane has a cranial concave free edge (Fig. 14–15A). Initially, most of its increase in length results from dilatation of the ventricles on each side of it (Fig. 14–15B); this produces an external *interventricular groove.* Later, there is active growth of septal tissue as the *muscular portion of the interventricular septum* forms.

A crescentic *interventricular foramen* between the free edge of the interventricular septum and the fused endocardial cushions permits communication between the right and left ventricles until about the end of the seventh week (Figs. 14–15 and 14–17).

The interventricular foramen usually closes around the end of the seventh week as the bulbar ridges fuse (Fig. 14–18). Closure results from the fusion of subendocardial tissue from three sources: (1) the right bulbar

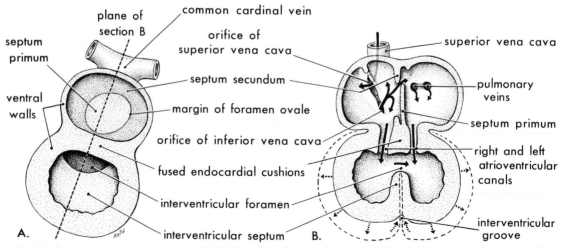

Figure 14–15 Schematic diagrams illustrating partitioning of the heart. *A,* Sagittal section late in the fifth week, showing the cardiac septa and foramina. *B,* Frontal section of slightly later stage, illustrating the directions of blood flow through the heart and the expansion of the ventricles.

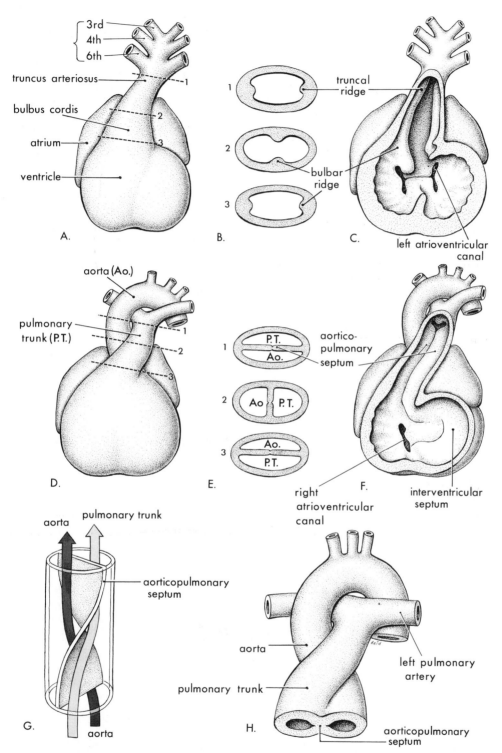

Figure 14–16 Schematic drawings illustrating partitioning of the bulbus cordis and truncus arteriosus. A, Ventral aspect of heart at five weeks. B, Transverse sections through the truncus arteriosus and bulbus cordis, illustrating the truncal and bulbar ridges. C, The ventral wall of the heart has been removed to demonstrate the ridges. D, Ventral aspect of heart after partitioning of the truncus arteriosus. E, Sections through the newly formed aorta (Ao.) and pulmonary trunk (P.T.), showing the aorticopulmonary septum. F, Six weeks. The ventral wall of the heart and pulmonary trunk have been removed to show the aorticopulmonary septum. G, Diagram illustrating the spiral form of the aorticopulmonary septum. H, Drawing showing the great arteries twisting around each other as they leave the heart.

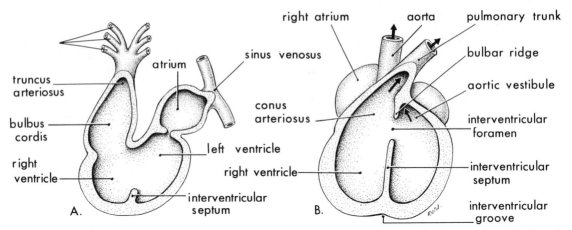

Figure 14–17 Sketches illustrating incorporation of the bulbus cordis into the ventricles and partitioning of the bulbus cordis and truncus arteriosus into the aorta and pulmonary trunk. *A*, Sagittal section at five weeks, showing the bulbus cordis as one of the five primitive chambers of the heart. *B*, Schematic frontal section at six weeks after the bulbus cordis has been incorporated into the ventricles to become the conus arteriosus of the right ventricle and the aortic vestibule.

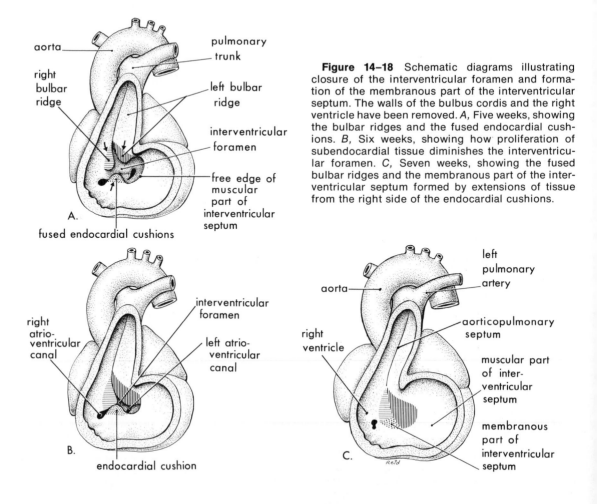

Figure 14–18 Schematic diagrams illustrating closure of the interventricular foramen and formation of the membranous part of the interventricular septum. The walls of the bulbus cordis and the right ventricle have been removed. *A*, Five weeks, showing the bulbar ridges and the fused endocardial cushions. *B*, Six weeks, showing how proliferation of subendocardial tissue diminishes the interventricular foramen. *C*, Seven weeks, showing the fused bulbar ridges and the membranous part of the interventricular septum formed by extensions of tissue from the right side of the endocardial cushions.

ridge; (2) the left bulbar ridge; and (3) the fused atrioventricular endocardial cushions.

The *membranous part of the interventricular septum* is derived from extensions of tissue from the right side of the fused endocardial cushions. This tissue fuses with the aorticopulmonary septum and the muscular part of the interventricular septum (Fig. 14–18C). After closure of the interventricular foramen, the pulmonary trunk is in communication with the right ventricle, and the aorta communicates with the left ventricle.

Partitioning and Fate of the Bulbus Cordis and Truncus Arteriosus (Fig. 14–

16). During the fifth week, bulges form in the walls of the bulbus cordis. These bulges, called *bulbar ridges,* are first filled with cardiac jelly but are later invaded by mesenchymal cells. Similar *truncal ridges* form in the truncus arteriosus and are continuous with the bulbar ridges. The spiral orientation of the ridges, possibly caused by the streaming of blood from the ventricles, results in a spiral *aorticopulmonary septum* when these ridges fuse (Fig. 14–16D to G). This septum divides the bulbus cordis and the truncus arteriosus into two channels, the *aorta* and the *pulmonary trunk.* Because of the spiral

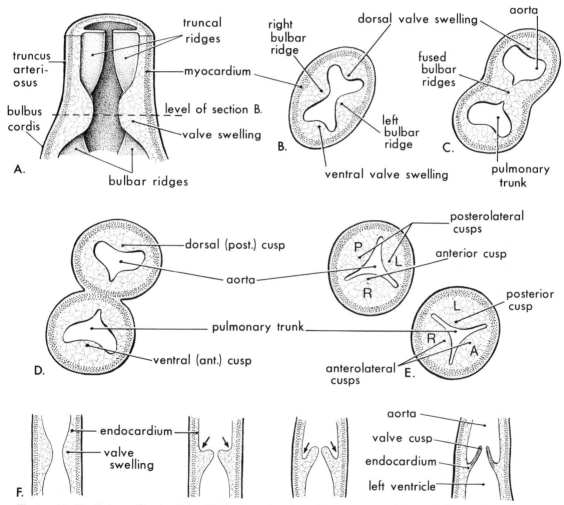

Figure 14–19 Schematic drawings illustrating development of the semilunar valves of the aorta and pulmonary trunk. *A,* Sketch of a sagittal section of the truncus arteriosus and bulbus cordis, showing the valve swellings. *B,* Transverse section through the bulbus cordis. *C,* Similar section after fusion of the bulbar ridges. *D,* Formation of the walls and valves of the aorta and pulmonary trunk. *E,* Rotation of the vessels has established the adult relations of the valves. *F,* Longitudinal sections of the aorticoventricular junction, illustrating successive stages in the hollowing (arrows) and thinning of the valve swellings to form valve cusps.

septum, the pulmonary trunk twists around the ascending aorta (Figs. 14–16*H* and 14–18).

The bulbus cordis is gradually incorporated into the walls of the venricles. In the adult right ventricle, it is represented by the *conus arteriosus*, which gives origin to the pulmonary trunk. In the adult left ventricle, the bulbus cordis forms the walls of the *aortic vestibule*, the part of the ventricular cavity just inferior to the aortic valve (Fig. 14–17*B*).

Development of the Cardiac Valves (Figs. 14–19 and 14–20). The *semilunar valves* develop from three *valve swellings*, or ridges, of subendocardial tissue at the orifices of both the aorta and the pulmonary trunk. These swellings become hollowed out and reshaped to form three thin-walled cusps or "pockets." The *atrioventricular valves* (tricuspid and mitral valves) develop similarly from localized proliferations of subendocardial tissue around the atrioventricular canals.

These swellings become hollowed out on their ventricular sides (Fig. 14–20*C* and *D*).

Development of the Ventricular Walls (Figs. 14–11 and 14–20). Cavitation of the ventricular walls forms a spongework of the muscle bundles. Some of these bundles remain as the *trabeculae carneae*, and others become transformed into the *papillary muscles* and the *chordae tendineae*, which connect the ventricular wall with the atrioventricular valves (Fig. 14–20*C* and *D*).

Development of the Conducting System. Initially the muscle layers of the atrium and ventricle are continuous. The primitive atrium acts as the temporary *pacemaker* of the heart, but the sinus venosus soon takes over this function.

The sinoatrial node is originally in the right wall of the sinus venosus, but it is incorporated with the sinus venosus into the wall of the right atrium and lies where the superior vena cava enters the atrium (Fig. 14–20*D*).

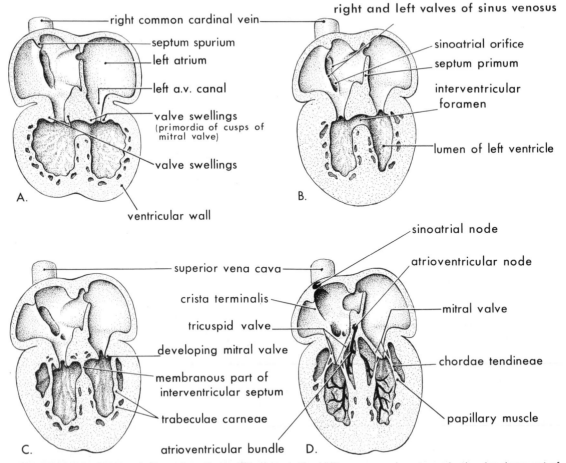

Figure 14–20 Schematic frontal sections of the heart, illustrating successive stages in the development of the atrioventricular valves, the chordae tendineae, and the papillary muscles. *A,* Five weeks. *B,* Six weeks. *C,* Seven weeks. *D,* Five months, also showing the conducting system of the heart (modified from Patten, 1968).

After incorporation of the sinus venosus, cells from the left wall of the sinus venosus are found in the base of the interatrial septum just anterior to the opening of the coronary sinus. They, together with cells from the atrioventricular canal region, make up the atrioventricular node and the bundle of His (Anderson and Taylor, 1972). This specialized tissue is normally the only pathway from the atria to the ventricles, because as the four chambers of the heart develop, a band of connective tissue grows in from the epicardium. This tissue subsequently separates the muscle of the atria from that of the ventricles and forms part of the adult cardiac skeleton.

The sinoatrial node, the atrioventricular node, and the atrioventricular bundle soon become richly supplied with nerves, but histological differentiation of these tissues is a continuing process up to, and after, birth.

There is evidence that abnormalities of the conducting tissue may cause unexpected death during infancy ("crib death"). Anderson and Ashley (1974) have observed conducting tissue abnormalities in the hearts of several infants who died unexpectedly and were classified as "crib deaths," or *Sudden Infant Death Syndrome (SIDS)*. There remains a lack of consensus that a single mechanism is responsible for the sudden and unexpected death of an apparently well infant. Currently, considerable research is being undertaken in an attempt to determine the pathogenic factors responsible for such deaths. *For a good discussion of SIDS*, see Vaughan et al. (1979).

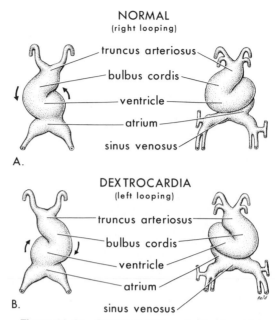

Figure 14–21 Sketches of the primitive heart tube during the fourth week. *A*, Normal bending to the right or right looping. *B*, Abnormal bending to the left or left looping, resulting in dextrocardia.

CONGENITAL MALFORMATIONS OF THE HEART AND GREAT VESSELS

Heart malformations are relatively common, constituting about 25 per cent of all congenital malformations. Their overall incidence is about 0.7 per cent of live births and about 2.7 per cent of stillbirths. Most heart malformations are believed to be determined by multiple factors, genetic and possibly also nongenetic, each of which has a minor effect.

Some types of congenital cardiac malformation cause very little disability, but others are incompatible with extrauterine life. Owing to recent advances in cardiovascular surgery, many types of congenital heart disease can be corrected surgically (Keith et al., 1978).

Not all congenital malformations of the heart and great vessels are described. Emphasis is placed on those that are compatible with life or amenable to surgery. Descriptions of uncommon and clinically unimportant malformations appear in small print.

ABNORMALITIES OF POSITION

Dextrocardia (Displacement of the Heart to the Right). If the heart tube bends to the left instead of to the right (Fig. 14–21), there is transposition in which the heart and its vessels are reversed left to right as in a mirror image. Dextrocardia is the most frequent positional abnormality of the heart, but it is still relatively uncommon. In *dextrocardia with situs inversus* (transposition of the viscera, the liver being on the left side and the heart on the right, and so forth) the incidence of accompanying cardiac defects is low, and such hearts, if there are no other associated vascular abnormalities, can function normally. In *isolated dextrocardia,* the abnormal position of the heart is not accompanied by displacement of other viscera, but accompanying cardiac defects are often present.

Ectopia Cordis (Fig. 14–22). In this extremely rare condition, the heart is partly or completely exposed on the surface of the thorax. It is usually associated with an open pericardium (pericardial sac) and widely separated halves of the sternum. Ectopia cordis results from faulty development of the sternum and pericardium, which may result from failure of complete fusion of the lateral folds in the thoracic region during the fourth week (see Fig. 5–1).

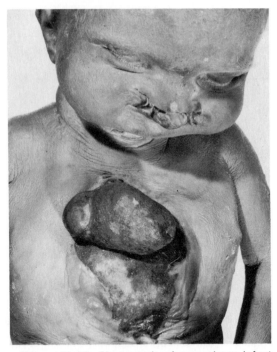

Figure 14–22 Photograph of a newborn infant with cleft sternum, extrathoracic ectopia cordis, and bilateral cleft lip. Fortunately, ectopia cordis is an extremely rare condition.

ABNORMALITIES OF THE INTERATRIAL SEPTUM

Atrial septal defect (ASD) is a common type of congenital heart abnormality. The common form of ASD is persistent or *patent* foramen ovale (Figs. 14–23, 14–24, 14–25, and 14–26).

Probe Patent Foramen Ovale (Fig. 14–23B). In up to 25 per cent of persons, a probe can be passed obliquely from one atrium to the other through the upper part of the floor of the fossa ovalis; the heart is otherwise normal. This defect is not considered a pathological occurrence, but a probe patent foramen ovale may be forced open as a result of other cardiac defects and contribute to functional pathology of the heart (Keith et al., 1978). Probe patent foramen ovale results from incomplete adhesion (i.e., imperfect sealing) between the septum primum and the septum secundum after birth.

Clinically Significant Types of Atrial Septal Defect (Figs. 14–24 and 14–25). There are *four main types of atrial septal defect (ASD):* (1) secundum type ASD; (2) endocardial cushion defect with primum type ASD; (3) sinus venosus type ASD; and (4) common atrium. The first two types are relatively common.

Secundum Type ASD (Figs. 14–24A to D and 14–25). Defects in the area of the foramen ovale are classified as secundum type ASD and include both septum primum and septum secundum defects. This is one of the most common congenital cardiac defects. The patent foramen ovale usually results from abnormal resorption of the septum primum during the formation of the foramen secundum (Fig. 14–12B). If resorption occurs in abnormal locations, the septum primum is fenestrated or net-like (Fig. 14–24A). If ex-

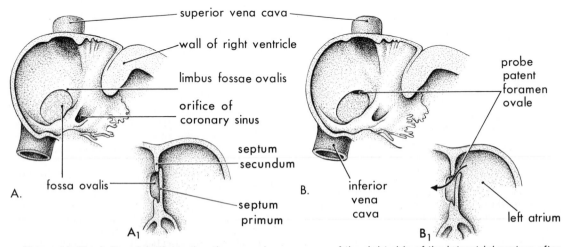

Figure 14–23 *A,* Drawing illustrating the normal appearance of the right side of the interatrial septum after adhesion of the septum primum to the septum secundum. *A₁,* Sketch of a frontal section of the interatrial septum illustrating formation of the fossa ovalis. Note that the floor of this fossa is formed by the septum primum. *B* and *B₁,* Similar views of a probe patent foramen ovale resulting from incomplete adhesion of the septum primum to the septum secundum.

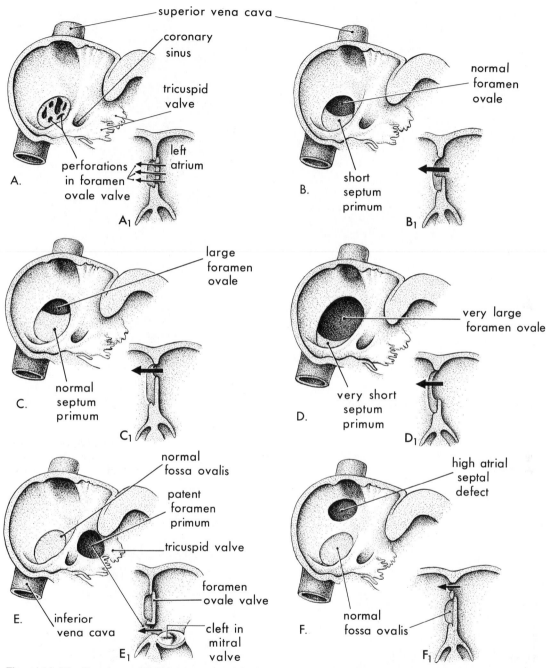

Figure 14–24 Drawings of the right aspect of the interatrial septum (*A* to *F*) and sketches of frontal sections through the septum (*A₁* to *F₁*) illustrating various types of atrial septal defect. *A,* Patent foramen ovale resulting from resorption of the septum primum in abnormal locations. *B,* Patent foramen ovale caused by excessive resorption of the septum primum, sometimes called the "short flap defect." *C,* Patent foramen ovale resulting from an abnormally large foramen ovale. *D,* Patent foramen ovale resulting from (1) an abnormally large foramen ovale, and (2) excessive resorption of the septum primum. *E* Endocardial cushion defect with primum type atrial septal defect. The frontal section *E₁* also shows the cleft in the septal leaflet of the mitral valve. *F,* High septal defect resulting from abnormal absorption of the sinus venosus into the right atrium. Note that in *E* and *F* the fossa ovalis has formed normally.

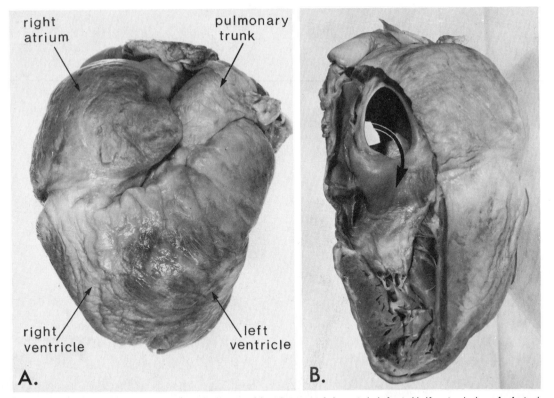

Figure 14–25 Photographs of an adult heart with a large atrial septal defect. Half actual size. *A,* Anterior view showing the large right ventricle, right atrium, and pulmonary trunk. *B,* Right atrial aspect showing the large atrial septal defect (arrow) resulting from an abnormally large foramen ovale and excessive resorption of the septum primum (see Figure 14–24*D*).

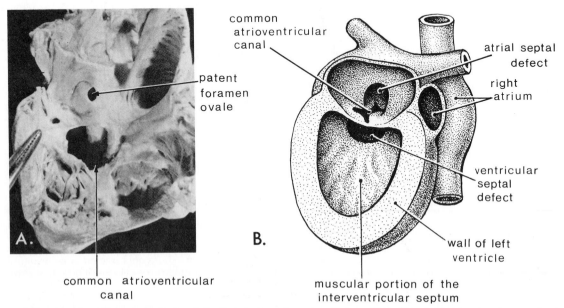

Figure 14–26 *A,* Photograph of an infant's heart, sectioned and viewed from the right side, showing a patent foramen ovale and a common atrioventricular canal. (From Lev, M.: *Autopsy Diagnosis of Congenitally Malformed Hearts.* 1953. Courtesy of Charles C Thomas, Publisher, Springfield, Ill.) *B,* Schematic drawing of a heart illustrating various defects of the cardiac septa.

cessive resorption of the septum primum occurs, the resulting short septum primum does not close the foramen ovale (Fig. 14–24*B*).

If an abnormally large foramen ovale results from defective development of the septum secundum, a normal septum primum will not close the abnormal foramen ovale at birth (Fig. 14–24*C*). Large atrial septal defects may result from a combination of excessive resorption of the septum primum and a large foramen ovale (Figs. 14–24*D* and 14–25*B*).

Endocardial Cushion Defect with Primum Type ASD (Fig. 14–24E). The incomplete form of endocardial cushion defect is relatively common. The septum primum does not fuse with the endocardial cushions, leaving a *patent foramen primum;* usually there is also a cleft in the anterior cusp of the mitral valve.

In the complete form of the endocardial cushion defect, fusion of the endocardial cushions does not occur. This produces a large hole in the center of the heart known as atrioventricularis communis (Fig. 14–26). *Atrioventricularis communis occurs in about 20 per cent of persons with the Down syndrome* (see Fig. 8–4); otherwise it is a relatively uncommon cardiac defect.

Sinus Venosus Type ASD (Fig. 14–24F). All defects of this type are located high in the interatrial septum, peripheral to the fossa ovalis. They may result from incomplete absorption of the sinus venosus into the right atrium and/or abnormal development of the septum secundum. This type of ASD is commonly associated with partial anomalous pulmonary venous connections. The sinus venosus defect is one of the rarest types of ASD.

Common Atrium. In this rare condition, the interatrial septum is completely absent. This defect results from failure of the septum primum and the septum secundum to develop.

ABNORMALITIES OF THE INTERVENTRICULAR SEPTUM

Ventricular septal defect (VSD) is the most common type of cardiac defect. Most septal defects involve the upper oval-shaped membranous part of the interventricular septum. Isolated ventricular septal defects are detected at a rate of 10 to 12 per 10,000 between birth and five years, and VSD occurs in about 20 per cent of children with

congenital heart disease (Keith et al., 1978).

Membranous Septal Defect (Fig. 14–26B). This is the most common type of VSD. Incomplete closure of the interventricular foramen and failure of the membranous part of the interventricular septum to develop result from failure of extensions of subendocardial tissue to grow from the right side of the fused endocardial cushions and fuse with the aorticopulmonary septum and the muscular part of the interventricular septum (Fig. 14–18*C*).

Muscular Septal Defect. This less common type of VSD may appear anywhere in the muscular part of the septum. Sometimes there are multiple defects, the so-called "Swiss cheese" type of VSD. Muscular septal defects probably result from excessive resorption of myocardial tissue during formation of the muscular part of the interventricular septum.

Absence of the Interventricular Septum. Complete failure of the septum to form is very rare and results in a three-chambered heart, called *cor triloculare biatriatum.*

ABNORMALITIES IN THE DIVISION OF THE TRUNCUS ARTERIOSUS

Persistent Truncus Arteriosus (Fig. 14–27). This malformation results from failure of the aorticopulmonary septum to develop and divide the truncus arteriosus into the aorta and pulmonary trunk. Usually there is also defective fusion of the bulbar ridges. The most common type is a single arterial vessel that gives rise to the pulmonary trunk and ascending aorta (Fig. 14–27*A* and *B*). In the next most common type, the right and left pulmonary arteries arise close together from the dorsal wall of the persistent truncus arteriosus (Fig. 14–27*C*). Two less common types of persistent truncus arteriosus are illustrated in Figure 14–27*D* and *E*.

Aorticopulmonary Septal Defect. This very rare condition consists of a round or oval opening, often called an *aortic window,* between the aorta and pulmonary trunk near the aortic valve. It results from a localized defect in the aorticopulmonary septum.

Complete Transposition of the Great Vessels (Fig. 14–28). In typical cases, the aorta lies anterior to the pulmonary trunk and arises from the right ventricle; the pulmonary trunk arises from the left ventricle. For survival, there must be an associated septal defect or patent ductus arteriosus to permit some interchange between the pulmo-

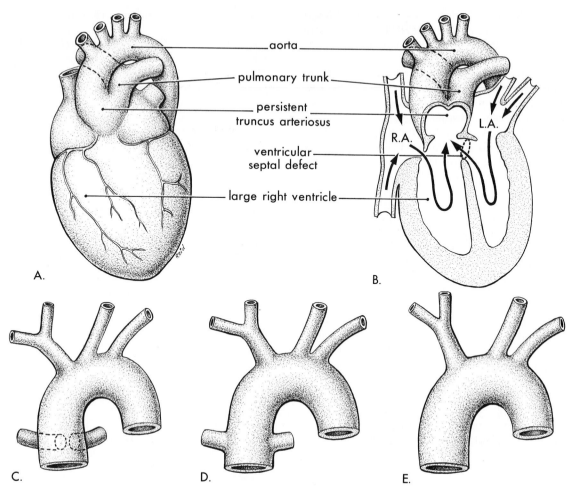

Figure 14–27 Drawings illustrating the main types of persistent truncus arteriosus. *A,* The common trunk divides into an aorta and short pulmonary trunk. *B,* Sketch showing circulation in this heart and a ventricular septal defect. *C,* The right and left pulmonary arteries arise close together from the truncus arteriosus. *D,* The pulmonary arteries arise independently from the sides of the truncus arteriosus. *E,* No pulmonary arteries are present. In such cases, the lungs are supplied by bronchial arteries.

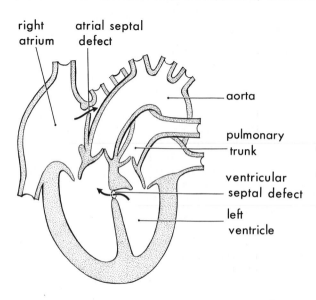

Figure 14–28 Diagram of a heart illustrating complete transposition of the great vessels (arteries). The ventricular and atrial septal defects allow mixing of the blood.

nary and systemic circulations. Although there have been many attempts to explain the embryological basis of transposition of the great arteries, the following *conal growth hypothesis* is favored by many investigators. During partitioning of the bulbus cordis and truncus arteriosus, the aorticopulmonary septum fails to pursue a spiral course. The straight septum probably results from failure of the conus arteriosus to develop normally during incorporation of the bulbus cordis into the ventricles.

Unequal Division of the Truncus Arteriosus (Fig. 14–29B and C). Rarely, partitioning of the truncus arteriosus above the valves is unequal, resulting in one great artery being large and the other small (stenotic). Often the aorticopulmonary septum is not aligned with the interventricular septum and a ventricular septal defect results. The

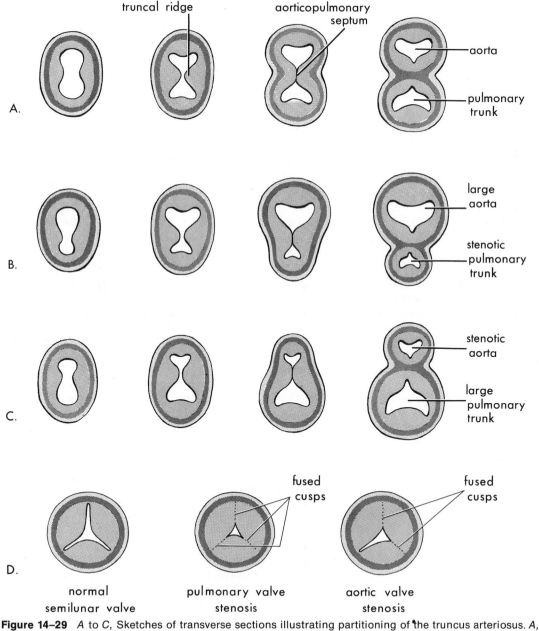

Figure 14–29 *A* to *C*, Sketches of transverse sections illustrating partitioning of the truncus arteriosus. *A*, Normal. *B*, Unequal partitioning, giving rise to a small pulmonary trunk. *C*, Unequal partitioning, resulting in a small aorta. *D*, Sketches illustrating a normal semilunar valve and stenotic pulmonary and aortic valves.

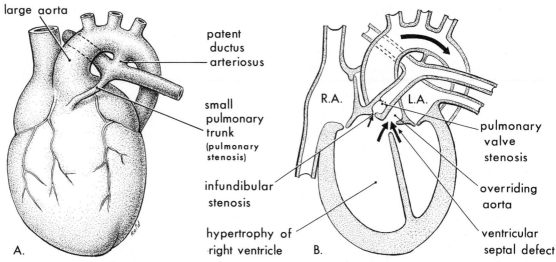

Figure 14–30 *A,* Drawing of an infant's heart showing a small pulmonary trunk (pulmonary stenosis) and a large aorta resulting from unequal partitioning of the truncus arteriosus. There is also hypertrophy of the right ventricle and a patent ductus arteriosus. *B,* Frontal section of a heart illustrating the tetralogy of Fallot. Observe the four cardiac abnormalities. *Note that the large aorta straddles, or overrides, the VSD.*

larger vessel (aorta or pulmonary trunk) usually lies over (overrides) the ventricular septal defect.

Pulmonary Stenosis (Figs. 14–29 and 14–30). In *pulmonary valve stenosis,* the pulmonary valve cusps are fused together to form a dome with a narrow central opening (Fig. 14–29*D*). In *infundibular pulmonary stenosis,* the conus arteriosus (infundibulum) of the right ventricle is underdeveloped. The two types of pulmonary stenosis may occur together or as separate entities. Depending upon the degree of obstruction to blood flow, there is a variable degree of hypertrophy of the right ventricle.

Tetralogy of Fallot (Fig. 14–30B). This is a classic and common group of cardiac defects, consisting of (1) pulmonary stenosis or narrowing of the region of the right ventricular outflow, (2) ventricular septal defect, (3) overriding aorta, and (4) hypertrophy of the right ventricle.

Pulmonary Atresia. If division of the truncus arteriosus is so unequal that the pulmonary trunk has no lumen, or there is no orifice at the level of the pulmonary valve, the malformation is called pulmonary atresia. There may or may not be an associated ventricular septal defect.

Aortic Stenosis and Atresia (Fig. 14–29D). In *aortic valve stenosis,* the edges of the valve are usually fused together to form a

dome with a narrow opening. This condition may be present at birth *(congenital)* or develop after birth *(acquired).* This valvular stenosis causes extra work for the heart and results in hypertrophy of the left ventricle and the production of a *heart murmur.*

In *subaortic stenosis,* there is often a band of fibrous tissue just below the aortic valve; this results from persistence of tissue that normally degenerates as the valve forms. When obstruction of the aorta or its valve is complete, the condition is called *aortic atresia.*

FORMATION AND DERIVATIVES OF THE AORTIC ARCHES

As the branchial arches develop during the fourth week (see Figs. 10–1 and 10–2), they receive arteries from the aortic sac, called *aortic arches* (Figs. 14–1 and 14–32). The aortic arches terminate in the dorsal aorta of the corresponding side. Although six pairs of aortic arches develop, they are not all present at the same time; e.g., by the time the sixth pair of aortic arches has formed, the first two pairs have involuted and disappeared (Fig. 14–31*C*).

During the sixth to eighth weeks, the primitive aortic arch pattern is transformed into the adult arterial arrangement.

The First Pair of Aortic Arches. These

vessels largely disappear, but the remaining parts form the maxillary arteries. These aortic arches may also contribute to the development of the external carotid arteries.

The Second Pair of Aortic Arches. Dorsal portions of these vessels persist and form the stems of the stapedial arteries.

The Third Pair of Aortic Arches. The proximal parts of these arteries form the *common carotid arteries,* and the distal portions join with the dorsal aortae to form the *internal carotid arteries*.

The Fourth Pair of Aortic Arches. The *left fourth aortic arch* forms part of the arch of the aorta. The proximal part of the arch of the aorta develops from the aortic sac and the distal part forms from the left dorsal aorta.

The *right fourth aortic arch* becomes the proximal portion of the *right subclavian artery*. The distal part of the subclavian artery forms from the right dorsal aorta and the right seventh intersegmental artery. The left subclavian artery is not derived from an aortic arch; it forms from the left seventh intersegmental artery (Fig. 14–32A). As development proceeds, differential growth shifts the origin of the left subclavian artery cranially, so it comes to lie close to the origin of the left common carotid artery (Fig. 14–32D).

The Fifth Pair of Aortic Arches. In about 50 per cent of embryos, the fifth pair of aortic arches are rudimentary vessels that soon degenerate, leaving no derivatives. In about as many embryos, these arteries never develop.

The Sixth Pair of Aortic Arches. The *left sixth aortic arch* develops as follows: (1) The proximal part persists as the proximal part of the left pulmonary artery; and (2) the distal part, which passes from the left pulmonary artery to the dorsal aorta, persists as a shunt called the *ductus arteriosus* (Figs. 14–32 and 14–33).

The *right sixth aortic arch* develops as follows: (1) The proximal part persists as the proximal part of the right pulmonary artery; and (2) the distal part degenerates.

The distal parts of the pulmonary arteries are derived from buds of the sixth aortic arches (Figs. 14–31C and 14–32A) which grow into the develop-

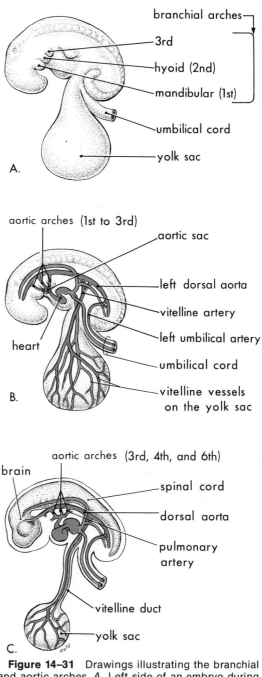

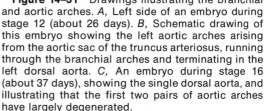

Figure 14–31 Drawings illustrating the branchial and aortic arches. *A,* Left side of an embryo during stage 12 (about 26 days). *B,* Schematic drawing of this embryo showing the left aortic arches arising from the aortic sac of the truncus arteriosus, running through the branchial arches and terminating in the left dorsal aorta. *C,* An embryo during stage 16 (about 37 days), showing the single dorsal aorta, and illustrating that the first two pairs of aortic arches have largely degenerated.

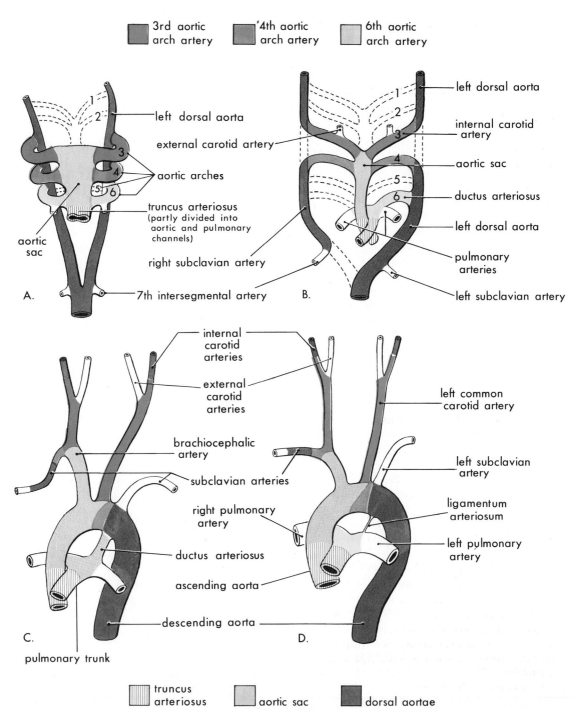

Figure 14-32 Schematic drawings illustrating the changes that result in transformation of the truncus arteriosus, aortic sac, aortic arches, and dorsal aortae into the adult arterial pattern. The vessels that are not shaded or colored are not derived from these structures. *A,* Aortic arches at six weeks; by this stage the first two pairs of aortic arches have largely disappeared. *B,* Aortic arches at seven weeks; the parts of the dorsal aortae and aortic arches that normally disappear are indicated with broken lines. *C,* Arterial arrangement at eight weeks. *D,* Sketch of the arterial vessels of a six-month-old infant. Note that the ascending aorta and the pulmonary arteries are considerably smaller in C than in D. This represents the relative flow through these vessels at the different stages of development. Observe the large size of the ductus arteriosus in C, and note that it is essentially a direct continuation of the pulmonary trunk.

ing lungs. Following partitioning of the truncus arteriosus, the pulmonary arteries arise from the pulmonary trunk.

Relations of the Aortic Arches and the Recurrent Laryngeal Nerves (Fig. 14–33).

The transformation of the sixth pair of aortic arches explains why the course of the recurrent laryngeal nerves differs on the two sides. These branches of the vagus nerves supply the sixth pair of branchial arches and hook around the sixth pair of aortic arches on their way to the larynx.

On the right, because the distal part of the right sixth aortic arch and the fifth aortic artery degenerate, the right recurrent laryngeal nerve moves up and hooks around the proximal part of the right subclavian artery (the derivative of the fourth aortic arch).

On the left, the left recurrent laryngeal nerve hooks around the ductus arteriosus (the persistent distal portion of the sixth aortic arch). When this vessel obliterates after birth, the nerve hooks around the ligamentum arteriosum (the derivative of the ductus arteriosus) and the arch of the aorta.

ABNORMAL TRANSFORMATION OF THE AORTIC ARCHES

Because of the many changes involved in transformation of the embryonic aortic arch system into the adult arterial pattern, it is understandable that variations may occur. Abnormalities result from the persistence of parts of aortic arches that normally disappear, from disappearance of parts that normally persist, or from both.

Coarctation of the Aorta (Fig. 14–34).

This relatively common malformation is characterized by a narrowing of the aorta, usually just above or below the ductus arteriosus. The classification into postductal and preductal coarctations is commonly used, but it must be stressed that in a significant number of instances the coarctation is directly opposite the ductus.

Postductal Coarctation (Fig. 14–34A and B). In this common type, the constriction is just inferior to the ductus arteriosus, which usually closes and forms the ligamentum arteriosum. This location allows for better development of collateral circulation during the fetal period, thus assisting with

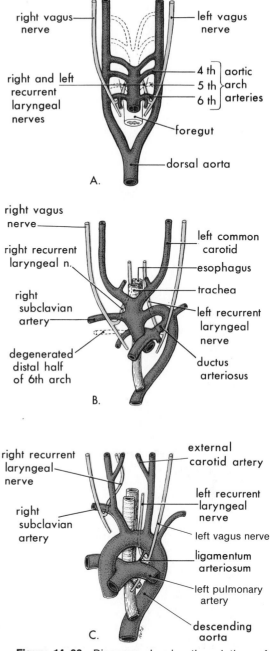

Figure 14–33 Diagrams showing the relations of the recurrent laryngeal nerves to the aortic arches and their derivatives. *A,* Six weeks, showing the recurrent laryngeal nerves hooked around the sixth pair of aortic arches. *B,* Eight weeks, showing the right recurrent laryngeal nerve hooked around the right subclavian artery, and the left recurrent laryngeal nerve hooked around the ductus arteriosus and the arch of the aorta. *C,* Adult, showing the left recurrent laryngeal nerve hooked around the ligamentum arteriosum and the arch of the aorta.

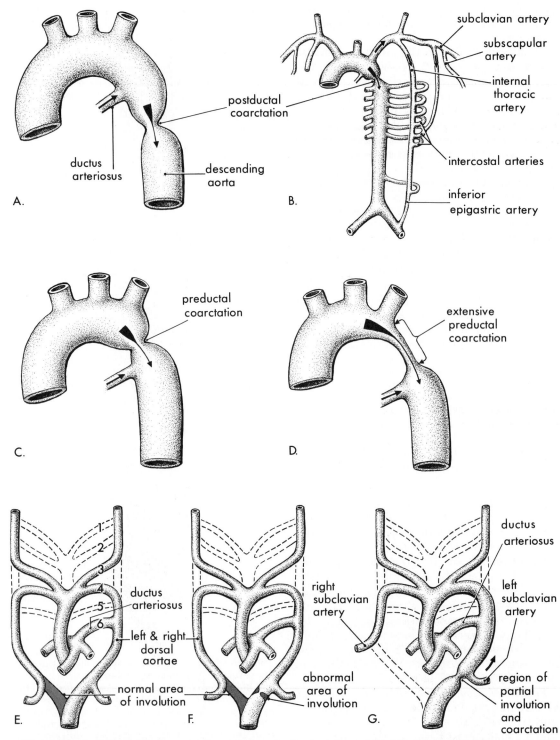

Figure 14–34 *A,* Postductal coarctation of the aorta, the commonest type. *B,* Diagrammatic representation of the common routes of collateral circulation that develop in association with coarctation of the aorta. *C* and *D,* Preductal coarctation. The type illustrated in *D* is usually associated with major cardiac defects. *E,* Sketch of the aortic arch pattern in a seven-week embryo, showing the areas that normally involute. Note that the distal segment of the right dorsal aorta normally involutes as the right subclavian artery develops. *F,* Localized abnormal involution of a small distal segment of the left dorsal aorta. *G,* Later stage showing the abnormally involuted segment appearing as a coarctation of the aorta. This moves (arrow) to the region of the ductus arteriosus with the left subclavian artery. These drawings illustrate one hypothesis on the embryological basis for coarctation of the aorta. For other hypotheses, see the text.

passage of blood to inferior parts of the body (Fig. 14–34*B*).

Preductal Coarctation (Fig. 14–34C and D). In this less common type, the constriction is just superior to the ductus arteriosus. The ductus usually remains open, providing a communication between the pulmonary artery and the descending aorta. The narrowed segment is occasionally extensive (Fig. 14–34*D*).

Causes of Coarctation. The causes of coarctation are not clearly understood. Coarctation is a common finding in cases of Turner syndrome, a condition caused by the loss of a sex chromosome (see Chapter 8). This and other observations suggest that genetic factors or environmental factors or both cause coarctation of the aorta (Warkany, 1971).

The embryological basis of coarctation of the aorta is unclear; there are three main views:

1. During formation of the aortic arch, muscle tissue of the ductus arteriosus may be incorporated into the wall of the aorta; then, when the ductus contracts at birth, the ductal muscle in the aorta also contracts and forms a coarctation.

2. There may be abnormal involution of a small segment of the left dorsal aorta (Fig. 14–34*F*). Later this stenotic segment (area of coarctation) moves cranially with the left subclavian artery to the region of the ductus arteriosus (Fig. 14–34*G*).

3. During fetal life the segment of the arch of the aorta between the left subclavian artery and the ductus arteriosus is normally narrow, because it carries little blood (Anderson and Ashley, 1974). Following normal closure of the ductus arteriosus, this region (called the *isthmus*) normally enlarges until it is the same diameter as the aorta. If the narrowing persists, a coarctation forms. Krediet (1965) believes that all types of coarctation are based on the existence of a stenosis in the isthmus.

Double Aortic Arch (Fig. 14–35). This rare abnormality is characterized by a *vascular ring* which may compress the trachea and esophagus. The ring results from failure of involution of the distal portion of the right dorsal aorta; thus both right and left aortic arches arise from the ascending aorta. The vascular ring is originally large but is reduced in size when (1) the descending aorta shifts to the left carrying the abnormal right aortic arch dorsally, and (2) the region of junction of the two dorsal aortae shifts upward. Usually the right aortic arch is larger and passes posterior to the trachea and esophagus.

Right Aortic Arch (Fig. 14–36). When the entire right dorsal aorta persists and the distal segment of the left dorsal aorta involutes, a right aortic arch results. There are two main types:

Right Aortic Arch Without a Retroesophageal Component (Fig. 14–36B). The ductus arteriosus (or ligamentum arteriosum) passes from the right pulmonary artery to the right aortic arch. Because no vascular ring is formed, this condition is usually asymptomatic. It is often associated with cardiac defects.

Right Aortic Arch with Retroesophageal Component (Fig. 14–36C). Originally there was proba-

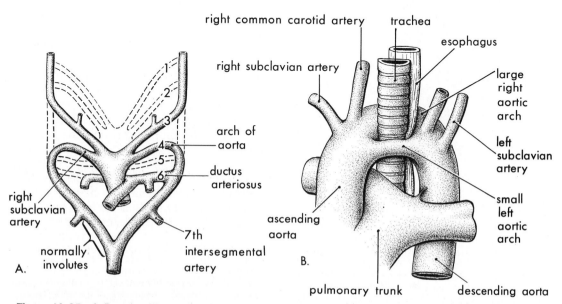

Figure 14–35 *A*, Drawing illustrating the embryological basis of double aortic arch. The distal portion of the right dorsal aorta persists and forms a right aortic arch. *B*, A large right aortic arch and a small left aortic arch arise from the ascending aorta and form a vascular ring around the trachea and esophagus. The right common carotid and subclavian arteries arise separately from the large right arch.

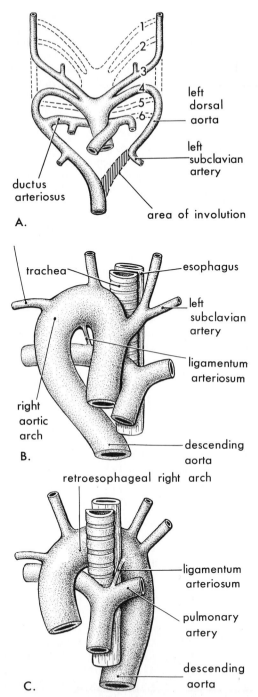

Figure 14–36 A, Sketch of the aortic arches showing abnormal involution of the distal portion of the left dorsal aorta and persistence of the entire right dorsal aorta and the distal portion of the right sixth aortic arch artery. B, Right aortic arch without retroesophageal component. C, Right aortic arch with retroesophageal component. The retroesophageal arch and the normal ductus arteriosus (ligamentum arteriosum) form a vascular ring that compresses the esophagus and the trachea.

bly a small left aortic arch (Fig. 14–35B) which then disappeared, leaving the right aortic arch posterior to the esophagus. The normal left ductus arteriosus (or ligamentum arteriosum) attaches to the descending aorta and forms a vascular ring, which may constrict the esophagus and trachea.

Abnormal Origin of the Right Subclavian Artery or Retroesophageal Subclavian Artery (Fig. 14–37). The right subclavian artery arises from the descending aorta and passes posterior to the trachea and esophagus to supply the right upper limb. This abnormal origin of the right subclavian artery occurs when the right fourth aortic arch and the right dorsal aorta involute cranial to the seventh intersegmental artery. As a result, the right subclavian artery forms from the right seventh intersegmental artery and the distal part of the right dorsal aorta. As development proceeds, differential growth shifts the origin of the right subclavian artery cranially so that it comes to lie close to the origin of the left subclavian artery. Although anomalous right subclavian artery is common and always forms a vascular ring, it is rarely clinically significant because the ring is usually not tight enough to constrict the esophagus and trachea.

FETAL AND NEONATAL CIRCULATION

The fetal cardiovascular system is designed to serve prenatal needs and to permit modifications at birth which establish the postnatal circulatory pattern. Most of the quantitative data relative to the distribution of blood flows has been obtained from studies on fetal lambs, but available human data suggest that the patterns are similar in the human fetus (Keith et al., 1978).

Good respiration in the newborn infant is dependent upon circulatory changes occurring at birth. Thus, to understand conditions such as the respiratory distress syndrome, in which *pulmonary hypoperfusion* plays such an important role, it is necessary to understand the fetal circulation and the changes that occur within the first hours after birth (Page et al., 1981).

Course of the Fetal Circulation (Fig. 14–38). Well-oxygenated blood returns from the placenta in the *umbilical vein*. This blood is about 80 per cent saturated with oxygen. The inferior vena eventually receives all the blood coming from the placenta.

About half the blood from the placenta passes through the *hepatic sinusoids,* whereas the remainder by-passes the liver,

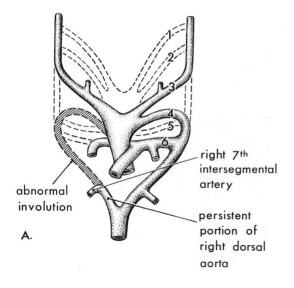

abnormal
involution

A.

right 7th
intersegmental
artery

persistent
portion of
right dorsal
aorta

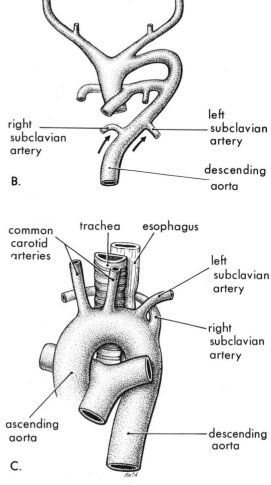

right
subclavian
artery

left
subclavian
artery

descending
aorta

B.

common
carotid
arteries

trachea   esophagus

left
subclavian
artery

right
subclavian
artery

ascending
aorta

descending
aorta

C.

Figure 14–37   Sketches illustrating the embryological basis of abnormal origin of the right subclavian artery. *A,* The right fourth aortic arch and the cranial portion of the right dorsal aorta have involuted. As a result, the right subclavian artery forms from the right seventh intersegmental artery and the distal segment of the right dorsal aorta. *B,* As the arch of the aorta forms, the right subclavian artery is carried cranially (arrows) with the left subclavian artery. *C,* The abnormal right subclavian artery arises from the aorta and passes posterior to the trachea and the esophagus.

going through the *ductus venosus* into the inferior vena cava. This blood flow is regulated by a *sphincter* in the ductus venosus close to the umbilical vein. When the sphincter relaxes, more blood passes through the ductus venosus; when the sphincter contracts, more blood is diverted through the *portal sinus* to the portal vein and hepatic sinusoids (Dickson, 1957). Although the presence of an *anatomical sphincter* in the ductus venosus is not universally accepted, there definitely appears to be a *physiological sphincter* which prevents overloading of the heart when venous flow in the umbilical vein is high, e.g., during a uterine contraction.

After a short course in the *inferior vena cava,* the blood enters the right atrium of the heart. Because the inferior vena cava contains deoxygenated blood from the lower limbs, abdomen, and pelvis, the blood entering the right atrium is not so well oxygenated as that in the umbilical vein, but it is still well-oxygenated blood (pO$_2$, 25 to 28 mm. Hg). The blood from the inferior vena cava is largely directed by the lower border of the septum secundum, the *crista dividens,* through the *foramen ovale* into the left atrium (Fig. 14–40); here it mixes with a relatively small amount of deoxygenated blood returning from the lungs via the pulmonary veins. The fetal lungs extract oxygen from the blood instead of providing it. From

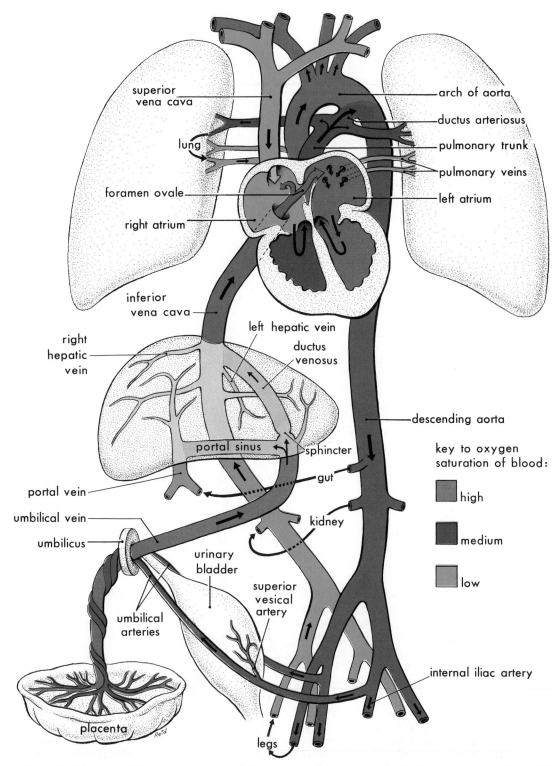

Figure 14–38 A simplified scheme of the fetal circulation. The colors indicate the oxygen saturation of the blood, and the arrows show the course of the fetal circulation. The organs are not drawn to scale. Observe that there are three shunts that permit most of the blood to bypass the liver and the lungs: (1) the *ductus venosus,* (2) the *foramen ovale,* and (3) the *ductus arteriosus.*

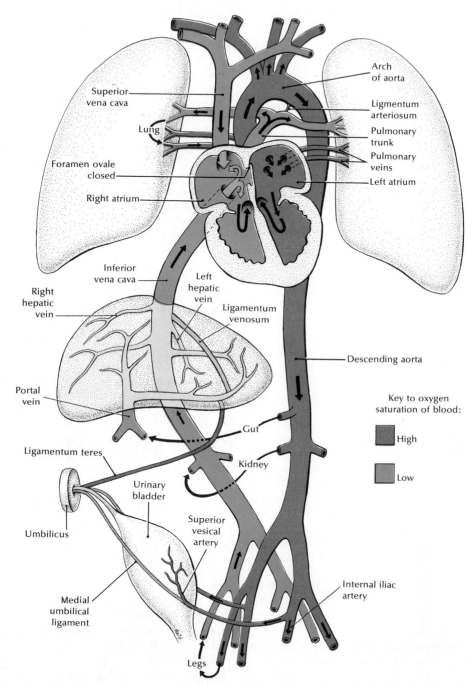

Figure 14–39 A simplified representation of the circulation after birth. The adult derivatives of the fetal vessels and structures that become nonfunctional at birth are also shown. The arrows indicate the course of the neonatal circulation. The organs are not drawn to scale. After birth, the three shunts that short-circuited the blood during fetal life cease to function, and the pulmonary and systemic circulations become separated.

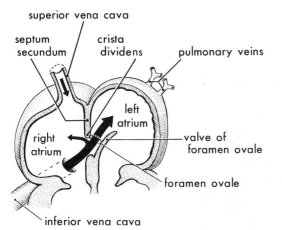

Figure 14–40 A schematic diagram illustrating how the crista dividens (lower edge of the septum secundum) separates the blood from the inferior vena cava into two streams. The larger stream passes through the foramen ovale into the left atrium, where it mixes with a small volume of deoxygenated blood from the pulmonary veins. The smaller stream remains in the right atrium and mixes with a large amount of deoxygenated blood from the superior vena cava and the coronary sinus.

the left atrium, the blood passes into the left ventricle and leaves via the ascending aorta. Consequently, the arteries to the heart, head and neck, and upper limbs receive well-oxygenated blood.

A small amount of oxygenated blood from the inferior vena cava is diverted by the *crista dividens* and remains in the right atrium (Fig. 14–40). This blood mixes with deoxygenated blood from the superior vena cava and coronary sinus and passes into the right ventricle. The blood leaves via the pulmonary trunk, and most of it passes through the *ductus arteriosus* into the aorta. Because of the high pulmonary vascular resistance in fetal life, pulmonary blood flow is low. Only 5 to 10 per cent of the cardiac output goes to the lungs; this is adequate because they are not functioning and so require little blood.

Forty to 50 per cent of the mixed blood in the descending aorta (about 58 per cent saturated with oxygen) passes into the umbilical arteries and is returned to the placenta for reoxygenation; the remainder circulates through the inferior half of the body.

Changes in the Cardiovascular System at Birth (Fig. 14–39). Important circulatory adjustments occur almost immediately at birth, when the circulation of fetal blood through the placenta ceases and the lungs begin to function. The three shunts that permitted much of the blood to bypass the liver and the lungs cease to function.

The foramen ovale, the ductus arteriosus, the ductus venosus, and the umbilical vessels are no longer needed. The sphincter in the ductus venosus constricts so that all blood reaching the liver must pass through the hepatic sinusoids. Occlusion of the placental circulation causes an immediate fall of blood pressure in the inferior vena cava and the right atrium. *Aeration of the lungs* is associated with a dramatic fall in pulmonary vascular resistance, a marked increase in pulmonary blood flow, and a progessive thinning of the walls of the pulmonary arteries. The thinning of the walls of these arteries results mainly from stretching as the lungs increase in size with the first few breaths. As a result of this increased pulmonary blood flow, the pressure in the left atrium is raised above that in the right atrium. This increased left atrial pressure closes the foramen ovale by pressing the valve of the foramen ovale, formed by the septum primum, against the septum secundum.

The right ventricular wall is thicker than the left ventricular wall in fetuses and in newborn infants because the right ventricle has been doing more work. Consequently, *right ventricular hypertrophy* is an inaccurate description of the otherwise normal heart of a newborn. By the end of the first month, the left ventricular wall is thicker than the right ventricular wall because the left ventricle is now working harder than the right one. Furthermore, the right ventricular wall becomes thinner, owing to the *atrophy* associated with its lighter workload.

The *ductus arteriosus* constricts at birth, but there is often a small shunt of blood from the aorta to the left pulmonary artery for a few days. The ductus arteriosus becomes functionally closed within 10 to 15 hours after birth, but in premature infants and in those with persistent hypoxia, it may remain open much longer (Page et al., 1981). Closure of the ductus arteriosus appears to be mediated by *bradykinin,* a substance released from the lungs during their initial inflation (Melmon et al., 1968). The action of this substance appears dependent on the high oxygen content of the aortic blood which normally results from ventilation of the lungs at birth.

The patency of the ductus arteriosus before birth is under the control of locally produced *prostaglandins* that act on the muscle cells in the wall of the ductus arteriosus, causing them to relax (Page et al., 1981). Inhibitors of prostaglandin synthesis, such as *indomethacin,* can pharmacologically cause constriction of a patent ductus arteriosus in premature infants (Vaughan et al., 1979).

The *umbilical arteries* constrict at birth, preventing loss of the infant's blood. If the cord is not tied for a minute or so, blood flow through the umbilical vein continues, transferring some blood from the placenta to the infant.

The change from the fetal to the adult pattern of circulation is not a sudden occurrence. It takes place over a period of days and weeks. During the transitional stage, there may be a right-to-left flow through the foramen ovale, and the ductus arteriosus usually remains patent for two or three months. *The closure of the fetal vessels and the foramen ovale is initially a functional change;* later there is anatomical closure resulting from proliferation of endothelial and fibrous tissues.

Adult Derivatives of Fetal Vessels and Structures Associated with the Fetal Circulation (Figs. 14–39, 14–41 and 14–42). Because of the changes in the cardiovascular system at birth, certain vessels and structures are no longer required. Their adult derivatives are:

1. The intra-abdominal portion of *the umbilical vein becomes the ligamentum teres*, which passes from the umbilicus to the porta hepatis where it attaches to the left branch of the portal vein (Fig. 14–41). The umbilical vein remains patent for some time and may be used for *exchange transfusions* during early infancy. This is done to prevent brain damage and death of anemic erythroblastotic infants. Most of the infant's blood is replaced with donor blood. See Allen and Umansky (1972) for a discussion of exchange transfusions and their use in the postnatal management of erythroblastosis fetalis. The lumen of the umbilical vein usually does not disappear completely, and so the ligamentum teres usually can be cannulated in adults, if necessary, for the injection of contrast medium or chemotherapeutic drugs (Kessler and Zimman, 1966). The potential patency of this vein may also be of functional significance in hepatic cirrhosis (Butler, 1952a).

2. *The ductus venosus becomes the ligamentum venosum,* which passes through the liver from the left branch of the portal vein to the

inferior vena cava to which it is attached (Fig. 14–41).

3. Most of the intra-abdominal portions of the *umbilical arteries become the medial umbilical ligaments;* the proximal parts of these vessels persist as the *superior vesical arteries.*

4. The *foramen ovale* normally closes functionally at birth. Later, anatomical closure results from tissue proliferation and adhesion of the septum primum (the valve of the foramen ovale) to the left margin of the septum secundum. Thus, *the septum primum forms the floor of the fossa ovalis* (Fig. 14–42). The lower edge of the septum secundum forms a rounded fold, the *limbus fossae ovalis (anulus ovalis),* which marks the former cranial boundary of the foramen ovale. In up to 25 per cent of persons, complete anatomical closure fails to occur. See previous description of probe patent foramen ovale and Figure 14–23B.

5. *The ductus arteriosus becomes the ligamentum arteriosum,* which passes from the left pulmonary artery to the arch of the aorta. Anatomical closure of the ductus normally occurs by the end of the third month.

Patent Ductus Arteriosus (Fig. 14–43). This common malformation is two to three times more frequent in females than in males. The reason for this is not known. *Patent ductus arteriosus is the most common congenital malformation associated with maternal rubella infection during early pregnancy* (see Chapter 8), but the mode of action of this teratogen (rubella virus) is unclear.

The embryological basis of patent ductus arteriosus is failure of the ductus arteriosus to involute after birth and form the ligamentum arteriosum. Failure of contraction of the muscular wall of the ductus arteriosus is the primary cause of patency (Gray and Skandalakis, 1972). There is some evidence that the low oxygen content of the blood in *neonatal respiratory distress* can adversely affect closure of the ductus arteriosus (Record and McKeown, 1953). There are also reports indicating that patent ductus arteriosus is found more often in populations living at high altitudes than in those living near sea level (Warkany, 1971). Large differences between aortic and pulmonary pressures can cause heavy flow through the ductus arteriosus, thereby preventing normal constriction. Such pressure differences may be caused by preductal coarctation of the aorta (Fig. 14–34C), transposition of the great vessels, or pulmonary stenosis and atresia (Fig. 14–30).

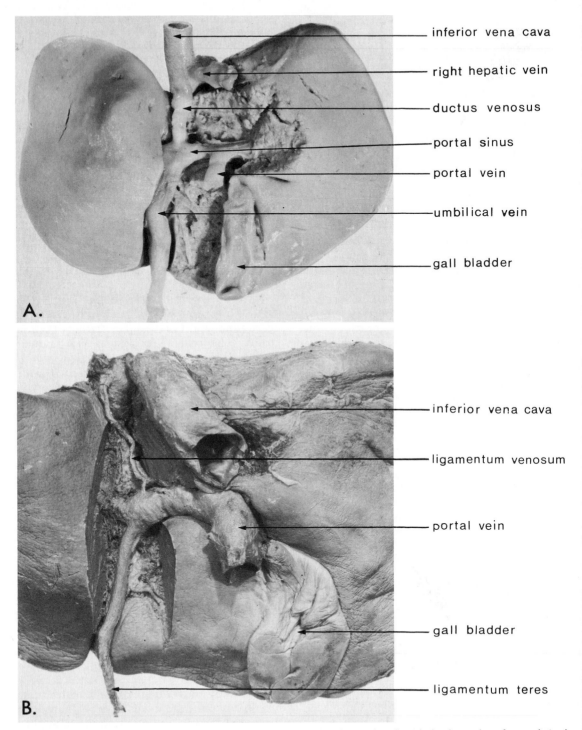

inferior vena cava

right hepatic vein

ductus venosus

portal sinus

portal vein

umbilical vein

gall bladder

inferior vena cava

ligamentum venosum

portal vein

gall bladder

ligamentum teres

A.

B.

Figure 14–41 Photographs of posterior views of dissected livers showing their visceral surfaces. Actual size. *A,* Fetal liver. *B,* Adult liver.

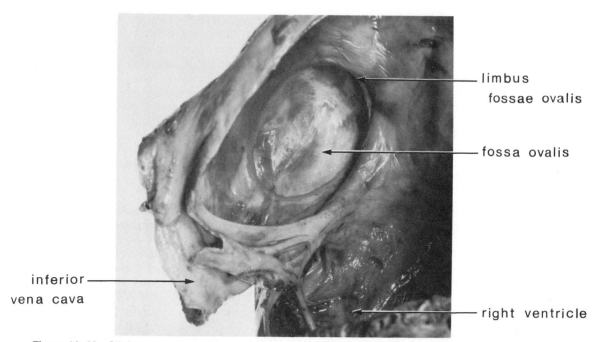

limbus
fossae ovalis

fossa ovalis

inferior
vena cava

right ventricle

Figure 14–42 Slightly retouched photograph of the right atrial aspect of an adult interatrial septum showing the fossa ovalis and the limbus fossae ovalis (×1.5). The floor of the fossa ovalis is derived from the septum primum, whereas the limbus fossae ovalis represents the free edge of the septum secundum.

THE LYMPHATIC SYSTEM

The lymph vascular system begins to develop at the end of the fifth week, about two weeks later than the cardiovascular system. Lymphatic vessels develop in a manner similar to that previously described for blood vessels (see Chapter 4), and make connections with the venous system. Another current view is that the earliest lymph vessels arise as capillary offshoots from the endothelium of veins (Yoffey and Courtice, 1970).

There are *six primary lymph sacs* (Fig. 14–44*A*): (1) two *jugular lymph sacs* near the junction of the subclavian veins with the anterior cardinal veins (the future internal jugular veins); (2) two *iliac lymph sacs* near the junction of the iliac veins with the posterior cardinal veins; (3) one *retroperitoneal*

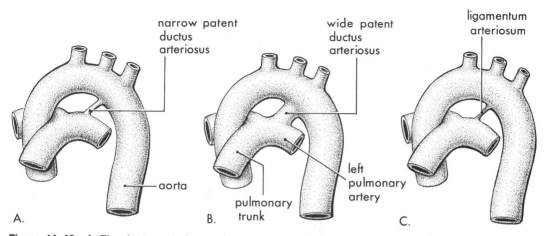

narrow patent
ductus
arteriosus

wide patent
ductus
arteriosus

ligamentum
arteriosum

aorta

left
pulmonary
artery

pulmonary
trunk

A.

B.

C.

Figure 14–43 *A,* The ductus arteriosus of a newborn infant. The ductus is normally patent for about two weeks after birth. *B,* Abnormal patent ductus arteriosus in a six-month-old infant. The ductus is nearly the same size as the left pulmonary artery. *C,* The ligamentum arteriosum, normal remnant of the ductus arteriosus, in a six-month-old infant.

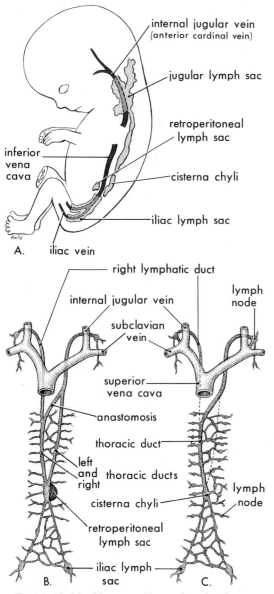

internal jugular vein
(anterior cardinal vein)

jugular lymph sac

retroperitoneal
lymph sac

inferior
vena
cava

cisterna chyli

iliac lymph sac

A. iliac vein

right lymphatic duct

lymph
node

internal jugular vein

subclavian
vein

superior
vena cava

anastomosis

thoracic duct

left
and thoracic ducts
right

cisterna chyli

lymph
node

retroperitoneal
lymph sac

iliac lymph
B.   sac   C.

Figure 14–44 Diagrams illustrating development of the lymphatic system in the human embryo. *A*, Left side of a seven-week embryo, showing the primary lymph sacs. *B*, Ventral view of the lymphatic system at about nine weeks, showing the paired thoracic ducts. *C*, Later stage, illustrating formation of the adult thoracic duct and the right lymphatic duct.

to the gut from the retroperitoneal lymph sac and the cisterna chyli.

Two large channels (right and left thoracic ducts) connect the jugular lymph sacs with the cisterna chyli; soon an anastomosis forms between these channels (Fig. 14–44*B*). The adult *thoracic duct* develops from (1) the caudal part of the right thoracic duct; (2) the anastomosis; and (3) the cranial part of the left thoracic duct. Because the primitive thoracic ducts are paired, there are many variations in the origin, course, and termination of the adult thoracic duct (Gray and Skandalakis, 1972). The *right lymphatic duct* is derived from the cranial part of the right thoracic duct (Fig. 14–44*C*). The thoracic and right lymphatic ducts become connected with the venous system at the angle between the internal jugular and subclavian veins (Fig. 14–44*C*). The superior portion of the embryonic cisterna chyli persists in the adult.

Development of Lymph Nodes. Except for the superior portion of the cisterna chyli, the lymph sacs become transformed into groups of lymph nodes during the early fetal period. Surrounding mesenchymal cells invade each lymph sac and break up the original cavity into a network of lymphatic channels, the *lymph sinuses*. Some mesenchymal cells give rise to the capsule and connective tissue framework of the lymph node. The *lymphocytes* that appear in lymph nodes before birth are derived from the *thymus gland,* a derivative of the third pair of pharyngeal pouches (see Fig. 10–7). Small lymphocytes leave the thymus and circulate to the other lymphoid organs. Later, some mesenchymal cells in the nodes differentiate into lymphocytes. Lymph nodules do not appear in the lymph nodes until just before and/or after birth. Lymph nodes also develop, as described, along the course of major lymph vessels.

The *spleen* develops from an aggregation of mesenchymal cells in the dorsal mesentery of the stomach (see Chapter 12). The *palatine tonsils* develop from the second pair of pharyngeal pouches (see Chapter 10). The *tubal tonsils* develop from aggregations of lymph nodules around the pharyngeal openings of the auditory (eustachian) tubes. The *pharyngeal tonsils* ("adenoids") develop from an aggregation of lymph nodules in the wall of the nasopharynx. The *lingual tonsil* develops from an aggregation of lymph nodules in the root of the tongue. Lymph nodules also develop in the mucosa of the respiratory and digestive systems.

lymph sac in the root of the mesentery on the posterior abdominal wall; (4) one *cisterna chyli* dorsal to the retroperitoneal lymph sac.

Lymphatic vessels grow out from these lymph sacs, principally along main veins, to the head, neck, and upper limbs from the jugular lymph sacs, to the lower trunk and lower limbs from the iliac lymph sacs, and

ABNORMALITIES OF THE LYMPHATIC SYSTEM

Congenital malformations of the lymphatic system are rare. There may be diffuse swelling of a part of the body or of an extremity *(congenital lymphedema)*. This condition may result from dilatation of primitive lymphatic channels or from a congenital hypoplasia of lymphatic vessels. More rarely, diffuse cystic dilatation of lymphatic channels involves widespread portions of the body.

Cystic Lymphangioma or Cystic Hygroma. These large swellings usually appear in the inferior third of the neck and consist of large single or multilocular fluid-filled cavities. Hygromas may be present at birth, but they often become evident during infancy. Most hygromas are probably derived from abnormal transformation of the jugular lymph sacs. Hygromas are believed to arise from (1) portions of the jugular lymph sac that are pinched off, or (2) lymphatic spaces that fail to establish connections with the main lymphatic channels.

SUMMARY

The cardiovascular system begins to develop during the third week from splanchnic mesoderm in the cardiogenic area. Paired endocardial heart tubes form and fuse into a single heart tube, the *primitive heart.* By the end of the third week, a functional cardiovascular system is present. As the heart tube grows, it bends to the right and soon acquires the general external appearance of the adult heart. The heart becomes partitioned into four chambers between the fourth and seventh weeks.

Three systems of paired veins drain into the primitive heart: (1) the *vitelline system,* which becomes the *portal system;* (2) the *cardinal veins,* which form the *caval system;* and (3) the *umbilical system,* which involutes after birth, except for a few adult derivatives.

As the branchial arches form during the fourth and fifth weeks they are penetrated by arteries, referred to as *aortic arches,* which arise from the *aortic sac.* During the sixth to eighth weeks, this primitive aortic arch pattern is transformed into the adult arterial arrangement of the carotid, subclavian, and pulmonary arteries.

The critical period of heart development is from about day 20 to day 50. There are numerous critical events during cardiac development, and deviation from the normal pattern at any one time may produce one or more cardiac defects. Because partitioning of the heart is complex, *defects of the cardiac septa are relatively common,* particularly ventricular septal defects. Some congenital malformations result from abnormal transformation of the aortic arches into the adult arterial pattern.

Because the lungs are nonfunctional during prenatal life, the fetal cardiovascular system is structurally designed so that blood is oxygenated in the placenta and largely bypasses the lungs. The modifications that establish the postnatal circulatory pattern at birth are not abrupt, but extend into infancy. Failure of the normal changes in the circulatory system to occur at birth results in two of the most common congenital abnormalities of the heart and great vessels: *patent foramen ovale* and *patent ductus arteriosus.*

The *lymphatic system* begins to develop during the fifth week in close association with the venous system. Six primary *lymph sacs* develop which later become interconnected by lymph vessels. Lymph nodes develop along the meshwork of lymphatic vessels; lymph nodules do not appear until just before and/or after birth. Sometimes a part of the jugular lymph sac becomes pinched off and may give rise to a tumor-like mass of dilated lymphatic vessels, called a *cystic hygroma.*

SUGGESTIONS FOR ADDITIONAL READING

Anderson, R. H., and Ashley, G. T.: Growth and development of the cardiovascular system: (a) anatomical development; in J. A. Davis and J. Dobbing (Eds.): *Scientific Foundations of Paediatrics.* Philadelphia, W. B. Saunders Co., 1974, pp. 165–198. A comprehensive discussion of early development of the cardiovascular system and a well-illustrated account of congenital abnormalities of the heart and great vessels.

Shinebourne, E. A.: Growth and development of the cardiovascular system: (b) functional development; in J. A. Davis and J. Dobbing (Eds.): *Scientific Foundations of Paediatrics,* Philadelphia, W. B. Saunders Co., 1974, pp. 198–213. A very good account by a consultant pediatric cardiologist of the changes that take place in the circulation at birth. Thorough discussions of the dramatic alterations in systemic and pulmonary vascular resistance as well as closure of the ductus arteriosus, foramen ovale, and ductus venosus.

CLINICALLY ORIENTED PROBLEMS

1. What is the *most common type of congenital cardiovascular malformation?* How common is it?

2. A female infant was born normally after a pregnancy complicated by a *rubella infection* (German measles) during the first trimester of pregnancy. She had *congenital cataracts* and *congenital heart disease.* A radiograph of the infant's chest at three weeks showed generalized *cardiac enlargement* with some increase in pulmonary vascularity. What congenital cardiovascular abnormality is commonly associated with *maternal rubella* during early pregnancy? What probably caused the cardiac enlargement?

3. In *tetralogy of Fallot* there are four cardiac abnormalities. Name them. What radiographic technique might be used to confirm a tentative diagnosis of this type of congenital heart disease?

4. A male infant was born after a full-term normal pregnancy. *Severe generalized cyanosis* was observed on the first day. A chest film revealed a *slightly enlarged heart* with a narrow base and increased pulmonary vascularity. A clinical diagnosis of *transposition of the great vessels* was made. What radiographic technique would likely be used to verify the diagnosis? What would this technique reveal in the present case? How was the infant able to survive after birth with this severe congenital abnormality of the great arteries?

5. During an autopsy on a 72-year-old man who had died following *chronic heart failure,* it was observed that the heart was very large and that the pulmonary artery and its main branches were dilated. Opening the heart revealed a large *atrial septal defect.* What type of ASD was probably present? Where would the defect likely be located? Explain why the pulmonary artery and its main branches were dilated.

The answers to these questions are given at the back of the book.

REFERENCES

Allen, F. H., Jr., and Umansky, I.: Erythroblastosis fetalis; *in* D. E. Reid, K. J. Ryan, and K. Benirschke: *Principles and Management of Human Reproduction.* Philadelphia, W. B. Saunders Co., 1972, pp. 811–832.

Anderson, R. C.: Causative factors underlying congenital heart malformations. *Pediatrics 14*:143, 1954.

Anderson, R. H., and Taylor, I. M.: Development of atrioventricular specialized tissue in human heart. *Br. Heart J. 34*:1205, 1972.

Arey, L. B.: *Developmental Anatomy: A Textbook and Laboratory Manual of Embryology.* Revised 7th ed. Philadelphia, W. B. Saunders Co., 1974, pp. 342–395.

Barnard, C. N., and Schire, V.: *The Surgery of the Common Congenital Cardiac Malformations.* Hoeber Medical Division, Harper & Row, Publishers, 1968.

Barry, A.: The aortic arch derivatives in the human adult. *Anat. Rec. 111*:221, 1951.

Bremer, J. L.: *Congenital Anomalies of the Viscera: Their Embryological Basis.* Cambridge, Harvard University Press, 1957.

Butler, H.: Gastro-esophageal haemorrhage in hepatic cirrhosis. *Thorax 7*:159, 1952a.

Butler, H.: An abnormal disposition of the pulmonary venous drainage. *Thorax 7*:249, 1952b.

Campbell, M.: Place of maternal rubella in the aetiology of congenital heart disease. *Br. Med. J. 1*:691, 1961.

Campbell, M.: Natural history of atrial septal defect. *Br. Heart J. 32*:820, 1970.

Campbell, M. S., and Sack, D. A.: Anomalies of the heart and great vessels; *in* A. A. Pearson and R. W. Sauter: *The Development of the Cardiovascular System.* Portland, University of Oregon Medical School Printing Department, 1968.

Collet, R. W., and Edwards, J. E.: Persistent truncus arteriosus: a classification according to anatomic types. *Surg. Clin. North Am. 29*:1245, 1949.

Dawes, G. S.: Foetal blood gas homeostasis; *in* G. E. W. Wolstenholme and M. O'Connor: *Foetal Autonomy.* London, J. & A. Churchill, 1969, pp. 162–185.

Dickson, A. D.: The development of the ductus venosus in man and the goat. *J. Anat. 91*:358, 1957.

Duckworth, J. W. A.: Embryology of congenital heart disease; *in* J. D. Keith, R. D. Rowe, and P. Vlad (Eds.): *Heart Disease in Infancy and Childhood.* 3rd ed. New York, The Macmillan Company, 1978.

Edwards, J. E., Dry, T. J., Parker, R. L., Burchell, H. B., Wood, E. H., and Bulbulian, A. H.: *An Atlas of Congenital Anomalies of the Heart and Great Ves-*

sels. Springfield, Ill., Charles C Thomas, Publisher, 1954.

Eidemiller, L. R., and Keane, J. M.: Development of the heart; *in* A. A. Pearson (Ed.): *The Development of the Cardiovascular System*. Portland, University of Oregon Medical School Printing Department, 1968, pp. 1–34.

Gray, S. W., and Skandalakis, J. E.: *Embryology for Surgeons. The Embryological Basis for the Treatment of Congenital Defects*. Philadelphia, W. B. Saunders Co., 1972, pp. 727–793.

Gunther, M.: The transfer of blood between baby and placenta in the minutes after birth. *Lancet 252*:1277, 1957.

Hamilton, W. J., Boyd, J. D., and Mossman, H. N.: *Human Embryology. Prenatal Development of Form and Function*. 4th ed. Cambridge, W. Heffer & Sons, Ltd., 1972.

Harley, H. R. S.: The sinus venosus type of interatrial septal defect. *Thorax 13*:12, 1958.

Hayek, H. von: Der functionelle Bau der Naberlaterien und des Ductus botalli. *Z. Anat. Entwicklungs-gesch. 105*:15, 1935.

Holmes, R. L.: Some features of the ductus arteriosus. *J. Anat. 92*:304, 1958.

Keith, J. D., Rowe, R. D., and Vlad, P.: *Heart Disease in Infancy and Childhood*. 3rd ed. New York, The Macmillan Company, 1978.

Kessler, R. E., and Zimman, D. S.: Umbilical vein angiography. *Radiology 87*:841, 1966.

Kleinman, C. S., Hobbins, C., Jaffe, C., Lynch, D. C., and Talner, N. S.: Echocardiographic studies of the human fetus: Prenatal diagnosis of congenital heart disease and cardiac dysrhythmias. *Pediatrics 65*:1059, 1980.

Krediet, P.: An hypothesis of the development of coarctation in man. *Acta Morphol. Neerlando Scand. 6*:207, 1965.

Lev, M.: *Autopsy Diagnosis of Congenitally Malformed Hearts*. Springfield, Ill., Charles C Thomas, Publisher, 1953.

Lind, J.: Normal perinatal circulation; *in* J. Emery (Ed.): The Anatomy of the Developing Lung. London, William Heinemann, Ltd., 1969, pp. 116–146.

Lucas, R. V., and Schmidt, R. E.: Anomalous venous connections, pulmonary and systemic; *in* A. J. Moss and F. H. Adams (Eds.): *Heart Disease in Infants, Children and Adolescents*. 2nd ed. Baltimore, The Williams & Wilkins Co., 1977.

Melmon, K. L., Cline, M. J., Hughes, T., and Nies, A. S.: Kinins: possible mediators of neonatal circulatory changes in man. *J. Clin. Invest. 47*:1295, 1968.

Mitchell, S. C.: The ductus arteriosus in the neonatal period. *J. Pediatr. 51*:12, 1957.

Mitchell, S. C., Horones, S. B., and Berendes, H. W.: Congenital heart disease in 56,109 births; incidence and natural history. *Circulation 43*:323, 1971.

Moss, A. J., and Adams, F. H. (Eds.): *Heart Diseases in Infants, Children and Adolescents*. 2nd ed. Baltimore, The Williams & Wilkins Co., 1977.

Moss, A. J., Emmanouilides, G. C., Adams, F. H., and Chuang, K.: Response of ductus arteriosus and pulmonary and systemic arterial pressure to changes in oxygen environment in newborn infants. *Pediatrics 33*:937, 1964.

Neill, C. A.: Development of the pulmonary veins. *Pediatrics 18*:880, 1956.

Page, E. W., Villee, C. A., and Villee, D. B.: *Human Reproduction. Essentials of Reproductive and Peri-*

natal Medicine. 3rd ed., Philadelphia, W. B. Saunders Co., 1981.

Patten, B. M.: *Human Embryology*. 3rd ed. New York, McGraw-Hill Book Co., 1968.

Paul, M. H.: Transposition of the great arteries; *in* A. J. Moss and F. H. Adams (Eds.): *Heart Disease in Infants, Children and Adolescents*. 2nd ed. Baltimore, The Williams & Wilkins Co., 1977.

Pearson, A. A., and Sauter, R. W.: Observation on the innervation of the umbilical vessels in human embryos and fetuses. *Anat. Rec. 160*:406, 1968.

Pearson, A. A., and Sauter, R. W.: *The Development of the Cardiovascular System*. Portland, University of Oregon Medical School Printing Department, 1968.

Peltonen, T., and Hirvonen, L.: Experimental studies on the fetal and neonatal circulation. *Acta Pediatr. Scand. Suppl.* 161:1965.

Record, R. G., and McKeown, T.: Observations relating to the aetiology of patent ductus arteriosus. *Br. Heart J.* 15:376, 1953.

Rosenquist, G. C., and Bergsma, D. (Eds.): *Morphogenesis and Malformation of the Cardiovascular System*. The National Foundation — March of Dimes. Birth Defects: Original Article Series. New York, Alan R. Liss Inc., Vol. XIV, No. 7, 1978.

Rubin, A. (Ed.): *Handbook of Congenital Malformations*. Philadelphia, W. B. Saunders Co., 1967, pp. 19–42.

Scopes, J. W.: Fetal circulation; *in* E. E. Philipp, J. Barnes, and M. Newton (Eds.): *Scientific Foundations of Obstetrics and Gynecology*. London, William Heinemann, Ltd., 1970, pp. 270–276.

Severn, C. B.: A morphological study of the development of the human liver. II. Establishment of liver parenchyma, extrahepatic ducts, and associated venous channels. *Am. J. Anat. 133*:85, 1972.

Taussig, H. B.: Cardiac abnormalities; *in* M. Fishbein (Ed.): *Birth Defects*. Philadelphia, J. B. Lippincott Co., 1963.

van Mierop, L. H. S.: Transposition of the great arteries. I. Clarification of further confusion. *Am. J. Cardiol. 28*:735, 1971.

van Praagh, R.: Malposition of the heart; *in* A. J. Moss and F. H. Adams (Eds.): *Heart Disease in Infants, Children and Adolescents*. 2nd ed. Baltimore, The Williams & Wilkins Co., 1977.

van Praagh, R.: Transposition of the great arteries. II. Transposition clarified. *Am. J. Cardiol. 28*:739, 1971.

van Praagh, R., van Praagh, S., Nebasar, R. A., Muster, A. J., Sinha, S. N., and Paul, M. H.: Tetralogy of Fallot: Underdevelopment of the pulmonary infundibulum and its sequelae. *Am. J. Cardiol. 26*:25, 1970.

Vaughan, V. C., McKay, R. J., and Behrman, R. E. (Eds.): *Nelson Textbook of Pediatrics*. 11th ed. Philadelphia, W. B. Saunders Co., 1979, pp. 546, 1980.

Warkany, J.: *Congenital Malformations: Notes and Comments*. Chicago, Year Book Medical Publishers, Inc., 1971.

Yoffey, J. M., and Courtice, F. C.: *Lymphatics, Lymph and Lymphomyeloid Complex*. London, Academic Press, Inc., 1970.

Young, I. M.: On being born; *in* R. Passmore and J. S. Robson (Eds.): *A Companion to Medical Studies*. Vol. I. *Anatomy, Biochemistry, Physiology and Related Subjects*. Oxford, Blackwell Scientific Publications, 1968, pp. 39.1–39.11.

15

THE ARTICULAR AND SKELETAL SYSTEMS

The articular and skeletal systems develop from mesoderm, the formation of which is described in Chapter 4. With the formation of the notochord and the neural tube, the *intraembryonic mesoderm* lateral to these structures thickens to form two longitudinal columns of *paraxial mesoderm*. The somites arise from the paraxial mesoderm, beginning during stage 9 (about 20 days). Externally, they appear as pairs of bead-like elevations along the dorsolateral surface of the embryo (see Fig. 5–9). The development and early differentiation of the *somites* are illustrated in Figure 15–1. Each somite becomes differentiated into a ventromedial part called the *sclerotome* and a dorsolateral part called the *dermomyotome*. Initially, the somites are composed of *compact aggregates of mesenchymal cells*.

Shortly after the somites form, their ventral and medial walls lose their organization and break up into sclerotomal cells, which are collectively referred to as the *sclerotomes* (Fig. 15–1). Sclerotomal cells soon surround the notochord and the neural tube, and form the primordia of the *vertebrae* and the ribs. Although it has been commonly stated that the sclerotomal cells migrate medially to surround the notochord, recent investigations indicate that these cells probably do not migrate medially, but their positions change relative to surrounding structures because of shifts brought about by the growth of these structures. (Blechschmidt and Gasser, 1978, and Gasser, 1979).

Cells of the sclerotomes give rise to the vertebral column and the ligaments associated with it. The *dermomyotome* (all but the sclerotome of a somite), gives rise to the dermis of the skin and to the dorsal musculature.

Mesodermal cells give rise to a loosely organized tissue known as *mesenchyme* (embryonic connective tissue). Mesenchymal cells have the ability to differentiate in many different ways, e.g., into fibroblasts, chon-

droblasts, or osteoblasts. In addition to mesenchyme of somite origin, there is mesenchyme that develops from the splanchnic and somatic mesoderms. Some mesenchyme in the head region is derived from neuroectoderm, e.g., *neural crest cells,* which give rise to bones of branchial arch origin (see Chapter 10).

DEVELOPMENT OF BONE AND CARTILAGE

Most bones first appear as condensations of mesenchymal cells that form hyaline cartilage models that become ossified by endochondral ossification. Some bones develop in mesenchyme by intramembranous bone formation.

Histogenesis of Cartilage. Cartilage develops from mesenchyme and first appears in embryos of about five weeks. In areas where cartilage is to develop, the mesenchyme condenses and the cells proliferate and become rounded. Subsequently, collagenous fibers and/or elastic fibers are deposited in the intercellular substance, or *matrix*. The *chondroblasts* form the collagenous fibrils and the ground substance of the matrix. Three types of cartilage (hyaline, fibrocartilage, and elastic) are distinguished according to the type of matrix formed. Of these, *hyaline cartilage is the most widely distributed type.*

Histogenesis of Bone. Bone always develops by transformation of pre-existing connective tissue; it develops in two types of connective tissue, mesenchyme and cartilage. Like cartilage, bone consists of cells and an organic intercellular substance called its *matrix,* which consists of collagen fibrils embedded in an amorphous component.

Intramembranous Ossification (Fig. 15–2). This type of bone formation occurs in mesenchyme. Because the mesenchyme

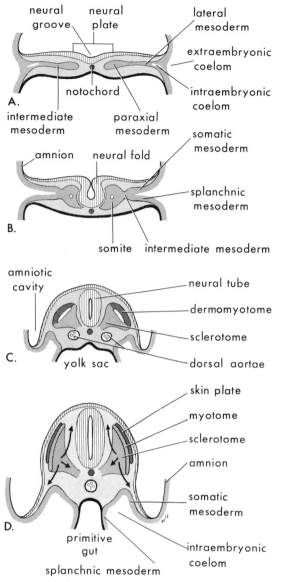

Figure 15–1 Transverse sections through embryos of various ages, illustrating the formation and early differentiation of somites. *A,* Presomite embryo during stage 8 (about 18 days), showing the paraxial mesoderm from which the somites are derived. *B,* Embryo during stage 10 (about 22 days). *C,* Embryo during stage 12 (about 26 days). The dermomyotome region of the somite gives rise to a myotome and a skin plate. *D,* Embryo during stage 13 (about 28 days). The arrows indicate apparent migration of cells from the sclerotome regions of the somites.

forms a layer, it was at first regarded as a membrane. This explains why ossification occurring in mesenchyme was called intramembranous ossification. The mesenchyme condenses and becomes highly vascular; some cells differentiate into *osteoblasts*

(bone-forming cells) and begin to deposit matrix, or intercellular substance, called *osteoid* (bone-like) *tissue,* or prebone. The osteoblasts are almost completely separated from each other, contact being maintained by a few tiny processes only. Calcium phosphate is deposited in the osteoid tissue. By some yet unknown mechanism, this tissue is organized into bone. Bone osteoblasts become trapped in the matrix and become *osteocytes.*

The new bone at first has no organized pattern, being formed in spicules. The spicules soon become organized and coalesce into lamellae (layers). Concentric lamellae form around blood vessels, forming *Haversian systems.* Some osteoblasts remain at the periphery of the developing bone and continue to lay down layers, forming plates of compact bone on the surfaces. Between the surface plates, the intervening bone remains spiculated or spongy. This spongy environment is somewhat accentuated by the action of cells of different origin, called *osteoclasts,* which absorb bone. In the interstices of spongy bone, the mesenchyme differentiates into bone marrow. During fetal and postnatal life, there is continual remodeling of bone by simultaneous action of osteoclasts and os-

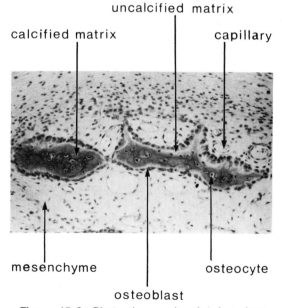

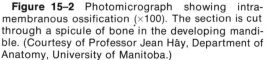

Figure 15–2 Photomicrograph showing intramembranous ossification (×100). The section is cut through a spicule of bone in the developing mandible. (Courtesy of Professor Jean Hay, Department of Anatomy, University of Manitoba.)

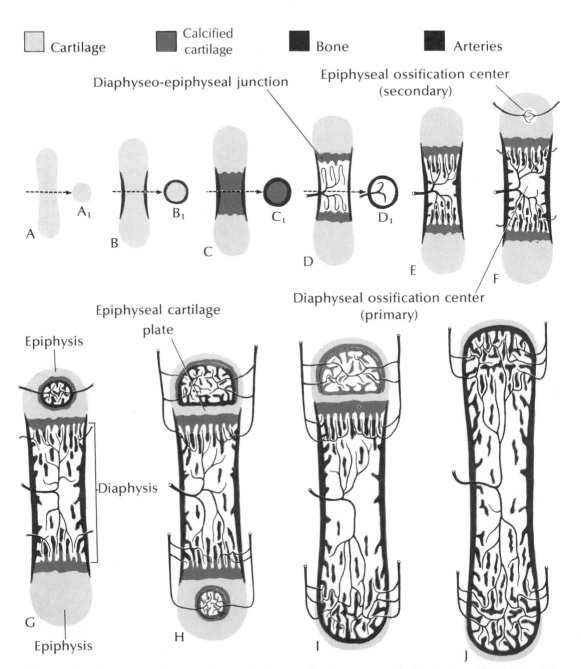

Figure 15–3 Schematic diagrams illustrating intracartilaginous, or endochondral, ossification and the development of a typical long bone. *A* to *J* are longitudinal sections, and *A*₁ to *D*₁ are cross sections at the levels indicated. *A,* Cartilage model of the bone. *B,* A subperiosteal ring of bone appears. *C,* Cartilage begins to calcify. *D,* Vascular mesenchyme enters the calcified cartilage. *E,* At each diaphyseo-epiphyseal junction, there is a zone of ossification. *F,* Blood vessels and mesenchyme enter the superior epiphyseal cartilage. *G,* The epiphyseal ossification center grows. *H,* A similar center develops in the inferior epiphyseal cartilage. *I,* The inferior epiphyseal plate is ossified. *J,* The superior epiphyseal plate ossifies, forming a continuous bone marrow cavity. When the epiphyseal plates ossify, the bone can no longer grow in length. (Modified from Bloom, W., and Fawcett, D. W.: *A Textbook of Histology.* 10th ed. Philadelphia, W. B. Saunders Co., 1975.)

teoblasts, but the activity of the osteoblasts slows down in females after the menopause.

Ossification begins at the end of the embryonic period and thereafter makes demands on the maternal supply of calcium and phosphorus. Pregnant women are advised to maintain adequate intake of these elements in order to preserve healthy bones and teeth.

Intracartilaginous (Endochondral) Ossification (Fig. 15–3). This type of bone formation occurs in pre-existing cartilaginous models. In a long bone, for example, the *primary ossification center* appears in the *diaphysis* (the portion of a long bone between the ends, or extremities), also called the *body,* or shaft. Here the cartilage cells increase in size, or hypertrophy, the matrix becomes calcified, and the cells die. Concurrently, a thin layer of bone is deposited under the *perichondrium* surrounding the diaphysis; thus, the perichondrium becomes the *periosteum.* Invasion of vascular connective tissue from the periosteum breaks up the cartilage. Some of these invading cells differentiate into hemopoietic cells of the bone marrow, and others differentiate into osteoblasts that deposit bone matrix on the spicules of calcified cartilage. This process continues toward the *epiphyses,* or ends of the bone. The spicules of bone are remodeled by the action of osteoclasts and osteoblasts.

Lengthening of long bones occurs at the *diaphyseal-epiphyseal junction.* The cartilage cells in this region proliferate by mitosis. Toward the diaphysis, the cartilage cells hypertrophy, and the matrix becomes calcified and broken up into spicules by vascular tissue from the marrow, or medullary cavity. Bone is deposited on these spicules; absorption of this bone keeps the spongy bone masses relatively constant in length and enlarges the marrow cavity. The region of bone formation at the center of the body of a long bone is called the *primary (diaphyseal) ossification center* (Fig. 15–3F). At birth, the diaphyses are largely ossified, but most *epiphyses* are still cartilaginous.

Most *secondary (epiphyseal) ossification centers* appear in the epiphyses during the first few years of the postnatal period. The epiphyseal cartilage cells hypertrophy, and there is invasion of vascular connective tissue. Ossification, as described previously, spreads in all directions, and only the articular cartilage and a transverse plate of cartilage, the *epiphyseal cartilage plate,* remain cartilaginous. Upon completion of growth, this epiphyseal plate is replaced by spongy bone, the epiphyses and the diaphysis are united, and further elongation of the bone does not occur. In most bones, the epiphyses have fused with the diaphysis by about the age of 20 years. Growth in diameter of the bone results from deposition of bone at the periosteum and from absorption on the medullary surface. The rate of deposition and absorption is balanced to regulate the thickness of the compact bone and the size of the marrow cavity. The internal reorganization of bone continues throughout life.

Development of irregular bones is similar to that of the epiphyses of long bones. Ossification begins centrally and spreads in all directions.

DEVELOPMENT OF JOINTS

The terms *articulation* and *joint* are used synonymously to refer to the structural arrangements that join two or more bones together at their place of meeting. Joints may be classified in several ways. Joints with little or no movement are classified according to the type of material holding the bones together, e.g., the bones involved in *fibrous joints* are joined by fibrous tissue (Fig. 15–4D).

Synovial Joints (Fig. 15–4B). The mesenchyme between the developing bones, known as the *interzonal mesenchyme,* differentiates as follows: (1) Peripherally, it gives rise to the capsular and other ligaments. (2) Centrally, it disappears and forms the joint cavity. (3) Where it lines the capsule and the articular surfaces, it forms the synovial membrane. Probably as a result of joint movement, the mesenchymal cells subsequently disappear from the surfaces of the articular cartilages. Examples of this type of joint are the knee and elbow joints.

Cartilaginous Joints (Fig. 15–4C). The interzonal mesenchyme between the developing bones differentiates into hyaline cartilage (e.g., the costochondral joints) or fibrocartilage (e.g., the symphysis pubis). Hyaline cartilage caps the bones participating in the joint.

Fibrous Joints (Fig. 15–4D). The interzonal mesenchyme between the developing bones differentiates into dense fibrous connective tissue, e.g., the sutures of the skull (Fig. 15–8).

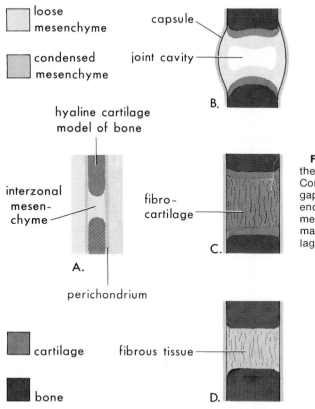

Figure 15–4 Schematic drawings illustrating the development of different types of joints. *A*, Condensed mesenchyme continues across the gap, or interzone, between the developing bones, enclosing some loose mesenchyme (the interzonal mesenchyme) between them. This primitive joint may differentiate into *B*, a synovial joint. *C*, a cartilaginous joint, or *D*, a fibrous joint.

THE AXIAL SKELETON

The axial skeleton includes the vertebral column, the twelve pairs of ribs, the sternum, and the skull.

DEVELOPMENT OF THE VERTEBRAL COLUMN

Precartilaginous Stage. During the fourth week, cells from the sclerotomes of the somites are found in three main areas (Figs. 15–1*D* and 15–5*A*):

1. *Surrounding the notochord.* In a frontal section through an embryo, the sclerotomes appear as paired condensations of mesenchymal cells that soon fuse in the midline around the notochord (Fig. 15–5*B*). Each sclerotome consists of loosely arranged cells cranially and densely packed cells caudally. Some of the densely packed cells move cranially opposite the center of the myotome and give rise to the *intervertebral disc* (Fig. 15–5*D*). The remaining densely packed cells fuse with the loosely arranged cells of the immediately caudal sclerotome to form the mesenchymal *centrum* of a vertebra. Thus, each centrum develops from two adjacent sclerotomes and so becomes an intersegmental structure. The nerves come to lie in close relationship to the intervertebral discs and the intersegmental (e.g., intercostal) arteries come to lie on each side of the vertebral bodies.

The *notochord* degenerates and disappears where it is surrounded by the developing vertebral body. Between the vertebrae, the notochord expands to form the gelatinous center of the intervertebral disc, called the *nucleus pulposus* (Fig. 15–5*D*). This nucleus is later surrounded by the circularly arranged fibers of the *anulus fibrosus*. These two structures together constitute the intervertebral disc. Remnants of the notochord (notochordal "rest" cells) may persist in any part of the axial skeleton and give rise to a *chordoma*. This slow-growing neoplasm occurs most frequently in the base of the skull and in the lumbosacral region.

2. *Surrounding the neural tube.* These mesenchymal cells will form the *vertebral arch*.

3. *In the body wall.* These mesenchymal cells will form the *costal processes*, which develop into ribs in the thoracic region.

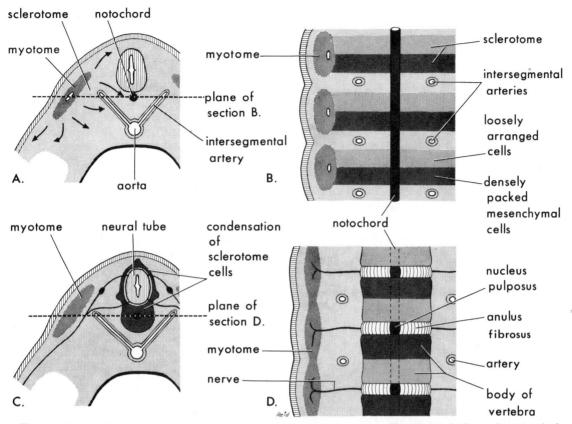

Figure 15–5 *A,* Partial transverse section through a four-week embryo. The arrows indicate the spread of mesenchymal cells from the sclerotome region of the somite on the right. *B,* Diagrammatic frontal section of this embryo, showing that the condensation of sclerotome cells around the notochord consists of a cranial area of loosely packed cells and a caudal area of densely packed cells. *C,*Partial transverse section through a five-week embryo showing the condensation of sclerotome cells around the notochord and the neural tube, forming a mesenchymal vertebra. *D,* Diagrammatic frontal section illustrating that the vertebral body forms from the cranial and caudal halves of two successive sclerotome masses. The intersegmental arteries now cross the bodies of the vertebrae, and the spinal nerves lie between the vertebrae. The notochord is degenerating except in the region of the intervertebral disc, where it forms the nucleus pulposus.

Chondrification (Fig. 15–6B). During the sixth week, chondrification centers appear in each mesenchymal vertebra. The two centers in each centrum fuse at the end of the embryonic period to form the cartilaginous centrum. Concomitantly, centers in the vertebral arches fuse with each other and with the centrum. The spinous and transverse processes are derived from extensions of chondrification centers in the vertebral arch.

Ossification of Typical Vertebrae. Ossification begins during the embryonic period and ends at about the twenty-fifth year.

Prenatal Period. At first, there are two *primary ossification centers,* ventral and dorsal (Fig. 15–6C), for the centrum. These differ from the chondrification centers that are side by side (Fig. 15–6B). The two primary ossification centers soon fuse to form one center. Hence, three primary centers appear by the end of the embryonic period: one in the centrum and one in each half of the vertebral arch (Fig. 15–6C). Ossification becomes evident in the vertebral arches around the eighth week. At birth, each vertebra consists of three bony parts connected by cartilage (Fig. 15–6D).

Postnatal Period. The halves of the vertebral arch usually fuse during the first three to five years. The laminae of the arches first unite in the lumbar region, and subsequent union progresses cranially. The vertebral arch articulates with the centrum at cartilaginous *neurocentral joints,* which permit the vertebral arches to grow as the spinal cord

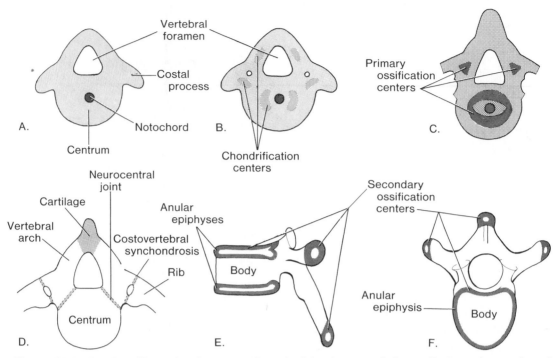

Figure 15–6 Drawings illustrating the stages of vertebral development. *A,* Precartilaginous (mesenchymal) vertebra at five weeks. *B,* Chondrification centers in a mesenchymal vertebra at six weeks. *C,* Primary ossification centers in a cartilaginous vertebra at seven weeks. *D,* A thoracic vertebra at birth, consisting of three bony parts. Note the cartilage between the halves of the vertebral (neural) arch and between the arch and the centrum (the neurocentral joint). *E* and *F,* Two views of a typical thoracic vertebra at puberty, showing the location of the secondary centers of ossification.

enlarges. These joints disappear when the vertebral arch fuses with the centrum during the third to sixth years.

During or shortly *after puberty,* five secondary centers appear: one for the tip of the spinous process, one for the tip of each transverse process, and two rim epiphyses (*anular epiphyses*), one on the superior and one on the inferior rim of the vertebral body (Fig. 15–6*E*).

The vertebral body is a composite of the superior and inferior anular epiphyses and the mass of bone between them. It includes the centrum, parts of the vertebral arch, and the facets for the heads of the ribs. The terms "body" and "centrum" are not, therefore, interchangeable. All secondary centers unite with the rest of the vertebra at about 25 years.

Ossification of Atypical Vertebrae. Exceptions to the typical ossification of vertebrae occur in the atlas, the axis, the seventh cervical vertebra, the lumbar vertebrae, the sacrum, and the coccyx. For details

of their ossification, consult Williams and Warwick (1980).

Variation in the Number of Vertebrae. About 95 per cent of normal people have 7 cervical, 12 thoracic, 5 lumbar, and 5 sacral vertebrae. About 3 per cent of people have one or two more vertebrae, and about 2 per cent have one less. To determine the number of vertebrae, it is necessary to examine the entire vertebral column, because an apparent extra (or absent) vertebra in one segment of the column may be compensated for by an absent (or extra) vertebra in an adjacent segment, e.g., 11 thoracic type vertebrae with 6 lumbar type vertebrae.

DEVELOPMENT OF RIBS

The ribs develop from the mesenchymal costal processes of the thoracic vertebrae (Fig. 15–6*A*). They become cartilaginous during the embryonic period and later ossify. The original union of the costal processes with the vertebra is replaced by a synovial joint (Fig. 15–6D).

DEVELOPMENT OF THE STERNUM

A pair of mesenchymal *sternal bands,* which at first are widely separated, develop ventrolaterally in the body wall and independent of the developing ribs. *Chondrification* occurs in these bands, forming two *sternal plates,* one on each side of the median plane. The cranial six or so costal cartilages become attached to them. The plates gradually fuse craniocaudally in the median plane to form cartilaginous models of the manubrium, the *sternebrae,* or segments of the body, and the xiphoid process. Centers of ossification appear craniocaudally before birth, except that for the xiphoid process, which appears during childhood.

DEVELOPMENT OF THE SKULL

The skull develops from mesenchyme around the developing brain. It consists of the *neurocranium,* a protective case for the brain, and the *viscerocranium,* the main skeleton of the jaws.

Cartilaginous Neurocranium, or Chondrocranium (Fig. 15–7). Initially, this consists of the cartilaginous base of the developing skull, which forms by fusion of several cartilages. Later, endochondral ossification of the chondrocranium forms the bones of the base of the skull.

The *parachordal cartilage,* or *basal plate,* forms around the cranial end of the notochord and fuses with the cartilages derived from the sclerotome regions of the occipital somites. This cartilaginous mass contributes to the base of the occipital bone; later, extensions grow around the cranial end of the spinal cord and form the boundaries of the *foramen magnum.* The *hypophyseal cartilages* form around the developing hypophysis, or pituitary gland, and fuse to form the body of the sphenoid bone. The *trabeculae cranii* fuse to form the body of the ethmoid bone. The *ala orbitalis* forms the lesser wing of the sphenoid bone. *Otic capsules* appear around the developing inner ears, or otic vesicles (see Chapter 19) and form the petrous and mastoid portions of the temporal bone. *Nasal capsules* develop around the nasal sacs (see Chapter 10) and contribute to the ethmoid bone.

Membranous Neurocranium (Figs. 15–7D and 15–8). Intramembranous ossification occurs in the mesenchyme at the sides and top of the brain, forming the cranial vault, or *calvaria.* During fetal life, the flat bones of the vault are separated by dense connective tissue membranes, or fibrous joints, called *sutures* (Fig. 15–8). The six large fibrous areas where several sutures meet, called *fontanelles,* are also present. The softness of the bones and their loose connections at the sutures enable the calvaria to undergo changes of shape, or *molding,* during birth. During this process, the frontal bone becomes flat, the occipital bone becomes drawn out, and one parietal bone slightly overrides, or overlaps, the other. The bony and cartilaginous base of the skull is deformed little, if any, during *parturition* (birth). Within a day or so after birth, the shape of the calvaria returns to normal.

Cartilaginous Viscerocranium. This consists of the cartilaginous skeleton of the first two pairs of *branchial arches* (see Chapter 10 and Fig. 10–4). After endochondral ossification, (1) the dorsal end of the *first arch cartilage* (Meckel's cartilage) forms two middle ear bones, the malleus and the incus (see Chapter 19); (2) the dorsal end of the *second arch cartilage* (Reichert's cartilage) forms the stapes of the middle ear and the styloid process of the temporal bone, and the ventral end ossifies to form the lesser cornu and superior part of the body of the hyoid bone.

The cartilages of the third, fourth, and sixth branchial arches are found only in the ventral portions of the arches. The third arch cartilages give rise to the greater cornua and inferior part of the body of the hyoid bone. The cartilages of the fourth and sixth branchial arches fuse to form the laryngeal cartilages, except for the epiglottis (see Fig. 10–4).

Membranous Viscerocranium. Intramembranous ossification occurs within the maxillary prominence of the first branchial arch and subsequently forms the maxillary, the zygomatic, and the squamous temporal bones. The squamous temporal bones later become part of the neurocranium. The mesenchyme in the mandibular prominence of this arch condenses around the first arch cartilage (Meckel's cartilage) and undergoes intramembranous ossification to form the mandible. Some endochondral ossification occurs at the center of the chin and at the mandibular condyle. This cartilage disap-

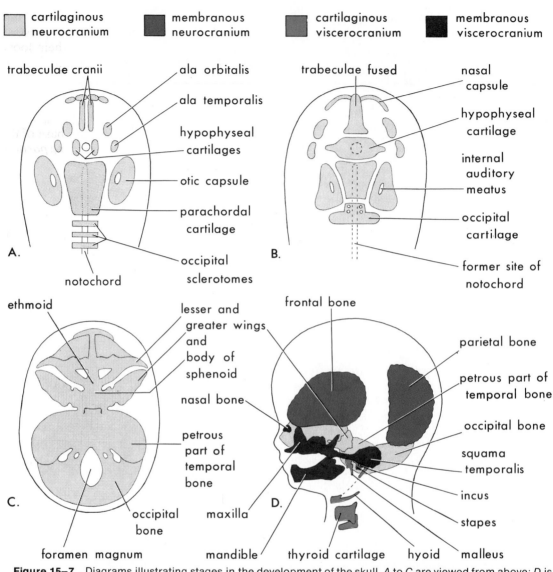

Figure 15–7 Diagrams illustrating stages in the development of the skull. *A* to *C* are viewed from above; *D* is a lateral view. *A*, Six weeks, showing the various cartilages that will fuse to form the chondrocranium. *B*, Seven weeks, after fusion of some of the paired cartilages. *C*, 12 weeks, showing the cartilaginous base of the skull or chondrocranium formed by the fusion of various cartilages. *D*, 20 weeks, indicating the derivation of the bones of the fetal skull.

pears ventral to the portion that forms the sphenomandibular ligament; thus, Meckel's cartilage does not form much of the adult mandible.

The Fetal and Newborn Skull (Fig. 15–8). The newborn skull, after recovering from molding (adaptation of the head to the birth canal), like the fetal skull, is rather round, and its bones are quite thin. The skull is large in proportion to the rest of the skeleton, and the face is relatively small compared with the calvaria. The small facial region results from the small size of the jaws, the virtual absence of paranasal air sinuses, and the general underdevelopment of the facial bones.

Postnatal Growth of the Skull. The fibrous sutures of the newborn calvaria (cranial vault) permit the skull to enlarge during infancy and childhood. The increase in the size of the calvaria is greatest during the first two years, the period of most rapid postnatal growth of the brain. A person's calvaria normally increases in capacity until

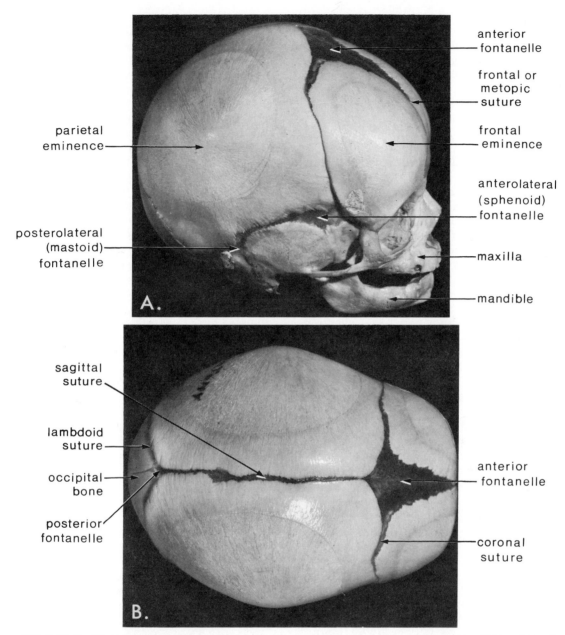

anterior fontanelle

frontal or metopic suture

frontal eminence

anterolateral (sphenoid) fontanelle

maxilla

mandible

parietal eminence

posterolateral (mastoid) fontanelle

A.

sagittal suture

lambdoid suture

occipital bone

posterior fontanelle

anterior fontanelle

coronal suture

B.

Figure 15–8 Photographs of a fetal skull showing the fontanelles, the bones, and the connecting sutures. *A,* Lateral view. *B,* Superior view. The posterior and anterolateral fontanelles disappear by growth of surrounding bones, within two or three months after birth, but they remain as junctions, or sutures, for several years. The posterolateral fontanelles disappear in a similar manner by the end of the first year, and the anterior fontanelle by the end of the second year. The two halves of the frontal bone normally begin to fuse during the second year, and the frontal or metopic suture is often obliterated by the eighth year. The other sutures begin to disappear during adult life, but the times when the sutures close are subject to wide variations.

about 15 or 16 years of age. After this, the calvaria usually increases slightly in size for three to four years because of thickening of its bones.

There is also rapid growth of the face and jaws, coinciding with the eruption of the dedicuous teeth; these changes are still more marked after the permanent teeth erupt (see Chapter 20). There is concurrent enlargement of the frontal and facial regions associated with the increase in the size of the paranasal air sinuses. Most of the *paranasal sinuses* (air-filled extensions of the nasal cavities that reach into certain cranial bones) are *rudimentary or absent at birth*. Growth of the paranasal sinuses is important in altering the shape of the face and in adding resonance to the voice.

MALFORMATIONS OF THE AXIAL SKELETON

Klippel-Feil Syndrome (Brevicollis). The main features of this syndrome are short neck, low hairline, and restricted neck movements. In most cases, the number of cervical vertebral bodies is less than normal, but the number of pedicles may be normal. In some cases, there is a lack of segmentation of several elements of the cervical region of the vertebral column. The number of cervical nerve roots may also be normal, but they are small, as are the intervertebral foramina. Patients with Klippel-Feil syndrome are often otherwise normal, but association of this malformation with other congenital abnormalities is not uncommon.

Spina Bifida Occulta (see Fig. 18–12A). This defect of the vertebral arch results from failure of fusion of the halves of the vertebral arch and is commonly observed in radiographs of the cervical, lumbar, and sacral regions. Frequently, only one vertebra is affected. *Spina bifida occulta of the first sacral vertebra occurs in about 10 per cent of people.* A common defect, spina bifida occulta usually causes no serious physiological disturbances (Dennison, 1971). *The spinal cord and spinal nerves are usually normal* and neurological symptoms are commonly absent. The skin over the bifid spine is intact and there may be no external evidence of the defect; sometimes, the malformation is indicated by a dimple or a tuft of hair. In about 3 per cent of normal adults, there is *spina bifida occulta of the atlas*. At other cervical levels this condition is rare, and, when present, it is sometimes accompanied by other abnormalities of the cervical region of the vertebral column.

Spina bifida cystica (Fig. 18–15), a severe type of spina bifida involving the spinal cord and the meninges, is discussed in Chapter 18. *Neurological symptoms are usually present.*

Rachischisis (Fig. 15–10). The term *rachischisis* (cleft of vertebral column) refers to the vertebral abnormalities encountered in a complex group of developmental malfor-

mations *(axial dysraphic disorders)* that affect primarily the axial structures of the body. In these cases, the neural folds fail to fuse, either because of faulty induction by the underlying notochord and its associated mesenchyme, or because of the action of teratogenic agents on the neuroepithelial cells making up the neural folds. The neural and vertebral defects may be extensive (see Fig. 18–19), or they may be restricted to a small area (see Fig. 18–17).

Accessory Ribs (Fig. 15–9A). Accessory ribs, which may be rudimentary or fairly well developed, result from the development of the costal processes of cervical or lumbar vertebrae. These processes form ribs in the thoracic region.

The most common type of accessory rib is a *lumbar rib,* but it causes no problems. *Cervical ribs* are less common, but are present in 0.5 to 1 per cent of people. A cervical rib is attached to the seventh cervical vertebra and may be unilateral or bilateral. Pressure of a cervical rib on the brachial plexus or on the subclavian vessels may produce symptoms.

Fused Ribs. Fusion of ribs occasionally occurs posteriorly when two or more ribs arise from a single vertebra. Fused ribs are often associated with a hemivertebra.

Hemivertebra (Fig. 15–9B). The developing vertebral bodies have two juxtaposed

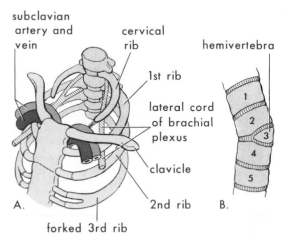

Figure 15–9 Drawings of vertebral and rib abnormalities. *A,* Cervical and forked ribs. Observe that the left cervical rib has a fibrous band passing posterior to the subclavian vessels and attaching to the sternum. Very likely, this condition produced neurovascular changes in the left upper limb. *B,* Anterior view of the vertebral column showing a hemivertebra (half vertebra). The right half of the third thoracic vertebra is absent. Note the associated lateral curvature, or scoliosis, of the vertebral column (spine).

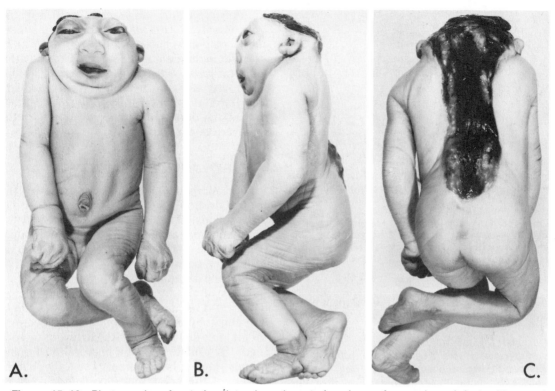

Figure 15–10 Photographs of anterior, lateral, and posterior views of a newborn infant with acrania (absence of cranial vault), anencephaly (absence of forebrain), rachischisis (extensive cleft in vertebral column), and myeloschisis (severe malformation of the spinal cord). Infants with these severe craniovertebral malformations involving the brain and spinal cord usually die within a few days after birth. For descriptions of anencephaly and spina bifida with myeloschisis, see Chapter 18.

chondrification centers (Fig. 15–6B) that soon unite. A hemivertebra results from failure of one of the chondrification centers to appear and subsequent failure of half of the vertebra to form. These defective vertebrae produce *scoliosis* (lateral curvature of the vertebral column).

Cleft Sternum. Minor sternal clefts (e.g., a notch, or foramen, in the xiphoid process) are common and are of little concern. Larger clefts are rare and are often associated with herniation of thoracic viscera, usually the heart (see Fig. 14–22). A *sternal foramen* of varying size and form occurs occasionally at the junction of the third and fourth sternebrae. This insignificant foramen *results from incomplete fusion of the cartilaginous sternal plates* during the embryonic period.

Skull Malformations. These abnormalities range from major defects that are incompatible with life to those that are minor and insignificant. With large defects, there is often herniation of the meninges and/or of the brain (see Figs. 18–31 and 18–32).

Acrania (Cranioschisis) (Fig. 15–10). In this condition, the cranial vault is almost absent, and an extensive defect of the vertebral column is often present. Acrania associated with anencephaly occurs about once in 1000 births and is incompatible with life. The malformation results from failure of the cranial end of the neural tube to close during the fourth week and subsequent failure of the cranial vault to form. For a discussion of anencephaly, see Chapter 18.

Craniosynostosis (Figs. 15–11 and 15–12). Several rare skull deformities result from premature closure of the skull sutures. Prenatal closure results in the most severe abnormalities. The cause of craniosynostosis is unknown, but genetic factors appear to be important. These abnormalities are much more common in males than in females, and they are often associated with other skeletal malformations. The type of deformed skull produced depends upon which sutures close prematurely. If the sagittal suture closes early, the skull becomes long, narrow, and wedge-shaped *(scaphocephaly,* Fig. 15–11); this type constitutes about half of the cases of craniosynostosis. Another 30 per cent of cases involve premature closure of the coronal suture. This results in a high, tower-like skull *(oxycephaly, or*

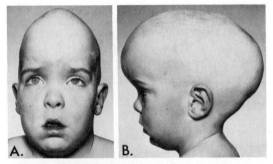

Figure 15–11 Photographs of a boy with a long, narrow, wedge-shaped skull, a condition called scaphocephaly, resulting from craniosynostosis, or premature closure of the sagittal suture. (From Laurence, K. M., and Weeks, R.: Abnormalities of the central nervous system; *in* A. P. Norman [Ed.]: *Congenital Abnormalities of Infancy*. 2nd ed. 1971. Courtesy of Blackwell Scientific Publications.)

turricephaly, Fig. 15–12*A*). If the coronal or the lambdoid suture closes prematurely on one side only, the skull is twisted and asymmetrical *(plagiocephaly*, Fig. 15–12*B*).

Microcephaly (see Fig. 18–33). Infants with this condition are born with a normal-sized or slightly small cranium. The fontanelles close during early infancy and the sutures close during the first year, but this abnormality is not caused by the premature closure of sutures. Microcephaly results from an abnormality of the central nervous system in which the brain and, consequently, the skull fail to grow. Generally, microcephalics are severely mentally retarded. This malformation is also discussed in Chapter 18.

Malformations at the Craniovertebral Junction. Congenital abnormalities at the craniovertebral junction are present in about 1 per cent of newborn infants, but they may not produce symptoms until adult life. The following are examples of these malfor-

mations: *basilar invagination* (upward displacement of the bone about the foramen magnum); *assimilation of the atlas* (nonsegmentation at the junction of the atlas and occipital bone); *atlantoaxial dislocation; Arnold-Chiari malformation* (see Chapter 18); and *separate odontoid process* (failure of the odontoid centers to fuse with the centrum of the axis).

THE APPENDICULAR SKELETON

The appendicular skeleton consists of the pectoral (shoulder) and pelvic girdles and the limb bones. The bones appear initially as mesenchymal condensation in the limb buds. During the sixth week, the mesenchymal primordia of bones in the limb buds undergo chondrification to form hyaline cartilage models of the future appendicular skeleton (Fig. 15–13*D* and *E*). The clavicle initially develops by intramembranous ossification, but it later develops growth cartilages at both ends. The models of the pectoral girdle and upper limb bones appear slightly before those of the pelvic girdle and lower limbs, and the bone models of each limb appear in a proximodistal sequence.

Ossification begins in the long bones by the end of the embryonic period and initially occurs in the diaphysis of a long bone, from a primary center of ossification. By 12 weeks, primary centers have appeared in nearly all bones of the limbs (Fig. 15–14). *The clavicles begin to ossify before any other bones in the body.* The femora are the next bones to show traces of ossification.

The first indication of ossification in the cartilaginous model of a long bone is visible near the center of the future body (shaft). This is called the *primary center of ossification*. Primary centers appear at different times in different developing bones, but most of them appear between the seventh and twelfth weeks. Virtually all of them are present by birth. The part of a bone ossified from a primary center is called the *diaphysis*.

The *secondary ossification centers* of the bones at the knee are the first to appear. The secondary centers for the distal end of the femur and the proximal end of the tibia usually appear during the ninth month of intrauterine life. Consequently, they may be present at birth, although most secondary centers of ossification appear after birth. The

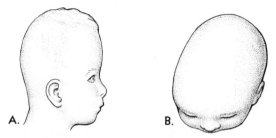

Figure 15–12 Drawings illustrating skull malformations. *A*, Oxycephaly, or turricephaly, showing the towerlike skull resulting from premature closure of the coronal suture. *B*, Plagiocephaly, illustrating a type of asymmetrical skull resulting from premature closure of the coronal and lambdoid sutures on the left side.

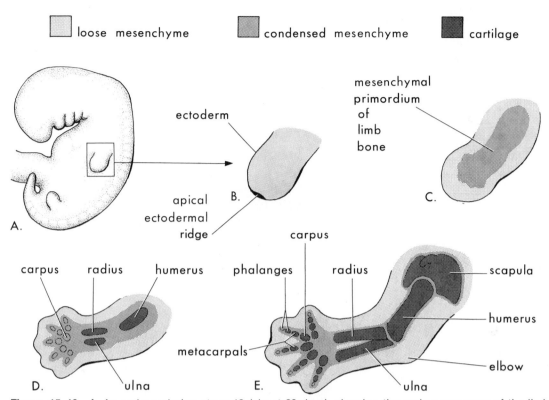

loose mesenchyme   condensed mesenchyme   cartilage

Figure 15–13 *A*, An embryo during stage 13 (about 28 days), showing the early appearance of the limb buds. *B*, Schematic drawing of a longitudinal section through an upper limb bud. The apical ectodermal ridge has an inductive influence on the loose mesenchyme in the limb bud; it promotes growth of the mesenchyme and appears to give it the ability to form specific cartilaginous elements. *C*, Similar sketch of an upper limb bud during stage 15 (about 33 days), showing the mesenchymal primordium of a limb bone. *D*, Upper limb at six weeks, showing the hyaline cartilage models of the various bones. *E*, Later in the sixth week, showing the completed cartilaginous models of the bones of the upper limb.

part of a bone ossified from a secondary center is called the *epiphysis*.

The bone formed from the primary center in the diaphysis does not fuse with that formed from the secondary centers in the epiphyses until the bone grows to its adult length; this enables lengthening of the bone to continue until that size is reached. During the growth of a bone, a plate of cartilage known as the *epiphyseal cartilage plate* intervenes between the diaphysis and the epiphysis (Fig. 15–3). The epiphyseal cartilage plate is eventually replaced by bone development at each of its two sides, diaphyseal and epiphyseal. When this occurs, growth of the bone ceases.

The development of irregular bones is similar to that of the primary center of long bones, and only one, the calcaneus in the foot, develops a secondary center of ossification.

Bone age is a good index of general maturation. Determination of the number, size, and fusion of epiphyseal centers from radiographs is a commonly used method. A radiologist determines the bone age of a person by assessing the ossification centers. Two criteria are used: (1) *The appearance of calcified material* in the diaphysis and/or the epiphysis. This is specific for each diaphysis and epiphysis for each bone for each sex. (2) The disappearance of the dark line representing the epiphyseal cartilage plate. This indicates that the epiphysis has fused with the diaphysis; this fusion occurs at specific times for each epiphysis. *Fusion occurs one to two years earlier in females than in males.*

Useful information may be obtained from study of ossification in the hand and wrist at all ages of childhood. Because girls are more advanced than boys in skeletal development, separate standards are used. For details of the expected times of appearance of various ossification centers and the evaluation of osseous maturation, see Bayer and Bayley (1959) and Vaughan et al. (1979).

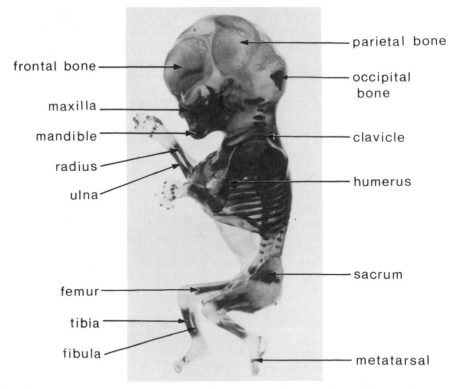

frontal bone

maxilla

mandible

radius

ulna

femur

tibia

fibula

parietal bone

occipital bone

clavicle

humerus

sacrum

metatarsal

Figure 15–14 Photograph of a 12-week fetus that has been cleared and stained with alizarin to show the developing skeleton. Observe the degree of progression of ossification from the primary centers, which is endochondral in the appendicular and axial skeletons, except for the clavicles and most of the cranial bones that are recognizable here. The part of the occipital bone visible here is preformed in cartilage and is undergoing endochondral ossification. *Actual size.* (Courtesy of Dr. Gary Geddes.)

GENERALIZED SKELETAL MALFORMATIONS

Achondroplasia, or Hypoplastic Chondrodystrophy (see Fig. 8–12). This condition, the most common cause of dwarfism, occurs about once in 10,000 births. The limbs are short because of a disturbance of endochondral ossification at the epiphyseal cartilage plates, particularly of long bones, during fetal life. The trunk is about normal length, but the head may be slightly enlarged. Achondroplasia is an *autosomal dominant disorder*. For details of its inheritance, see Thompson and Thompson (1980).

Hyperpituitarism. Congenital infantile hyperpituitarism is rare; it causes the infant to grow at an abnormallly rapid rate. This may result in *gigantism* (excessive height and body proportions) or, in adults, *acromegaly* (enlargement of the soft tissues and bones of the face, hands, and feet). Both gigantism and acromegaly result from an excessive secretion of growth hormone arising from eosinophilic hyperplasia or from an eosinophilic, amphophilic, or chromophobic adenoma of the anterior pituitary (Gardner, 1975). Acromegaly results from an excess of growth hormone after the primary and secondary ossification centers have closed, i.e., after adolescence.

Hypothyroidism and Cretinism. Deficiency of fetal thyroid hormone production results in cretinism, a condition characterized by mental deficiency, skeletal abnormalities, and auditory and neurological disorders. Cretinism is now very rare except in areas where there is a lack of iodine in the soil and water; for details, see Koenig (1975) and Warkany (1971). Agenesis of the thyroid gland will also result in cretinism.

There is a variety of other rare generalized skeletal disorders; for descriptions, see Vaughan et. al (1979) and Warkany (1971).

Congenital malformations of the limbs are described and illustrated in Chapter 17.

SUMMARY

The skeletal system develops from mesenchyme. In most bones, such as the long bones in the limbs, the condensed mesenchyme undergoes chondrification to form hyaline cartilage models of bones. Ossification centers appear in these models by the end of the embryonic period, and the bones ossify by *endochondral ossification*. Some bones, e.g., the flat bones of the skull, develop by *intramembranous ossification.* The vertebral column and ribs develop from the sclerotomes of the somites. Each vertebra is formed by fusion of a condensation of the caudal half of one pair of *sclerotomes* with the cranial half of the subjacent pair of sclerotomes.

The developing skull consists of a neurocranium and a viscerocranium, each of which has membranous and cartilaginous components.

The appendicular skeleton develops from endochondral ossification of the cartilaginous models of the bones, which form from the mesenchyme in the developing limbs. Although there are numerous types of skeletal malformations, most of these, except for spina bifida occulta and accessory ribs, are rare.

Joints are classified as (1) fibrous joints, (2) cartilaginous joints, and (3) synovial joints (Fig. 15–4). They develop from the *interzonal mesenchyme* between the primordia of bones. In a fibrous joint, the intervening mesenchyme differentiates into dense fibrous connective tissue. In a cartilaginous joint, the mesenchyme between the bones differentiates into cartilage. In a synovial joint, a *synovial cavity* is formed within the intervening mesenchyme by breakdown, or cavitation, of the cells. The mesenchyme also gives rise to the synovial membrane and the capsular and other ligaments of the joint.

SUGGESTIONS FOR ADDITIONAL READING

Lloyd-Roberts, G. C.: Orthopaedic abnormalities; *in* E. P. Norman (Ed.): *Congenital Abnormalities in Infancy.* 2nd ed. Oxford, Blackwell Scientific publications, 1971, pp. 256–269. The clinical aspects of abnormalities of the skeletal system are described by an orthopedic surgeon.

Royer, P.: Growth and development of bony tissues; *in* J. A. Davis and J. Dobbing (Eds.): *Scientific Foundations of Paediatrics.* Philadelphia, W. B. Saunders Co., 1974, pp. 376–399. A comprehensive account of perinatal and postnatal growth and development of the skeleton by a research professor of pediatrics.

Sperber, G. H.: *Craniofacial Embryology.* 2nd ed. Chicago, Year Book Medical Publishers, Inc., 1976. A concise, well-written text on the developmental anatomy of the face and jaws, written by a professor with a dual background of dentistry and anatomy. This book would be of special interest to dental students and practitioners.

CLINICALLY ORIENTED PROBLEMS

1. What is the most common congenital malformation of the vertebral column? Where is the defect usually located? Does this congenital malformation usually cause symptoms (e.g., back problems)?
2. Occasionally, rudimentary ribs develop on the seventh cervical vertebra. Are these accessory ribs of clinical importance? What is the embryological basis of a cervical rib?
3. What vertebral defect can produce scoliosis? Define this condition. What is the embryological basis of this vertebral defect?
4. What is meant by the term *craniosynostosis*? What results from this developmental abnormality? Give a common example.
5. A child presented with characteristics of Klippel-Feil syndrome. What are the main features of this condition? What vertebral abnormalities are usually present?

The answers to these questions are given at the back of the book.

REFERENCES

Bagnall, K. M., Harris, P. F., and Jones, P. R.: A radiographic study of the human fetal spine. *J. Anat. 123*:777, 1977.

Bayer, L. M., and Bayley, N.: *Growth Diagnosis. Selected Methods for Interpreting and Predicting Physical Development From One Year to Maturity.* Chicago, The University of Chicago Press, 1959.

Blechschmidt, E., and Gasser, R. F.: *Biokinetics and Biodynamics of Human Differentiation.* Springfield, Ill., Charles C Thomas, Publisher, 1978.

Bloom, W., and Fawcett, D. W.: *A Textbook of Histology.* 10th ed. Philadelphia, W. B. Saunders Co., 1975, pp. 244–287.

Dennison, W. M.: Spina bifida; *in* J. C. Mustardé: *Plastic Surgery in Infancy and Childhood.* Edinburgh, E. & S. Livingstone, Ltd., 1971, pp. 381–385.

Ford, E. H. R.: The growth of the foetal skull. *J. Anat.* *90*:63, 1956.

Gardner, E., and Gray, D. J.: Prenatal development of the human hip joint. *Am. J. Anat.* *87*:163, 1950.

Gardner, L. I. (Ed.): *Endocrine and Genetic Diseases of Childhood and Adolescence.* 2nd ed. Philadelphia, W. B. Saunders Co., 1975.

Gasser, R. F.: Evidence that sclerotomal cells do not migrate medially during normal embryonic development of the rat. *Am. J. Anat.* *154:*509, 1979.

Gilbert, P. W.: Origin and development of head cavities in the human embryo. *J. Morph.* *90*:149, 1952.

Gray, D. J., and Gardner, E.: Prenatal development of the human knee and superior tibiofibular joints. *Am. J. Anat.* *86*:235, 1950.

Gray, D. J., Gardner, E., and O'Rahilly, R.: The prenatal development of the skeleton and joints of the human hand. *Am. J. Anat.* *101*:169, 1957.

Grüneberg, H.: *The Pathology of Development. A Study of Inherited Skeletal Disorders in Animals.* Oxford, Blackwell Scientific Publications, 1963.

Haines, R. W.: The development of joints. *J. Anat.* *81*:33, 1947.

Johnson, W. H., and Kennedy, J. A.: *Radiographic Anatomy of the Human Skeleton.* Edinburgh, E. & S. Livingstone, Ltd., 1961.

Koenig, M. P.: Endemic goiter and endemic cretinism; *in* L. I. Gardner (Ed.): *Endocrine and Genetic Diseases of Childhood and Adolescence.* 2nd ed. Philadelphia, W. B. Saunders Co., 1975, pp. 260–269.

Laurence, K. M., and Weeks, R.: Abnormalities of the central nervous system; *in* A. P. Norman (Ed.): *Congenital Abnormalities in Infancy.* 2nd ed. Oxford, Blackwell Scientific Publications, 1971, pp. 25–86.

Maroteaux, P., and Lamy, M.: Achondroplasia in man and animal. *Clin. Orthop.* *33*:91, 1964.

Mutch, J., and Walmsley, R.: The aetiology of cleft vertebral arch in spondylolisthesis. *Lancet* *270*:74, 1956.

Noback, C. R., and Robertson, G. G.: Sequences of appearance of ossification centers in the human skeleton during the first five prenatal months. *Am. J. Anat.* *89*:1, 1951.

Peacock, A.: Observations on the prenatal development of the intervertebral disc in man. *J. Anat.* *85*:260, 1951.

Rubin, A., and Friedenberg, Z. B.: Musculoskeletal system; *in* A. Rubin (Ed.): *Handbook of Congenital Malformations.* Philadelphia, W. B. Saunders Co., 1967, pp. 51–85.

Sensenig, E. C.: The early development of the human vertebral column. *Contr. Embryol. Carneg. Instn.* *33*:21, 1949.

Smith, D. W.: *Recognizable Patterns of Human Malformation: Genetic, Embryologic, and Clinical Aspects.* 2nd ed. Philadelphia, W. B. Saunders Co., 1976, pp. 347–349.

Swinyard, C. A. (Ed.): *Limb Development and Deformity: Problems of Evaluation and Rehabilitation.* Springfield, Ill., Charles C Thomas, Publisher, 1969.

Thompson, J. S., and Thompson, M. W.: *Genetics in Medicine.* 3rd ed. Philadelphia, W. B. Saunders Co., 1980.

Vaughan, V. C., McKay, R. J., and Behrman, R. E.: *Nelson Textbook of Pediatrics.* 11th ed. Philadelphia, W. B. Saunders Co., 1979.

Warkany, J.: *Congenital Malformations: Notes and Comments:* Chicago, Year Book Medical Publishers, Inc., 1971.

Warkany, J., and Kirkpatrick, J. A., Jr.: Skeletal defects; *in* V. C. Vaughan. and R. J. McKay (Eds.): *Nelson Textbook of Pediatrics.* 10th ed. Philadelphia, W. B. Saunders Co., 1975, pp. 1470–1490.

Wells, L. H.: Congenital deficiency of the vertebral pedicle. *Anat. Rec.* *145*:193, 1963.

Williams, P. L., and Warwick, R.: *Gray's Anatomy.* 36th ed. (British). Philadelphia, W. B. Saunders Co., 1980.

16

THE MUSCULAR SYSTEM

The muscular system develops from mesoderm, except for the muscles of the iris, which develop from the ectoderm of the optic cup (see Chapter 19). Muscle tissue develops from primitive cells called *myoblasts,* which are derived from mesenchyme (embryonic connective tissue).

STRIATED SKELETAL MUSCLE

The myoblasts that form the striated skeletal *muscles of the trunk* are derived from myotomal mesoderm of the somites (see Fig. 15–1). The *limb muscles* develop from mesenchyme in the limb buds (see Chapter 17). *Tongue muscles* are formed from head mesenchyme, and many *muscles of the face,* jaws, neck, and shoulders develop from mesenchyme in the *branchial arches* (see Chapter 10).

The first indication of muscle development is the elongation of the nuclei and cell bodies of the mesenchymal cells to form *myoblasts.* Soon, these cells begin to fuse with one another to form elongated, multinucleated, cylindrical structures called *myotubes.* Growth occurs by continued fusion of myoblasts and myotubes. The specialized myofilaments develop in the cytoplasm of the myotubes during or soon after fusion of the myoblasts. Soon, the *myofibrils* and other organelles characteristic of striated muscle fibers (cells) develop. Because muscle cells are long and narrow, it is customary to call them fibers (Ham and Cormack, 1979).

As the myotubes differentiate into muscle fibers, they become invested with external lamina, individually or in groups, which segregates them from the surrounding connective tissue.

Most striated skeletal muscle fibers develop before birth, and almost all the remaining ones are formed by the end of the first year. Increase in the size of a muscle occurs as the result of an increase in the diameter of the fibers through the formation of more myofilaments.

After birth, muscles must increase in length and width in order to grow with the skeleton. Their ultimate size depends on the amount of exercise that is performed. Not all embryonic muscle fibers persist; many of them fail to establish themselves as necessary units of the muscle and therefore degenerate.

Typical Myotomes. Each myotome divides into a small dorsal *epaxial division* and a larger ventral *hypaxial division* (Fig. 16–1*B*). Each developing spinal nerve also divides and sends a branch to each division, the *dorsal primary ramus* supplying the epaxial division and the *ventral primary ramus,* the hypaxial division. Some muscles remain segmentally arranged like the somites (e.g., the intercostals), but most myoblasts migrate away from the myotomes and form nonsegmented muscles.

Derivatives of the Epaxial Divisions (Fig 16–2). Myoblasts from these parts of the myotomes form the extensor muscles of the neck, the extensor muscles of the vertebral column, and the lumbar extensor musculature. The extensor muscles derived from the caudal sacral and the coccygeal myotomes degenerate; their adult derivatives are the dorsal sacrococcygeal ligaments.

Derivatives of the Hypaxial Divisions (Fig. 16–2). Myoblasts from the cervical myotomes form the scalene, prevertebral, geniohyoid, and infrahyoid muscles. The thoracic myotomes form the lateral and ventral flexor muscles of the vertebral column, and the lumbar myotomes form the quadratus lumborum muscle. The sacrococcygeal myotomes form the muscles of the pelvic diaphragm and probably the striated muscles of the anus and the sex organs.

The Branchial Arches (see Fig. 10–5). The migration of myoblasts from these arches to form the muscles of mastication, of facial expression, and of the pharynx and larynx is described in Chapter 10 (see Table 10–1). These muscles are innervated by bran-

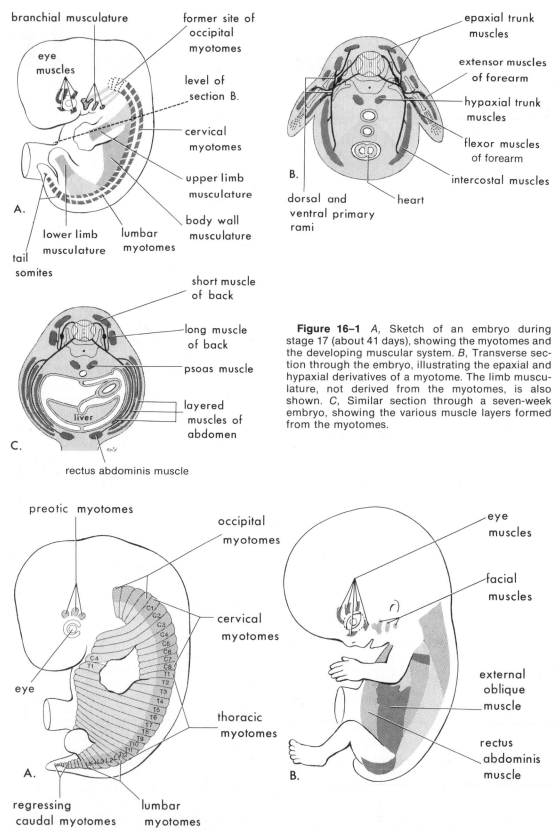

Figure 16–1 *A,* Sketch of an embryo during stage 17 (about 41 days), showing the myotomes and the developing muscular system. *B,* Transverse section through the embryo, illustrating the epaxial and hypaxial derivatives of a myotome. The limb musculature, not derived from the myotomes, is also shown. *C,* Similar section through a seven-week embryo, showing the various muscle layers formed from the myotomes.

Figure 16–2 Drawings illustrating the developing muscular system: *A,* Six-week embryo, showing the myotome regions of the somites that give rise to most skeletal muscles. *B,* Eight-week embryo, showing the developing superficial trunk musculature.

chial arch nerves (cranial nerves V, VII, IX, and X).

The Ocular Muscles (Figs. 16–1 and 16–2). The origin of the extrinsic eye muscles is unclear, but it has been suggested that they may be derived from mesenchymal cells near the prochordal plate (Gilbert, 1957). The mesoderm in this area may give rise to the three *preotic myotomes* (Fig. 16–2A). Myoblasts differentiate from mesenchymal cells. Groups of myoblasts, each supplied by its own cranial nerve (III, IV, or VI), form the extrinsic muscles of the eye.

The Tongue Muscles (see Figs. 10–5 and 10–10). Initially, there are four indistinct *occipital myotomes;* the first pair disappears. Myoblasts from the remaining three myotomes form the tongue muscles, innervated by the hypoglossal or twelfth cranial nerve.

The Limb Muscles (see Figs. 15–13 and 16–1). The musculature of the limbs develops in situ from the mesenchyme surrounding the developing bones. This mesenchyme is derived from the somatic layer of lateral mesoderm (see Fig. 15–1).

SMOOTH MUSCLE

Smooth muscle fibers differentiate from the splanchnic mesenchyme surrounding the endoderm of the primitive gut and its derivatives (see Fig. 15–1). The smooth muscle in the walls of many blood and lymphatic vessels arises from somatic mesoderm. The muscles of the iris (the sphincter and dilator pupillae) and the myoepithelial cells of mammary and sweat glands are thought to be derived from mesenchymal cells that originate from ectoderm.

The first evidence of the differentiation of smooth muscle is the development of elongated nuclei in spindle-shaped cells, called *myoblasts.* During early development, new myoblasts continue to differentiate from mesenchymal cells. During later development, division of existing myoblasts gradually replaces the differentiation of myoblasts in the production of new smooth muscle tissue.

As smooth muscle cells differentiate, filamentous *contractile elements* develop in their cytoplasm, and the external surface of each cell acquires a surrounding *external lamina.* As the smooth muscle fibers develop into sheets or bundles, the fibroblasts and/or muscles cells synthesize and lay down collagenous, elastic, and reticular fibers.

STRIATED CARDIAC MUSCLE

Cardiac muscle develops from splanchnic mesenchyme surrounding the endocardial heart tube (see Fig. 14–7). Cardiac myoblasts differentiate from the *myoepicardial mantle* and form the *myocardium,* or heart muscle. Cardiac muscle fibers arise by differentiation and growth of single cells, unlike striated skeletal muscle fibers, which develop by fusion of cells. Growth of cardiac muscle fibers results from the formation of new myofilaments. The myoblasts adhere to each other as in developing skeletal muscle, but the intervening cell membranes do not disintegrate; these areas of adhesion give rise to the *intercalated discs.*

Late in the embryonic period, special bundles of muscle cells develop with relatively few myofibrils and relatively larger diameters than typical cardiac muscle fibers. These atypical cardiac muscle cells, called *Purkinje fibers,* later form the conducting system of the heart (see Fig. 14–20D).

CONGENITAL MALFORMATIONS OF MUSCLES

Congenital Absence of Muscles. Absence of one or more muscles is more common than is generally recognized. In rare instances, failure of normal muscle development may be widespread, leading to immobility of multiple joints, a condition known as *arthrogryposis multiplex congenita* (Vaughan et al., 1979). Persons with this disorder have congenital stiffness of one or more joints associated with hypoplasia of the associated muscles.

Usually, only a single muscle is absent on one side of the body, or, at times, only a portion of the muscle fails to develop. Occasionally, the same muscle or muscles may be absent on both sides of the body. Any muscle in the body may occasionally be absent; common examples are the sternocostal head of the pectoralis major (Fig. 16–3), the palmaris longus, the trapezius, the serratus anterior, and the quadratus femoris. The failure of these muscles to develop is usually of little consequence, and the individual may not even be aware of the deficiency.

Absence of the pectoralis major is occasionally associated with absence of the mammary gland and/or hypoplasia of the nipple (Fig. 16–3), syndactyly (see Chapter 17), microdactyly, webbed fingers, and scoliosis (see Fig. 15–9).

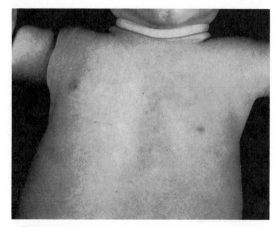

Figure 16–3 Photograph of the chest of an infant with congenital absence of the left pectoralis major muscle. Note the absence of an anterior axillary fold on the left and the low location of the left nipple. (From Vaughan, V. C., McKay, R. J., and Behrman, R. E.: *Nelson Textbook of Pediatrics.* 11th ed. Philadelphia, W. B. Saunders Co., 1979.)

Some muscular deficiencies are of a more serious nature. These include congenital absence of the diaphragm, which is usually associated with severe pulmonary atelectasis and pneumonia, and absence of muscle(s) of the anterior abdominal wall, which may be associated with severe gastrointestinal and genitourinary malformations (see Fig. 13–17). Occasionally, individuals with a congenitally absent muscle develop muscular dystrophy in later life. The most common association is between congenital absence of the pectoralis major muscle and the Landouzy-Déjérine facioscapulohumeral form of muscular dystrophy (Mastaglia, 1974). At present, the relationship between congenital muscle abnormalities and muscular dystrophy is obscure.

It is not certain whether congenital deficiencies of skeletal muscles result from failure of the muscle cells to develop, or from a pathological process affecting the muscles or their nerve supply during intrauterine life.

Variations in Muscles. *All muscles are subject to a certain amount of variation,* but some are affected more often than others. Certain muscles are functionally vestigial (e.g., those of the external ear and scalp). Some muscles present in other primates appear in only some humans (e.g., the sternalis). Variations in the form, position, and attachments of muscles are common and are usually functionally insignificant.

The sternocleidomastoid muscle is sometimes injured at birth, resulting in a condition known as *congenital torticollis* (Moore, 1980). There is fixed rotation and tilting of the head, owing to *fibrosis and shortening of the sternocleidomastoid* muscle on one side (Fig. 16–4). Most cases of torticollis result from tearing of fibers of the sternocleidomastoid muscle during delivery of the infant. Bleeding into the muscle often occurs in a localized area, forming a small swelling called a *hematoma*. Later, a mass often develops, owing to necrosis of muscle fibers and fibrosis (formation of fibrous tissue). Shortening of the muscle usually follows; this causes lateral bending of the head to the affected side, and a slight turning away of the head from the side of the short muscle.

SUMMARY

Most skeletal muscle is derived from the myotome regions of the somites, but some head and neck muscles are derived from branchial arch mesoderm, and the limb musculature develops from mesenchyme derived

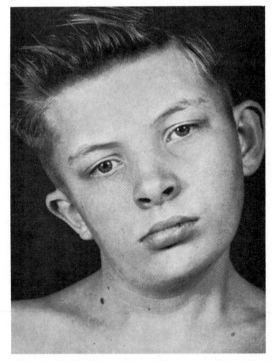

Figure 16–4 Photograph of the head and neck of a 12-year-old boy with congenital torticollis (wryneck). Shortening, or contracture, of the right sternocleidomastoid muscle has caused tilting of the head to the right and turning of the chin to the left. There is also asymmetric development of the face and skull. (From Vaughan, V. C., McKay, R. J., and Behrman, R. E.: *Nelson Textbook of Pediatrics.* 11th ed. Philadelphia, W. B. Saunders Co., 1979.)

from the somatic mesoderm. Cardiac muscle and most smooth muscle are derived from splanchnic mesoderm. Absence or variation of muscles is fairly common and is usually of little consequence.

SUGGESTIONS FOR ADDITIONAL READING

Adams, R. D., Denny-Brown, D., and Pearson, C. M. (Eds.): *Diseases of Muscle. A Study In Pathology.* New York, Harper & Row, Inc., 1962, pp. 3–61. A thorough account of the histogenesis of skeletal muscle with a good discussion of the regeneration of skeletal muscle and descriptions of congenital abnormalities of muscle.

Mastaglia, F. L.: The growth and development of the skeletal muscles; *in* J. A. Davis and J. Dobbing (Eds.): *Scientific Foundations of Paediatrics.* Philadelphia, W. B. Saunders Co., 1974, pp. 348–375. A well-illustrated account of the histogenesis and cytogenesis of skeletal muscle. The innervation of muscle, electrophysiological changes during development, and abnormalities of skeletal muscle development are discussed. Many excellent electron-micrographs are included.

CLINICALLY ORIENTED PROBLEMS

1. An infant presented with absence of the left anterior axillary fold, and the left nipple was much lower than usual. Absence of which muscle probably caused these unusual observations? Would the infant be likely to suffer any disability?

2. A patient was concerned when he learned that he had only one palmaris longus muscle. Is this a common occurrence? Does the absence of this muscle cause a disability?

3. The parents of a 4-year-old girl observed that she always held her head slightly tilted to the right side and that one of her neck muscles was more prominent than the others. Name the muscle that was prominent. Did it pull the child's head to the right side? What is this deformity called? What probably caused the muscle shortening that resulted in this condition?

4. Failure of striated muscle to develop in the midline of the anterior abdominal wall is associated with the formation of a severe congenital abnormality of the urinary system. What is this malformation called? What is the embryological basis of the failure of muscle formation?

The answers to these questions are given at the back of the book.

REFERENCES

Arey, L. B.: The history of the first somite in human embryos. *Contr. Embryol. Carneg. Instn. 27*:235, 1938.

Bates, M. N.: The early development of the hypoglossal musculature of the cat. *Am. J. Anat. 83*:329, 1948.

Copenhaver, W. M., Kelly, D. E., and Wood, R. L.: *Bailey's Textbook of Histology.* 17th ed. Baltimore, The Williams & Wilkins Co., 1978.

Deuchar, E. M.: Experimental demonstration of tongue muscle origin in chick embryos. *J. Embryol. Exp. Morphol. 6*:527, 1958.

Dubowitz, V.: *Muscle Disorders in Childhood.* Philadelphia, W. B. Saunders Co., 1978.

Enesco, M.: Increase in the number of nuclei in various striated muscles of the growing rat. *Anat. Rec. 139*:225, 1961.

Gasser, R. F.: The development of the facial muscles in man. *Am. J. Anat. 120*:357, 1967.

Gilbert, P. W.: The origin and development of the human extrinsic ocular muscles. *Contr. Embryol. Carneg. Inst. 36*:59, 1957.

Godman, J. C.: On the regeneration and redifferentiation of mammalian striated muscles. *J. Morphol. 100*:27, 1957.

Ham, A. W., and Cormack, D. H.: *Histology.* 8th ed. Philadelphia, J. B. Lippincott Co., 1979.

Kamieniecka, Z.: The stages of development of human foetal muscles with reference to some muscular diseases. *J. Neurol. Sci. 7*:319, 1968.

Moore, K. L.: *Clinically Oriented Anatomy.* Baltimore, The Williams & Wilkins Co., 1980.

Rubin, A., and Friedenberg, Z. B.: Musculoskeletal system; *in* A. Rubin (Ed.): *Handbook of Congenital Malformations.* Philadelphia, W. B. Saunders Co., 1967, pp. 51–85.

Vaughan, V. C., McKay, R. J., and Behrman, R. E.: *Nelson Textbook of Pediatrics.* 11th ed. Philadelphia, W. B. Saunders Co., 1979.

Walker, B. E.: The origin of myoblasts in normal and dystrophic mice. *Anat. Rec. 142*:289, 1962.

17

THE LIMBS

LIMB DEVELOPMENT

The general features of limb development are described and illustrated in Chapter 5. Development of the limb bones is described in Chapter 15, and formation of the limb musculature is outlined in Chapter 16. The purpose of this chapter is to consolidate this material.

The *limb buds* first appear as small elevations on the ventrolateral body wall toward the end of the fourth week (Fig. 17–1). The upper limb buds appear disproportionately low on the embryo's trunk because development of the head and neck occurs in advance of the rest of the embryo. The early stages of limb development are alike for the upper and

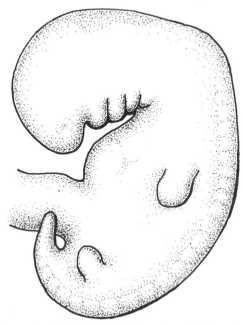

Figure 17–1 Drawing of an embryo during stage 13 (about 28 days), showing the paddle-shaped appearance of the early limb buds. At this stage, the limb buds consist of a core of mesenchyme and a covering layer of ectoderm. Early development of the upper and lower limb buds is similar, except that development of the upper limb bud precedes that of the lower limb bud by a few days.

lower limbs (Figs. 17–1 and 17–2), except that development of the *upper limb buds* precedes that of the *lower limb buds* by a few days. The upper limb buds develop opposite the caudal cervical segments, and the lower limb buds form opposite the lumbar and upper sacral segments (Fig. 17–4). Each limb bud consists of a mass of mesenchyme derived from the somatic mesoderm and is covered by a layer of ectoderm.

The *apical ectodermal ridge* (see Fig. 15–13) exerts an inductive influence on the limb mesenchyme that promotes growth and development of the limbs. The distal ends of the flipper-like limb buds soon flatten into paddle-like hand or foot plates (Fig. 17–2). By the end of the sixth week, the mesenchymal tissue in the periphery of the hand plates condenses to form *digital rays* (finger rays), which outline the pattern of the digits. In the seventh week, similar rays develop in the foot plates. Soon, the intervening regions of tissue break down, forming notches between the digital rays. This tissue breakdown progresses inward from the circumference, producing the fingers and toes. If this process is incomplete or is arrested, varying degrees of webbing, or syndactyly, result (Fig. 17–9).

As the limbs elongate and the bones form, myoblasts aggregate and develop into a large muscle mass in each limb bud. In general, this muscle mass separates into dorsal (extensor) and ventral (flexor) components. *The limb musculature develops in situ* from the mesenchyme surrounding the developing bones. The limbs receive no mesenchymal contribution from the myotome regions of the somites (O'Rahilly, 1967).

Early in the seventh week, the limbs extend ventrally, and then the developing upper and lower limbs rotate in opposite directions and to different degrees (Fig. 17–3). Originally, the flexor aspect of the limbs is ventral and the extensor aspect dorsal, and the preaxial and postaxial borders are cranial and caudal, respectively (Fig. 17–4*A* and *D*). The upper limbs rotate laterally through 90

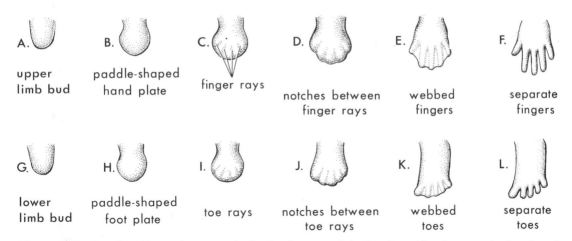

A. upper limb bud
B. paddle-shaped hand plate
C. finger rays
D. notches between finger rays
E. webbed fingers
F. separate fingers

G. lower limb bud
H. paddle-shaped foot plate
I. toe rays
J. notches between toe rays
K. webbed toes
L. separate toes

Figure 17–2 Drawings illustrating stages in the development of the hands and feet between the fourth and eighth weeks. The early stages of limb development are alike, except that development of the hands precedes that of the feet by a few days.

degrees on their longitudinal axes; thus, the future elbows point backward, or posteriorly, and the extensor muscles come to lie on the lateral and posterior aspects of the limb. The lower limbs rotate medially through almost 90 degrees; thus, the future knees face forward, or anteriorly, and the extensor muscles lie on the anterior aspect of the lower limb. It should also be clear that the radius and the tibia are homologous bones, as are the ulna and the fibula, just as the thumb and the great toe are homologous digits.

Dermatomes and Cutaneous Innervation of the Limbs (Fig. 17–4). Because of its relationship to the growth and rotation of the limbs, the cutaneous segmental nerve supply of the limbs is considered in this chapter rather than in the chapter dealing with the nervous system or the integumentary system.

A *dermatome is defined as the area of skin supplied by a single spinal nerve and its spinal ganglion.* The peripheral nerves grow from the limb plexuses (brachial and lumbosacral) into the mesenchyme of the limb buds during the fifth week. The spinal nerves are distributed in segmental bands and supply both dorsal and ventral surfaces of the limb buds. As the limbs elongate, the cutaneous distribution of the spinal nerves migrates along the limbs and no longer reaches the surface in the distal part of the limbs. Although the original dermatomal pattern changes during growth of the limbs, an orderly sequence of distribution can still be recognized in the adult (Fig. 17–4C and F).

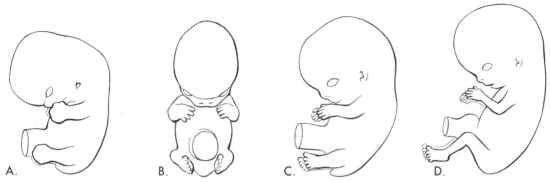

A.   B.   C.   D.

Figure 17–3 Drawings illustrating positional changes of the developing limbs. *A,* During stage 19 (about 48 days), showing the limbs extending ventrally and the hand and foot plates facing each other. *B,* During stage 20 (about 51 days), showing the upper limbs bent at the elbows and the hands curved over the thorax. *C,* During stage 22 (about 54 days), showing the soles of the feet facing medially. *D,* About 56 days. Note that the elbows now point caudally and the knees cranially.

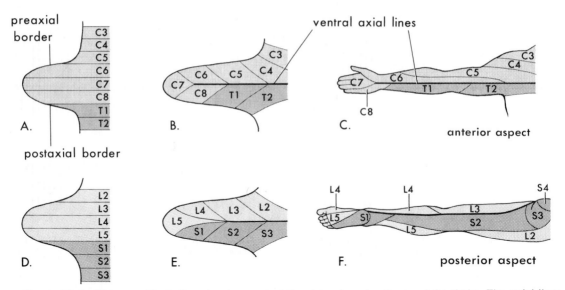

Figure 17–4 Diagrams illustrating development of the dermatomal patterns of the limbs. The *axial lines* indicate where there is no sensory overlap. *A* and *D*, Ventral aspect of the limb buds early in the fifth week. At this stage, the dermatomal patterns show the primitive segmental arrangement. *B* and *E*, Similar views later in the fifth week, showing the modified arrangement of dermatomes. *C* and *F*, The dermatomal patterns in the adult upper and lower limbs. The primitive dermatomal pattern has disappeared, but an orderly sequence of dermatomes can still be recognized. In *F*, note that most of the original ventral surface of the lower limb lies on the back of the adult limb. This results from the medial rotation of the lower limb that occurs toward the end of the embryonic period. In the upper limb, the ventral axial line extends along the anterior surface of the arm and forearm. In the lower limb, the ventral axial line extends along the medial side of the thigh and knee to the posteromedial aspect of the leg to the heel.

In the upper limb, observe that the areas supplied by C5 and C6 adjoin the areas supplied by T2, T1, and C8, but the overlap between them is minimal at the *ventral axial line* (Fig. 17–4C).

A *cutaneous nerve area is the area of skin supplied by a peripheral nerve;* both cutaneous nerve areas and dermatomes show considerable overlapping. It should be emphasized that the dermatomal patterns indicate only that if the dorsal root of that segment is cut, there may be a slight deficit in the area indicated; however, because there is overlapping of dermatomes, a particular area is not exclusively innervated by a single segmental nerve. The limb dermatomes may be traced progressively down the lateral aspect of the upper limb and back up its medial aspect.

A comparable distribution of dermatomes occurs in the lower limbs, which may be traced down the ventral and then up the dorsal aspect of the lower limb. When the limbs descend, they carry their nerves with them; this explains the oblique course of the nerves of the brachial and lumbosacral plexuses.

LIMB MALFORMATIONS

Minor limb defects are relatively common, but major limb malformations are generally rare. Abnormalities of the fingers and toes are relatively common, but they can usually be corrected surgically. Although minor malformations are usually of no serious medical consequence, they may serve as indicators of more serious abnormalities; i.e., they may be part of a specific pattern of malformations (Smith, 1976). Many severe limb deformities occurred from 1957 to 1962 as a result of maternal ingestion of thalidomide (Fig. 17–5). This drug, widely used as a sedative and antinauseant, was withdrawn from the market in December 1961.

The terminology used to describe limb deficiencies in this book follows the commonly used international nomenclature in which only two basic descriptive terms are used (Swinyard, 1969): (1) *amelia*, complete absence of a limb or limbs (Fig. 17–5A), and (2) *meromelia* (Gr. *meros*, part, and *melos*, extremity), partial absence of a limb or limbs (Fig. 17–5B and C). Descriptive terms such as hemimelia, peromelia, ectromelia, phocomelia,

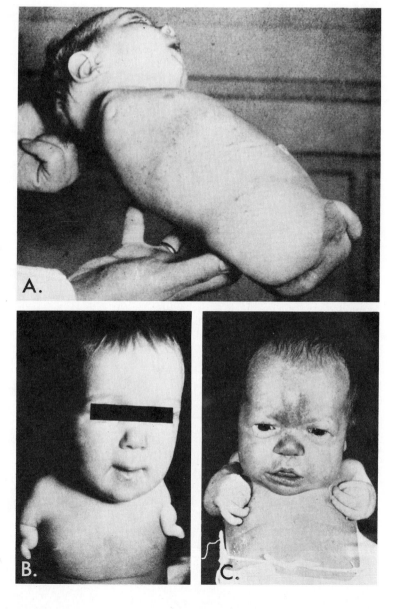

Figure 17–5 Limb malformations caused by thalidomide. *A,* Quadruple amelia. The upper and lower limbs are absent. *B,* Meromelia of the upper limbs. The upper limbs are represented by rudimentary stumps. *C,* Meromelia, with the rudimentary upper limbs attached directly to the trunk. (From Lenz, W., and Knapp, K.: *Ger. Med. Mon. 7*:253, 1962.) See also Figure 8–18.

and so forth are not used in current nomenclature because of their imprecision.

Cleft Hand and Cleft Foot (Lobster-Claw Deformities, Figs. 17–6*E* and *F* and 17–7). In these rare deformities, there is absence of one or more central digits, resulting from failure of development of one or more digital rays; thus the hand or foot is divided into two parts that oppose each other like lobster claws. The remaining digits are partially or completely fused (syndactyly).

Clubhand (Congenital Absence of the Radius). The radius is partially or completely absent. The hand deviates radially and the ulna bows with the concavity on the lateral side of the forearm. The deformity is often inherited.

Brachydactyly (Fig. 17–8*A*). Shortness of the fingers or toes is uncommon and results from reduction in the size of the phalanges. It is usually inherited as a dominant trait and is often associated with shortness of stature.

Polydactyly (Fig. 17–8*C* and *D*). Supernumerary fingers or toes are common. Often, the extra digit is incompletely formed and lacks proper muscular development; it is thus useless. If the hand is affected, the extra digit is most commonly medial or lateral rather than central. In the foot, the extra toe

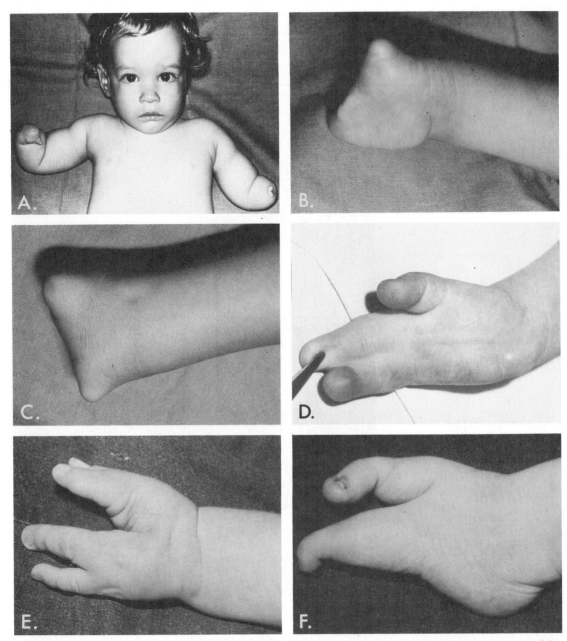

Figure 17–6 Photographs illustrating various types of meromelia. *A*, Absence of the hands and most of the forearms. *B*, Absence of the phalanges. *C*, Absence of the hand. *D*, Absence of the fourth and fifth phalanges and metacarpals. There is also syndactyly. *E*, Absence of the third phalanx, resulting in a cleft hand (lobster claw). *F*, Absence of the second and third toes, resulting in a cleft foot. (*D* is from Swenson, O.: *Pediatric Surgery.* 1958. Courtesy of Appleton-Century-Crofts, Inc.)

is usually on the lateral side. Polydactyly is inherited as a dominant trait.

Syndactyly (Fig. 17–9). Fusion of the fingers or toes is the most common limb malformation and is more frequent in the foot than in the hand. Syndactyly results from a lack of differentiation between two or more digits. Normally, the mesenchyme in the periphery of the hand and foot plates condenses to form the primordia of the fingers and toes, and the thinner tissue between them breaks down. Webbing of the skin between the fingers or toes results from failure of this tissue breakdown to occur (Fig.

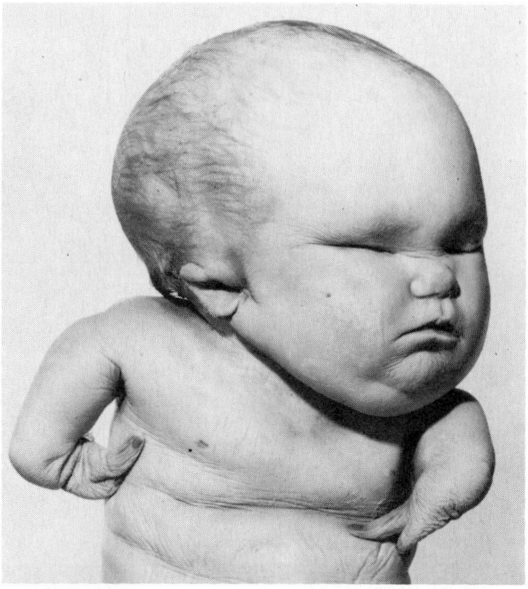

Figure 17–7 A fetus with absence of central digits, resulting in the division of the hands into two parts that oppose each other like lobster claws. This condition is often referred to as cleft hand or the lobster-claw deformity. This fetus also has an abnormally large head and low-set, malformed auricles of the external ears.

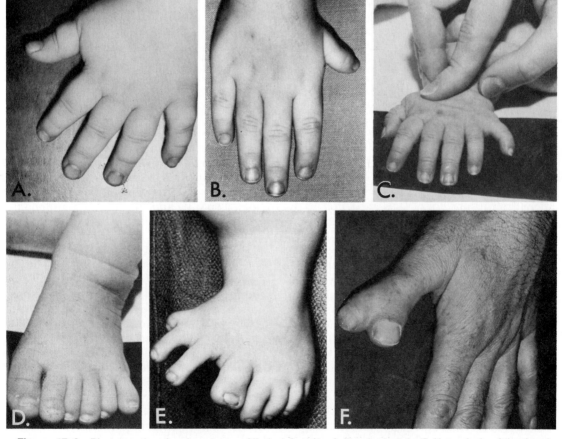

Figure 17–8 Photographs of various types of limb deformity. *A,* Brachydactyly. *B,* Hypoplasia of the thumb. *C,* Polydactyly, showing a supernumerary finger. *D,* Polydactyly, showing a supernumerary toe. *E,* Partial duplication of the foot. *F,* Partial duplication of the thumb. (*C* and *D* are from Swenson, O.: *Pediatric Surgery.* 1958. Courtesy of Appleton-Century-Crofts, Inc.)

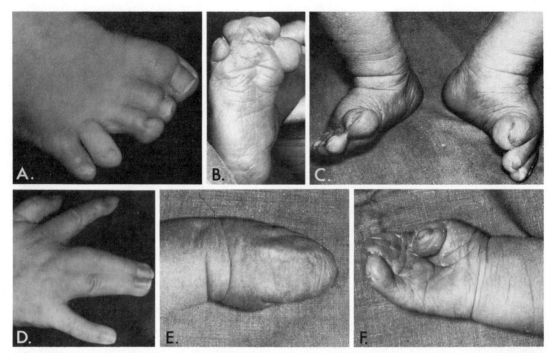

Figure 17–9 Photographs of various types of limb deformity. *A,* Syndactyly, showing skin webs between the first and second and the second and third toes. *B,* Syndactyly involving fusion of all the toes except the fifth. *C,* Syndactyly associated with clubfoot (talipes equinovarus). *D,* Syndactyly involving webbing of the third and fourth fingers. *E* and *F,* Dorsal and palmar views of a child's right hand, showing syndactyly, or fusion, of the second to fifth fingers. (*A* and *D* are from Swenson, O.: *Pediatric Surgery.* 1958. Courtesy of Appleton-Century-Crofts, Inc.)

17–2*E* and *K*). In some cases, there is also fusion of the bones (*synostosis*). Syndactyly is most frequently observed between the third and fourth fingers and between the second and third toes. It is inherited as a simple dominant or simple recessive trait.

Congenital Clubfoot (Fig. 17–9C). Any deformity of the foot involving the talus (ankle bone) is called clubfoot, or talipes (L. *talus,* heel, ankle + *pes,* foot). Clubfoot is a common deformity, occurring about once in 1000 births. It is characterized by an abnormal position of the foot that prevents normal weight bearing. As the child develops, he or she tends to walk on the ankle rather than on the sole of the foot; hence, the name *talipes.*

Talipes equinovarus (Fig. 17–9C) is the most common type of clubfoot, and occurs about twice as frequently in males. The sole of the foot is turned medially, and the foot is adducted and plantar flexed at the midtarsal joint. There is much uncertainty about the cause of clubfoot (Warkany, 1971). Although it is commonly stated that this condition

results from abnormal positioning or restricted movement of the fetus's lower limbs in utero, the evidence for this is inconclusive. Hereditary factors are involved in some cases, and it appears that environmental factors are also involved in most cases. *Clubfoot appears to follow a multifactorial pattern of inheritance* (see Chapter 8). Hence, any intrauterine position that results in abnormal positioning of the feet may cause clubfeet if the fetus is genetically predispositioned to this deformity.

Congenital Dislocation of the Hip. This deformity occurs in about 1 of every 1500 newborn infants and is more common in females than in males. The capsule of the hip joint is very relaxed at birth, and there is underdevelopment of the acetabulum of the hip bone and the head of the femur. The actual dislocation almost always occurs after birth.

Two causative factors are commonly suggested: (1) *Abnormal development of the acetabulum.* About 15 per cent of infants with congenital dislocation of the hip are breech deliveries, which suggests that breech posture in prenatal life may result in abnormal development of the acetabulum and the head of the femur. (2) *Generalized joint laxity* appears to be associated with dislocation of the hip. Joint laxity is often a dominantly inherited condition. Hence, congenital dislocation of the hip follows a multifactorial pattern of inheritance (see Chapter 8).

Causes of Limb Malformations. Abnormalities originate at different stages of development. Early suppression of limb development results in amelia (Fig. 17–5A). Later arrest or disturbance of differentiation or growth of the limbs results in various types of meromelia (Fig. 17–5B and C, and 17–6). Like other malformations, some limb deformities are caused by genetic factors, e.g., chromosomal abnormalities, as in trisomy 18 (see Fig. 8–5), or mutant genes, as in brachydactyly (Fig. 17–8A); by environmental factors, e.g., thalidomide (Fig. 17–5); or by a combination of genetic and environmental factors (*multifactorial inheritance*), e.g., congenital dislocation of the hip.

Experimental studies lend support to the suggestion that mechanical influences during intrauterine development may cause some limb malformations. A reduced quantity of amniotic fluid (oligohydramnios) is commonly associated with limb deformities (Dunn, 1976). However, the significance of mechanical influences in the uterus on congenital postural deformities is still open to question (McKeown, 1976).

SUMMARY

The limbs begin to appear toward the end of the fourth week as slight elevations of the ventrolateral body wall. The upper limb buds develop slightly before the lower limb buds. The tissues of the limb buds are derived from two main sources: somatic mesoderm and ectoderm.

Initially, the developing limbs are directed caudally; later, they project ventrally, and finally they rotate on their longitudinal axes. The upper and lower limbs rotate in opposite directions and to different degrees.

The majority of limb malformations appear to be caused by genetic factors; however, many deformities probably result from an interaction of genetic and environmental factors (multifactorial inheritance). Relatively few congenital malformations of the limbs can be attributed to specific environmental teratogens, except those resulting from thalidomide.

SUGGESTIONS FOR ADDITIONAL READING

Lloyd-Roberts, G. C.: Orthopedic abnormalities; *in* A. P. Norman (Ed.): *Congenital Abnormalities in Infancy,* 2nd ed. Oxford, Blackwell Scientific Publications, 1971, pp. 269–287. A clinically oriented discussion of the many abnormalities of the limbs. Treatment of the deformities and the expected results are also discussed.

Trueta, J.: The growth and development of bones and joints: Orthopaedic aspects; *in* J. A. Davis and J. Dobbing (Eds.): *Scientific Foundations of Paediatrics.* Philadelphia, W. B. Saunders Co., 1974, pp. 399–419. A comprehensive, well-illustrated account of the growth and development of bones and joints. Conditions caused by disturbed growth of bones and joints are discussed.

CLINICALLY ORIENTED PROBLEMS

1. Do more female infants have *congenital dislocation of the hip* than male infants? Are the hip joints of these infants usually dislocated at birth?
2. Are limb malformations similar to those caused by the drug thalidomide common? What was the characteristic *malformation syndrome* produced by thalidomide? Name the limb defects associated with this syndrome.
3. What is the most common type of clubfoot (talipes)? How common is it? Describe the feet of infants born with this malformation.
4. Is *syndactyly* common? Does it occur more often in the hands than in the feet? What is the embryological basis of syndactyly?

The answers to these questions are given at the back of the book.

REFERENCES

Barsky, A. J.: The upper extremity. The lower extremity; *both in* J. C. Mustardé (Ed.): *Plastic Surgery in Infancy and Childhood.* Edinburgh, E. & S. Livingstone, Ltd., 1971, pp. 463–509.

Blechschmidt, E.: The early stages of human limb development; *in* C. A. Swinyard (Ed.): *Limb Development and Deformity: Problems of Evaluation and Rehabilitation.* Springfield, Ill., Charles C. Thomas, Publisher, 1969, pp. 24–55.

Dunn, P. M.: Congenital postural deformities. *Br. Med. Bull. 32*:65, 1976.

Frantz, C. H., and O'Rahilly, R.: Congenital skeletal limb deficiencies. *J. Bone Joint Surg., 43A*:1202, 1961.

Keegan, J. J., and Garrett, F. D.: The segmental distribution of the cutaneous nerves in the limbs of man. *Anat. Rec. 102*:409, 1948.

Lamy, M., and Maroteaux, P.: The genetic study of limb malformations; *in* C. A. Swinyard (Ed.): *Limb Development and Deformity: Problems of Evaluation and Rehabilitation.* Springfield, Ill., Charles C Thomas, Publisher, 1969, pp. 170–174.

Lenz, W., and Knapp, K.: Foetal malformations due to thalidomide. *Ger. Med. Mon. 7*:253, 1962.

McKeown, T.: Human malformations: introduction; *in* C. L. Berry (Ed.): Human Malformations. *Br. Med. Bull. 32*:1, 1976.

Moore, K. L.: The vulnerable embryo: causes of malformation in man. *Manit. Med. Rev. 43*:306, 1963.

O'Rahilly, R.: Normal development of the human embryo; *in* C. H. Frantz (Ed.): *Normal and Abnormal Embryological Development.* Washington, National Research Council, 1967, pp. 3–15.

O'Rahilly, R., and Gardner, E.: The timing and sequence of events in the development of the limbs in the human embryo. *Anat. Embryol. 148*:1, 1975.

Rubin, A., and Friedenberg, Z. B.: Musculoskeletal system; *in* A. Rubin (Ed.): *Handbook of Congenital Malformations.* Philadelphia, W. B. Saunders Co., 1967, pp. 51–85.

Saunders, J. W.: Control of growth patterns in limb development; *in* C. H. Frantz (Ed.): *Normal and Abnormal Embryological Development.* Washington, National Research Council, 1967, pp. 16–26.

Smith, D. W.: *Recognizable Patterns of Human Malformation: Genetic, Embryologic and Clinical Aspects.* 2nd ed. Philadelphia, W. B. Saunders Co., 1976, pp. 166–177.

Swinyard, C. A. (Ed.): *Limb Development and Deformity: Problems of Evaluation and Rehabilitation.* Springfield, Ill., Charles C Thomas, Publisher, 1969.

Warkany, J.: *Congenital Malformations: Notes and Comments.* Chicago, Year Book Medical Publishers, Inc., 1971.

Warkany, J., and Kirkpatrick, J. A., Jr.: Skeletal defects; *in* V. C. Vaughan and R. J. McKay (Eds.):*Nelson Textbook of Pediatrics.* 10th ed. Philadelphia, W. B. Saunders Co., 1975, pp. 1470–1490.

Wolpert, L.: Mechanisms of limb development and malformation. *Br. Med. Bull. 32*:65, 1976.

Zwilling, E.: Abnormal morphogenesis in limb development; *in* C. A. Swinyard (Ed.): *Limb Development and Deformity: Problems of Evaluation and Rehabilitation.* Springfield, Ill., Charles C Thomas, Publisher, 1969, pp. 100–118.

THE NERVOUS SYSTEM

The nervous system develops from a thickened area of embryonic ectoderm called the *neural plate* (Fig. 18–1*A*), which appears during developmental stage 8 (about 18 days). Experimental studies in lower vertebrates have shown that the underlying notochord and paraxial mesoderm constitute a *chordamesoderm field* of organizer action that induces the overlying ectoderm to differentiate into the neural plate.

Formation of the *neural tube* and *neural crest* from the neural plate is described in Chapter 4 and is illustrated in Figure 18–1. The neural tube differentiates into the *central nervous system,* consisting of the brain and spinal cord, and the neural crest gives rise to most of the *peripheral nervous system,* consisting of cranial, spinal, and autonomic ganglia and nerves (Fig. 18–8). In addition, *neural crest cells* differentiate into Schwann cells, pigment cells, odontoblasts, meninges, and bone of branchial arch origin (see Chapters 4 and 10).

THE CENTRAL NERVOUS SYSTEM

Formation of the neural tube begins during stage 10 of development (22 to 23 days) in the region of the fourth to sixth somites (see Fig. 5–8*A*), which represents the future cervical region. At this stage, the cranial two thirds of the neural plate and neural tube, as far caudal as the fourth pair of somites, represent the *future brain,* and the caudal one third of the neural tube and neural plate represents the *future spinal cord.* Fusion of the neural folds proceeds in a somewhat irregular fashion in cranial and caudal directions. The neural tube is temporarily open at both ends, where it communicates freely with the amniotic cavity (see Figs. 5–9*B,* 18–1, and 18–2). The cranial opening, called the *rostral neuropore,* closes on about the twenty-fifth day during developmental stage 11 (see Fig. 5–10), and the *caudal neuropore*

closes during the next stage, about two days later (Fig. 18–2*D*).

The walls of the neural tube become thickened to form the brain and the spinal cord (Fig. 18–3). The lumen of the neural tube is converted into the *ventricular system of the brain* and the central canal of the spinal cord.

THE SPINAL CORD

The neural tube caudal to the fourth pair of somites develops into the spinal cord (Figs. 18–3 and 18–4). The lateral walls of the neural tube thicken until only a minute *central canal* is present at 9 to 10 weeks (Fig. 18–4*C*). The wall of the neural tube is initially composed of a thick, pseudostratified, columnar neuroepithelium (Fig. 18–4*A* and *D*). These neuroepithelial cells constitute the *ventricular zone (ependymal layer)* and give rise to all neurons and macroglial cells of the spinal cord (Fig. 18–5). Macroglial cells, or macroglia, are the larger types of neuroglial cells (astrocytes and oligodendrocytes). Soon, a *marginal zone* composed of the outer parts of the neuroepithelial cells becomes recognizable (Fig. 18–4*E*). This zone gradually becomes the white matter of the cord as axons grow into it and over it from nerve cell bodies in the spinal cord, in the spinal ganglia, and in the brain.

Some dividing neuroepithelial cells in the ventricular zone differentiate into primitive neurons called *neuroblasts.* The neuroblasts arise exclusively by division of neuroepithelial cells. These cells form an *intermediate zone (mantle layer)* between the ventricular and marginal zones (Fig. 18–4*E*). Neuroblasts become *neurons* as cytoplasmic processes develop (Fig. 18–5). The longest cytoplasmic processes are axons.

The primitive supporting cells of the central nervous system, the *glioblasts* (spongioblasts) differentiate from neuroepithelial

375

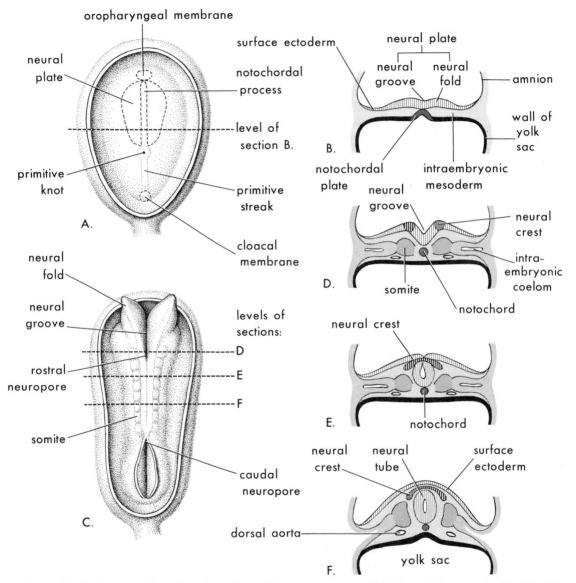

Figure 18-1 Diagrams illustrating formation of the neural crest and folding of the neural plate into the neural tube. *A,* Dorsal view of an embryo during developmental stage 8 (about 18 days), exposed by removing the amnion. *B,* Transverse section of this embryo showing the neural plate and early development of the neural groove. The developing notochord is also shown (for details of its development, see Chapter 4). *C,* Dorsal view of an embryo during developmental stage 10 (about 22 days). The neural folds have fused opposite the somites but are widely spread out at both ends of the embryo. The rostral and caudal neuropores are indicated. *D, E,* and *F,* Transverse sections of this embryo at the levels shown in *C,* illustrating formation of the neural tube and its detachment from the surface ectoderm. Note that some neuroectodermal cells are not included in the neural tube but remain between it and the surface ectoderm as the neural crest. These cells first appear as paired columns on the dorsolateral aspect of the neural tube, but they soon become broken up into a series of segmental masses. See Figure 18-8 for derivatives of neural crest cells.

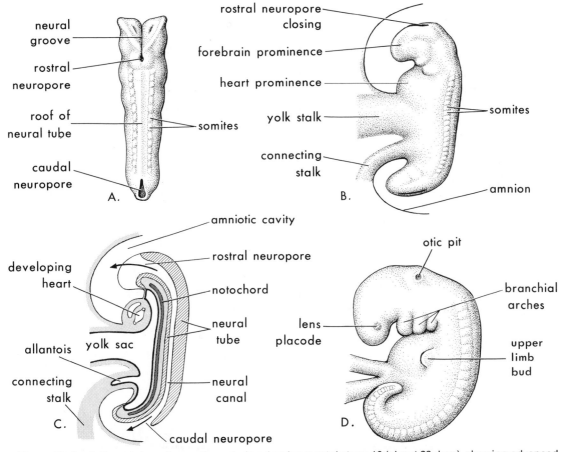

neural groove

rostral neuropore

roof of neural tube

somites

caudal neuropore

A.

rostral neuropore closing

forebrain prominence

heart prominence

yolk stalk

connecting stalk

somites

amnion

B.

amniotic cavity

rostral neuropore

notochord

neural tube

neural canal

developing heart

yolk sac

allantois

connecting stalk

C.

caudal neuropore

otic pit

branchial arches

lens placode

upper limb bud

D.

Figure 18–2 *A,* Dorsal view of an embryo during developmental stage 10 (about 23 days), showing advanced fusion of the neural folds. *B,* Lateral view of an embryo during developmental stage 11 (about 24 days), showing the forebrain prominence and closing of the rostral neuropore. *C,* Sagittal section of this embryo showing the transitory communication of the neural canal with the amniotic cavity (arrows). *D,* Lateral view of an embryo during developmental stage 12 (about 26 days). The neuropores are closed and the three primary brain vesicles, illustrated in Figure 18–3*A,* are present but are not visible.

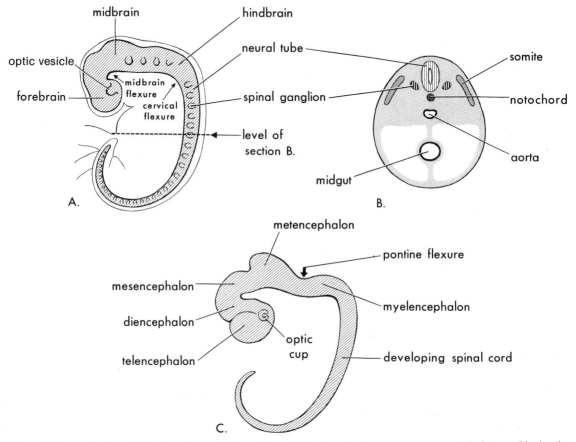

Figure 18-3 *A*, Schematic lateral view of an embryo during developmental stage 13 (about 28 days), showing the three primary brain vesicles. The two flexures demarcate the primary divisions of the brain. *B*, Transverse section of this embryo, showing the neural tube that will develop into the spinal cord in this region. The spinal ganglia derived from the neural crest are also shown. *C*, Schematic lateral view of the central nervous system of a six-week embryo, showing the secondary brain vesicles and the pontine flexure. The flexures occur as the brain grows rapidly and are important factors in determining the final shape of the brain.

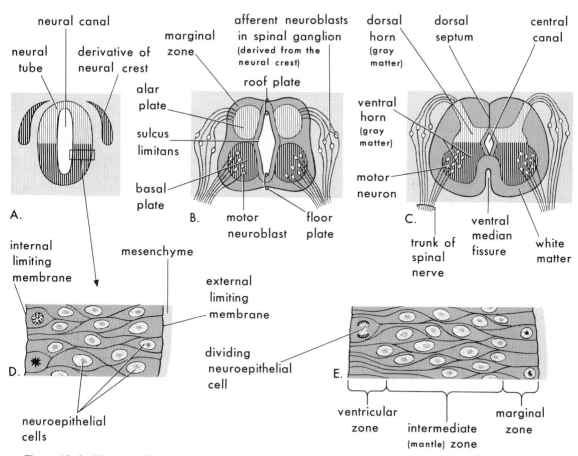

Figure 18–4 Diagrams illustrating development of the spinal cord. *A,* Transverse section through the neural tube of an embryo during developmental stage 10 (about 23 days). *B* and *C,* Similar sections at six and nine weeks, respectively. *D,* Section through the wall of the early neural tube. *E,* Section through the wall of the developing spinal cord showing the three different zones. In drawings A to C, note that the neural canal of the neural tube is converted into the central canal of the spinal cord.

cells mainly after neuroblast formation has ceased. The glioblasts migrate from the ventricular zone into the intermediate and marginal zones; some become *astroblasts* and then *astrocytes,* whereas other glioblasts become *oligodendroblasts* and, later, *oligodendrocytes.* When the neuroepithelial cells cease producing neuroblasts and glioblasts, they differentiate into *ependymal cells,* which give rise to the *ependyma,* or ependymal epithelium, lining the central canal of the spinal cord.

The *microglial cells (microglia),* which are scattered through the gray and white matter, are small neuroglial cells that are derived from *mesenchymal cells* (Fig. 18–5). These cells invade the central nervous system rather late in the fetal period, after it has been penetrated by blood vessels. In view of

their mesodermal origin and their role as scavenger cells, *microglia are classified as part of the reticuloendothelial system,* with properties similar to those of histiocytes in connective tissue (Barr, 1979).

Proliferation and differentiation of the neuroepithelial cells in the developing spinal cord produce thick walls and thin roof and floor plates (Fig. 18–4*B*). Differential thickening of the lateral walls soon produces a shallow longitudinal groove called the *sulcus limitans* (Figs. 18–4*B* and 18–6), which separates the dorsal part, or *alar plate* (lamina), from the ventral part, or *basal plate* (lamina). The alar and basal plates produce longitudinal bulges extending through most of the developing spinal cord's length on each side in its lateral walls. This regional separation is of fundamental importance be-

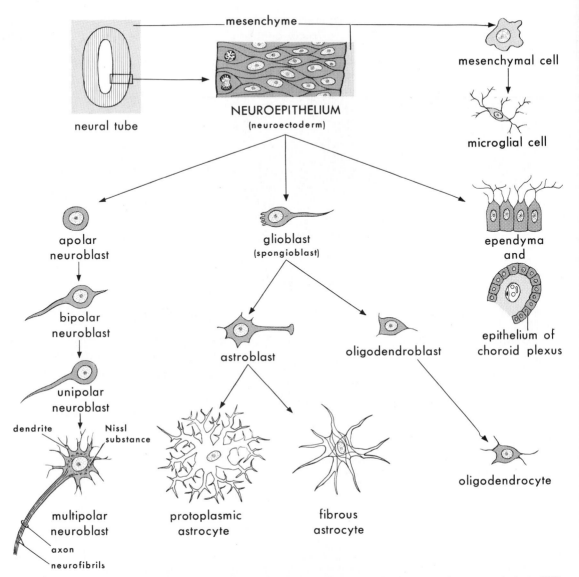

Figure 18–5 Schematic diagram illustrating the histogenesis of cells in the central nervous system. With further development, the multipolar neuroblast (lower left) becomes a nerve cell, or neuron. Neuroepithelial cells give rise to all neurons and macroglial cells. Microglial cells are derived from mesenchymal cells that invade the developing nervous system with the developing blood vessels.

cause the alar and basal plates are later associated with afferent and efferent functions, respectively.

The Alar Plates (Figs. 18–4, 18–6, and 18–7). Cell bodies in the alar plates form the dorsal gray matter in columns that extend the length of the spinal cord. In cross sections, these columns are called *dorsal horns.* Neurons in these columns constitute afferent nuclei, and groups of these nuclei form the dorsal gray columns. As the alar plates enlarge, the *dorsal septum* forms.

The Basal Plates (Figs. 18–4, 18–6, and 18–7). Cell bodies in the basal plates form the ventral and lateral gray columns. In transverse sections of the spinal cord, these columns are commonly called *ventral* and *lateral horns,* respectively. Axons of ventral horn cells grow out of the spinal cord and then receive sheaths formed by Schwann cells, derivatives of neural crest cells (Figs. 18–8 and 18–11). The nerve fibers thus formed are grouped into large bundles called *ventral roots* of the spinal nerves. As the

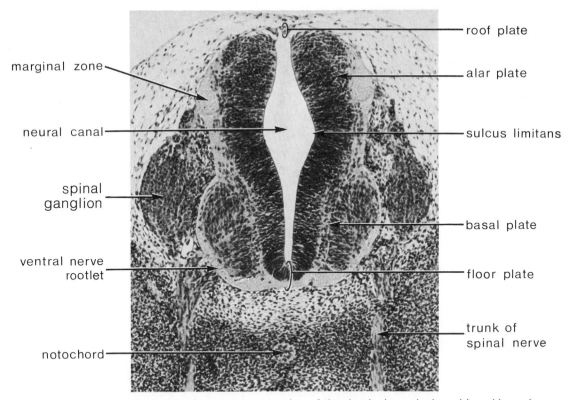

marginal zone

neural canal

spinal ganglion

ventral nerve rootlet

notochord

roof plate

alar plate

sulcus limitans

basal plate

floor plate

trunk of spinal nerve

Figure 18–6 Photomicrograph of a transverse section of the developing spinal cord in a 14-mm human embryo during stage 18 of development, about 44 days (× 75). The dorsal wall (roof plate) and the ventral wall (floor plate) contain no neuroblasts and are relatively thin. (Courtesy of Dr. J. W. A. Duckworth, Professor Emeritus of Anatomy, University of Toronto.)

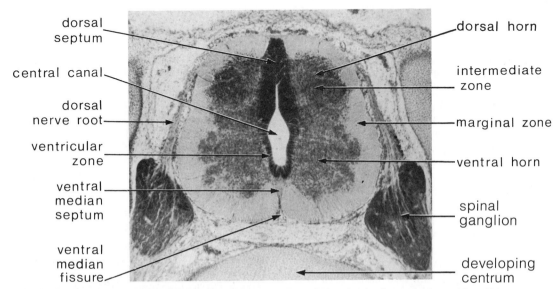

dorsal septum

central canal

dorsal nerve root

ventricular zone

ventral median septum

ventral median fissure

dorsal horn

intermediate zone

marginal zone

ventral horn

spinal ganglion

developing centrum

Figure 18–7 Photomicrograph of a transverse section of the developing spinal cord in a 20-mm human embryo during stage 20, about 50 days (× 60).

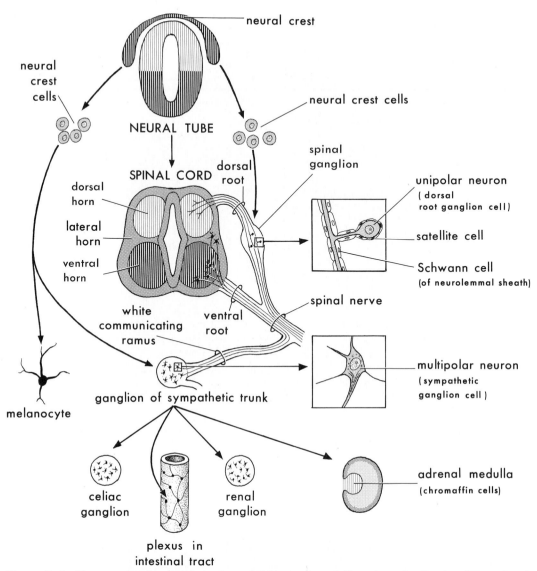

Figure 18–8 Diagram showing the derivatives of the neural crest. Neural crest cells also differentiate into cells of the afferent ganglia of cranial nerves. Formation of a spinal nerve is also illustrated.

basal plates enlarge, they bulge ventrally on each side of the midline and produce the *ventral median septum* and a deep longitudinal groove on the ventral surface of the spinal cord known as the *ventral median fissure* (Fig. 18–7).

Spinal Ganglia (Figs. 18–6 and 18–7). Unipolar neurons in the *spinal ganglia* (dorsal root ganglia) are derived from neural crest cells (Figs. 18–1*F,* 18–8, and 18–9). Their axons divide in a T-shaped fashion into central and peripheral processes. Both processes of spinal ganglion cells have the structural characteristics of axons, but the peripheral one is a dendrite in that there is conduction *toward* the cell body. The central

branches or processes enter the spinal cord and constitute the *dorsal roots* of spinal nerves (Figs. 18–7 and 18–8). Some central processes (axons) of spinal ganglion cells

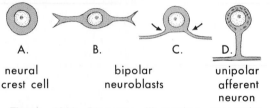

Figure 18–9 Diagrams illustrating successive stages in the differentiation of a neural crest cell into a unipolar afferent neuron of a spinal ganglion (dorsal root ganglion).

end in the dorsal gray column of the spinal cord, whereas others ascend to the brain in the dorsal white columns of the cord. These white columns lie between the right and left dorsal gray columns. The peripheral processes of spinal ganglion cells pass in the spinal nerves (Figs. 18–6 and 18–8) to sensory endings in somatic or visceral structures.

Spinal Meninges. The mesenchyme surrounding the neural tube condenses to form a covering, or membrane, called the *primitive meninx*. The outermost layer of the primitive meninx thickens to form the *dura mater;* the innermost layer remains thin and becomes the *pia-arachnoid*. It is composed of pia mater and arachnoid, which together constitute the *leptomeninges*. Fluid-filled spaces appear within the leptomeninges and soon coalesce to form the *subarachnoid space* (Fig. 18–10). The origin of the pia mater and arachnoid from a single layer is indicated in the adult by the numerous delicate strands of connective tissue *(arachnoid trabeculae)* passing between them.

Positional Changes of the Spinal Cord (Fig. 18–10). In the embryo, the spinal cord extends the entire length of the vertebral canal and the spinal nerves pass through the intervertebral foramina at their levels of origin (Fig. 18–10*A*). Because the vertebral column and dura mater grow more rapidly than the spinal cord, this relationship does not persist. Consequently, the caudal end of the spinal cord gradually comes to lie at relatively higher levels. At six months, it lies at the level of the first sacral vertebra (Fig. 18–10*B*); in the newborn infant, it terminates at the level of the second or third lumbar vertebra (Fig. 18–10*C*); in the adult, it usually terminates at the inferior border of the first lumbar vertebra (Fig. 18–10*D*). This is an average level because the caudal end of the cord may be as high as the twelfth thoracic vertebra or as low as the third lumbar vertebra. As a result, the spinal nerve roots, especially those of the lumbar and sacral segments, run obliquely from the spinal cord to the corresponding level of the vertebral column. The dorsal and ventral nerve roots inferior to the end of the cord *(the conus medullaris)* form a sheaf of nerve roots called the *cauda equina* (Fig. 18–10*D*). Although the dura extends the en-

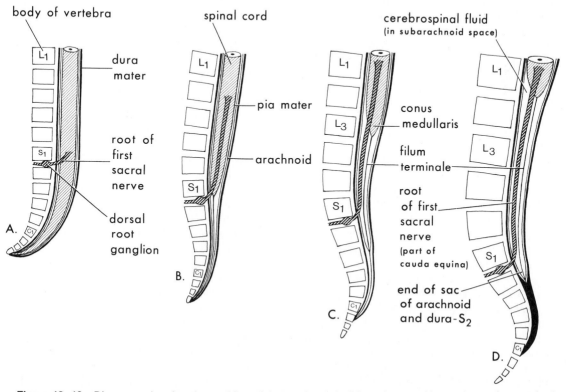

Figure 18–10 Diagrams showing the position of the caudal end of the spinal cord in relation to the vertebral column and the meninges at various stages of development. The increasing inclination of the root of the first sacral nerve is also illustrated. *A,* Eight weeks. *B,* 24 weeks. *C,* Newborn. *D,* Adult.

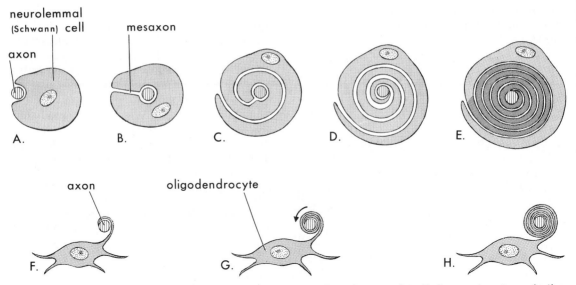

neurolemmal
(Schwann) cell

mesaxon

axon

A.   B.   C.   D.   E.

axon

oligodendrocyte

F.   G.   H.

Figure 18–11 Diagrammatic sketches illustrating myelination of axons. *A* to *E,* Successive stages in the myelination of a peripheral nerve fiber, or axon, by a Schwann cell. The axon first indents the cell, then the Schwann cell rotates around the axon as the mesaxon (site of invagination) elongates. The cytoplasm between the layers of cell membrane gradually condenses. Cytoplasm remains on the inside of the sheath between the myelin and axon. *F* to *H,* Successive stages in the myelination of a nerve fiber in the central nervous system by an oligodendrocyte. A process of the neuroglial cell wraps itself around an axon, and the intervening layers of cytoplasm move to the body of the cell. Myelination in the brain begins in the brain stem and reaches the level of the cerebral hemispheres by birth.

tire length of the vertebral column in the adult, the other layers of the meninges do not. The pia mater beyond the caudal end of the spinal cord forms a long fibrous thread, the *filum terminale,* which extends from the conus medullaris and attaches to the periosteum of the first coccygeal vertebra in the adult. The filum terminale also indicates the line of regression of the embryonic spinal cord.

A portion of the subarachnoid space, from which cerebrospinal fluid may be removed without damaging the spinal cord, extends below the cord. The removal of cerebrospinal fluid by insertion of a needle between certain lumbar vertebrae and into the subarachnoid space is known as *lumbar puncture.* Moore (1980) illustrates this technique. There is little danger that the needle will damage the spinal cord if it is inserted between the spinous processes of L3/L4 or L4/L5 vertebrae.

Myelination (Fig. 18–11). Myelin formation begins in the spinal cord during midfetal life and continues during the first postnatal year. In general, fiber tracts appear to become completely myelinated at about the time they become fully functional.

The myelin sheath surrounding nerve fibers in the spinal cord is formed by *oligo-*

dendrocytes. The lipoprotein plasma membrane of an oligodendrocyte is wrapped around the axon in a number of layers (Fig. 18–11*F* to *G*).

The myelin sheath is formed around axons of peripheral nerve fibers by the plasma membranes of neurolemma cells, or *Schwann cells* (Fig. 18–11*E*). Schwann cells are derived from neural crest cells (Fig. 18–8) that migrate peripherally and wrap themselves around the axons of somatic motor neurons and preganglionic autonomic motor neurons as they pass out of the central nervous system. These cells also wrap themselves around both the central and the peripheral processes of the somatic and visceral sensory neurons, as well as around the axons of postganglionic autonomic motor neurons (Fig. 18–11*A* to *F*).

Electron microscopy has shown that myelin is not only formed by the Schwann cells but consists of its plasma membrane wrapped around the axon (Fig. 18–11*A* to *E*). Hence, myelin is merely a series of condensed cell membranes after the cytoplasm is squeezed back into the body of the cell.

Beginning at about 20 weeks, the nerve fibers begin to have a whitish appearance, owing to the *deposition of myelin.*

CONGENITAL MALFORMATIONS OF THE SPINAL CORD AND/OR THE MENINGES

Most congenital malformations of the spinal cord result from defective closure of the caudal neuropore toward the end of the fourth week of development. Severe *neural tube defects* also involve the tissues overlying the spinal cord (meninges, vertebral arch, dorsal muscles, and skin; Figs. 18–12*B* to *D* and 18–17).

Malformations involving the caudal end of the neural tube and the vertebral arches are referred to as *spina bifida* (divided spine). This term describes nonfusion of the vertebral arches common to all types of spina bifida (Fig. 18–12). These conditions range from clinically significant types to minor, clinically unimportant types.

Spina Bifida Occulta (Fig. 18–12A). This is a *vertebral defect* resulting from failure of the two halves of the vertebral arch to fuse, usually in the sacral, lumbar, and cervical regions (see also Chapter 15). *This defect is in L5 or S1 in about 10 per cent of people.* In its most minor form, there is no defect in the skin, and the only evidence of its presence may be a small dimple with a tuft of hair. *Spina bifida occulta produces no clinical symptoms.*

Spinal Dermal Sinus (Figs. 18–13 and 18–14). A skin dimple in the sacral region may be associated with spina bifida occulta. Such dimples indicate the region of closure of the caudal neuro-

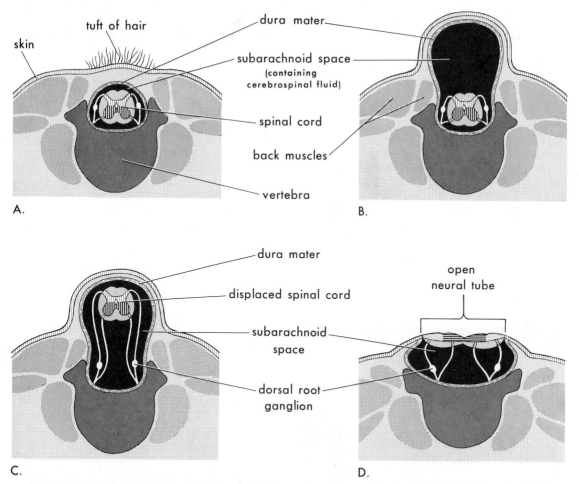

Figure 18–12 Diagrammatic sketches illustrating various types of spina bifida and the commonly associated malformations of the nervous system. *A,* Spina bifida occulta. About 10 per cent of people have this vertebral defect in L5 and/or S1. It causes no back problems. *B,* Spina bifida with meningocele. *C,* Spina bifida with meningomyelocele. *D,* Spina bifida with myeloschisis. (Modified from Patten, B. M.: *Human Embryology.* 2nd ed. Copyright © 1968 by McGraw-Hill, Inc. Used by permission of McGraw-Hill Book Company.)

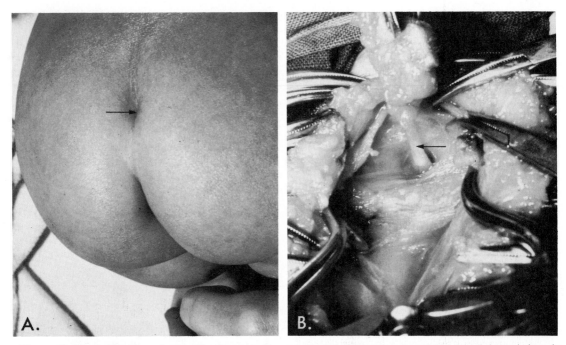

Figure 18–13 *A,* Photograph of a skin dimple in the sacral region. The opening of the spinal dermal sinus is indicated by an arrow. *B,* Photograph taken during removal of this sinus, showing the cord (arrow) connecting the dimple to the spinal dura mater. (Courtesy of Dr. Dwight Parkinson, Children's Centre, Winnipeg, Canada.)

pore at the end of the fourth week, and therefore represent the last place of separation between the surface ectoderm and the neural tube. In some cases, the dimple is connected with the dura mater by a fibrous cord (Fig. 18–13*B*).

A rare but more serious condition exists when a sinus, or fistula, connects the subarachnoid space with the exterior through the dimple on the skin (Fig. 18–14). This fistula may provide a source of recurring meningitis.

Very rarely, some surface ectodermal cells are incorporated into the neural tube during closure of the caudal neuropore, and these may give rise to a

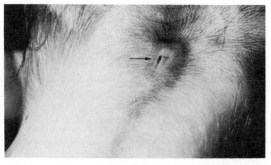

Figure 18–14 Photograph of a cervical dermal sinus (fistula) connecting the exterior (arrow) with the subarachnoid space around the upper cervical region of the spinal cord. (Courtesy of Dr. Dwight Parkinson, Children's Centre, Winnipeg, Canada.)

type of spinal cord tumor called an *intramedullary dermoid* (List, 1941).

Spina Bifida Cystica (Figs. 18–12 and 18–15 to 18–17). Severe types of spina bifida, involving protrusion of the spinal cord and/or the meninges through the defect in the vertebral arch, are often referred to collectively as *spina bifida cystica* because of the cyst-like protrusion, or sac, that is associated with these malformations. When the sac contains meninges and cerebrospinal fluid, the condition is called *spina bifida with meningocele* (Fig. 18–12*B*). The spinal cord and spinal roots are in their normal position, but there may be spinal cord abnormalities. If the spinal cord and/or nerve roots are included in the sac, the malformation is called *spina bifida with meningomyelocele* (Fig. 18–12*C*, 18–15, and 18–16). The spinal cord used to be called the spinal medulla; thus, *myelo* (Gr. *myelos,* medulla) is included in the name of this malformation.

In meningomyeloceles, there is often a marked *neurological deficit* inferior to the level of the protruding sac. This deficit occurs because nervous tissue is often incorporated in the wall of the sac, a condition that impairs early development of nerve fibers. Meningomyeloceles may be covered

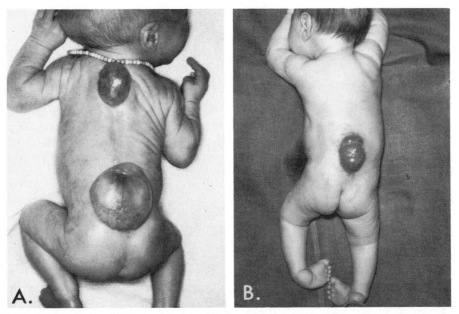

Figure 18–15   Photographs of infants with spina bifida cystica. *A,* Spina bifida with meningomyelocele in the thoracic and lumbar regions. *B,* Spina bifida with myeloschisis in the lumbar region (see also Figs. 18–12*D* and 18–17). Note the nerve involvement affecting the lower limbs. (Courtesy of Dr. Dwight Parkinson, Children's Centre, Winnipeg, Canada.)

by skin or by a thin, easily ruptured membrane. *Spina bifida cystica occurs about once in every 1000 births.*

Meningomyelocele is a more common and a very much more severe malformation than meningocele. Meningoceles and meningomyeloceles may occur anywhere along the spinal axis, but they are most common in the lumbar region. Some cases of meningomyelocele are associated with *craniolacunia,* or

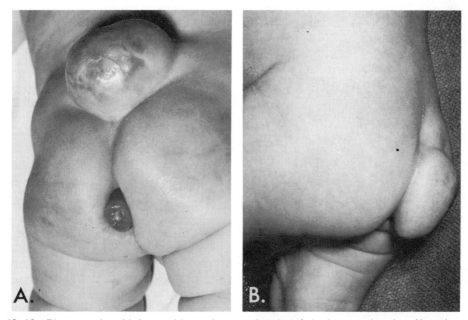

Figure 18–16   Photographs of infants with meningomyeloceles. *A,* In the sacral region. Note the prolapsed rectum. *B,* In the sacrococcygeal region. Note the good covering of skin over this neural tube defect. (Courtesy of Dr. Dwight Parkinson, Children's Centre, Winnipeg, Canada.)

defective development of the calvaria, which results in depressed nonossified areas on the inner surfaces of the bones of the calvaria.

Spina bifida cystica shows considerable geographical variation in incidence; in the British Isles, for example, the incidence varies from 4.2 per 1000 newborn infants in South Wales to 1.5 per 1000 in southeastern England (Laurence and Weeks, 1971).

The most severe type of spina bifida cystica is called spina bifida with myeloschisis (Figs. 18–12D, 18–17, and 18–19). In these cases, the neural plate is open because the neural folds failed to meet and fuse. As a result, the spinal cord in the area concerned is represented by a flattened mass of nervous tissue. Extensive *myeloschisis associated with rachischisis,* as shown in Figure 18–19, is rare. It is much more usual for a short part of the neural tube to fail to form (Figs. 18–12D

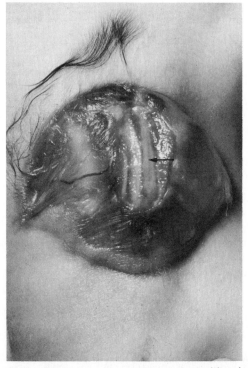

Figure 18–17 Photograph of an infant with spina bifida with myeloschisis in the lumbar region. The neural plate is surrounded by a delicate semitransparent membrane. The arrow indicates the open neural tube. The central canal of the spinal cord opens into the sac. Note the tufts of hair on the surrounding skin. (From Laurence, K. M., and Weeks, R.: Abnormalities of the central nervous system; *in* A. P. Norman [Ed.]: *Congenital Abnormalities in Infancy.* 2nd ed. 1971. Courtesy of Blackwell Scientific Publications.)

and 18–17). As the spinal cord is examined longitudinally, it can be observed to spread out and to enter a *cystic dilation of the subarachnoid space,* across which the roots of the spinal nerves pass. In these cases, the central canal of the spinal cord opens into the cystic dilation of the subarachnoid space.

There is suggestive evidence that myeloschisis and the resulting spina bifida are caused by local overgrowth of the neural plate (Fig. 18–18). Infants with severe forms of myeloschisis, as illustrated in Figure 18–19, have no chance of survival.

Spina bifida cystica shows varying degrees of neurological deficit, depending on the position and extent of the lesion. There is usually a corresponding dermatome loss of sensation, along with complete or partial skeletal muscle paralysis (Fig. 18–15B). The level of the lesion determines the area of anesthesia (area of skin without sensation) and the muscles affected. *Sphincter paralysis* (bladder and/or anal sphincters) is common with lumbosacral meningomyeloceles (Fig. 18–16A). There is almost invariably a *saddle anesthesia* when the sphincters are involved, that is, loss of sensation in the region that impinges on the saddle during riding, corresponding roughly to the area of the buttocks, perineum, and inner aspects of the thighs.

Diastematomyelia is associated with spina bifida in rare cases. The spinal cord is split into halves by a bony spicule or a fibrous band. Each half of the spinal cord is surrounded by a dural sac. Diagnosis of the condition is based on radiographs of the vertebral column and on myelograms. The spicule of bone or fibrous band can be removed surgically, and the prognosis is good if the operation is performed early (Perret, 1960).

Spina bifida cystica and/or anencephaly (Fig. 18–19) is strongly suspected in utero when there is a high level of alpha fetoprotein in the amniotic fluid. Alpha fetoprotein may also be elevated in the maternal blood serum in these cases, and methods are being developed that will allow screening of all pregnant women for elevated serum alpha fetoprotein levels (Thompson and Thompson, 1980). An *amniocentesis* would be performed on pregnant women with high levels of alpha fetoprotein, if the patient approved, for the determination of the alpha fetoprotein level in the amniotic fluid. An ultrasound scan would also be requested to try to confirm the presence of a neural tube defect. The

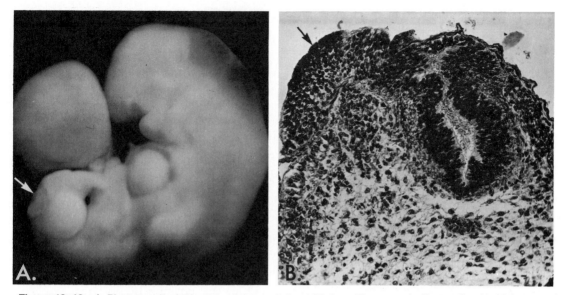

Figure 18–18   *A,* Photograph of a human embryo of about 28 days. The arrow indicates the site of the *neural tube defect* resulting from failure of closure of the caudal neuropore. Normally the caudal neuropore closes on day 27. *B,* Photomicrograph of a transverse section through the defect. The arrow indicates an abnormal fold of neural tissue extending over the left side of the embryo. It appears that this overgrown neural fold has prevented closure of the neural tube. (From Lemire, R. J., Shepard, T. H., and Alvord, E. J., Jr.: *Anat. Rec. 152*:9, 1965.)

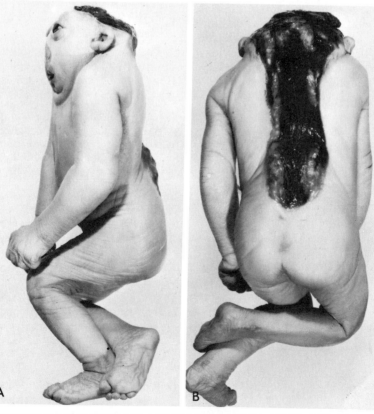

Figure 18–19   Photographs of an infant with acrania (absence of the cranial vault), anencephaly (absence of the forebrain), rachischisis (failure of fusion of several vertebral arches), and spina bifida with myeloschisis (failure of closure of the neural folds). These infants are usually born dead (stillborn) or die shortly after birth.

fetal vertebral column can be detected by ultrasound at 10 to 12 weeks' gestation, and, if present, a *spina bifida cystica,* such as meningocele or meningomyelocele, is sometimes visible as a cystic mass adjacent to the affected area of the vertebral column (Page et al., 1981).

THE BRAIN

The neural tube cranial to the fourth pair of somites develops into the brain (Gr. *enkephalos,* brain). The adult brain is described as consisting of a number of regions; the relation of these divisions to each other will be better understood after development of the brain has been considered.

Fusion of the neural folds in the cranial region and *closure of the rostral neuropore* result in the formation of three primary brain vesicles (Fig. 18–20), from which the brain develops.

The Brain Vesicles (Figs. 18–3 and 18–20). During the fourth week *three primary brain vesicles* form: the *forebrain,* or prosencephalon, the *midbrain,* or mesencephalon, and the *hindbrain,* or rhombencephalon (Figs. 18–3A and 18–20). During the fifth week, the forebrain partly divides into two vesicles, the *telencephalon* and the *diencephalon,* and the hindbrain partly divides into the *metencephalon* and the *myelencephalon.* As a result, there are *five secondary brain vesicles* (Fig. 18–20B).

The Brain Flexures (Figs. 18–3 and 18–21). During the fourth week, the brain grows rapidly and bends, or flexes, ventrally with the head fold (see Chapter 5). This produces the *midbrain flexure* in the midbrain region and the *cervical flexure* at the junction of the hindbrain and the spinal cord. Later, between these flexures, unequal growth in the hindbrain produces the *pontine flexure* in the opposite direction (Figs. 18–3C and 18–21A). This flexure results in thinning of the roof of the hindbrain (Fig. 18–21D).

Initially, the developing brain has the same basic structure as the developing spinal cord; however, the brain flexures produce considerable variation in the outline of transverse sections at different levels of the brain and in the position of the gray and white matter. The *sulcus limitans* extends cranially to the junction between the midbrain and the forebrain, and the alar and basal plates are recognizable only in the midbrain and the hindbrain.

THE HINDBRAIN (RHOMBENCEPHALON)

The cervical flexure demarcates the hindbrain vesicle from the developing spinal cord (Figs. 18–3A and 18–21A). Later, this junction is arbitrarily defined as the level of the

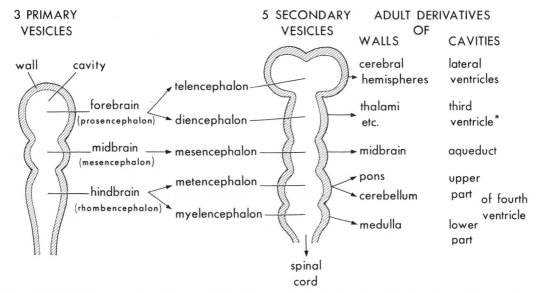

3 PRIMARY VESICLES

wall   cavity

forebrain (prosencephalon)

midbrain (mesencephalon)

hindbrain (rhombencephalon)

5 SECONDARY VESICLES

telencephalon

diencephalon

mesencephalon

metencephalon

myelencephalon

spinal cord

ADULT DERIVATIVES OF

WALLS   CAVITIES

cerebral hemispheres   lateral ventricles

thalami etc.   third ventricle*

midbrain   aqueduct

pons   upper part of fourth ventricle
cerebellum

medulla   lower part

Figure 18–20 Diagrammatic sketches of the brain vesicles, indicating the adult derivatives of their walls and cavities. *The anterior part of the third ventricle forms from the cavity of the telencephalon; most of the third ventricle is derived from the cavity of the diencephalon.

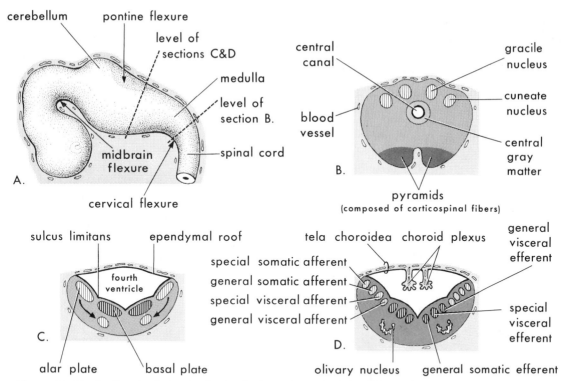

Figure 18–21   *A,* Sketch of the developing brain at the end of the fifth week, showing the three primary divisions of the brain and the brain flexures. *B,* Transverse section through the caudal part of the myelencephalon (developing closed part of the medulla). *C* and *D,* Similar sections through the rostral part of the myelencephalon (developing "open" part of the medulla), showing the position and successive stages of differentiation of the alar and basal plates. The arrows in *C* show the pathway taken by neuroblasts from the alar plates to form the olivary nuclei.

superior rootlet of the first cervical nerve, roughly at the foramen magnum. The pontine flexure appears in the hindbrain vesicle in the future pontine region. The pontine flexure divides the hindbrain into caudal (myelencephalon) and rostral (metencehpalon) parts. The myelencephalon becomes the *medulla* (oblongata), and the metencephalon gives rise to the *pons* and *cerebellum.* The cavity of the hindbrain becomes the fourth ventricle and the central canal of the caudal part of the medulla.

The Myelencephalon (Fig. 18–21). The caudal part of the myelencephalon (closed portion of medulla) resembles the spinal cord both developmentally and structurally. The lumen of the neural tube becomes a small central canal. Unlike those of the spinal cord, neuroblasts from the alar plates migrate into the marginal zone and form isolated areas of gray matter, the *gracile nucleus* medially and the *cuneate nucleus* laterally. These nuclei are associated with correspond-

ingly named tracts that enter the medulla from the spinal cord. The ventral area of the medulla contains a pair of fiber bundles, called the *pyramids,* consisting of corticospinal fibers descending from the developing cerebral cortex.

The rostral part of the myelencephalon ("open" portion of medulla) is wide and rather flat, especially opposite the pontine flexure (Fig. 18–21C and D). The pontine flexure causes the lateral walls of the medulla to move outward like the pages of an opening book. It also causes the roof plate to become stretched and greatly thinned. In addition, the cavity of this part of the myelencephalon (part of future fourth ventricle) becomes somewhat rhomboidal, or diamond-shaped. The walls of the medulla move laterally, and the alar plates come to lie lateral to, rather than dorsal to, the basal plates. As the positions of the plates change, the motor nuclei generally develop medial to sensory nuclei (Fig. 18–21C).

Neuroblasts in the basal plates of the medulla, like those of the spinal cord, develop into motor nuclei and organize into three cell columns, or nuclei, on each side (Fig. 18–21D). From medial to lateral, they are (1) *general somatic efferent,* represented by neurons of the hypoglossal nerve; (2) *special visceral (branchial) efferent,* represented by the neurons innervating muscles derived from the branchial arches (see Chapter 10); and (3) *general visceral efferent,* represented by some neurons of the vagus and glossopharyngeal nerves.

Neuroblasts of the alar plates form four columns, or nuclei, on each side. From medial to lateral, they are (1) *general visceral afferent,* receiving impulses from the viscera; (2) *special visceral afferent,* receiving taste fibers; (3) *general somatic afferent,* receiving impulses from the surface of the head; and (4) *special somatic afferent,* receiving impulses from the ear. Some neuroblasts from the alar plates migrate ventrally and form the olivary nuclei (Fig. 18–21C and D).

The Metencephalon (Fig. 18–22). The

walls of the metencephalon form the pons and the cerebellum, and its cavity forms the superior part of the fourth ventricle. As in the rostral part of the myelencephalon, the pontine flexure causes divergence of the lateral walls of the pons and spreads the gray matter in the floor of the fourth ventricle. As in the myelencephalon, neuroblasts in each basal plate develop into motor nuclei and organize into three columns, or nuclei, on each side.

The *cerebellum* develops from symmetrical thickenings of dorsal parts of the alar plates. Initially, the cerebellar swellings project as small bulges into the fourth ventricle (Fig. 18–22B). As the cerebellar swellings enlarge and fuse in the midline, they soon overgrow the rostral half of the fourth ventricle and overlap the pons and medulla (Fig. 18–22D). Some neuroblasts in the intermediate (mantle) zone of the alar plates migrate

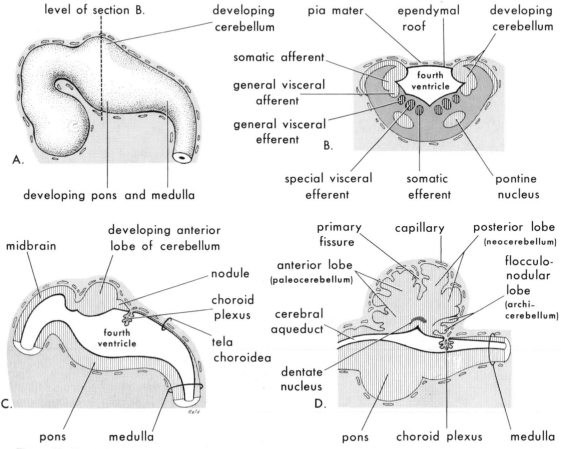

Figure 18–22 *A,* Sketch of the developing brain at the end of the fifth week. *B,* Transverse section through the metencephalon (developing pons and cerebellum), showing the derivatives of the alar and basal plates. *C* and *D,* Sagittal sections of the hindbrain at about 6 and 17 weeks, respectively, showing successive stages of development of the pons and cerebellum.

to the marginal zone and form the *cerebellar cortex;* other neuroblasts from these plates give rise to the central nuclei, the largest of which is the *dentate nucleus.* Cells from the alar plates also give rise to the *pontine nuclei,* the *cochlear* and *vestibular nuclei,* and the *sensory nuclei of the trigeminal nerve.*

Divisions of the Cerebellum (Fig. 18–22C and D). The structure of the human cerebellum reflects its phylogenetic development. The *archicerebellum* (flocculonodular lobe), the oldest part phylogenetically, has many connections with the vestibular apparatus. The *paleocerebellum* (vermis and anterior lobe) is of more recent development and is associated with sensory data from the limbs. The *neocerebellum* (posterior lobe), the newest part phylogenetically, is concerned with selective control of limb movement.

Nerve fibers connecting the cerebral and cerebellar cortices with the spinal cord pass through the marginal layer of the ventral region of the metencephalon. This region of the brain stem is called the *pons* (L., a bridge) because of the robust band of nerve fibers that crosses the median plane and forms a bulky ridge on its anterior and lateral aspects.

Choroid Plexuses and Cerebrospinal Fluid (Figs. 18–21, 18–22, and 18–26). The thin ependymal roof of the fourth ventricle is covered externally by vascular *pia mater,* derived from mesenchyme associated with the hindbrain. This vascular pia mater, together with the ependymal roof, forms the *tela choroidea.* Because of active proliferation of the vascular pia mater, the tela choroidea invaginates into the fourth ventricle and differentiates into the tufted *choroid plexus.* Similar plexuses develop in the roof of the third ventricle and in the medial walls of the lateral ventricles. Four choroid plexuses are formed; these are responsible for the secretion of ventricular fluid that becomes *cerebrospinal fluid* when additions are made to it from the surfaces of the brain and spinal cord, and from the pia-arachnoid layer of the meninges.

During the mid-fetal period, the thin roof of the fourth ventricle bulges outward in three locations and ruptures to form foramina. The median and lateral apertures (foramen of Magendie and foramina of Luschka, respectively) permit the ventricular fluid to enter the *subarachnoid space* from the fourth ventricle.

The main site of absorption of cerebrospinal fluid into the venous system is through the *arachnoid villi,* which are protrusions of the arachnoid that extend into the *dural venous sinuses.* These villi consist of a thin cellular layer derived from the epithelium of the arachnoid and the endothelium of the sinus.

THE MIDBRAIN (MESENCEPHALON)

The midbrain undergoes less change than any other part of the developing brain, except the most caudal part of the hindbrain. The neural canal narrows to form the *cerebral aqueduct* (Fig. 18–22D), a canal that joins the third and fourth ventricles.

Neuroblasts migrate from the alar plates into the roof, or *tectum,* and aggregate to form four large groups of neurons, the paired *superior* and *inferior colliculi* (concerned with visual and auditory reflexes, respectively).

Neuroblasts from the basal plates give rise to groups of neurons in the *tegmentum* (red nuclei, nuclei of the third and fourth cranial nerves, and the reticular nuclei). Fibers growing from the cerebrum form the cerebral peduncles anteriorly.

The *substantia nigra* (black nucleus), a broad layer of gray matter adjacent to the cerebral peduncle, is also believed to differentiate from the basal plate. The cerebral peduncles become progressively more prominent as more descending fiber groups (corticopontine, corticobulbar, and corticospinal) pass through the developing midbrain on their way to the brain stem and the spinal cord (Fig. 18–23 E).

THE FOREBRAIN (PROSENCEPHALON)

Before closure of the rostral neuropore, two lateral outgrowths, or diverticula, called *optic vesicles* (Fig. 18–3A) appear, one on each side of the forebrain. The optic vesicles are the primordia of the *retinae* and *optic nerves* (see Chapter 19). A second pair of diverticula soon arise more dorsally and rostrally; these are called the *telencephalic vesicles* (Fig. 18–23C). They are the primordia of the *cerebral hemispheres,* and their cavities become the *lateral ventricles.*

The anterior part of the forebrain, including the primordia of the cerebral hemispheres, is known as the *telencephalon,* and the posterior part of the forebrain is called

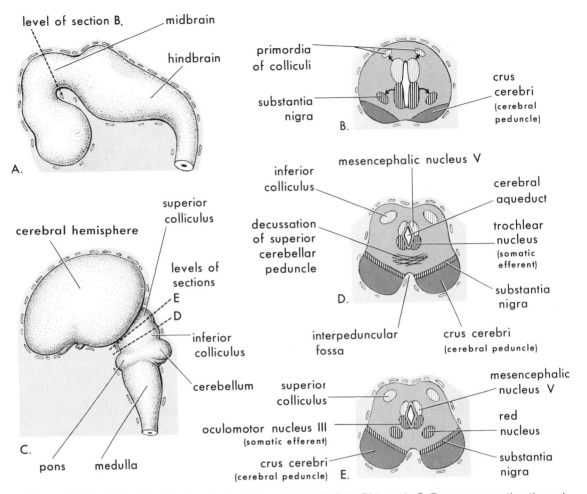

Figure 18–23 *A,* Sketch of the developing brain at the end of the fifth week. *B,* Transverse section through the mesencephalon (developing midbrain), showing early migrations of cells from the basal and alar plates. *C,* Sketch of the developing brain at about 11 weeks. *D* and *E,* Transverse sections of the developing midbrain at the level of the inferior and superior colliculi, respectively.

the *diencephalon.* The cavities of the telencephalon and diencephalon both contribute to the formation of the *third ventricle,* although the cavity of the diencephalon contributes more.

The Diencephalon (Fig. 18–24). Three swellings develop in the lateral walls of the third ventricle; these later become the *epithalamus,* the *thalamus,* and the *hypothalamus.* The thalamus is separated from the epithalamus by the *epithalamic sulcus* and from the hypothalamus by the *hypothalamic sulcus.* The hypothalamic sulcus is not a continuation of the sulcus limitans into the forebrain and does not, like the sulcus limitans, divide sensory and motor areas.

The thalamus on each side develops rapidly and bulges into the cavity of the third ventricle, reducing it to a narrow cleft. The thalami meet in about 80 per cent of brains and fuse in the midline, forming a bridge of gray matter across the third ventricle called the *massa intermedia,* or interthalamic adhesion.

The *hypothalamus* arises by proliferation of neuroblasts in the intermediate zone of the diencephalic walls ventral to the hypothalamic sulci. Later, a number of nuclei concerned with endocrine activities and homeostasis develop. A pair of nuclei, the *mammillary bodies,* form pea-sized swellings on the ventral surface of the hypothalamus (Fig. 18–24*C*).

The *epithalamus* develops from the roof

and dorsal portion of the lateral wall of the diencephalon. Initially, the epithalamic swellings are large, but they later become relatively small. The *pineal body* develops as a midline diverticulum of the caudal part of the diencephalic roof (Fig. 18–24C and D).

Proliferation of cells in its walls soon converts it into a solid cone-shaped organ.

The *paraphysis* sometimes develops as an evagination of the diencephalic roof dorsal to the *interventricular foramen*. This vestigial structure may persist postnatally and give rise to colloid

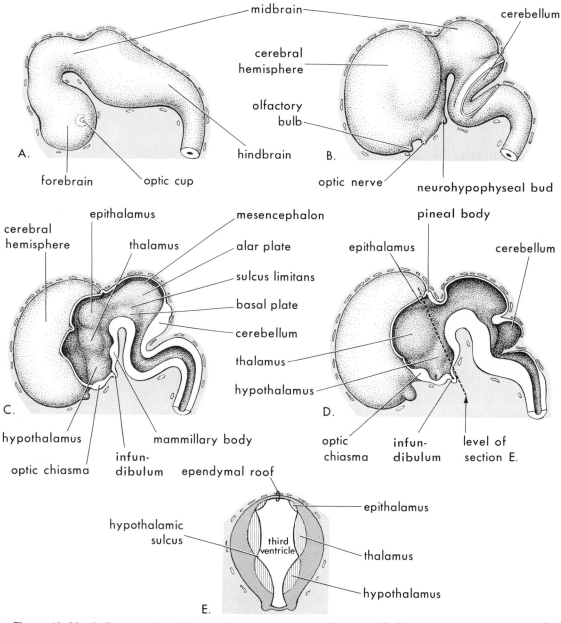

Figure 18–24 *A,* External view of the brain at the end of the fifth week. *B,* Similar view at seven weeks. *C,* Median sagittal section of this brain, showing the medial surface of the forebrain and midbrain. *D,* Similar section at eight weeks. *E,* Transverse section through the diencephalon, showing the epithalamus dorsally, the thalamus laterally, and the hypothalamus ventrally.

cysts (Bull and Sutton, 1949). Colloid cysts usually obstruct the interventricular foramina between the lateral and third ventricles, producing *hydrocephalus* of the lateral ventricles.

The Pituitary Gland (Hypophysis Cerebri) (Figs. 18–25, 18–26, and Table 18–1). This gland develops from two different sources: an upgrowth from the ectoderm of the stomodeum and a downgrowth from the neuroectoderm of the diencephalon. This double origin explains why the pituitary gland is composed of two completely different types of tissue. The *adenohypophysis* (glandular portion) arises from the oral ec-

toderm, and the *neurohypophysis* (nervous portion) originates from the neuroectoderm.

During developmental stage 11 (about 24 days), a diverticulum called *Rathke's pouch* arises from the roof of the *stomodeum* (primitive mouth cavity) and grows toward the brain (Fig. 18–25*A* and *B*). By the fifth week, this pouch has elongated and become constricted at its attachment to the oral epithelium, giving it a nipple-like appearance (Fig. 18–25*C*). By this stage, it has come into contact with the *infundibulum,* a ventral downgrowth, or diverticulum, of the diencephalon (Figs. 18–24 to 18–26).

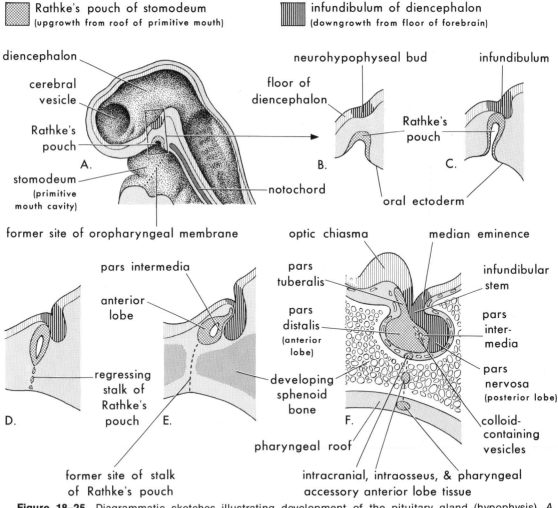

Figure 18–25 Diagrammatic sketches illustrating development of the pituitary gland (hypophysis). *A,* Sagittal section of the cranial end of an embryo during developmental stage 11 (about 24 days), showing Rathke's pouch as an upgrowth from the stomodeum and the neurohypophyseal bud from the forebrain. *B* to *D,* Successive stages of the developing pituitary gland. By eight weeks, Rathke's pouch loses its connection with the oral cavity and is in close contact with the infundibulum, the primordium of the stalk and the posterior lobe of the pituitary (neurohypophysis). *E* and *F,* Later stages, showing proliferation of the anterior wall of Rathke's pouch forming the anterior lobe of the pituitary (adenohypophysis).

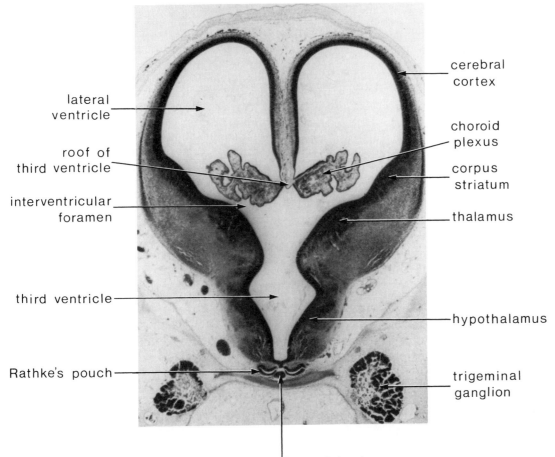

lateral ventricle

roof of third ventricle

interventricular foramen

third ventricle

Rathke's pouch

cerebral cortex

choroid plexus

corpus striatum

thalamus

hypothalamus

trigeminal ganglion

neurohypophyseal bud

Figure 18–26 Photomicrograph of a transverse section through the diencephalon and the cerebral vesicles of a 20-mm human embryo during developmental stage 20 (about 50 days) at the level of the interventricular foramina (× 20). This section is through the plane indicated in Figure 18–28A. The choroid fissure is located at the junction of the choroid plexus and the medial wall of the lateral ventricle.

Adenohypophysis. The parts of the pituitary gland that develop from the ectoderm of the stomodeum (pars anterior, pars intermedia, and pars tuberalis; Table 18–1) are often referred to as the *adenohypophysis.*

The stalk of Rathke's pouch passes between the chondrification centers of the developing presphenoid and basisphenoid bones of the skull (Fig. 18–25E). During the sixth week, the connection of Rathke's pouch with the oral cavity disappears (Fig. 18–25D and E). A remnant of this stalk may persist and give rise to a *pharyngeal hypophysis* in the pharyngeal roof (Fig. 18–

TABLE 18–1   DERIVATION AND TERMINOLOGY OF THE PITUITARY GLAND

Oral Ectoderm			
(From roof of stomodeum) ⟶	Adenohypophysis (glandular portion)	Pars distalis Pars tuberalis Pars intermedia	Anterior lobe
Neuroectoderm			Posterior lobe
(From floor of diencephalon) ⟶	Neurohypophysis (nervous portion)	Pars nervosa Infundibular stem Median eminence	

25*F*). Very rarely, accessory masses of anterior lobe tissue may occur outside the capsule of the gland but within the sella turcica of the sphenoid bone, or in the substance of this bone.

A remnant of the site of the stalk of Rathke's pouch, called the *basipharyngeal canal,* is visible in sections of the newborn sphenoid bone in about 1 per cent of cases. It can also be identified in a small number of skull radiographs of newborn infants (usually those with skull abnormalities). Occasionally, *craniopharyngiomas* develop in the pharynx or in the basisphenoid, but most often they form in and/or above the sella turcica.

During subsequent development, cells of the anterior wall of Rathke's pouch proliferate actively and give rise to the *pars distalis* of the pituitary gland. Later, a small extension, the *pars tuberalis,* grows around the *infundibular stem.* The extensive proliferation of the anterior wall of Rathke's pouch reduces the lumen to a narrow residual cleft (Fig. 18–25*E*); it is usually not recognizable in the adult gland and is represented by a zone of cysts. In humans, cells of the posterior wall of Rathke's pouch do not proliferate; they give rise to the thin, poorly defined *pars intermedia,* which becomes an inconspicuous, discontinuous layer (Fig. 18–25*F*).

Neurohypophysis. The part of the pituitary gland that develops from neuroectoderm (*infundibulum*) is often referred to as the *neurohypophysis.* The infundibulum gives rise to the *median eminence,* the *infundibular stem,* and the *pars nervosa* (Fig. 18–25*F*). Initially, the walls of the infundibulum are thin, like the floor plate of the diencephalon, but the distal end of the infundibulum soon becomes solid as the neuroepithelial cells proliferate. These cells later differentiate into *pituicytes* resembling neuroglial cells. Nerve fibers grow into the pars nervosa from the hypothalamic area, to which the infundibular stem is attached.

The Telencephalon (Figs. 18–27 to 18–30). The telencephalon consists of a median part and two lateral diverticula, the cerebral vesicles, which are the primordia of the *cerebral hemispheres* (Figs. 18–20, 18–24*B*, and 18–25*A*). The cavity of the median portion forms the extreme anterior part of the third ventricle. At first, the cerebral vesicles are in wide communication with the cavity of the third ventricle through the *interventricular foramina* (Fig. 18–26).

Along a line known as the *choroid fissure* (Fig. 18–29*A*), part of the medial wall of the developing cerebral hemisphere becomes

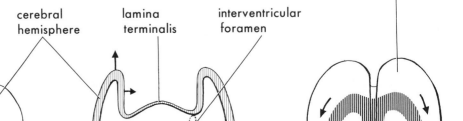

Figure 18–27 *A,* Sketch of the dorsal surface of the forebrain, indicating how the ependymal roof of the diencephalon is carried out to the dorsomedial surface of the cerebral hemispheres. *B,* Diagrammatic section of the forebrain, showing how the developing cerebral hemispheres grow from the lateral walls of the forebrain and expand in all directions until they cover the diencephalon. The arrows indicate some directions in which the hemisphere expands. The rostral wall of the forebrain, the lamina terminalis, is very thin. *C,* Sketch of the forebrain, as viewed anteriorly, showing how the ependymal roof is finally carried down into the temporal lobes as a result of the C-shaped growth pattern of the cerebral hemispheres.

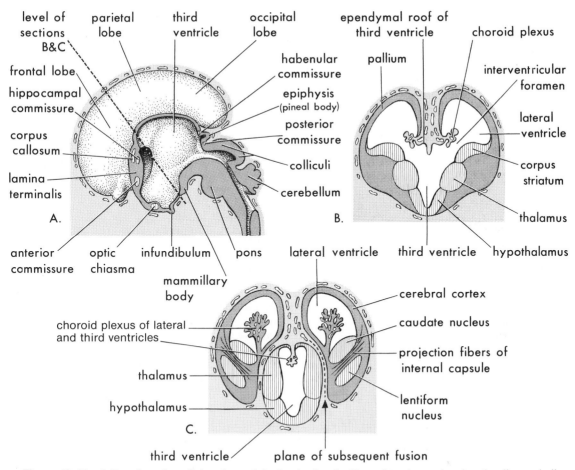

Figure 18–28 *A,* Drawing of medial surface of the forebrain of a 10-week embryo, showing the diencephalic derivatives, the main commissures, and the expanding cerebral vesicles (hemispheres). *B,* Transverse section through the forebrain at the level of the interventricular foramen showing the corpus striatum and the choroid plexus of the lateral ventricles. *C,* Similar section at about 11 weeks, showing division of the corpus striatum into caudate and lentiform nuclei by the internal capsule. The developing relationship of the cerebral hemispheres to the diencephalon is also illustrated.

very thin. Initially, this thin ependymal portion lies in the roof of the hemisphere and is continuous with the ependymal roof of the third ventricle (Fig. 18–27*A*). The choroid plexus of the lateral ventricle later forms at this site (Figs. 18–26 and 18–28). As the hemispheres expand, somewhat like inflating balloons, they cover successively the diencephalon, the midbrain, and the hindbrain. The hemispheres eventually meet each other in the midline, flattening their medial surfaces. The mesenchyme trapped in the longitudinal fissure between them gives rise to the *falx cerebri.*

The *corpus striatum* appears during the sixth week as a prominent swelling in the floor of each cerebral hemisphere (Figs. 18–26 and 18–28*B*). The floor of the hemi-

sphere expands more slowly than the thin cortical wall because it contains the rather large corpus striatum. Consequently, the cerebral hemispheres assume a C-shape (Fig. 18–29). The growth and curvature of the hemispheres also affect the shape of the lateral ventricles within them, forming anterior and inferior horns to their bodies. The caudal end of the hemisphere turns downward and then forward, forming the temporal lobe. In so doing, it carries the ventricle (forming the inferior horn) and the choroid fissure with it (Fig. 18–29). Here, the thin medial wall of the hemisphere is invaginated along the choroid fissure by vascular pia mater to form the *choroid plexus of the inferior horn* (Figs. 18–26 and 18–28*B*).

As the cerebral cortex differentiates,

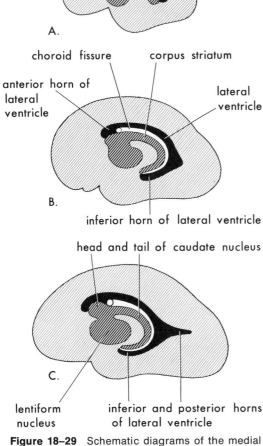

Figure 18–29 Schematic diagrams of the medial surface of the developing right cerebral hemisphere, showing development of the lateral ventricle, the choroid fissure, and the corpus striatum. *A,* 13 weeks. *B,* 21 weeks. *C,* 32 weeks.

Cerebral Commissures (Fig. 18–28). As the cortex develops, groups of fibers, or commissures, connect corresponding areas of the cerebral hemispheres with one another. The most important of these fibers cross in the lamina terminalis, the rostral end of the forebrain (Fig. 18–28*A*). This lamina extends from the roof plate of the diencephalon to the optic chiasma and is the natural pathway from one hemisphere to the other.

The first commissures to form, the *anterior commissure* and the *hippocampal commissure,* are small fiber bundles that connect phylogenetically older parts of the brain. The anterior commissure connects the olfactory bulb and related brain areas of the one hemisphere with those of the opposite side. The hippocampal commissure connects the hippocampal formations.

The largest commissure is the *corpus callosum* (Fig. 18–28*A*), connecting neocortical areas. The corpus callosum initially lies within the lamina terminalis, but fibers are added to it as the cortex enlarges, and it gradually extends beyond the lamina terminalis. The rest of the lamina terminalis lies between the corpus callosum and the fornix. It becomes stretched out to form the thin septum pellucidum. By birth, the corpus callosum extends over the roof of the diencephalon. The *optic chiasma* develops in the ventral part of the lamina terminalis (Fig. 18–28*A*) and consists of fibers from the medial halves of the retinae, which cross to join the optic tract of the opposite side.

Differentiation of the Cerebral Cortex. The walls of the developing cerebral hemispheres initially show the three typical zones of the neural tube (ventricular, intermediate, and marginal). Later, a fourth one, the subventricular zone, appears. Cells of the intermediate zone migrate into the marginal zone and give rise to the cortical layers. Thus, the gray matter is located peripherally, and axons from its cell bodies pass centrally to form the *medullary center.*

Initially, the surface of the hemispheres is smooth (Fig. 18–30*A*), but as growth proceeds, a complex pattern of sulci and gyri develops. These permit a considerable increase in the area of the cerebral cortex without requiring an extensive increase in cranial size. As a hemisphere grows, the cortex lying over the outer surface of the corpus striatum grows relatively slowly and is soon overgrown (Fig. 18–30*C*). This buried cortex, hidden from view in the depths of the lateral sulcus (fissure) of the cerebral hemisphere, is known as the *insula.*

fibers passing to and from it pass through the corpus striatum and divide it into *caudate* and *lentiform nuclei.* This fiber pathway, called the *internal capsule* (Fig. 18–28*C*), bcomes C-shaped as the hemisphere assumes this form. The caudate nucleus becomes elongated and horseshoe-shaped, conforming to the outline of the lateral ventricle (Fig. 18–29). Its pear-shaped head and elongated body lie in the floor of the anterior horn and body of the lateral ventricle, whereas its tail makes a U-shaped turn to gain the roof of the inferior horn.

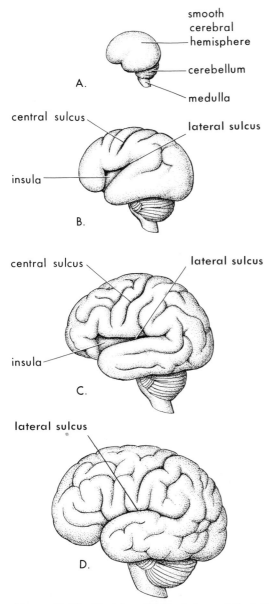

smooth
cerebral
hemisphere

cerebellum

medulla

A.

central sulcus

lateral sulcus

insula

B.

central sulcus

lateral sulcus

insula

C.

lateral sulcus

D.

Figure 18–30 Sketches of lateral views of the left cerebral hemisphere, showing successive stages in the development of sulci and gyri. *Half actual size.* Note the gradual narrowing of the lateral sulcus (fissure) and burying of the insula. *A,* 13 weeks. *B,* 26 weeks. *C,* 35 weeks. *D,* Newborn. At birth, the brain weighs about 400 grams. Note that the surface of the cerebral hemispheres grows rapidly during the fetal period, forming many convolutions (gyri), which are separated by many grooves (sulci).

CONGENITAL MALFORMATIONS OF THE BRAIN AND/OR THE MENINGES

Abnormal development of the brain is not uncommon, owing to the complexity of its embryological history. Most major congenital malformations of the brain result from *defective closure of the rostral neuropore* during the fourth week and involve the overlying tissues (future meninges and calvaria). The factors causing the faulty development may be primarily either genetic or environmental in nature.

Congenital abnormalities of the brain can result from alterations in the morphogenesis or the histogenesis of the nervous tissue, or they can result from developmental failures occurring in associated structures (notochord, somites, mesenchyme, and skull). Faulty development or histogenesis of the cerebral cortex can result in various types of congenital mental retardation. Prenatal factors are known to be involved in the development of *cerebral palsy*, but it is most often due to a normal fetus being damaged at birth (Warkany, 1971).

Defects in the formation of the cranium *(cranium bifidum)* are often associated with congenital malformations of the brain and/or meninges. Such defects of the cranium are usually in the median plane and usually in the cranial vault. The defect is often in the squamous part of the occipital bone and may include the posterior lip of the foramen magnum.

When the defect in the cranium is small, usually only the meninges herniate, and the malformation is called a *meningocele* (Fig. 18–31*B*). When the cranial defect is large, the meninges and part of the brain (Gr. *enkephalos*) herniate, forming a *meningoencephalocele* (Fig. 18–31*C*). If the protruding part of the brain contains part of the ventricular system, the malformation is called a *meningohydroencephalocele* (Figs. 18–31*D* and 18–32). The part of the brain in the sac is dependent on the location of the cranial defect. Cranium bifidum associated with herniation of the brain and/or its meninges occurs about once in every 2000 births.

Exencephaly and Anencephaly (Fig. 18–19). These severe malformations of the brain result from failure of the rostral neuropore to close properly during the fourth week. As a result, the forebrain primordium is abnormal, and development of the calvaria is defective. Most of the embryo's brain is exposed or extruding from the skull, a condition known as *exencephaly*. Exencephalus is occasionally observed in aborted human embryos, but not at birth.

Owing to the abnormal structure and vas-

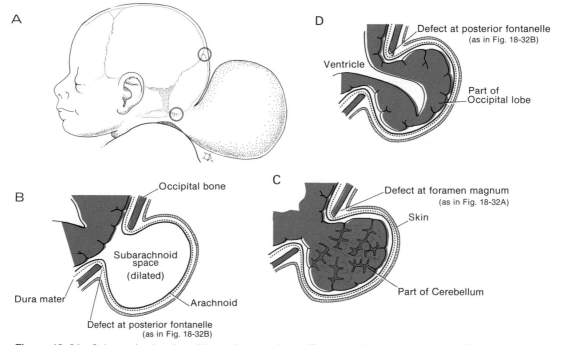

Figure 18–31   Schematic drawings illustrating cranium bifidum and the various types of herniation of the brain and/or cranial meninges associated with it. *A*, Sketch of the head of a newborn infant with a large protrusion from the occipital region of the skull, similar to that shown in Figure 18–32. The *upper red circle* indicates a cranial defect at the posterior fontanelle, and the *lower red circle* indicates a cranial defect at the foramen magnum. *B, Meningocele* consisting of a protrusion of the cranial meninges that is filled with cerebrospinal fluid. *C, Meningoencephalocele* consisting of a protrusion of part of the cerebellum that is covered by cranial meninges and skin. *D, Meningohydroencephalocele* consisting of a protrusion of part of the occipital lobe that contains part of the posterior horn of a lateral ventricle.

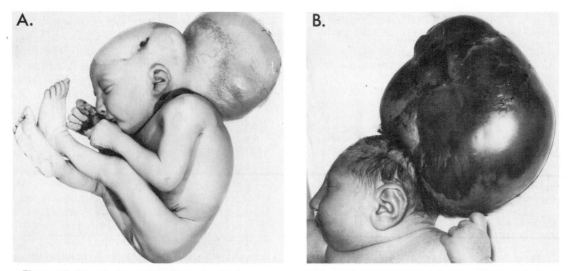

Figure 18–32   Photographs of infants with large meningoencephaloceles. *A*, Occipital area. *B*, Occipital and parietal areas. The infant shown in *B* also has microcephaly (small cranium). (Courtesy of Dr. Dwight Parkinson, Children's Centre, Winnipeg, Canada.)

cularization of the embryonic exencephalic brain, the nervous tissue undergoes degeneration until most of it is replaced in fetuses by a spongy, vascular mass consisting mostly of hindbrain structures. Although this condition is called *anencephaly* (Gr. *an*, without + *enkephalos*, brain), a rudimentary brain stem and traces of the basal ganglia are usually present.

Anencephaly is a common malformation, occurring about once in every 1000 births, and it is about four times more common in females than in males. It is always associated with *acrania* and may be associated with *rachischisis* (Fig. 18–19) when defective neural tube closure is extensive.

Anencephaly accounts for about one half of the severe *neural tube defects* in Great Britain, and it is the most common serious malformation seen in stillborn fetuses (Laurence and Weeks, 1971). *Sustained extrauterine life is impossible in infants born with anencephaly*. Infants afflicted with this defect survive for a few hours after birth, at most.

Anencephaly or another type of neural tube defect is suspected in utero when there is an *elevated level of alpha fetoprotein* in the amniotic fluid. Anencephaly can be easily diagnosed by ultrasonography, fetoscopy, and radiography because extensive parts of the brain and skull are absent. Anencephaly has been detected as early as 14 weeks in utero (Page et al., 1981). Usually, a *therapeutic abortion* is performed, if the mother requests it, when continuation of the pregnancy will result in the birth of a child with severe malformations, such as anencephaly, that are incompatible with life after birth.

Exencephaly and anencephaly can be readily induced experimentally in rats by various teratogenic agents (Giroud, 1960). Studies of exencephalic human abortuses suggest that the process is similar in humans. Genetic factors are certainly involved because of the well-established familial incidence of anencephaly. Anencephaly has a *multifactorial inheritance* (see Chapter 8). An excess of amniotic fluid (*polyhydramnios*) is often associated with anencephaly, possibly because the fetus lacks the neural control for swallowing amniotic fluid. Thus, it does not pass into the intestines for absorption and subsequent transfer to the placenta for disposal.

For a mother who has had one anen-

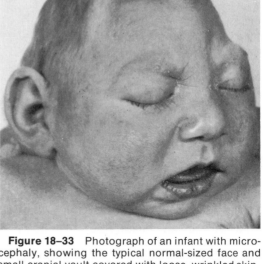

Figure 18–33 Photograph of an infant with microcephaly, showing the typical normal-sized face and small cranial vault covered with loose, wrinkled skin. (From Laurence, K. M., and Weeks, R.: Abnormalities of the central nervous system; *in* A. P. Norman [Ed.]: *Congenital Abnormalities in Infancy*. 2nd ed. 1971. Courtesy of Blackwell Scientific Publications.)

cephalic infant, the *recurrence risk* of one of the neural tube defects (anencephaly or spina bifida cystica) is about 10 per cent for each subsequent pregnancy.

Microcephaly (Fig. 18–33). In this uncommon condition, the calvaria is small, but the face is normal-sized. Generally, these infants are grossly mentally retarded because the brain is small and underdeveloped, a condition known as *microencephaly*. The cause of this condition is often uncertain; some cases appear to be genetic in origin, and others seem to be associated with environmental factors. Exposure to large amounts of ionizing radiation during the embryonic period and to infectious agents during the fetal period are possible contributing factors (see Chapter 8). Microcephaly can be detected in utero by ultrasound. Successive *ultrasound scans* carried out over the period of gestation are helpful in assessing the rate of growth of the fetal cranium (Page et al., 1981).

Microcephaly (Gr. *mikros* small + *kephale*, head) results from *microencephaly* (Gr. *mikros*, small + *enkephalos*, brain) because growth of the *calvaria*, or *cranial vault*, is largely due to pressure from the growing brain. *Microencephaly* also results in a thick cranial vault with few or no convolutional markings of the brain on its inner surface.

A small head may result from *premature synostosis* (osseous union) of all the cranial sutures (see Chapter 15), but the cranial vault will be thin with exaggerated convolutional markings. *Premature synostosis of all cranial sutures occurs rarely.* Premature synostosis of one suture produces an abnormality in the shape of the skull (see Figs. 15–11 and 15–12).

Hydranencephaly. In this extremely rare malformation, the cerebral hemispheres are partially or completely absent. Only the basal nuclei and remnants of the midbrain are found rostral to the hindbrain. These infants generally appear normal at birth, but mental development fails to occur after birth. The cause of this malformation is uncertain, but there is some evidence that it may be caused by an early obstruction of blood flow to the areas supplied by the internal carotid arteries (Laurence and Weeks, 1971).

Agenesis of the Corpus Callosum. In this condition, there is complete or partial absence of the corpus callosum. The condition may be asymptomatic, but seizures and mental deficiency are common. In two sisters with agenesis of the corpus callosum, the only symptoms were seizures: recurrent in one, but only occasional and minor in the other. *Their I.Q.'s were average.* The cause of this malformation is unknown; there is no evidence that it is inherited (Shager et al., 1957).

Hydrocephalus (Fig. 18–34). Overproduction of cerebrospinal fluid (CSF), obstruction of its flow, or interference with its absorption results in *an excess of CSF,* a condition known as hydrocephalus (Gr. *hydōr,* water + *kephalē,* head). Hydrocephalus often results from *congenital aqueductal stenosis,* in which the cerebral aqueduct is narrow or consists of several minute channels. Blockage of CSF circulation results in dilation of the ventricles superior to the obstruction and in pressure on the cerebral hemispheres. This squeezes the brain between the ventricular fluid and the bones of the cranium. In infants, the internal pressure pressure results in expansion of the brain and the calvaria because the sutures and fontanelles are still open (see Fig. 15–8).

Hydrocephalus usually refers to internal hydrocephalus, in which all or part of the ventricular system is enlarged. All ventricles

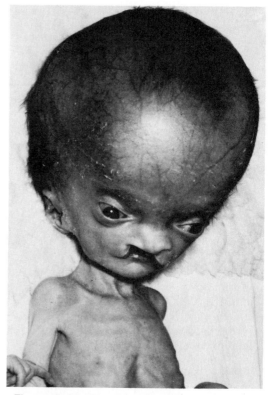

Figure 18–34 Photograph of an infant with hydrocephalus, bilateral cleft lip, and deformed limbs. (Courtesy of Dr. T. V. N. Persaud, Professor of Anatomy and Head of the Anatomy Department, University of Manitoba, Winnipeg, Canada.)

are enlarged if the apertures of the fourth ventricle or the subarachnoid spaces are blocked, whereas the lateral and third ventricles are dilated when the cerebral aqueduct is obstructed. Although rare, obstruction of one interventricular foramen can produce dilation of one ventricle.

Hydrocephalus is often associated with spina bifida cystica, although the hydrocephalus may not be obvious at birth. Hydrocephalus often produces thinning of the bones of the calvaria, prominence of the forehead, atrophy of the cerebral cortex and white matter, and compression of the basal nuclei and diencephalon.

Arnold-Chiari or Chiari Malformation. A tongue-like projection of the medulla and cerebellum herniates into the foramen magnum. It occurs about once in every 1000 births and is frequently associated with spina bifida with meningomyelocele, spina bifida with myeloschisis, and hydrocephaly. Downward displacement of the hindbrain obstructs the flow of cerebrospinal fluid through the foramina of the fourth ventricle.

The cause of the Arnold-Chiari malformation is uncertain, but Barry et al. (1957) believe that it results primarily from overgrowth of the cerebellum and medulla. In some, the posterior cranial fossa is abnormally small. These abnormalities result in herniation of parts of the cerebellum and medulla into the vertebral canal in the cervical region.

Mental Retardation. Congenital impairment of intelligence may result from various genetically determined conditions. It is well known that mental retardation may result from the action of a single mutant gene, or a chromosomal abnormality. Chromosomal abnormalities and mental deficiency are discussed in Chapter 8. Disorders of protein, carbohydrate, or fat metabolism may also cause mental retardation. Maternal and fetal infections (syphilis, rubella virus, toxoplasmosis, and cytomegalic inclusion disease) and cretinism are commonly associated with mental retardation (Clayton, 1973). Inadequate mental development throughout the postnatal growth period can result from birth injuries, cerebral infections, cerebral trauma, and poisoning. For good discussions of mental retardation and its many causes, see Warkany (1971) and Vaughan et al. (1979).

THE PERIPHERAL NERVOUS SYSTEM

The peripheral nervous system consists of the cranial, spinal, and visceral nerves and the cranial, spinal, and autonomic ganglia. The peripheral nervous system develops from various sources. All sensory cells (both somatic and visceral) of the peripheral nervous system are derived from *neural crest cells*. The cell bodies of these sensory cells are located outside the central nervous system. With the exception of the cells in the spiral ganglion of the cochlea and in the vestibular ganglion of the eighth cranial nerve (vestibulocochlear), all the peripheral sensory cells are at first bipolar, but the two processes soon unite to form a single process and a unipolar type of neuron (Fig. 18–9*D*). This process has peripheral and central branches, or processes. The peripheral process terminates in a sensory ending, whereas the central process enters the spinal cord or brain (Fig. 18–8). The sensory cells of the ganglion of cranial nerve VIII (vestibulocochlear) remain bipolar.

The cell body of each afferent neuron is closely invested by a capsule of *satellite cells* (Fig. 18–8), also derived from neural crest cells. This capsule is continuous with the neurolemmal sheath of Schwann cells, also derived from the neural crest, that surrounds the axons of afferent neurons. External to the satellite cells is a layer of connective tissue that is continuous with the endoneurial sheath of the nerve fibers. This connective tissue and the endoneurial sheath are derived from mesenchyme.

The neural crest cells in the brain region migrate to form sensory ganglia only in relation to the trigeminal (CN V), the facial (CN VII), the vestibulocochlear (CN VIII), the glossopharyngeal (CN IX), and the vagus (CN X).

Cells of the neural crest also differentiate into multipolar neurons of the *autonomic ganglia* (Fig. 18–8), including ganglia of the sympathetic trunks along the sides of the vertebral bodies; collateral, or prevertebral, ganglia in plexuses of the thorax and abdomen (e.g., the cardiac, celiac, and mesenteric plexuses); and parasympathetic, or terminal, ganglia in or near the viscera (e.g., the submucosal, or Meissner's, plexus). Cells of the paraganglia, called *chromaffin cells,* are also derived from the neural crest. The term *paraganglia* includes several widely scattered groups of cells that are similar in many ways to medullary cells of the adrenal, or suprarenal, glands. The cell groups largely lie retroperitoneally, often in association with sympathetic ganglia. The carotid and aortic bodies also have small islands of chromaffin cells associated with them. These widely scattered groups of chromaffin cells constitute the *chromaffin system*.

Cells of the neural crest also give rise to melanoblasts (the precursors of the melanocytes) and to cells of the adrenal medulla (see Fig. 13–19).

The Spinal Nerves (Figs. 18–4, 18–6, and 18–8). Motor nerve fibers in the spinal cord region begin to appear at the end of the fourth week. They arise from cells in the *basal plates* of the developing spinal cord and emerge as a continuous series of rootlets along its ventrolateral surface. The fibers destined for a particular developing muscle group become arranged in a bundle, forming a *ventral nerve root*. The myelin sheaths of nerve fibers inside the spinal cord are formed by *oligodendroglia cells,* whereas outside the spinal cord the myelin sheaths are formed by *Schwann cells* (Figs. 18–8 and 18–11).

The *dorsal nerve root* (sensory) is formed by axons of neural crest cells that migrate to the dorsolateral aspect of the spinal cord, where they become the cells of the *spinal ganglion* (Figs. 18–4 and 18–7 to 18–9). The central processes of neurons in this spinal ganglion collect into a single bundle that grows into the spinal cord, opposite the apex of the *dorsal horn* of gray matter (Fig. 18–4*B* and *C*). The distal processes of spinal ganglion cells grow towards the ventral nerve root (motor) and eventually join with it to form a *spinal nerve* (Figs. 18–4, 18–6, and 18–8).

Immediately after being formed, a *mixed spinal nerve* divides into dorsal and ventral primary rami (L., branches). The *dorsal primary ramus,* the smaller division, innervates the dorsal axial musculature (see Fig. 16–1), the vertebrae, the posterior intervertebral joints, and part of the skin of the back. The *ventral primary ramus,* the major division of each spinal nerve, contributes to the innervation of the limbs and the ventrolateral parts of the body wall. *The major plexuses (cervical, brachial, and lumbosacral) are formed by the ventral primary rami.*

As each limb bud develops, the nerves of the segments opposite to it elongate and grow into its mesenchyme (see Fig. 17–4), and are distributed to its muscles, which differentiate from the mesenchyme. The skin of the developing limbs is also supplied in a segmental manner. Early in development, successive ventral primary rami are joined by connecting loops of nerve fibers, especially those supplying the limbs (e.g., the *brachial plexus*). The dorsal divisions of the trunks of these plexuses supply the extensor muscles and the extensor surface of the limbs, and the ventral divisions of the trunks supply the flexor muscles and the flexor surface. The *dermatomes* and the cutaneous innervation of the limbs are described in Chapter 17.

The Cranial Nerves (see Figs. 10–6 *A* and 18–35). Twelve pairs of cranial nerves form during the fifth and sixth weeks of development. They are classified into three groups, according to their embryological origins.

The Somatic Efferent Cranial Nerves (Fig. 18–35). The *trochlear* (CN IV), *abducens* (CN VI), *hypoglossal* (CN XII), and the greater part of the *oculomotor* (CN III) nerves are homologous with the ventral roots of spinal nerves. The cells of origin of these

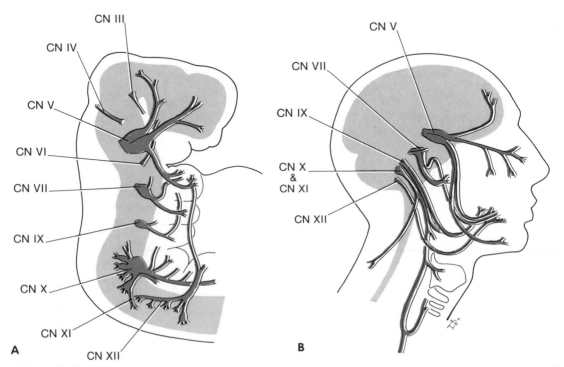

Figure 18–35 *A,* Schematic drawing of a 5-week-old embryo, showing the distribution of most of the cranial nerves, especially those supplying branchial arches. *B,* Schematic drawing of an adult, showing the general distribution of most of the cranial nerves.

nerves are located in the *somatic efferent column* (basal plates) of the brain stem (Fig. 18–22), and their axons are distributed to the muscles derived from the *head myotomes* (preotic and occipital; see Fig. 16–2).

The hypoglossal nerve (CN XII) resembles a spinal nerve more than do the other somatic efferent cranial nerves (CN III, CN IV, and CN VI), so it will be described first. The hypoglossal nerve develops by the fusion of the ventral root fibers of three or four occipital nerves (Fig. 18–35A). Sensory roots, corresponding to the dorsal roots of spinal nerves, are absent. The somatic motor fibers originate from the *hypoglossal nucleus*, consisting of motor cells resembling those of the ventral horn of the spinal cord. These fibers leave the ventrolateral wall of the medulla in several groups, the *hypoglossal nerve roots*, which converge to form the common trunk of CN XII (Fig. 18–35B). They grow rostrally and eventually innervate the *muscles of the tongue*, which are thought to be derived from the *occipital myotomes* (see Fig. 16–2). With the development of the neck, the hypoglossal nerve comes to lie at a progressively higher level (Fig. 18–35B).

The abducens nerve (CN VI) arises from nerve cells in the basal plates of the metencephalon (Fig. 18–22) and passes from its ventral surface to the posterior of the three preotic myotomes (Fig. 16–2A), from which the *lateral rectus muscle* of the eye is thought to originate.

The trochlear nerve (CN IV) arises from nerve cells in the somatic efferent column in the posterior part of the midbrain (Fig. 18–23). Although a motor nerve, it emerges from the brain stem dorsally and then passes ventrally to supply the *superior oblique muscle* of the eye.

The oculomotor nerve (CN III) supplies most of the muscles of the eye, i.e., *the superior, inferior, and medial recti* and the *inferior oblique muscles* of the eye, which are thought to be derived from the first preotic myotomes.

The Nerves of the Branchial Arches (see Figs. 10–6A and 18–35A). Cranial nerves V, VII, IX, and X supply the embryonic branchial arches; thus, the structures that develop from these arches are innervated by these cranial nerves (see Chapter 10).

The trigeminal nerve (CN V) is the nerve of the first branchial arch (see Figs. 10–6A and 18–35A), but it has an ophthalmic division that is not a branchial component. *CN V is chiefly sensory* and is the principal sensory nerve for the head. Its large ganglion lies beside the rostral end of the pons, and its cells are derived from the most anterior part of the *neural crest*. The central processes of cells in this ganglion form the large sensory root of CN V, which enters the lateral portion of the pons. The peripheral processes of cells in this ganglion separate into three large divisions (the *ophthalmic*, the *maxillary*, and the *mandibular nerves*). Their sensory fibers supply the skin of the face (see Figs. 10–6B and 18–35B) as well as the lining of the mouth and the nose. The *motor fibers of CN V* arise from cells in the most anterior part of the *special visceral efferent column* in the metencephalon (Fig. 18–22). The motor nucleus of CN V lies at the midlevel of the pons. The fibers leave the pons at the site of the entering sensory fibers and pass to the muscles of *mastication* and to other muscles derived from this arch, which develop in the mandibular prominence of the first branchial arch (see Fig. 10–5 and Table 10–1). The *mesencephalic nucleus of CN V* (Fig. 18–23D) differentiates from cells in the midbrain that extend forward from the metencephalon.

The facial nerve (CN VII) is the nerve of the second branchial arch. It consists mostly of motor fibers that arise principally from a nuclear group in the *special visceral efferent column* in the caudal part of the pons (Fig. 18–22). These fibers are distributed to the *muscles of facial expression* and to other muscles that develop in the mesenchyme of the second branchial arch (see Fig. 10–5 and Table 10–1). The small general visceral efferent component of CN VII terminates in the peripheral autonomic ganglia of the head. The *sensory fibers of CN VII* arise from the cells of the *geniculate ganglion*. The central processes of these cells enter the pons, and the peripheral processes pass to the *greater superficial petrosal nerve* and, via the *chorda tympani nerve*, to the taste buds in the anterior two thirds of the tongue (see Fig. 10–18).

The glossopharyngeal nerve (CN IX) is the nerve of the third branchial arch. Its motor fibers arise from the special and, to a lesser extent, the general visceral efferent columns of the anterior part of the myelencephalon (Fig. 18–21). *CN IX forms from several rootlets* that arise from the medulla, just caudal to the developing inner ear, or *auditory vesicle*. All the fibers from the special visceral efferent column are distributed to the stylopharyngeus muscle, which is derived from mesenchyme in the third branchial arch (see Fig. 10–5 and Table 10–1). The general efferent fibers are distributed to the *otic ganglion,* from which postganglionic fibers pass to the parotid and posterior lingual glands. The *sensory fibers of CN IX* are distributed as general sensory and special (taste) visceral afferent fibers to the posterior part of the tongue (see Fig. 10–18).

The vagus nerve (CN X) is formed by fusion of the nerves of the fourth, fifth, and sixth branchial arches. It has large visceral efferent and visceral afferent components that are distributed to the heart, to the foregut and its derivatives, and to a considerable part of the midgut (see Chapter 12).

The nerve of the fourth branchial arch becomes the *superior laryngeal nerve,* which supplies the cricothyroid muscle and the constrictors of the pharynx derived from mesenchyme in this arch. The nerve of the sixth branchial arch becomes the *inferior laryngeal nerve,* which supplies various laryngeal muscles. The nerve of the fifth branchial arch cannot be identified after the arch disappears.

The accessory nerve (CN XI) has two separate origins (Fig. 18–35). The *cranial root* (part) is a posterior extension of CN X, and the *spinal root* (part) arises from the upper five or six cervical segments of the spinal cord. The fibers of the cranial root emerge from the lateral surface of the medulla, where they join the vagus nerve and supply the muscles of the soft palate and the intrinsic muscles of the larynx. The fibers of the spinal root supply the sternocleidomastoid and trapezius muscles.

The Special Sensory Nerves. The *olfactory nerve (CN I)* arises from the *olfactory bulb* (see Chapter 10). The olfactory cells are bipolar neurons that differentiate from cells in the epithelial lining of the primitive *nasal sac* (see Fig. 10–21). The axons of the *olfactory cells* are collected into 18 to 20 bundles, around which the *cribriform plate* of the ethmoid bone develops. These unmyelinated nerve fibers end in the olfactory bulb.

The optic nerve (CN II) is formed by more than one million nerve fibers that grow into the brain from neuroblasts in the primitive retina (see Fig. 19–2). Because the optic nerve develops from the evaginated wall of the forebrain, it actually represents a fiber tract of the brain. Development of the optic nerve is described in more detail in Chapter 19.

The vestibulocochlear nerve (CN VIII) consists of two kinds of sensory fiber in two bundles; these fibers are known as the vestibular and cochlear nerves. The *vestibular nerve* originates in the semicircular ducts (see Fig. 19–15), and the *cochlear nerve* proceeds from the cochlear duct, in which the *organ of Corti* develops.

The bipolar neurons of the vestibular nerve have their cell bodies in the *vestibular ganglion*. The central processes of these cells terminate in the *vestibular nuclei* in the floor of the fourth ventricle. The bipolar neurons of the *cochlear nerve* have their cell bodies in the *spiral ganglion* (see Fig. 19–15*I*). The central processes of these cells end in the ventral and dorsal *cochlear nuclei* in the medulla.

THE AUTONOMIC NERVOUS SYSTEM

Functionally, the autonomic system can be divided into sympathetic (thoracolumbar) and parasympathetic (craniosacral) parts.

The Sympathetic Nervous System. During the fifth week, *neural crest cells* in the thoracic region migrate along each side of the spinal cord, where they form paired masses dorsolateral to the aorta (see Figs. 13–20 and 18–8). These segmentally arranged *sympathetic ganglia* are connected in a bilateral chain by longitudinal nerve fibers. These ganglionated cords, called *sympathetic trunks,* are located on each side of the vertebral bodies.

Some neural crest cells migrate ventral to the aorta and form neurons in the *preaortic ganglia*, such as the celiac and mesenteric ganglia (Fig. 18–8). Other neural crest cells migrate to the area of the heart, lungs, and gastrointestinal tract, where they form terminal ganglia in *sympathetic organ plexuses,* located near or even within these organs.

After the sympathetic trunks have formed, axons of sympathetic neurons located in the *intermediolateral cell column* (lateral horn) of the thoracolumbar segments of the spinal cord pass via the ventral root of a spinal nerve and a *white ramus communicans* (connecting branch) to a *paravertebral ganglion* (Fig. 18–8). Here they may synapse with the neurons or ascend or descend in the sympathetic trunk to synapse at other levels. Other *preganglionic fibers* pass through the paravertebral ganglia without synapsing, forming splanchnic nerves to the viscera. The *postganglionic fibers* course through a *gray ramus communicans*, passing from a sympathetic ganglion into a spinal nerve. Hence, the sympathetic trunks are composed of ascending and descending fibers.

The Parasympathetic Nervous System. The preganglionic parasympathetic fibers arise from *neurons in nuclei of the brain stem and in the sacral region of the spinal cord.* The fibers from the brain stem leave via the oculomotor (CN III), facial (CN VII), glossopharyngeal (CN IX), and vagus (CN X) nerves. The postganglionic neurons are located in peripheral ganglia or in plexuses near or within the structure being innervated (e.g., the pupil of the eye, the salivary glands, the heart, and the gastrointestinal tract).

Congenital Aganglionic Megacolon (Hirschsprung's Disease). This condition of extreme dilation and hypertrophy of the colon results from *failure of neural crest cells to migrate into the wall of the colon* and to differentiate into parasympathetic ganglion cells. Histologically, there is an absence of ganglion cells in the *myenteric plexus* (Auerbach's plexus) in the affected region of the bowel, usually the rectum and/or sigmoid colon. In less common cases, the *aganglionosis* also affects proximal parts of the colon and even the small bowel (Nixon and Wilkinson, 1971).

Congenital aganglionic megacolon (Gr. *megas*, big) is characterized clinically by *fecal retention* and abdominal distention starting at birth. The functional abnormality is a decrease in muscular tone and contractile activity of an aganglionic segment of bowel, without the reciprocal relation necessary to produce the onward movement of feces by peristalsis of the colon proximal to the aganglionic segment. *Congenital aganglionic megacolon accounts for about 20 per cent of the cases of intestinal obstruction in infants* (Vaughan et al., 1979).

SUMMARY

The central nervous system develops from a dorsal thickening of ectoderm known as the *neural plate.* This plate appears around the middle of the third week and becomes infolded to form a *neural groove* and *neural folds.* When the neural folds fuse to form the *neural tube* in the fourth week, some neuroectodermal cells are not included but remain between the neural tube and the surface ectoderm as the *neural crest.* Development of the neural plate and neural tube is believed to be induced by the notochord and the paraxial mesoderm.

The cranial end of the neural tube forms the brain, consisting of the forebrain, the midbrain, and the hindbrain. The forebrain gives rise to the cerebral hemispheres and the diencephalon; the midbrain becomes the adult midbrain; and the hindbrain gives rise to the pons, cerebellum, and medulla oblongata. The remainder of the neural tube becomes the spinal cord.

The lumen of the neural tube becomes the ventricles of the brain and the central canal of the spinal cord. The walls of the neural tube become thickened by proliferation of neuroepithelial cells, which give rise to all nerve and macroglial cells in the central nervous system. The microglia are believed to differentiate from mesenchymal cells that enter the central nervous system with the blood vessels.

The *pituitary gland* develops from two completely different parts: (1) an ectodermal upgrowth from the stomodeum, known as *Rathke's pouch,* and (2) a neuroectodermal downgrowth from the diencephalon called the *neurohypophyseal bud.* The *adenohypophysis* arises from the oral ectoderm, and the *neurohypophysis* arises from the neuroectoderm (see Table 18–1).

Cells in the cranial, spinal, and autonomic ganglia are derived from the neural crest. *Schwann cells,* which myelinate the axons, also arise from the neural crest. Similarly, most of the autonomic nervous system and all chromaffin tissue, including the adrenal medulla, develop from the neural crest.

Congenital malformations of the central nervous system are common. Defects of closure of the neural tube (*neural tube defects*) account for most abnormalities (e.g., *spinal bifida cystica* and *anencephaly*). The malformations may be limited to the nervous system, or they may include overlying tissues (bone, muscle, and connective tissue). Some malformations are caused by genetic abnormalities; others result from environmental factors such as infectious agents, drugs, and metabolic disease. Most malformations are probably caused by a combination of genetic and environmental factors.

Most gross abnormalities (e.g., anencephaly) are incompatible with life. Other severe malformations (e.g., spina bifida cystica) often cause functional disability (e.g., muscle paralysis).

Severe abnormalities may result from congenital malformations of the ventricular system. There are two main types of *hydrocephalus:* internal, or noncommunicating hydrocephalus (blockage of cerebrospinal fluid flow in the ventricular system) and communicating hydrocephalus (blockage of cerebrospinal fluid in the subarachnoid space).

Mental retardation may result from chromosomal abnormalities arising during gametogenesis or from metabolic disorders or maternal and fetal infections occurring during prenatal life. Various postnatal conditions (e.g., cerebral infection or trauma) may also cause abnormal mental development.

SUGGESTIONS FOR ADDITIONAL READING

Crelin, E. S.: *Development of the Nervous System. A Logical Approach to Neuroanatomy.* Ciba Clinical Symposia. Vol. 26. A succinct account of the normal development of the nervous system, beautifully illustrated by Dr. F. Netter. Dr. Crelin shows clearly that the key to understanding the complexities of the nervous system is a good knowledge of how it develops from the neural plate.

Dobbing, J.: *Late development of the brain and its vulnerability; in* J. A. Davis and J. Dobbing (Eds.): *Scientific Foundations of Paediatrics.* Philadelphia, W. B. Saunders Co., 1974, pp. 565–576. The author emphasizes that the developing brain — from the time of birth until maturity — is subject to a number of different types of pathological conditions. He also discusses the brain growth spurt, emphasizing the desirability of promoting good body growth during the vulnerable period of brain growth after birth.

Lemire, R. J., Loeser, J. D., Leech, R. W., and Alvord, E. C.: *Normal and Abnormal Development of the Human Nervous System.* Hagerstown, Harper & Row, Publishers, 1975. The authors consolidate currently available information on the normal development of the nervous system and analyze the various diseases of the developing nervous system. They thoroughly discuss possible causes of abnormalities of the nervous system and the times at which these etiological factors could produce abnormal development.

CLINICALLY ORIENTED PROBLEMS

1. A pregnant woman developed polyhydramnios over the course of a few days *(acute hydramnios).* Using ultrasonography, a radiologist reported that the fetus had *acrania* and was probably anencephalic. How soon can anencephaly be detected by *ultrasound scanning* of the abdomen? Why is polyhydramnios associated with *anencephaly?* What other techniques could be used to confirm the diagnosis of anencephaly?

2. A male infant was born with a large *lumbar meningomyelocele* that was covered with a thin membrane. Within a few days, the sac ulcerated and began to leak. A marked *neurological deficit* was detected inferior to the level of the sac. What is the basis of the neurological deficit? What structures would likely be affected?

3. *A CT scan* of an infant with an enlarged head showed dilation of the lateral and third ventricles. What is this condition called? Where would the block most likely be to produce this abnormal *dilation of the ventricles?* Is this condition usually recognizable before birth? How do you think this condition might be treated surgically?

4. Is an enlarged head in an infant synonymous with *hydrocephalus?* What condition is usually associated with an abnormally small head? Is growth of the skull dependent on growth of the brain? What environmental factors are known to cause *microencephaly?*

5. A radiologist reporting on a *pneumoencephalogram* stated that the patient's ventricles were dilated posteriorly, and that the lateral ventricles were also widely separated by a dilated third ventricle. *Agenesis of the corpus callosum* was diagnosed. What is the common symptom associated with agenesis of the corpus callosum? Are some patients asymptomatic? What is the basis of the dilated third ventricle?

The answers to these questions are given at the back of the book.

REFERENCES

Angevine, J. B., Jr., Bodian, D., Coulombre, A. J., Edds, M. V., Jr., Hamburger, V., Jacobson, M., Lyser, K. M., Prestige, M. C., Sidman, R. L., Varon, S., and Weiss, P. A.: Embryonic vertebrate central nervous system: revised terminology. *Anat. Rec.* *166*:257, 1970.

Ariens Kappers, J.: Development of the human paraphysis. *J. Comp. Neurol. 102*:425, 1955.

Barr, M. L.: *The Human Nervous System. An Anatomic Viewpoint.* 3rd ed. Hagerstown, Harper & Row, Publishers, 1979.

Barry, A., Patten, B. M., and Stewart, B. H.: Possible factors in the development of the Arnold-Chiari malformation. *J. Neurol. 14*:285, 1957.

Beks, J. W. F.: Defects of the lumbar spinal axis; *in* A. J. C. Huffstadt (ed.): *Congenital Malformations.* Amsterdam, Excerpta Medica, 1980, pp. 91–97.

Boyd, J. D.: Observations on the human pharyngeal hypophysis. *J. Endocrinol. 14*:66, 1956.

Bull, J. W. D., and Sutton, D.: Diagnosis of paraphyseal cysts. *Brain 72*:487, 1949.

Clayton, B. E. (Ed.): *Mental Retardation: Environmental Hazards.* London, Butterworth & Co. (Publishers) Ltd., 1973.

Conel, J. L.: *Postnatal Development of the Human Cerebral Cortex.* Cambridge, Mass., Harvard University Press, 1959.

Corliss, C. E.: *Patten's Human Embryology. Elements of Clinical Development.* New York, McGraw-Hill Book Co., 1976, pp. 199–235.

Dennison, W. M.: Spina bifida; *in* J. C. Mustardé (Ed.): *Plastic Surgery in Infancy and in Childhood.* Edinburgh, E. & S. Livingston, Ltd., 1971, pp. 381–385.

Giroud, A.: Causes and morphogenesis of anencephaly; *in* G. E. W. Wolstenholme and C. M. O'Connor (Eds.): *Ciba Foundation Symposium on Congenital Malformations.* London, J. & A. Churchill, 1960, pp. 199–212.

Groff, R. A., and Pitts, F. W.: Nervous system; *in* A. Rubin (Ed.): *Handbook of Congenital Malformations.* Philadelphia, W. B. Saunders Co., 1967, pp. 86–102.

Hamilton, W. J., Boyd, J. D., and Mossman, H. W.: *Human Embryology.* 4th ed. Cambridge, W. Heffer & Sons Ltd., 1972.

Holmes, R. L., and Sharp, J. A.: *The Human Nervous System. A Developmental Approach.* London, J. & A. Churchill, 1969.

Jacobson, M.: *Developmental Neurobiology.* 2nd ed. New York, Plenum Press, 1978.

Jacobson, S.: Neuroembryology; *in* B. A. Curtis, S. Jacobson, and E. M. Marcus (Eds.): *An Introduction to the Neurosciences.* Philadelphia, W. B. Saunders Co., 1972, pp. 20–35.

Jost, A.: Hormonal factors in the development of the fetus. *Cold Spring Harbor Symp. Quant. Biol. 19*:167, 1954.

Kalter, H.: *Teratology of the Central Nervous System.* Chicago, University of Chicago Press, 1968.

Langman, J.: *Medical Embryology. Human Development — Normal and Abnormal.* 4th ed., Baltimore, The Williams & Wilkins Co., 1981.

Langman, J., Guerrant, R. L., and Freeman, B. G.: Behavior of neuroepithelial cells during closure of the neural tube. *J. Comp. Neurol. 127*:399, 1966.

Larroche, J. C.: Part II: The development of the central nervous system during intrauterine life; *in* F. Falkner (Ed.): *Human Development.* Philadelphia, W. B. Saunders Co., 1966, pp. 257–276.

Laurence, K. M.: The pathology of hydrocephalus. *Ann. R. Coll. Surg. Engl. 24*:388, 1958.

Laurence, K. M., Carter, C. O., and David, P. A.: The major central nervous system malformations in South Wales. I. Incidence, local variations and geographical factors. *Br. J. Prev. Soc. Med. 22*:146, 1968.

Laurence, K. M., Carter, C. O., and David, P. A.: The major central nervous system malformations in South Wales. II. Pregnancy factors, seasonal variations and social class effects. *Br. J. Prev. Soc. Med. 22*:212, 1968.

Laurence, K. M., and Weeks, R.: Abnormalities of the central nervous system; *in* A. P. Norman (Ed.): *Congenital Abnormalities in Infancy.* 2nd ed. Oxford, Blackwell Scientific Publications, 1971, pp. 25–86.

Lemire, R. J.: Variations in development of the caudal neural tube in human embryos (Horizons XIV-XXI). *Teratology 2*:361, 1969.

Lemire, R. J.: Embryology of the nervous system; *in* J. A. Davis and J. Dobbing (Eds.): *Scientific Foundations of Paediatrics.* Philadelphia, W. B. Saunders Co., 1974, pp. 547–564.

Lemire, R. J., Shepard, T. H., and Alvord, E. C., Jr.: Caudal myeloschisis (lumbo-sacral spina bifida cystica) in a five millimeter (Horizon XIV) human embryo. *Anat. Rec. 152*:9, 1965.

List, C. F.: Intraspinal epidermoids, dermoids, and dermal sinuses. *Surg. Gynecol. Obstet. 73*:525, 1941.

Loggie, J. M. H.: Growth and development of the autonomic nervous system; *in* J. A. Davis and J. Dobbing (Eds.): *Scientific Foundations of Paediatrics.* Philadelphia, W. B. Saunders Co., 1974, pp. 640–648.

MacLean, N., Mitchell, J. M., Harnden, D. G., Williams, J., Jacobs, P. A., Buckton, K. A., Baikie, A. G., Court Brown, W. M., McBride, J. A., Strong, J. A., Close, H. G., and Jones, D. C.: A survey of sex-chromosome abnormalities among 4514 mental defectives. *Lancet 1*:293, 1962.

Marshal, W. A.: *Development of the Brain.* Edinburgh, Oliver and Boyd, 1968.

Moore, K. L.: *Clinically Oriented Anatomy.* Baltimore, The Williams & Wilkins Co., 1980, p. 661.

Milhorat, T. H.: *Hydrocephalus and the Cerebrospinal Fluid.* Baltimore, The Williams & Wilkins Co., 1972.

Nixon, H. H., and Wilkinson, A. W.: Abnormalities of the gastrointestinal system; *in* A. P. Norman (Eds.): *Congenital Abnormalities in Infancy.* 2nd ed. Oxford, Blackwell Scientific Publications, 1971, pp. 244–246.

O'Rahilly, R., and Gardner, E.: The timing and sequence of events in the development of the human nervous system during the embryonic period proper. *Z. Anat. Entwicklungsgesch. 134*:1, 1971.

Padget, D. H.: Neuroschisis and human embryonic maldevelopment. New evidence on anencephaly, spina bifida and diverse mammalian defects. *J. Neuropathol. Exp. Neurol. 29*:192, 1970.

Page, E. W., Villee, C. A., and Villee, D. B.: *Human Reproduction. Essentials of Reproductive and Perinatal Medicine.* 3rd ed. Philadelphia, W. B. Saunders Co., 1981.

Perret, G.: Symptoms and diagnosis of diastematomyelia. *Neurology 10*:1, 1960.

Parkinson, D.: The meningomyeloceles and allied malformations. *Manit. Med. Rev. 43*:76, 1963.

Penrose, L. S.: Genetics of anencephaly. *J. Ment. Defic. Res. 1*:4, 1957.

Peters, P. W., Dormans, J. A., and Geelan, J. A.: Light microscopic and ultrastructural observations in advanced stages of induced exencephaly and spina bifida. *Teratology 19*:183, 1979.

Rodier, P. M., Reynolds, S. S., and Roberts, W. N.: Behavioral consequences of interference with CNS development in the early fetal period. *Teratology 19*:327, 1979.

Russel, D. S. P.: *Observations on the Pathology of Hydrocephalus*. Med. Res. Council Report No. 265, London, 1949.

Shager, N. T., Kelly, A. B., and Wagner, J. A.: Congenital absence of the corpus callosum. *N. Engl. J. Med. 256*:1171, 1957.

Smith, D. W., and Gong, B. T.: Scalp-hair patterning as a clue to early fetal brain development. *J. Pediatr. 83*:375, 1973.

Thompson, J. S., and Thompson, M. W.: *Medical Genetics*. 3rd ed. Philadelphia, W. B. Saunders Co., 1980, p. 270.

Touwen, B. C. L.: Neurological development of the infant; *in* J. A. Davis and J. Dobbing (Eds.): *Scientific Foundations of Paediatrics*. Philadelphia, W. B. Saunders Co., 1974, pp. 587–614.

Vaughan, V. C., McKay, R. J., and Behrman, R. E. (Eds.): *Nelson Textbook of Pediatrics*. 11th ed. Philadelphia, W. B. Saunders Co., 1979.

Warkany, J.: *Congenital Malformations. Notes and Comments*. Chicago, Year Book Medical Publishers, Inc., 1971, pp. 189–352.

THE EYE AND THE EAR

THE EYE

The *visual organs,* or eyes, develop from three sources: (1) *neuroectoderm* of the forebrain, (2) *surface ectoderm* of the head, and (3) *mesoderm* between the two aforementioned layers.

Eye formation is first evident during stage 10 of development (about 22 days) when a pair of grooves called *optic sulci* appear in the neural folds at the cranial end of the embryo (Fig. 19–1*A* and *B*). As the neural folds fuse to form the prosencephalon, or *forebrain,* the optic sulci evaginate to form a pair of hollow diverticula called *optic vesicles,* which project from the sides of the forebrain into the adjacent mesenchyme (see Figs. 18–3*A* and 19–1*C*). The formation of the optic vesicles is induced by the mesenchyme adjacent to the developing brain.

As the bulb-like optic vesicles grow laterally, their distal ends expand and their connections with the forebrain constrict to form *optic stalks* (Fig. 19–1*D*). As the optic vesicles grow outward, displacing the mesenchyme, their external surfaces become flattened.

Concurrently, the surface ectoderm adjacent to the optic vesicles thickens and forms *lens placodes* (Fig. 19–1*C*). The formation of lens placodes is induced by the optic vesicles. The inducing agent is probably a chemical substance produced by the lens vesicles. The central region of each lens placode soon invaginates (Fig. 19–1*D*) and sinks below the surface, forming a *lens pit* (fovea lentis). The edges of this pit gradually approach each other and fuse to form a spherical *lens vesicle* (Fig. 19–1*F*), which becomes pinched off from the surface ectoderm. It is totally surrounded by mesenchyme. *The lens vesicle develops into the lens of the eye.*

As the lens vesicles are developing, the optic vesicles invaginate and become double-walled, cup-like structures called *optic cups* (see Figs. 18–3*C* and 19–1*H*). The opening of the optic cup is large at first, but later the

rim of the optic cup bends inward and converges (Figs. 19–3 and 19–4). By this stage, the lens vesicles have separated from the surface ectoderm and have come to lie within the optic cups. Linear grooves called choroid fissures, or *optic fissures* (Fig. 19–1*E* to *H*), develop on the inferior surface of the optic cups and along the optic stalks. Hyaloid blood vessels develop in the mesenchyme in these fissures. The *hyaloid artery,* a branch of the *ophthalmic artery,* supplies the inner layer of the optic cup, the lens vesicle, and the mesenchyme in the optic cup (Fig. 19–1*H*). The *hyaloid vein* returns blood from these structures. As the edges of the optic fissure come together and fuse, the hyaloid vessels are enclosed within the optic nerve (Fig. 19–2*E* and *F*). The distal portions of the hyaloid vessels eventually degenerate, but their proximal portions persist as the *central artery and vein of the retina.*

The Retina (Figs. 19–3 to 19–5).

The retina develops from the walls of the optic cup, which is an outgrowth of the brain (*diencephalon*). The outer, thinner layer of the optic cup becomes the pigment epithelium, and the inner, thicker layer differentiates into the complex neural layer of the retina. During the embryonic and early fetal periods, the two retinal layers are separated by an intraretinal space representing the cavity of the original optic vesicle. This space gradually disappears as the retina forms.

The pigment epithelium becomes firmly fixed to the choroid, but its attachment to the neural layer of the retina is not so firm. Hence, detachment of the retina may follow a blow to the eye. It often occurs during fixation and preparation of an eye for histological study. The detachment consists of separation of the pigment epithelium from the neural layer of the retina, i.e., at the site of embryonic adherence of the outer and inner layers of the optic cup.

413

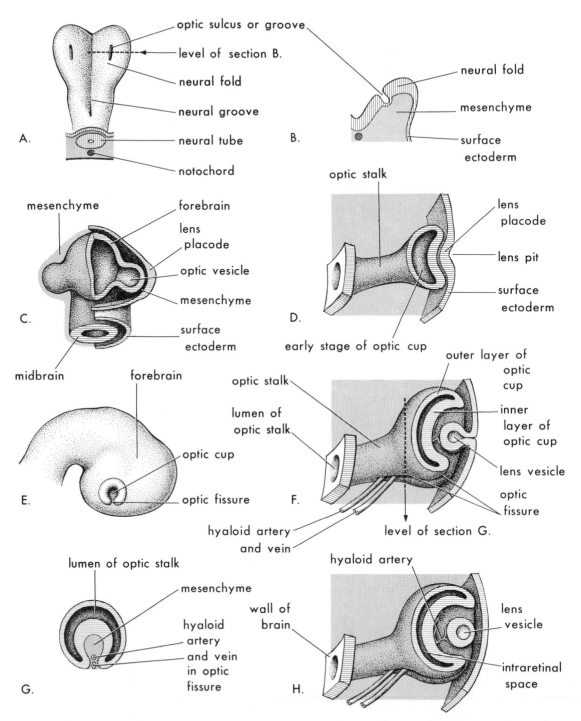

Figure 19–1 Drawings illustrating early eye development. *A,* Dorsal view of the cranial end of an embryo during stage 10 of development (about 22 days), showing the first indication of eye development. Note that the neural folds have not fused to form the primary brain vesicles at this stage. *B,* Transverse section through an optic sulcus. *C,* Schematic drawing of the forebrain, its covering layers of mesoderm and surface ectoderm from an embryo during stage 13 of development (about 28 days), *D, F,* and *H,* Schematic sections of the developing eye illustrating successive stages in the development of the optic cup and the lens vesicle. *E,* Lateral view of the brain of an embryo during stage 14 of development (about 32 days), showing the external appearance of the optic cup. *G,* Transverse section through the optic stalk showing the optic fissure and its contents. Note that the edges of the optic fissure grow together and fuse, thereby completing the optic cup and enclosing the central artery and vein of the retina in the cup and the optic nerve.

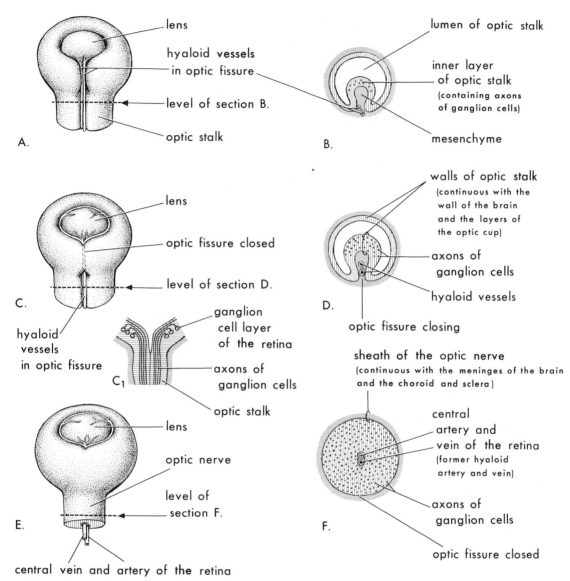

A.

lens

hyaloid vessels
in optic fissure

level of section B.

optic stalk

B.

lumen of optic stalk

inner layer
of optic stalk
(containing axons
of ganglion cells)

mesenchyme

C.

lens

optic fissure closed

level of section D.

hyaloid
vessels
in optic fissure

C₁

ganglion
cell layer
of the retina

axons of
ganglion cells

optic stalk

D.

walls of optic stalk
(continuous with the
wall of the brain
and the layers of
the optic cup)

axons of
ganglion cells

hyaloid vessels

optic fissure closing

E.

lens

optic nerve

level of
section F.

central vein and artery of the retina

F.

sheath of the optic nerve
(continuous with the meninges of the brain
and the choroid and sclera)

central
artery and
vein of the retina
(former hyaloid
artery and vein)

axons of
ganglion cells

optic fissure closed

Figure 19–2   Diagrams illustrating closure of the optic fissure and formation of the optic nerve. *A, C,* and *E,* Views of the inferior surface of the optic cup and stalk, showing progressive stages in the closure of the optic fissure. *C₁,* Schematic sketch of a longitudinal section of a portion of the optic cup and optic stalk, showing axons of ganglion cells of the retina growing through the optic stalk to the brain. *B, D,* and *F,* Transverse sections through the optic stalk, showing successive stages in the closure of the optic fissure and in formation of the optic nerve. Note that the lumen of the optic stalk is gradually obliterated as axons of ganglion cells accumulate in the inner layer of the optic stalk. The changes in the optic stalk that result in the formation of the optic nerve occur between the sixth and eighth weeks.

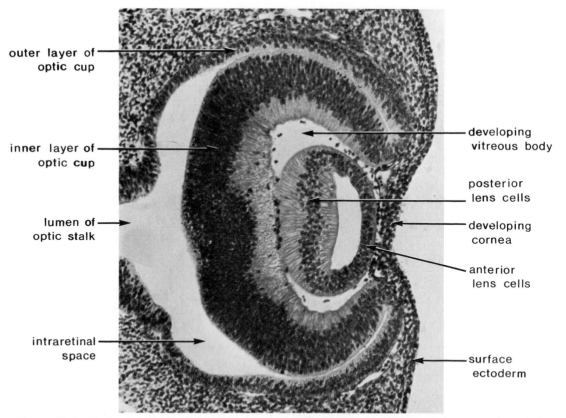

Figure 19–3 Photomicrograph of a sagittal section through the developing eye of a human embryo during stage 17 of development (about 41 days) (×200). The intraretinal space represents the cavity of the original optic vesicle. (Courtesy of Dr. J. W. A. Duckworth, Professor Emeritus of Anatomy, University of Toronto.)

Labels on figure:
- outer layer of optic cup
- inner layer of optic cup
- lumen of optic stalk
- intraretinal space
- developing vitreous body
- posterior lens cells
- developing cornea
- anterior lens cells
- surface ectoderm

Because the optic vesicle grows outward from the forebrain, the layers of the optic cup are continuous with the wall of the brain. Under the influence of the lens, the inner layer of the optic cup proliferates and forms a thick neuroepithelium. Subsequently, the cells of this layer differentiate into rods and cones, bipolar cells, and ganglion cells.

Because the optic vesicle invaginates as it forms the optic cup, the neural layer of the retina is "inverted"; i.e., light-sensitive parts of the photoreceptor cells are adjacent to the pigment epithelium. As a result, light must pass through most of the retina before reaching the receptors; however, because the retina is thin and transparent, it does not produce a significant barrier to light. The neural layer of the developing retina is continuous with the inner layer of the optic stalk (Figs. 19–1F and G and 19–2D). Consequently, axons of the ganglion cells pass into the inner wall of the optic stalk and gradually convert it into the *optic nerve* (Fig. 19–2B, D, and F).

Myelination of the optic nerve fibers is incomplete at birth. After the eyes have been exposed to light for about ten weeks, myelination of the fibers is complete, but the process normally stops short of the optic disc (Kwitko, 1979).

The Ciliary Body (Figs. 19–5 and 19–6). The pigmented portion of the epithelium of the ciliary body is derived from the outer layer of the optic cup, and it is continuous with the pigment epithelium of the retina. The nonpigmented portion of the ciliary epithelium represents the forward prolongation of the neural layer of the retina, in which no neural elements differentiate. The ciliary muscle and connective tissue develop from mesenchyme that indents the edge of the optic cup.

The Iris (Figs. 19–5 and 19–6). The iris develops from the edge of the optic cup, which bends inward and partially covers the lens. The epithelium of the iris represents both layers of the optic cup and is continuous with the double-layered epithelium of the ciliary body and with the pigment epithelium and the neural layer of the retina.

The eyes of most Caucasians are blue at birth because the small amount of dark melanin pigment in the epithelial layers on the posterior aspect of the iris appears blue through the stroma anterior to it. Pigment begins to form in the stroma during the first few days after birth. The final color of the eye depends on the density of the stroma and on how much pigment is deposited in it and in the two layers of epithelial cells at the posterior aspect of the iris. If very little pigment is deposited, the eye remains blue. When the stroma is dense, the eye appears gray. If more pigment is deposited, the eye appears brown. The eyes of deeply-pigmented races may look blue, hazel, or brown at birth, depending upon the amount of pigment present in the iris. They become darker during the first few days as more pigment is deposited in the iris, and changes in eye color may be noticeable for several weeks.

Pigmentation of the border layer of the iris, a condensation of the iris stroma, is acquired during the first few years after birth. As a result, the final color of the eyes is not determined until this time. The border layer is thin and lightly pigmented in blue-eyed persons, and thick and heavily pigmented in dark-eyed persons.

The dilator and sphincter pupillae muscles of the iris are derived from the neuroectoderm of the outer layer of the optic cup. These smooth muscles are derived from the pigment epithelium by a transformation of

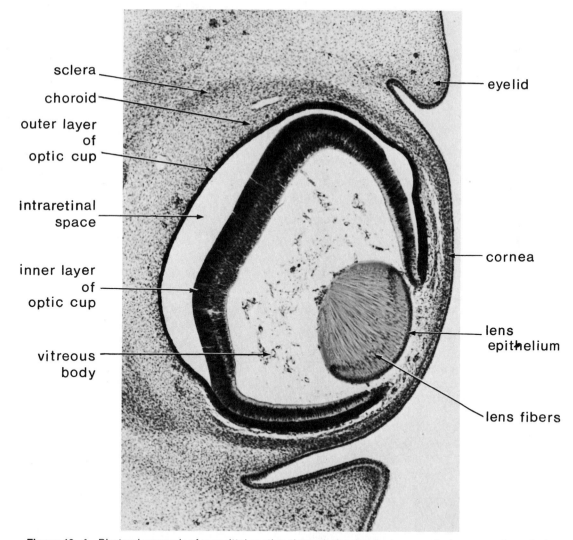

sclera

choroid

outer layer of optic cup

intraretinal space

inner layer of optic cup

vitreous body

eyelid

cornea

lens epithelium

lens fibers

Figure 19–4 Photomicrograph of a sagittal section through the developing eye of a human embryo during stage 20 of development (about 50 days) (×75). The intraretinal space, representing the cavity of the optic vesicle, gradually disappears as the inner and outer layers of the optic cup fuse to form the retina (Fig. 19–5).

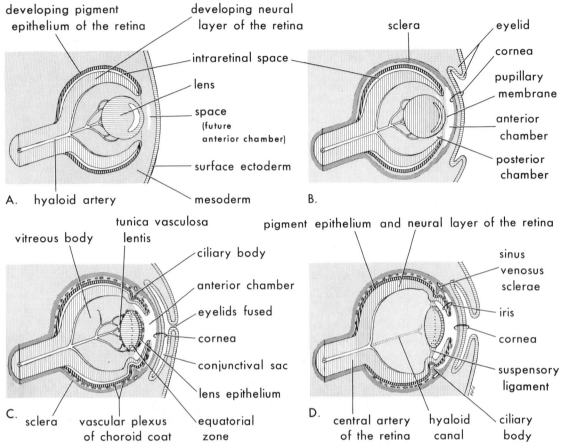

Figure 19–5 Drawings of sagittal sections of the eye, showing successive developmental stages. *A*, Five weeks. *B*, Six weeks. *C*, 20 weeks. *D*, Newborn. Note that the layers of the optic cup are fused and form the pigment epithelium and neural layer of the retina and that they extend anteriorly as the double epithelium of the ciliary body and the iris. The retina and the optic nerve are formed from the optic cup and the optic stalk, outgrowths of the brain (Fig. 19–1C).

epithelial cells into smooth muscle fibers. Electron microscopic studies confirm this unusual event. The vascular connective tissue of the iris is derived from mesenchyme located anterior to the rim of the optic cup.

The Lens (Fig. 19–5). The lens develops from the hollow lens vesicle, a derivative of the surface ectoderm (Fig. 19–1D and F). The anterior wall of this vesicle, composed of low columnar cells, does not change appreciably as it becomes the *anterior epithelium* of the adult lens. The cells forming the posterior wall of the lens vesicle lengthen considerably to form *lens fibers* that grow into and gradually obliterate the cavity of the lens vesicle (Fig. 19–5A to C). The first lens fibers, which form by elongation and differentiation of the posterior epithelial cells of the lens vesicle, are oriented in an anteropos-

terior direction. Succeeding fibers are derived superficially by mitosis, elongation, and differentiation of cells at the *equator of the lens*. Hence, these lens fibers are arranged meridionally in concentric layers. The older and deeper lens fibers lose their nuclei, but the epithelial cells in the *equatorial zone* (Fig. 19–5C) continue to multiply and differentiate into new lens fibers. Consequently, the concentric layers of lens fibers show varying degrees of differentiation. Fiber formation continues throughout life.

The lens is enclosed by a specialized elastic capsule of intercellular substance, *the lens capsule*, which is produced by the underlying epithelial cells (Willis et al., 1969). The lens grows rapidly during the first few years after birth; thereafter, it grows more slowly.

ace that forms in the mesenchyme posteri-
to the developing iris and anterior to the
eveloping lens. When the pupillary mem-
ane disappears and the pupil forms, the
nterior and posterior chambers of the eye
ommunicate with each other.

**The Sclera and Choroid (Figs. 19–4 and
9–5).** The mesenchyme surrounding the
otic cup differentiates into an inner vas-
ilar layer, the choroid, and an outer fibrous
yer, the sclera. Toward the margin of the
otic cup, the choroid becomes modified to
rm the cores of the ciliary processes, con-
sting chiefly of capillaries supported by
elicate connective tissue. The sclera is con-
nuous with the substantia propria of the
ornea (Fig. 19–5D).

At the attachment of the optic nerve to the eye,
e choroid is continuous with the pia-arachnoid
the brain, which forms a sheath around the
otic nerve. The sclera is continuous with the
ira mater of the brain, which also forms a sheath
ound this nerve. The continuity of these layers
understandable when it is recalled that the eyes
velop from outgrowths of the brain. The sub-
achnoid space around the brain also extends
ound the optic nerves as far as their attachment
the eyes.

The Eyelids (Fig. 19–5). The eyelids develop
om two ectodermal, or cutaneous, folds con-
ning cores of mesenchyme. The eyelids meet
d adhere by about the tenth week and remain
herent until about the twenty-sixth week (see
apter 6). While the eyelids are adherent, a
osed conjunctival sac exists anterior to the
ornea; when the eyes open, the *conjunctiva*
vers the "white" of the eye and lines the
elids. The eyelashes and glands are derived
om the surface ectoderm in a manner similar to
at described for other parts of the integument
e Chapter 20). The muscles and tarsal plates
velop from mesenchyme in the cores of the
elids.

Although it is generally stated that the upper
d lower eyelids are separated by 26 weeks,
n et al. (1973) concluded that the eyelids begin
separate from each other by about 21 weeks.

The Lacrimal Glands. At the superolateral
gles (superior fornices) of the conjunctival
s, the lacrimal glands develop from a number
solid buds from the surface ectoderm. These
anch and become canalized to form the ducts
l alveoli of the glands. The lacrimal glands are
all at birth and do not function fully for about
weeks. Hence, the newborn infant does not
oduce tears when it cries.

The Postnatal Eyes. The size of the
es of a newborn infant is about two thirds
t of the eyes of an adult. The eyes grow
ly rapidly in the first year and then grow

more slowly until puberty; after this, additional growth is negligible.

*Infants do not see well for the first few
weeks.* They have a vacant stare, mainly
because of incomplete development of the
macula lutea and fovea centralis, the area of
the retina responsible for visual acuity.
These structures are not completely developed until about a month after birth. Persistence of a vacant stare beyond the first few
weeks may indicate mental retardation or a
visual defect such as bilateral *optic atrophy*
(Kwitko, 1979).

CONGENITAL MALFORMATIONS
OF THE EYE

Because of the complexity of eye development, many congenital abnormalities may
occur, but most of them are relatively rare.
The type and severity of the malformation
depend upon the embryonic stage during
which development is disturbed. The critical
period of human eye development is during
developmental stages 10 to 20 (22 to 50
days).

**Congenital Colobomas (Figs. 19–7 and
19–8).** Defect of the eyelid, *palpebral coloboma,* is uncommon and is usually characterized by a small notch in the upper eyelid.
Rarely, colobomas are in the lower eyelid
and may then be associatd with the *Treacher
Collins syndrome* (see Chapter 10). Palpebral
colobomas probably result from a local developmental disturbance in the growth of the
eyelid.

In *coloboma iridis,* there is an inferior
defect in the iris, giving the pupil a keyhole
appearance (Fig. 19–8). The gap, or notch,
may be limited to the iris, or it may extend

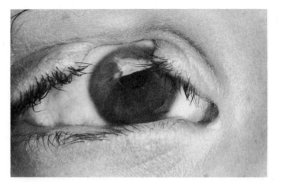

Figure 19–7 Photograph of the eye of a child
with a coloboma of the iris and upper eyelid. (From
Brown, C. A.: Abnormalities of the eyes and associ-
ated structures; in A. P. Norman [Ed.]: *Congenital
Abnormalities in Infancy.* 2nd ed. 1971. Courtesy of
Blackwell Scientific Publications.)

nonpigmented portion of
the ciliary epithelium
(continuous with the
neural layer of the retina)

pigment
the cili
(continuous
epitheliu

ciliary
processes

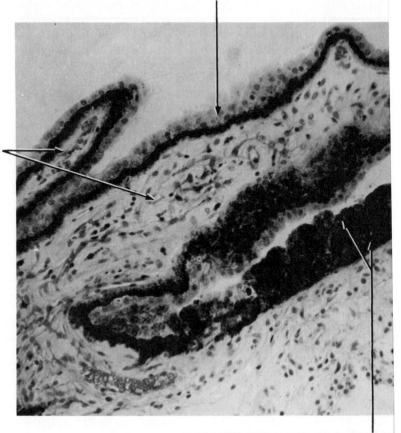

double-layered epithelium
(continuous with the neural and pigmented laye

Figure 19-6 Photomicrograph of the root of the adult iris (right) and ciliary processes, a
and iridial parts of the retina (×215). (From Leeson, T. S., and Leeson, C. R.: *Histology.* 3rd
W. B. Saunders Co., 1976.)

The developing lens is invested by a vascular
mesenchymal layer, the *tunica vasculosa lentis;*
the anterior portion of this capsule is called the
pupillary membrane (Fig. 19–5B and C). The
portion of the hyaloid artery that supplies the
tunica vasculosa lentis disappears during the late
fetal period. As a result, the tunica vasculosa
lentis and the pupillary membrane degenerate
(Fig. 19–5D), but the lens capsule produced by the
lens epithelium persists. The former site of the
hyaloid artery is indicated by the *hyaloid canal* in
the vitreous body (Fig. 19–5D). It is usually
inconspicuous in the living eye.

The *vitreous body* forms within the cavity of the
optic cup and is derived partly from the neuroectoderm of the optic cup and partly from mesenchyme that enters the optic cup.

The Aqueous Chambers
the Cornea (Figs. 19–3 to
anterior chamber develops fr
space that forms in the mese
ed between the developing le
face ectoderm (Fig. 19–5A to
enchyme superficial to this s
substantia propria of the c
mesothelium of the anterior
the lens is established, it fr
inductor and influences the su
to develop into the epitheliun
and the conjunctiva. The mes
to the developing anterior cha
stroma of the iris.

The posterior chamber de

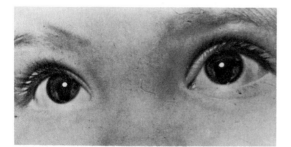

Figure 19–8 Photograph of the eyes of a child, showing typical bilateral coloboma of the iris. (From Rahn, E. K., and Scheie, H. G.: The eye; *in* A. Rubin [Ed.]: *Handbook of Congenital Malformations.* Philadelphia, W. B. Saunders Co., 1967.)

deeper and involve the ciliary body and retina. Typical colobomas of the iris result from failure of closure of the optic fissure. This may be genetically determined or be caused by environmental factors (Warkany, 1971).

Congenital coloboma of the retina is characterized by a localized gap in the retina, usually inferior to the optic disc. The defect also results from *defective closure of the optic fissure.* The condition is bilateral in more than one half of cases.

Congenital Glaucoma (Fig. 19–9). High intraocular pressure and enlargement of the eye result from abnormal development of the drainage mechanism of the aqueous humor. *Intraocular tension* rises as a result of imbalance between production of aqueous humor and its outflow. This probably results from absence of or abnormal development of the *sinus venosus sclerae* (Fig. 19–5D). Congenital glaucoma is usually caused by recessive mutant genes, but the condition sometimes results from maternal rubella infection during early pregnancy (see Fig. 8–19B).

Congenital Cataract (see Fig. 8–18A). In this condition, the lens is opaque and frequently appears grayish-white. Many lens opacities are inherited, but some are caused by noxious agents (particularly the *rubella virus*) that affect early development of the lens. *Rubella cataract and other ocular abnormalities caused by the rubella virus could be completely prevented* if immunity to rubella were conferred on all women of reproductive age (Warkany, 1981).

Another cause of cataract is the *enzymatic deficiency of congenital galactosemia.* Cataracts are not present at birth, but they appear as early as the second week after birth. Owing to the enzyme deficiency, large amounts of galactose from milk accumulate in the infant's blood and tissues, causing injury to the lens that results in cataract formation (Crowley, 1974). Hence, the cataracts are not congenital, but the enzyme deficiency causing them is.

Congenital Ptosis (Fig. 19–10). Drooping of one or both upper eyelids at birth is relatively common. It results from abnormal development or *failure of development of the levator palpebrae superioris muscle.* Congenital ptosis may also result from incomplete innervation of this muscle. If ptosis is associated with inability to move the eyeball upward, there is also failure of the superior rectus muscle of the eye to develop. Congenital ptosis is hereditary, and the isolated defect is usually transmitted as an autosomal dominant trait (Brown, 1971).

Persistent Pupillary Membrane. Remnants of the pupillary membrane, which normally covers the anterior surface of the lens during the early fetal period (Fig. 19–5B), commonly persist as strands of connective tissue over the pupil. They seldom interfere with vision and usually are of no consequence. Very rarely, the entire pupillary membrane persists at birth, giving rise to a

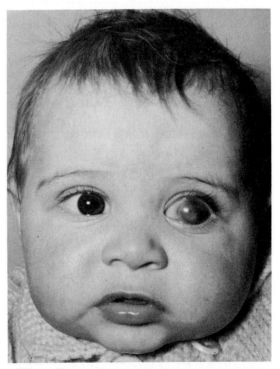

Figure 19–9 Photograph of a child with congenital glaucoma (buphthalmos), showing enlargement of the left eye. (Courtesy of Dr. C. A. Brown, Consultant Ophthalmologist, Bristol Eye Hospital, England.)

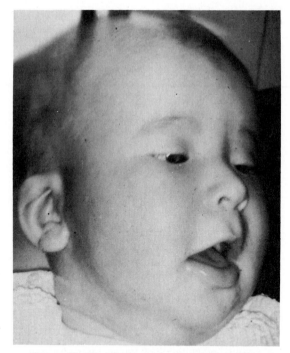

Figure 19–10 Photograph of an infant with congenital bilateral ptosis. Unilateral ptosis is more common. Drooping of the upper eyelid results from abnormal development or failure of development of the levator palpebrae superioris, the muscle that elevates the eyelid. In bilateral cases, the infant contracts the frontalis muscle in an attempt to raise the upper eyelids. (From Schaffer, A. J., and Avery, M. E.: *Diseases of the Newborn.* 4th ed. Philadelphia, W. B. Saunders Co., 1977.)

condition known as *congenital atresia of the pupil.*

Persistence of the Hyaloid Artery. The distal portion of this artery normally degenerates. If a small part of this portion persists, it may appear as a freely moving vessel, or cord, projecting from the optic disc into the vitreous body. In some cases, this remnant may form a cyst. Rarely, the entire distal portion of the artery persists and extends from the optic disc through the vitreous body to the lens. In most of these cases, the eye is microphthalmic (very small), but in some cases the eye is otherwise normal (Mann, 1957).

Congenital Detachment of the Retina. This condition, in which the pigmented and neural layers of the retina are separated, results when the inner and outer layers of the optic cup fail to fuse to form the retina and obliterate the intraretinal space (Fig. 19–4). The separation may be partial or complete. This abnormality may result from unequal rates of growth of the two layers of the optic cup; as a result, these layers are not in perfect apposition. Sometimes, the layers of the optic cup appear to have fused and separated later; such secondary detachments usually occur in association with other malformations of the eye and head.

Congenital Aphakia. Congenital absence of the lens is extremely rare and results from failure of the lens placode to form, a condition that is probably caused by failure of lens induction by the optic vesicle. This condition may also result from degeneration of the lens during the fetal period as the result of a pathological process.

Cryptophthalmos. This very rare condition results from failure of the eyelids to develop. As a result, the skin of the forehead continues over the eye(s). The eyeball is small and defective, and the cornea and conjunctiva usually do not develop.

Microphthalmos. The eye on one side may be very small and may be associated with gross ocular abnormalities, or it may be a normal-appearing miniature eye. The affected side of the face is underdeveloped and the orbit is small. Microphthalmos may be associated with other congenital abnormalities (e.g., facial cleft) or it may be part of a syndrome (e.g., trisomy 13).

Microphthalmos results from arrested development of the eye after the optic vesicle has formed. If the interference with development occurs before the embryonic optic fissure closes, microphthalmos is associated with gross ocular defects. When eye development is arrested later in the embryonic or early fetal periods, *simple microphthalmos* results (small eye without gross ocular abnormalities).

Some cases of microphthalmos are inherited. The hereditary pattern may be recessive or sex-linked with low penetrance (Smith and Guberina, 1979). Most cases of so-called simple microphthalmia are probably caused by infectious agents (e.g., rubella virus, *Toxoplasma gondii,* and herpes simplex virus) that cross the placental membrane during the fetal period (Rahn and Scheie, 1967).

Anophthalmos (Fig. 19–11). The eyelid forms, but no eyeball is present in this condition. In some cases, eye tissue may be recognizable histologically. Anophthalmia may be unilateral or bilateral. Usually, there are no other defects.

In *primary anophthalmos,* eye development is arrested early and results from failure of the optic vesicle to form. In *secondary anophthalmos,* the entire forebrain is suppressed, and absence of the eyes is one of several malformations. In consecutive anophthalmos, or *degenerative anophthalmos,* the optic vesicle develops and then undergoes degeneration, probably as the result of a teratogenic agent.

Cyclopia (Fig. 19–12). In this very rare condition, the eyes are partially or completely fused into a single *median eye* enclosed in a single orbit. Usually there is a tubular nose (proboscis) above the eye. The abnormality is frequently associated with other severe craniocerebral malformations that are *incompatible with life.* Cyclopia appears to result from suppression of midline cerebral structures that develop from the cranial part of the neural plate (Rahn and Scheie, 1967). Cyclopia is transmitted by recessive inheritance.

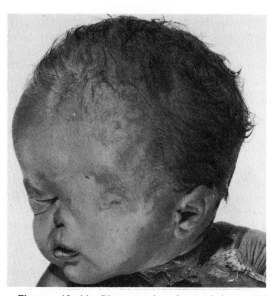

Figure 19–11 Photograph of an infant with apparent anophthalmia and a single nostril. Without microscopic examination, the clinical distinction between anophthalmia and microphthalmia is difficult.

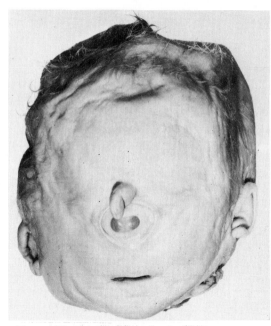

Figure 19–12 Photograph of an infant with cyclopia. The fused eyes are situated in a single midline orbit that is surmounted by a proboscis, a tubular structure representing the nose. Owing to gross deformities of the skull, brain, and other organs, these malformed infants do not survive long after birth.

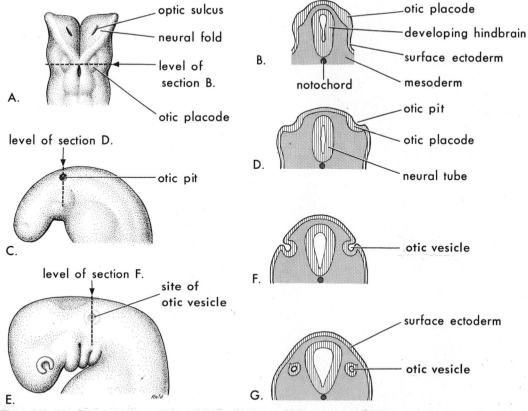

Figure 19–13 Drawings illustrating early development of the inner ear. *A,* Dorsal view of an embryo during stage 10 of development (about 22 days), showing the otic placodes. *B, D, F,* and *G,* Schematic sections illustrating successive stages in the development of otic vesicles. *C* and *E,* Lateral views of the cranial region of embryos during developmental stages 11 and 13 (about 24 and 28 days, respectively).

THE EARS

The ear consists of three anatomical parts: external, middle, and internal. The external and middle parts are concerned mainly with the transference of sound waves from the exterior to the internal ear, which contains the *vestibulocochlear organ* concerned with equilibration and hearing.

The Internal Ear (Figs. 19–13, 19–14, and 19–15). This is the first of the three anatomical divisions of the ear to appear. Early in the fourth week, a thickened plate of surface ectoderm, the *otic placode,* appears on each side of the developing hindbrain. Each placode soon invaginates and sinks below the surface ectoderm into the underlying mesenchyme to form an *otic pit.* The edges of the pit come together and fuse to form an *otic vesicle* (otocyst), the primordium of the *membranous labyrinth.* The otic vesicle soon loses its connection with the surface ectoderm. A hollow diverticulum grows out from the otic vesicle and elongates to form the *endolymphatic duct and sac* (Figs. 19–14 and 19–15A to E).

Two regions of each otic vesicle soon become recognizable: a dorsal *utricular portion,* from which the *utricle, semicircular ducts,* and *endolymphatic duct* arise; and a ventral *saccular portion,* which gives rise to the *saccule* and the *cochlear duct.*

Three flat disc-like diverticula grow out from the utricular portion, and soon the central portions of the walls of these diverticula fuse and then disappear (Fig. 19–15B to E).

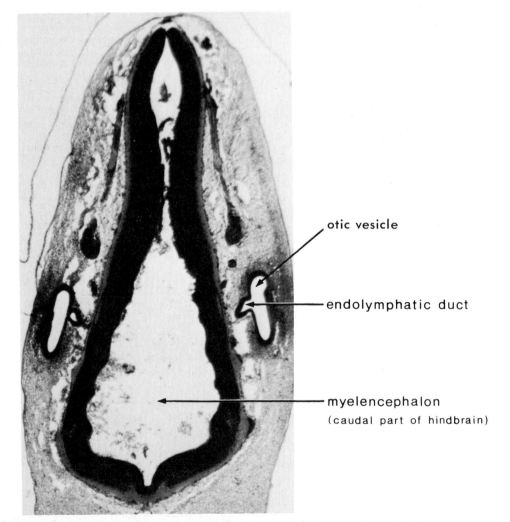

otic vesicle

endolymphatic duct

myelencephalon
(caudal part of hindbrain)

Figure 19–14 Transverse section through the head region of a human embryo during stage 14 of development (about 32 days), showing the otic vesicle (×27).

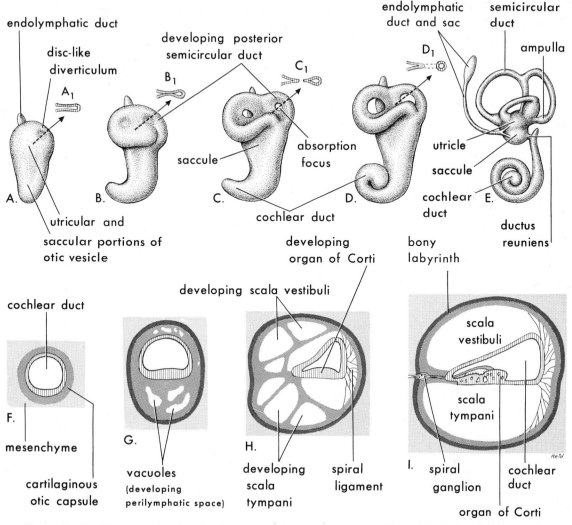

Figure 19–15 Diagrams showing development of the membranous and bony labyrinths of the internal ear. *A* to *E*, Lateral views showing successive stages in the development of the otic vesicle into the membranous labyrinth from the fifth to eighth weeks. A_1 to D_1, Diagrammatic sketches illustrating the development of a semicircular duct. *F* to *I*, Sections through the cochlear duct showing successive stages in the development of the organ of Corti and the perilymphatic space from the eighth to the twentieth weeks.

The peripheral unfused portions of the diverticula become the *semicircular ducts,* which are attached to the utricle and enclosed in the *semicircular canals* of the bony labyrinth.

Localized dilatations, the *ampullae,* develop at one end of each canal. Sensory nerve endings differentiate in the ampullae (*cristae ampullares*) and in the utricle and saccule (*maculae utriculi* and *sacculi*).

From the ventral saccular portion of the otocyst, a tubular diverticulum, the *cochlear duct,* grows and coils to form the *cochlea* (Fig. 19–15C to *E*). The connection of the

cochlea with the *saccule* becomes constricted to form the narrow *ductus reuniens.*

The *organ of Corti* differentiates from cells in the wall of the cochlear duct (Fig. 19–15F to *I*). Ganglion cells of the eighth cranial nerve migrate along the coils of the cochlea and form the *cochlear (spiral) ganglion,* from which nerve processes grow to the organ of Corti, where they terminate on the *hair cells.* The cells of the cochlear ganglion retain their embryonic bipolar condition (see Fig. 18–9B) and do not become unipolar like spinal ganglion cells (see Fig. 18–9D).

The mesenchyme around the otic vesicle

condenses and differentiates into a cartilaginous *otic capsule* (Fig. 19–15*F*). As the membranous labyrinth enlarges, vacuoles appear in the cartilaginous otic capsule and soon coalesce to form the *perilymphatic space*. The membranous labyrinth is now suspended in fluid, the *perilymph,* in the perilymphatic space. The perilymphatic space related to the cochlear duct develops in two divisions, the *scala tympani* and the *scala vestibuli* (Fig. 19–15*H* and *I*). The cartilaginous otic capsule ossifies to form the *bony labyrinth* of the internal ear.

The Middle Ear (see Figs. 10–4 and 19–16). Development of the tubotympanic recess from the first pharyngeal pouch is described in Chapter 10. The distal portion of this recess expands and becomes the *tympanic cavity*. The proximal unexpanded portion becomes the *auditory tube*. As the tym-

panic cavity expands, its endodermal epithelium gradually envelops the middle ear bones, or *auditory ossicles* (malleus, incus, and stapes), their tendons and ligaments, and the *chorda tympani nerve*. All these structures receive a more or less complete epithelial investment. During the late fetal period, expansion of the tympanic cavity gives rise to the *mastoid antrum*.

Some mastoid air cells begin to develop during fetal life, but most of them form after birth. The epithelial lining of the middle ear first induces erosion of the surrounding bone and then lines the spaces thus formed. This process of invasion of bone by epithelium to form air sacs is called *pneumatization*. The mucous membrane of the air cells is continuous with that of the antrum and, through it, with the tympanic cavity. Air cell development is a continuing process. The *mastoid process* is absent at birth, but it forms a distinct prominence on the temporal bone by the end of

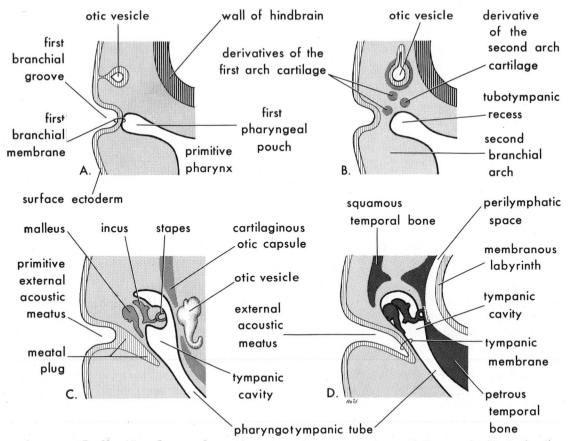

Figure 19–16 Schematic drawings showing development of the middle ear. *A,* Four weeks, illustrating the relation of the otic vesicle to the branchial apparatus. *B,* Five weeks, showing the tubotympanic recess and branchial arch cartilages. *C,* Later stage, showing the tubotympanic recess (future tympanic cavity) beginning to envelop the ossicles. *D,* Final stage of ear development, showing the relationship of the middle ear to the perilymphatic space and the external acoustic meatus.

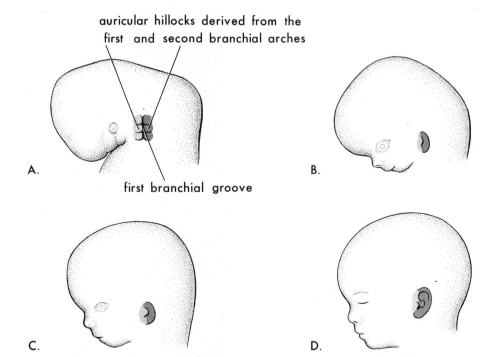

auricular hillocks derived from the
first and second branchial arches

A.   B.

first branchial groove

C.   D.

Figure 19–17 Drawings illustrating development of the auricle of the external ear. *A*, Six weeks. *B*, Eight weeks. *C*, Ten weeks. *D*, 32 weeks. As the auricles develop, they move from the neck to the side of the head.

the first year. The greatest growth of the mastoid process occurs between the third year and puberty.

The development of the middle ear bones is described in Chapter 10 (see Fig. 10–4). The muscle attached to the malleus, the *tensor tympani*, is derived from mesenchyme of the first branchial arch and is innervated by CN V (see Fig. 18–35), the nerve of that arch. The *stapedius muscle* is derived from the second branchial arch and is therefore supplied by CN VII, the nerve of this arch.

The External Ear (Figs. 19–16 and 19–17). The *external acoustic meatus* develops from the dorsal end of the first branchial groove. The ectodermal cells at the bottom of this funnel-shaped tube proliferate and extend inward as a solid epithelial plate called the *meatal plug*. Late in the fetal period, the central cells of this plug degenerate, forming a cavity that becomes the inner part of the external acoustic meatus.

The early *tympanic membrane* is represented by the first branchial membrane, which separates the first branchial groove and the first pharyngeal pouch (Fig. 19–16A). As development proceeds, mesenchyme grows between the branchial membrane and later differentiates into the fibrous stratum of the tympanic membrane. Thus, the *tympanic membrane* develops from ectoderm of the meatal plug, endoderm of the tubotympanic recess, and mesenchyme of the first and second branchial arches.

The *auricle* develops from six swellings called *auricular hillocks*, which arise around the margins of the first branchial groove (Fig. 19–17A). The swellings are produced by proliferation of mesenchyme from the first and second branchial arches. As the auricle grows, the contribution of the first branchial arch becomes relatively reduced (Fig. 19–17B to D). The lobule is the last part of the auricle to develop.

The external ears begin to develop in the upper part of the future neck region (Fig. 19–17A), but as the mandible develops, the auricles move to the side of the head and ascend to the level of the eyes (Fig. 19–17B to D).

The parts of the auricle derived from the first branchial arch are supplied by its nerve, the mandibular branch of the trigeminal, but the parts derived from the second branchial arch are supplied by cutaneous branches of the cervical plexus, especially the lesser occipital and the greater auricular nerves. The facial nerve of the second arch has few cutaneous branches, but some fibers contribute to the sensory innervation of the skin in the mastoid region and probably on both aspects of the auricle.

CONGENITAL MALFORMATIONS OF THE EAR

Congenital Deafness. Because the formation of the inner ear is independent of the development of the middle ear, congenital impairment of hearing may be the result of maldevelopment of the sound-conducting apparatus of the middle ear or of the neurosensory (perceptive) structures of the inner ear.

Most types of congenital deafness are caused by genetic factors (Konigsmark and Gorlin, 1976). In *deaf-mutism,* the ear abnormality is usually perceptive in type. This condition may occur with several other head and neck abnormalities as a part of the *first arch syndrome* (see Fig. 10–12).

Rubella infection during the critical period of embryonic development of the ear, particularly during the seventh and eighth weeks, can also cause maldevelopment of the organ of Corti (Gray, 1959). Congenital deafness may be associated with maternal goiter, which may result in fetal hypothyroidism (Thould and Scowen, 1964).

Congenital fixation of the stapes results in severe conductive deafness in an otherwise normal ear. Failure of differentiation of the anular ligament results in fixation of the stapes to the bony labyrinth. Defects of the malleus and incus are often associated with the *first arch syndrome* (see Fig. 10–12).

Auricular Abnormalities. There is a wide normal variation in the shape of the auricle. Minor variations of the auricle may be clues to serious congenital malformations, e.g., renal disorders (Smith, 1976). The auricles are often abnormal in shape and low-set in malformed infants (see Figs. 10–12 and 17–7); in infants afflicted with chromosomal syndromes (see Table 8–1 and Fig. 8–4); and in infants affected by maternal ingestion of certain drugs (see Fig. 8–16C).

Auricular appendages, or *tags* (Fig. 19–18), are relatively common and result from the development of accessory auricular hillocks. They usually appear anterior to the auricle, more often unilaterally than bilaterally. The appendages consist of skin, but they may also contain cartilage.

Absence and hypoplasia of the auricle are rare and are usually associated with the first arch syndrome. *Anotia* results from failure of the auricular hillocks to develop, and *microtia* results from suppressed development of the auricular hillocks. Atresia of the external

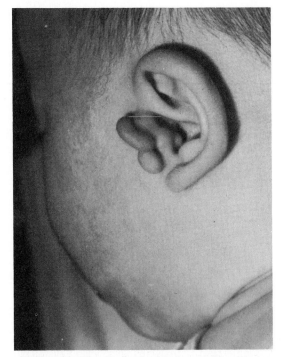

Figure 19–18 Photograph of a child with two auricular appendages, or tags, which result from the formation of accessory auricular hillocks. (From Swenson, O.: *Pediatric Surgery.* 1958. Courtesy of Appleton-Century-Crofts, Inc.)

acoustic meatus and middle ear abnormalities are usually also present.

Auricular Sinus and Fistula (see Fig. 10–9F). These sinuses are usually preauricular and are commonly located in a triangular area anterior to the auricle. The sinuses are usually narrow tubes or pits that end blindly and have pinpoint external openings. Some sinuses may contain a vestigial cartilaginous mass at their base.

The embryological basis of these sinuses is uncertain, but some may be related to defective closure of the dorsal part of the first branchial groove. Other auricular sinuses appear to represent ectodermal folds that are sequestered during formation of the auricle (Gray and Skandalakis, 1972). Auricular fistulas connecting the exterior with the tympanic cavity or the intratonsillar cleft are extremely rare.

Atresia of the External Acoustic Meatus. Blockage of the meatus results from failure of the meatal plug to canalize. Usually, the deep part of the meatus is open, but the superficial part is blocked by bone or fibrous tissue. Most cases are associated with the *first arch syndrome.* Often, abnormal development of both the first and second branchial arches is involved. The auricle is also usually severely affected and malformations of the middle or the inner ear are sometimes present. Atresia of the external acoustic meatus usually results from autosomal dominant inheritance, but it may also be caused by environmental factors.

SUMMARY

The eyes and the ears begin to develop during the fourth week. These special sense organs, particularly the eyes, are very sensitive to the teratogenic effects of conditions such as viral infections. The most serious defects result from disturbances of development during the fourth to eighth weeks, but defects of sight and hearing may result from infection of tissues and organs by certain microorganisms during the fetal period.

The Eye. The first indications of the eyes are the *optic sulci*. Soon, these sulci form a pair of *optic vesicles* on each side of the forebrain. The optic vesicles contact the surface ectoderm and induce development of the lens placodes, the primordia of the lenses. As the *lens placodes* invaginate to form *lens vesicles,* the optic vesicles invaginate to form *optic cups*. The retina forms from the two layers of the optic cup.

The retina, the optic nerve fibers, the iris muscles, and the epithelium of the iris and ciliary body are derived from the *neuroectoderm*. The *surface ectoderm* gives rise to the lens, the epithelium of the lacrimal glands and ducts, the eyelids, the conjunctiva, and the cornea. The *mesoderm* gives rise to the eye muscles (except those of the iris), and to all connective and vascular tissues of the cornea, iris, ciliary body, choroid, and ·sclera. The sphincter and dilator muscles of the iris develop from the ectoderm of the optic cup.

There are many congenital *ocular malformations,* but most of them are rare. Some malformations are caused by defective closure of the optic fissure. Congenital cataract and glaucoma may result from intrauterine infections (e.g., *rubella virus*) or from a *genetic disorder*.

The Ear. The *surface ectoderm* gives rise to the otic vesicle, which becomes the *membranous labyrinth* of the internal ear. The otic vesicle divides into (1) a dorsal *utricular portion,* which gives rise to the *utricle,* the semicircular ducts, and the endolymphatic duct; and (2) a ventral *saccular portion,* which gives rise to the *saccule* and the cochlear duct. The *organ of Corti* develops from the cochlear duct. The *bony labyrinth* develops from the surrounding mesenchyme.

The epithelium lining the tympanic cavity, the mastoid antrum, the mastoid air cells, and the auditory tube are derived from the endoderm of the tubotympanic recess of the first pharyngeal pouch. The auditory ossicles (malleus, incus, and stapes) develop from the cartilages of the first two branchial arches.

The epithelium of the *external acoustic meatus* develops from ectoderm of the first branchial groove. The *tympanic membrane* is derived from (1) the endoderm of the first pharyngeal pouch, (2) the ectoderm of the first branchial groove, and (3) the mesenchyme between these layers. The auricle develops from six *auricular hillocks,* or swellings, around the first branchial groove. These fuse to form the definitive auricle.

Congenital deafness may result from abnormal development of the membranous labyrinth and/or the bony labyrinth, as well as from abnormalities of the ossicles. Recessive inheritance is the most common cause of congenital deafness, but prenatal rubella virus infection is a major environmental factor known to cause defective hearing. There are many minor, clinically unimportant anomalies of the auricle. Low-set, malformed ears are often associated with chromosomal abnormalities, particularly *trisomy 18* and *trisomy 13* (see Chapter 8).

SUGGESTIONS FOR ADDITIONAL READING

Harcourt, B.: The Visual System; *in* J. A. Davis and J. Dobbing (Eds.): *Scientific Foundations of Paediatrics.* Philadelphia, W. B. Saunders Co., 1974, pp. 649–660. A detailed account of ocular organogenesis, with a summary of the chronology of events during early ocular development. The development of visual function and the effects of adverse influences upon the development of vision are also discussed.

Mann, I. C.: *The Development of the Human Eye.* 3rd ed. London, British Medical Assoc., 1974. A classic monograph on the embryology of the eye, written by a world-renowned ophthalmologist.

Toronto Hospital for Sick Children (Ophthalmologic Staff): *The Eye in Childhood.* Chicago, Year Book Medical Publishers, Inc. 1967. A comprehensive textbook by 14 contributors on the eye diseases of children. Included are complete discussions of all clinically significant congenital malformations of the eye and their treatment.

Wright, I.: Hearing and balance; *in* J. A. Davis and J. Dobbing (Eds.): *Scientific Foundations of Paediatrics.* Philadelphia, W. B. Saunders Co., 1974, pp. 661–680. An excellent description of the development of the inner and middle ear. Maldevelopment of the ear leading to congenital deafness is also discussed.

CLINICALLY ORIENTED PROBLEMS

1. *An infant was born blind and deaf* and with congenital heart disease. The mother had a *viral infection* early in her pregnancy. Considering the congenital malformations present, name the virus that was probably involved. What is the common *congenital cardiovascular lesion* found in infants whose mothers have this infection early in pregnancy? Is the history of a rash during the first trimester an essential factor in the development of *embryopathy?*

2. An infant was born with *bilateral ptosis, or drooping of the upper eyelids.* What is the embryological basis of this condition? Are hereditary factors involved? Injury to what nerve could also cause ptosis?

3. An infant had small, multiple *calcifications in the brain,* microcephaly, and microphthalmia. The mother was known to have a fondness for raw and rare meat. What *protozoon* might be involved? What is the embryological basis of the infant's congenital malformations? What advice might the doctor give the mother concerning future pregnancies?

4. A mentally retarded female infant had *low-set malformed ears,* a prominent occiput, and rocker-bottom feet. A *chromosomal abnormality* was suspected. What type of aberration was probably present? What is the usual cause of this abnormality? How long would the infant likely survive?

5. An infant was born with partial *detachment of the retina* in one eye. The eye was microphthalmic (small), and there was persistence of the posterior end of the *hyaloid artery.* What is the embryological basis of detachment of the retina? What is the usual fate of the hyaloid artery?

The answers to these questions are given at the back of the book.

REFERENCES

Anson, B. J., and Bast, T. H.: The development of the stapes of the human ear. *Quart. Bull. Northwestern Univ. Med. Sch. 33*:110, 1959.

Anson, B. J., Hanson, J. S., and Richany, S. F.: Early embryology of the auditory ossicles and associated structures in relation to certain anomalies observed clinically. *Ann. Otol. 69*:427, 1960.

Ashton, N.: Retinal angiogenesis in the human embryo. *Br. Med. Bull. 26*:103, 1970.

Balfour, H. H., Jr., Groth, K. E., and Edelman, C. K.: Ra 27/3 rubella vaccine. *Am. J. Dis. Child. 134*:350, 1980.

Barber, A. N.: *Embryology of the Human Eye.* St. Louis, The C. V. Mosby Co., 1955.

Brown, C. A.: Abnormalities of the eyes and associated structures; *in* A. P. Norman (Ed.): *Congenital Abnormalities in Infancy.* 2nd ed. Oxford, Blackwell Scientific Publications, 1971, pp. 147–198.

Crowley, L. V.: *An Introduction to Clinical Embryology.* Chicago, Year Book Medical Publishers, Inc., 1974.

Fraser, G. R.: A study of causes of deafness amongst 2,355 children in special schools; *in* L. Fisch (Ed.): *Research in Deafness in Children.* Oxford, Blackwell Scientific Publications, 1964.

Gray, J. E.: Rubella in pregnancy; fetal pathology in the internal ear. *Ann. Otol. 68*:170, 1959.

Gray, S. W., and Skandalakis, J. E.: *Embryology for Surgeons. The Embryological Basis for the Treatment of Congenital Defects.* Philadelphia, W. B. Saunders Co., 1972, pp. 15–61.

Hanson, J. R., Anson, B. J., and Strickland, E. M.: Branchial sources of the auditory ossicles in man. *Arch. Otolaryngol. 76*:200, 1962.

Harley, R. D., and Martyn, L. J.: Pediatric ophthalmology; *in* V. C. Vaughan, R. J. McKay, and R. E. Behrman (Eds.): *Nelson Textbook of Pediatrics.* 11th ed. Philadelphia, W. B. Saunders Co., 1979, pp. 1939–1979.

Hay, E. D., and Meier, S.: Stimulation of corneal differentiation by interaction between cell surface and extracellular matrix. II. Further studies on the nature and site of transfilter "induction." *Dev. Biol. 52*:141, 1976.

Haydon, G. D., and Arnold, G. G.: The ear; *in* E. L. Kendig, Jr. and V. Chernick (Eds.): *Disorders of the Respiratory Tract in Children.* 3rd ed. Philadelphia, W. B. Saunders Co., 1977.

Hough, J. V. D.: Congenital malformations of the middle ear. *Arch. Otolaryngol. 78*:335, 1963.

Jain, K. K., Bhandari, G. J., and Koronne, S. P.: Histogenesis of the human eyelid. *East. Arch. Ophthal. 3*:8, 1973.

Kanagasuntheram, R.: A note on the development of the tubotympanic recess in the human embryo. *J. Anat. 101*:731, 1967.

Konigsmark, B. W., and Gorlin, R. J.: *Genetic and Metabolic Deafness.* Philadelphia, W. B. Saunders Co., 1976.

Kwitko, M. L. (Ed.): *Surgery of the Infant Eye.* New York, Appleton-Century-Crofts, 1979.

Langman, J.: The first appearance of specific antigens during induction of the lens. *J. Embryol. Exp. Morphol.* 7:264, 1959.

Mann, I. C.: *Developmental Abnormalities of the Eye.* 2nd ed. Philadelphia, J. B. Lippincott Co., 1957.

O'Rahilly, R.: The early development of the otic vesicle in staged human embryos. *J. Embryol. Exp. Morphol.* 11:741, 1963.

O'Rahilly, R.: The early development of the eye in staged human embryos. *Contr. Embryol. Carneg. Instn. 38*:1, 1966.

O'Rahilly, R.: The prenatal development of the human eye. *Exp. Eye Res. 21*:93, 1975.

Papolczy, F.: Congenital cyclopia and orbital cyst together with other developmental anomalies on the same side of the face. *Br. J. Ophthalmol. 32*:439, 1948.

Pearson, A. A.: *The Development of the Eye.* Portland, University of Oregon Medical School Printing Department, 1964.

Pearson, A. A., and Jacobson, A. D.: *The Development of the Ear.* Portland, University of Oregon Medical School Printing Department, 1967.

Proctor, B.: The development of the middle ear spaces and their surgical significance. *J. Laryngol. 78*:631, 1964.

Rahn, E. K., and Scheie, H. G.: The eye; *in* A. Rubin (Ed.): *Handbook of Congenital Malformations.* Philadelphia, W. B. Saunders Co., 1967, pp. 170–228.

Sharp, H. S.: Abnormalities of the ear, nose and throat; *in* A. P. Norman (Ed.): *Congenital Abnormalities in Infancy.* 2nd ed. Oxford, Blackwell Scientific Publications, 1971, pp. 131–146.

Silcox, L. E.: The ear; *in* A. Rubin (Ed.): *Handbook of Congenital Malformations.* Philadelphia, W. B. Saunders Co., 1967, pp. 229–247.

Smith, B., and Guberina, C.: Congenital ocular anomalies; *in* M. L. Kwitko (Ed.): *Surgery of the Infant Eye.* New York, Appleton-Century-Crofts, 1979.

Smith, D. W.: *Recognizable Patterns of Human Malformation: Genetic, Embryologic, and Clinical Aspects.* 2nd ed. Philadelphia, W. B. Saunders Co., 1976.

Stevenson, A. C., and Cheeseman, E. A.: Hereditary deaf-mutism. *Ann. Hum. Genet. 20*:177, 1956.

Tamura, T., and Smelser, J. K.: Development of the sphincter and dilator muscles of the iris. *Arch. Ophthalmol. 89*:332, 1973.

Thould, A. K., and Scowen, E. F.: The syndrome of congenital deafness and goiter. *J. Endocrinol. 30*:69, 1964.

Tuchmann-Duplessis, H., and Mercier-Parot, L.: Production of congenital eye malformations, particularly in rat fetuses; *in* G. K. Smelser (Ed.): *The Structure of the Eye.* New York, Academic Press, Inc., 1961, p. 507.

Vaughan, V. C., McKay, R. J., and Behrman, R. E. (Eds.): *Nelson Textbook of Pediatrics.* 11th ed. Philadelphia, W. B. Saunders Co., 1979.

Warkany, J.: *Congenital Malformations: Notes and Comments.* Chicago, Year Book Medical Publishers, Inc., 1971, pp. 355–385.

Warkany, J.: Prevention of congenital malformations. *Teratology 23*:175, 1981.

Willis, N. R., Hollenberg, M. J., and Braekevelt, C. R.: The fine structure of the lens of the fetal rat. *Can. J. Ophthalmol. 4*:307, 1969.

Young, R. W., and Ocumpaugh, D. E.: Autoradiographic studies on the growth and development of the lens capsule in the rat. *Invest. Ophthalmol. 5*:583, 1966.

Zeiter, H. J.: Congenital microphthalmos. *Am. J. Ophthalmol. 55*:910, 1963.

THE INTEGUMENTARY SYSTEM

The Skin, the Cutaneous Appendages, and the Teeth

SKIN

The skin consists of two morphologically different layers that are derived from two different germ layers. The more superficial layer, the *epidermis,* is a specialized epithelial tissue derived from *surface ectoderm* (Fig. 20–1). The deeper and thicker layer, the *dermis,* is composed of vascular dense connective tissue derived from *mesenchyme.*

Epidermis (Fig. 20–1). The surface ectodermal cells proliferate and form a superficial protective layer of simple squamous epithelium, the *periderm.* The cells of this layer continually undergo keratinization and desquamation and are replaced by cells arising from the *basal layer.* The exfoliated cells form part of the *vernix caseosa,* a white, cheese-like, protective substance that covers the fetal skin. The other components of vernix caseosa are sebum from the sebaceous glands, fetal hair, and desquamated cells from the amnion. The basal layer of the epidermis is later called the *stratum germinativum* because it produces new cells that are displaced into the layers superficial to it.

By about 11 weeks, cells from the stratum germinativum have formed an intermediate layer (Fig. 20–1C). All layers of the adult epidermis are present at birth (Fig. 20–1D). Replacement of peridermal cells continues until about the twenty-first week; thereafter, the periderm gradually disappears. The desquamated epidermal cells then form part of the vernix caseosa; this substance persists until birth.

Proliferation of cells in the stratum germinativum also forms downgrowths called *epidermal ridges,* which extend into the developing dermis (Fig. 20–1C). These ridges are permanently established by the seventeenth week (Hale, 1952).

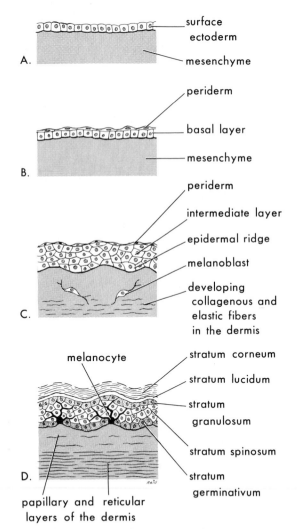

Figure 20–1 Drawings illustrating successive stages in the development of thick skin. *A,* Four weeks. *B,* Seven weeks. *C,* 11 weeks. *D,* Newborn. Note the position of the melanocytes in the basal layer of the epidermis and the way their branching processes extend between the epidermal cells to supply them with melanin.

The epidermal ridges produce ridges and grooves on the surface of the palms, including the fingers, and on the soles of the feet, including the toes. The type of pattern is determined genetically and constitutes the basis for using prints in criminal investigations and medical genetics. The study of the patterns of the epidermal ridges of the skin is called *dermatoglyphics*. The presence of abnormal chromosome complements affects the development of the ridge patterns; for example, infants with Down syndrome have distinctive patterns on the hands and feet that are of diagnostic value (see Chapter 8). For details about the use of dermatoglyphics in medical genetics, see Uchida and Summit (1979).

During the early fetal period, *melanoblasts* migrate from the *neural crest* to the dermoepidermal junction, where they differentiate into *melanocytes* (see Figs. 18–8 and 20–1D). The melanocytes, unlike the relatively round melanoblasts and *premelanocytes*, have several long processes. The cell bodies of the melanocytes are usually confined to the basal layers of the epidermis, but the processes of melanocytes extend for some distance between the epidermal cells (Fig. 20–1D).

The melanocytes begin producing melanin before birth and distribute it to the epidermal cells. Active pigmentary activity can be observed prenatally in the epidermis of dark-skinned races, but there is little evidence of such activity in white fetuses.

Dermis (Fig. 20–1). The dermis is derived from the mesenchyme underlying the surface ectoderm. Most of the mesenchyme of the dermis originates from the somatic layer of lateral mesoderm (see Chapter 4), but some mesenchyme is derived from the dermomyotome regions of the somites (see Chapter 15).

By 11 weeks, the mesenchymal cells have begun to produce collagenous and elastic connective tissue fibers (Fig. 20–1D). As the epidermal ridges form, the dermis projects upward into the epidermis and forms *dermal papillae*. Capillary loops develop in some dermal papillae, and sensory nerve endings form in others.

The first blood vessels in the dermis begin as simple endothelium-lined structures that differentiate from the mesenchyme. As the skin grows, new capillaries grow (bud) out from the simple vessels. Some of these capillaries acquire muscular coats through differentiation of myoblasts from the surrounding mesenchyme and become arterioles and arteries; others, through which a return flow of blood is established, acquire muscular coats and become venules and veins. As new blood vessels form, some transitory ones normally disappear.

CONGENITAL ABNORMALITIES OF SKIN

Disorders of Keratinization. *Ichthyosis* (Gr. *ichthys,* fish) is a general term applied to a group of disorders characterized by dryness and *fishskin-like scaling of the skin.* Scaling is often pronounced, involving the entire body surface.

A *harlequin fetus* results from a very rare keratinizing disorder that is inherited as an autosomal recessive trait (Vaughan et al., 1979). The skin is markedly thickened, ridged, and cracked. Affected infants have a grotesque appearance, and most of them die within the first week of life.

A *collodion baby* is covered at birth by a thick, taut membrane resembling collodion. This membrane cracks with the first respiratory efforts and begins to fall off in large sheets, but complete shedding may take several weeks.

Lamellar ichthyosis (Fig. 20–2), an autosomal recessive disorder, is present at birth. A newborn infant with this condition may first appear to be a "collodion baby," but the scaling persists and involves the entire body. Growth of hair may be curtailed, and development of the sweat glands is often impeded. Affected infants often suffer in hot weather, owing to their inability to sweat.

Congenital Ectodermal Dysplasia. In this rare hereditary disorder, there is partial failure of the epidermis and its appendages to develop. In severe cases, there are dental abnormalities and absence of hairs (Moynahan, 1971).

Angiomatous Malformations of the Skin. These *vascular malformations* present developmental flaws in which some transitory and/or surplus primitive vessels persist and, possibly, enlarge. These angiomatous malformations are often called *angiomas,* even though they may not be true tumors. Those composed of blood vessels may be mainly arterial, mainly venous, or mainly cavernous, but most often they are of a mixed type. Ones composed of lymphatics are often called *lymphangiomas,* or cystic *hygromas* (see Chapter 14). *True angiomas*

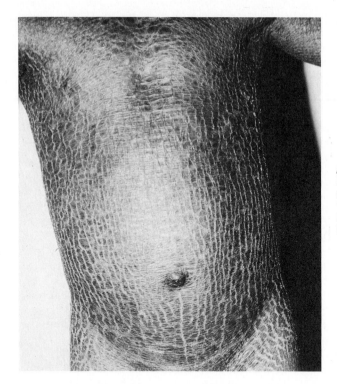

Figure 20–2 Photograph of infant with *lamellar ichthyosis*, a congenital disorder of keratinization characterized by pronounced scaling that involves the entire body. (From Vaughan, V. C., McKay, R. J., and Behrman, R. E.: *Nelson Textbook of Pediatrics*. 11th ed. Philadelphia, W. B. Saunders Co., 1979.)

are benign tumors of endothelial cells, usually composed of solid or hollow cords; the hollow cords contain blood.

Various terms are used to describe angiomatous malformations. *Nevus flammeus* is used to denote a flat, pink or red, flame-like blotch that often appears on the posterior surface (nape) of the neck. Nevus is not a good term because it is derived from the Latin word meaning mole, or *birthmark,* which may or may not be an angioma. It may be a melanoma.

A *port-wine stain* is a larger and darker angioma than the type previously described and is nearly always anterior or lateral, not posterior. It is sharply demarcated by the midline when it occurs near the median plane, whereas the common angioma (pinkish-red blotch on the nape) may cross the midline of the neck. A port-wine stain in the area of distribution of the trigeminal nerve (CN V) is sometimes associated with a similar type of angioma on the surface of the brain (*Sturge-Weber syndrome*).

Albinism. In generalized albinism, which is an autosomal recessive trait, the skin, hair, and retina lack pigment, but the iris usually shows some pigmentation. The condition results when the melanocytes fail to produce melanin because of lack of the enzyme tyrosinase, except for small amounts in the iris. In localized albinism, or *piebaldism,* which is an autosomal dominant trait, there is a lack of melanin in patches of skin and/or hair.

Absence of Skin. In rare cases, small areas of skin fail to form, giving the appearance of ulcers. The area usually heals by scarring unless a skin graft is performed. Absence of patches of skin is most common in the scalp.

HAIR

Hairs begin to develop early in the fetal period, but they do not become readily visible until about the twentieth week (see Fig. 6–1). Hairs are first recognizable on the eyebrows, upper lip, and chin.

A hair follicle begins as a solid downgrowth of the stratum germinativum of the epidermis and extends into the underlying dermis (Fig. 20–3*A*). The deepest part of the *hair bud* soon becomes club-shaped, forming a *hair bulb* (Fig. 20–3*B*). The epithelial cells of the hair bulb constitute the *germinal matrix,* which later gives rise to the hair. The hair bulb is then invaginated by a small mesenchymal *hair papilla* (Fig. 20–3*C*). The peripheral cells of the developing hair follicle form the *epithelial root sheath.* The surrounding mesenchymal cells differentiate

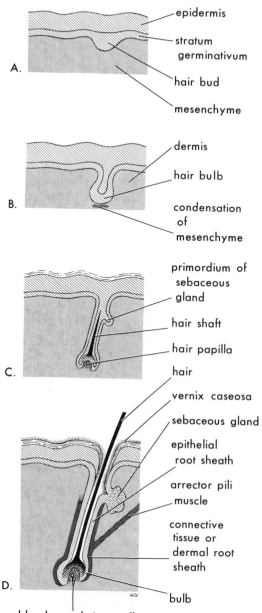

A. epidermis
stratum germinativum
hair bud
mesenchyme

B. dermis
hair bulb
condensation of mesenchyme

C. primordium of sebaceous gland
hair shaft
hair papilla
hair

D. vernix caseosa
sebaceous gland
epithelial root sheath
arrector pili muscle
connective tissue or dermal root sheath
bulb
blood vessels in papilla

Figure 20–3 Drawings showing successive stages in the development of a hair and its associated sebaceous gland. *A*, 12 weeks. *B*, 14 weeks. *C*, 16 weeks, *D*, 18 weeks.

into the *dermal (connective tissue) root sheath* (Fig. 20–3*D*).

As the cells in the *germinal matrix* proliferate, they are pushed upward and become keratinized to form the *hair shaft* (Fig. 20–3*C*). The hair grows through the epidermis and protrudes above the surface of the skin. Melanoblasts migrate into the hair bulb and

differentiate into melanocytes. Melanin produced by these cells is transferred to the hair-forming cells in the germinal matrix before birth.

The first hairs that appear, called *lanugo* (L. *lana,* fine wool), are fine and colorless. These hairs are replaced during the perinatal period by coarser hairs, called *vellus* (L. *vellus,* fleece, or coarse wool). This hair persists over most of the body, except in the axillary and pubic regions, where it is replaced at puberty by coarse *terminal hairs*. In males, similar terminal hairs also appear on the face and often on the chest. The *arrector pili muscle,* a small bundle of smooth muscle fibers, differentiates from the surrounding mesenchyme and becomes attached to the connective tissue sheath of the follicle and the papillary layer of the dermis. The arrector pili muscles, *the erectors of the hairs,* are poorly developed in the hairs of the axilla and in certain parts of the face. The hairs forming the eyebrows and the cilia forming the eyelashes have no arrector pili muscles.

CONGENITAL ABNORMALITIES OF HAIR

Congenital Alopecia (Atrichia Congenita). Fetal absence or loss of hair may occur alone or with other abnormalities of the skin and its derivatives. The hair loss may be caused by failure of hair follicles to develop or may result from follicles producing poor quality hairs.

Hypertrichosis. Excessive hairiness results from the development of supernumerary hair follicles, or from the persistence of hairs that normally disappear during the fetal period. Localized hypertrichosis is often associated with spina bifida occulta (see Fig. 18–12*A*).

GLANDS OF THE SKIN

Two kinds of glands occur in the skin: sebaceous glands and sweat glands.

Sebaceous Glands (Fig. 20–3). Most of these glands develop as buds from the side of the developing epithelial root sheath of a hair follicle (Fig. 20–3*C*). The glandular buds grow into the surrounding connective tissue and branch to form the primordia of several alveoli and their associated ducts (Fig. 20–3*D*). The central cells of the alveoli subsequently break down, forming an oily secretion called *sebum.* The sebum is extruded into the hair follicle and onto the surface of the skin, where it mixes with desquamated peridermal cells to form vernix caseosa. Sebaceous glands independent of hair follicles (e.g., in the glans penis and labia minora) develop in a similar manner from buds of the epidermis.

Sweat Glands (Fig. 20–4). Most *sweat glands* develop as solid epidermal downgrowths

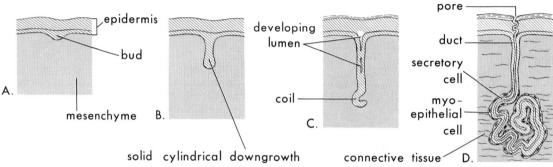

Figure 20–4 Diagrams illustrating successive stages in the development of a sweat gland.

that extend into the underlying dermis. As the bud elongates, its end becomes coiled, forming the primordium of the secretory portion of the gland, and the epithelial attachment of the developing gland to the epidermis forms the primordium of the duct. The central cells of these primordia degenerate, forming a lumen. The peripheral cells of the secretory portion of the gland differentiate into *secretory* and *myoepithelial cells*. These cells, derived from ectoderm, are thought to be specialized smooth muscle cells that aid in expelling sweat from the glands.

The distribution of the *large sweat glands* in humans is very limited; these glands are mostly confined to the axilla, the pubic region, and the areolae of the breasts. They develop from the downgrowths of the stratum germinativum of the epidermis that give rise to hair follicles. As a result, the ducts of these glands open, not onto the skin surface as do ordinary sweat glands, but into hair follicles above the openings of the sebaceous glands.

NAILS

Toenails and fingernails begin to develop at the distal ends of the digits at about 10 weeks (Fig. 20–5). Development of the fingernails precedes that of the toenails. The nails first appear as thickened areas of the developing epidermis on the dorsal aspect of each digit. These *nail fields*

are surrounded laterally and proximally by folds of epidermis called *nail folds*. Cells from the proximal nail fold grow over the nail field and become keratinized to form the *nail*, or *nail plate*. At first, the developing nail is covered by superficial layers of epidermis called the *eponychium*. This later degenerates, except at the base of the nail, where it persists and is sometimes called the cuticle. The skin under the free margin of the nail is called the *hyponychium*. The fingernails reach the fingertips by about 32 weeks; the toenails reach the toe tips by about 36 weeks.

CONGENITAL ABNORMALITIES OF NAILS

Anonychia. Partial or complete absence of nails may occur, but this condition is extremely rare. Anonychia results from failure of the nail fields to form or from failure of the nail folds to give rise to nails. The abnormality is permanent. It may be associated with congenital absence or extremely poor development of the hair, and with anomalies of the teeth.

MAMMARY GLANDS

The mammary glands first begin to develop during the sixth week as solid downgrowths of the epidermis that extend into the underlying mesenchyme (Fig. 20–6C).

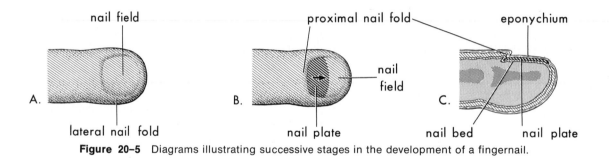

Figure 20–5 Diagrams illustrating successive stages in the development of a fingernail.

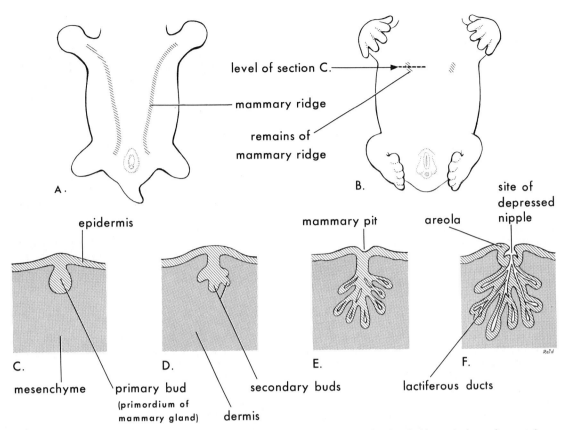

level of section C.

mammary ridge

remains of
mammary ridge

A.

B.

site of
depressed
nipple

epidermis

mammary pit

areola

C.

D.

E.

F.

mesenchyme

primary bud
(primordium of
mammary gland)

secondary buds

dermis

lactiferous ducts

Figure 20–6 Drawings illustrating development of the mammary glands. *A*, Ventral view of an embryo during stage 13 (about 28 days), showing the mammary ridges. *B*, Similar view at six weeks, showing the remains of these ridges. *C*, Transverse section through the trunk and the mammary ridge at the site of a developing mammary gland. *D, E,* and *F,* Similar sections showing successive stages of development between the twelfth week and birth.

These downgrowths occur along the *mammary ridges*, two thickened strips of ectoderm that extend from the axillary to the inguinal regions. These epithelial ridges appear during the fourth week, but disappear in humans except in the pectoral area.

Each primary mammary bud soon gives rise to several secondary buds that develop into *lactiferous ducts* and their branches (Fig. 20–6*D* and *E*). The fibrous connective tissue and fat develop from the surrounding mesenchyme. During the late fetal period, the epidermis at the origin of the mammary gland becomes depressed, forming a shallow *mammary pit* (Fig. 20–6*E*). The nipple forms during the perinatal period as the result of proliferation of mesenchyme underlying the areola, a circular area of skin around the nipple. The nipples are poorly formed and often depressed in newborn infants. Soon after birth, the nipples are raised from the shallow mammary pits by the proliferation of the surrounding connective tissue.

The mammary glands of newborn males and females are often enlarged, and some secretion, often called "witch's milk," may be produced. These transitory changes are caused by maternal hormones passing into the fetal circulation through the placental membrane. Only the main ducts are formed at birth, and the mammary glands remain underdeveloped until puberty. In females, the glands then enlarge rapidly (Fig. 20–7), mainly because of fat and other connective tissue development. Growth of the duct system also occurs under the influence of estrogen and progesterone secreted by the ovaries at puberty.

If pregnancy occurs, the glandular tissue becomes completely developed, and the intralobular ducts undergo rapid development, forming buds that become alveoli. The mammary glands in males normally undergo little postnatal development. Estrogen given to males tends to make their rudimentary mammary glands develop into the feminine type. *Gynecomastia* (Gr. *gynē*, woman + *mastos,* breast) may also be associated with *Klinefelter syndrome* (see Chapter 8).

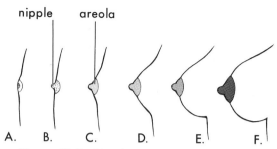

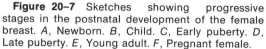

Figure 20–7 Sketches showing progressive stages in the postnatal development of the female breast. *A,* Newborn. *B,* Child. *C,* Early puberty. *D,* Late puberty. *E,* Young adult. *F,* Pregnant female.

CONGENITAL ABNORMALITIES OF MAMMARY GLANDS

Absence of the Nipple (Athelia) and Absence of the Breast (Amastia). These rare congenital abnormalities may occur bilaterally or unilaterally. They result from failure of development or from complete disappearance of the mammary ridge(s). These conditions may also result from failure of a mammary bud to form.

Aplasia of the Breast. The breasts of a post-pubertal female often differ somewhat in size. Slight differences in size are not congenital malformations, but marked differences are regarded as deformities because both glands are exposed to the same hormones at puberty. In these cases, there is often associated rudimentary development of muscles, usually the pectoralis major.

Supernumerary Breasts and Nipples (Figs. 20–8 and 20–9). An extra breast (*polymastia*) or nipple (*polythelia*) occurs in about 1 per cent of the female population and is an inheritable condition (Haagensen, 1967). Supernumerary nipples are also relatively common in males. Often, they are mistaken for moles (nevi). An extra breast or

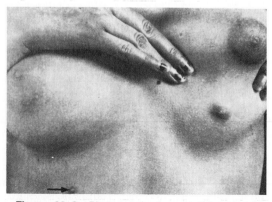

Figure 20–8 Photograph of an adult female with a supernumerary nipple on the right (arrow) and a supernumerary breast below the normal left one. (From Haagensen, C. D.: *Diseases of the Breast.* 2nd ed. Philadelphia, W. B. Saunders, Co., 1971.)

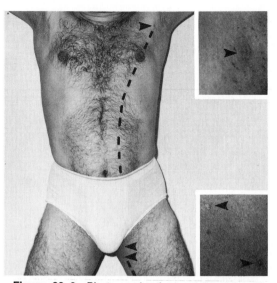

Figure 20–9 Photograph of an adult male with bilateral nipples in the axillary and thigh regions. The inset photographs are enlargements of the nipples on the left (arrows). The broken line indicates the original position of the left mammary ridge, along which the extra nipples have developed. (Courtesy of Dr. Kunwar Bhatnagar, Department of Anatomy, University of Louisville, Louisville, Ky.)

nipple usually develops just inferior to the normal breast.

Less commonly, extra mammary glands appear in the axillary or abdominal regions. In these positions, the extra nipples or breasts develop from extra mammary buds along the mammary ridges. They usually do not become obvious until pregnancy occurs. About one third of affected persons have two extra nipples or breasts. Very rarely, supernumerary mammary tissue occurs in a location other than along the course of the mammary ridges and probably develops from tissue that was displaced from the mammary ridge.

Inverted Nipples. Sometimes, the mammary pit in which the nipple is located prenatally (Fig. 20–7) fails to become elevated above the skin surface. This inverted nipple may make breast feeding of an infant difficult.

TEETH

Two sets of teeth normally develop: the *primary dentition*, or deciduous teeth, and the *secondary dentition*, or permanent teeth.

Each tooth develops from ectoderm and mesoderm. The enamel is derived from ectoderm of the oral cavity; all other tissues differentiate from the associated mesenchyme.

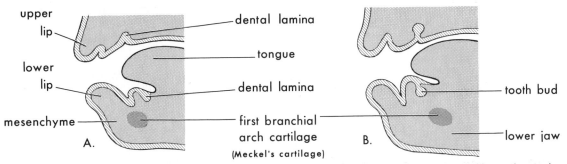

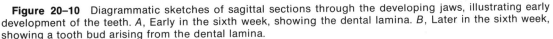

Figure 20–10 Diagrammatic sketches of sagittal sections through the developing jaws, illustrating early development of the teeth. *A*, Early in the sixth week, showing the dental lamina. *B*, Later in the sixth week, showing a tooth bud arising from the dental lamina.

Tooth development appears to be initiated by the mesenchyme's inductive influence on the overlying ectoderm. Present evidence indicates that this mesenchyme is of neural crest origin. *Tooth development is a continuous process*, but it is usually divided into stages (bud, cap, and bell stages) based upon the appearance of the developing tooth. Not all teeth begin to develop at the same time. The first tooth buds appear in the anterior mandibular region; later, tooth development occurs in the anterior maxillary region, and then it progresses posteriorly in both jaws. Tooth development continues for a number of years after birth (Table 20–1).

The Dental Lamina and the Bud Stage. The first indications of tooth development appear in histological sections early in the sixth week as thickenings of the oral epithelium, a derivative of the surface ectoderm (Figs. 20–10*A* and 20–11*A*). The epithelium folds into the underlying mesenchyme. These U-shaped bands, called *dental laminae,* follow the curve of the primitive jaws. Localized proliferations of cells in the dental laminae produce round or oval swellings called *tooth buds* (Figs. 20–10*B* and 20–11*B*), which grow into the mesenchyme. These tooth buds, sometimes called *term germs*, develop into the first teeth, called *deciduous teeth* because they are shed during childhood. There are 10 tooth buds in each jaw, one for each deciduous tooth.

The tooth buds for the permanent teeth with deciduous predecessors begin to appear at about 10 fetal weeks from deeper continuations of the dental lamina, and they lie lingual to the deciduous tooth buds (Fig. 20–11*D*). The permanent molars that have no deciduous predecessors develop as buds from backward extensions of the dental la-

minae. The tooth buds for the permanent teeth appear at different times, mostly during the fetal period. The buds for the second and third permanent molars, however, appear after birth (about the fourth month and fifth year, respectively).

The Cap Stage (Figs. 20–11*C* and 20–12). The deep surface of each tooth bud soon becomes slightly invaginated by a mass of condensed mesenchyme called the *dental papilla*, on which the tooth bud sits like a cap. The mesenchyme of the dental papilla gives rise to the *dentin* and the *dental pulp*. The ectodermal portion of this cap-shaped developing tooth is called an *enamel organ* because it later produces enamel. The outer cellular layer of the enamel organ is called the *outer enamel epithelium*, and the inner cellular layer lining the "cap" is called the *inner enamel epithelium*. The central core of loosely arranged cells between the layers of enamel epithelium is called the *enamel reticulum* (stellate reticulum).

As the enamel organ and the dental papilla form, the mesenchyme surrounding them condenses and forms a capsule-like structure called the *dental sac*, or follicle (Fig. 20–11*E* and *F*), which will give rise to the cementum and the periodontal ligament.

The Bell Stage (Figs. 20–11*D* and *E*, 20–12, and 20–13). As the enamel organ differentiates, the developing tooth assumes a bell shape. The mesenchymal cells in the dental papilla adjacent to the inner enamel epithelium differentiate into *odontoblasts*. These cells produce *predentin* and deposit it adjacent to the inner enamel epithelium. Later, the predentin calcifies and becomes *dentin*. As the dentin thickens, the odontoblasts regress toward the center of the dental papilla, but cytoplasmic processes of

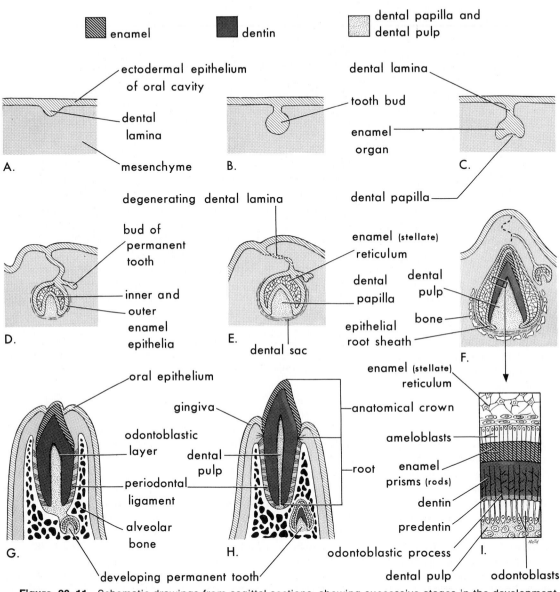

Figure 20–11 Schematic drawings from sagittal sections, showing successive stages in the development and eruption of an incisor tooth. *A*, Six weeks, showing the dental lamina. *B*, Seven weeks, showing the bud stage of tooth development. *C*, Eight weeks, showing the cap stage of development of the enamel organ. *D*, 10 weeks, showing the early bell stage of the enamel organ of the deciduous tooth and the bud stage of the developing permanent tooth. *E*, 14 weeks, showing the advanced bell stage of the enamel organ. Note that the connection (dental lamina) of the tooth to the oral epithelium is degenerating. *F*, 28 weeks, showing the enamel and dentin layers. *G*, Six months *postnatal*, showing early tooth eruption. *H*, 18 months *postnatal*, showing a fully erupted deciduous incisor tooth. The permanent incisor tooth now has a well-developed crown. *I*, Section through a developing tooth, showing the ameloblasts (enamel producers) and the odontoblasts (dentin producers).

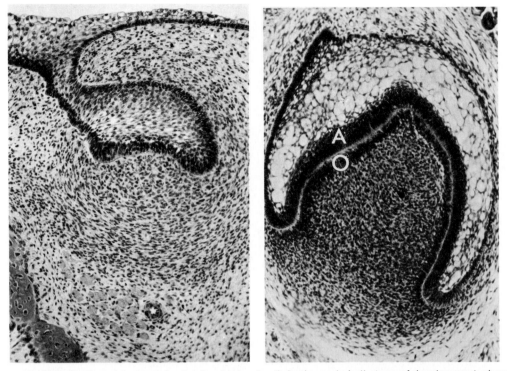

Figure 20–12 *Left:* Photomicrograph of a developing tooth in the early bell stage of development, showing the enamel organ attached to the oral mucosa by the dental lamina (×100). Compare with Figure 20–11C and D. *Right:* Photomicrograph of a developing tooth in the late bell stage with ameloblasts (inner enamel epithelium, *A*) differentiated and in contact with odontoblasts (*O*). Compare with Figure 20–11D. (From Leeson, C. R., and Leeson, T. S.: *Histology.* 4th ed. Philadelphia, W. B. Saunders Co., 1981.)

the odontoblasts, called *odontoblastic processes*, remain embedded in the dentin (Fig. 20–11F and I). These processes are also called *Tomes' dentinal fibers*, or *processes*.

Cells of the inner enamel epithelium adjacent to the dentin differentiate into *ameloblasts*. These cells produce *enamel* in the form of prisms (rods) over the dentin. As the enamel increases, the ameloblasts regress toward the outer enamel epithelium. Enamel and dentin formation begins at the tip (cusp) of the tooth and progresses toward the future root.

The development of the *root* begins after dentin and enamel formation is well advanced. The inner and outer enamel epithelia come together in the neck region of the tooth and form an epithelial fold called the *epithelial root sheath*. This sheath grows into the mesenchyme and initiates root formation. The odontoblasts adjacent to this sheath form dentin continuous with that of the crown. As the dentin increases, it reduces

the pulp cavity to a narrow canal through which the vessels and nerves pass.

The inner cells of the dental sac differentiate into *cementoblasts*, which produce *cementum*. This is deposited over the dentin of the root and meets the enamel at the neck of the tooth *(the cementoenamel junction)*.

As the teeth develop and the jaws ossify, the outer cells of the dental sac also become active in bone formation. Each tooth soon becomes surrounded by bone, except over its crown. The tooth is held in its bony socket *(dental alveolus)* by the *periodontal ligament,* a derivative of the dental sac. (Fig. 20–11G). Some parts of the fibers of this ligament are embedded in the cementum; other parts are embedded in the bony wall of the socket.

Tooth Eruption (Figs. 20–11G and 20–14). The mandibular teeth usually erupt before the maxillary teeth, and girls' teeth usually erupt sooner than boys' teeth.

As the root of the tooth grows, the crown

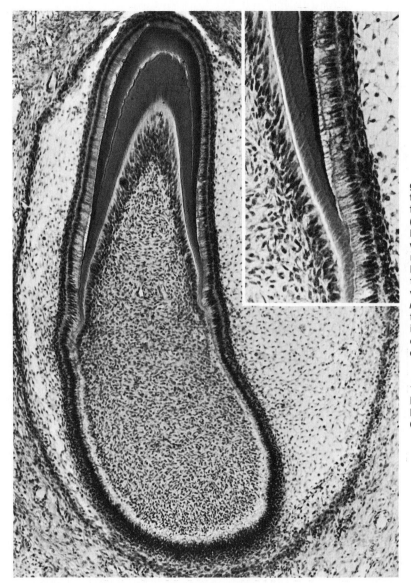

Figure 20-13 Photomicrograph of a developing tooth at the stage during which crown formation is well advanced. (Compare with Figure 20-11F.) Enamel and dentin are present with a thin layer of predentin in relation to the odontoblasts. Note the connective tissue dental sac enveloping the entire developing tooth (×75). *Upper right insert*: A higher magnification of part of the tooth showing, from left to right, pulp, odontoblasts, predentin, dentin, enamel (*black*), ameloblasts, stratum intermedium, and enamel (stellate) reticulum (× 175). (From Leeson, C. R., and Leeson, T. S.: *Histology*. 4th ed. Philadelphia, W. B. Saunders Co., 1981.)

gradually erupts through the oral mucosa. The part of the oral mucosa around the erupted crown becomes the gum, or *gingiva*. Eruption of the deciduous teeth usually occurs between the sixth and twenty-fourth months after birth (Table 20-1).

The permanent teeth develop in a manner similar to that just described for deciduous teeth. As a permanent tooth grows, the root of the corresponding deciduous tooth is gradually resorbed by osteoclasts. Consequently, when the deciduous tooth is shed, it consists only of the crown and the uppermost portion of the root. The permanent teeth usually begin to erupt during the sixth year and continue to appear until early adulthood.

The development of the face is determined by the development of the *paranasal air sinuses* (see Chapter 10) and by the growth of the maxilla and the mandible to accommodate the teeth. It is the lengthening of the bony sockets for the teeth that results in the increase in the depth of the face during childhood.

CONGENITAL ABNORMALITIES OF TEETH

The most common dental abnormality present at birth is *premature eruption of one or more of the deciduous teeth*, usually the mandibular incisors. Most congenital abnor-

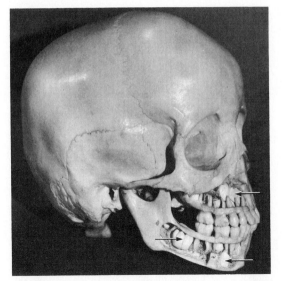

Figure 20–14 Photograph of the skull of a child in the fourth year. The jaws have been dissected to show the relations of the developing permanent teeth (arrows) to the erupted deciduous teeth.

malities of teeth are not visible at birth because the teeth usually do not begin to erupt until several months after birth (Table 20–1).

Enamel Hypoplasia (Fig. 20–15). Defective enamel formation results in grooves, pits, or fissures on the enamel surface. These defects result from temporary disturbances in enamel formation. Various factors may injure the ameloblasts (e.g., nutritional deficiency, tetracycline therapy, and diseases such as measles).

Rickets during the critical period of permanent tooth development is the most common known cause of enamel hypoplasia. *Rickets is a condition caused by a deficiency of vitamin D*, especially during infancy and childhood, and is characterized by disturbance of normal ossification.

Abnormalities in Shape (Fig. 20–15A to G). Abnormally shaped teeth are relatively common. Occasionally, spherical masses of enamel called *enamel pearls,* or drops, are attached to the tooth. They are formed by aberrant groups of ameloblasts. The maxillary lateral incisor teeth may assume a slender, tapering shape (peg-shaped lateral incisors).

Congenital syphilis affects the differentiation of the permanent teeth, resulting in screwdriver-shaped incisors, with central notches in their incisive edges.

Numerical Abnormalities (Fig. 20–15H and I). One or more *supernumerary teeth* may develop, or the normal number of teeth may fail to form. Supernumerary teeth usually appear in the area of the maxillary incisors, where they disrupt the position and eruption of normal teeth. The extra teeth commonly erupt posterior to the normal ones.

In *partial anodontia*, one or more teeth are absent. Congenital absence of one or more teeth is often a familial trait. In *total anodontia*, no teeth develop; this very rare condition is usually associated with congenital ectodermal dysplasia.

Abnormal Size of Teeth. Disturbances during the differentiation of teeth may result in gross alterations of dental morphology, e.g., *macrodontia* (large teeth) and *microdontia* (small teeth).

Natal Teeth and Caps. It is not uncommon for one or two mandibular incisors to be erupted at birth: Premature eruption of teeth occurs in about 1 of every 2000 newborn infants. Sometimes, the gingiva (gum) grows over these teeth within two or three weeks. Often, prematurely

TABLE 20–1 ORDER AND TIME OF ERUPTION OF TEETH AND TIME OF SHEDDING OF DECIDUOUS TEETH*

	Medial Incisor	Lateral Incisor	Canine	First Molar	Second Molar
	Deciduous Teeth				
Eruption (months)	6 to 8	8 to 10	16 to 20	12 to 16	20 to 24
Shedding (years)	6 to 7	7 to 8	10 to 12	9 to 11	10 to 12

	Medial Incisor	Lateral Incisor	Canine	First Premolar	Second Premolar	First Molar	Second Molar	Third Molar
	Permanent Teeth							
Eruption (years)	7 to 8	8 to 9	10 to 12	10 to 11	11 to 12	6 to 7	12	13 to 25

*(From Moore, K. L. *Clinically Oriented Anatomy.* © 1980. Courtesy of Williams & Wilkins Co., Baltimore.)

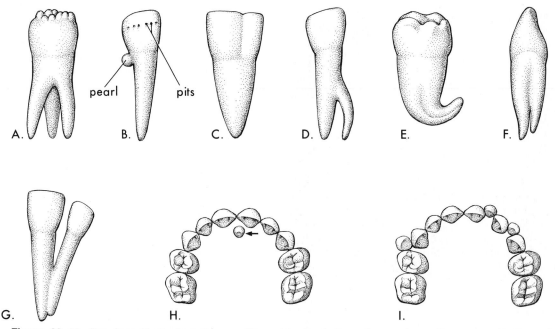

Figure 20–15 Drawings illustrating abnormalities of teeth. *A*, Irregular raspberry-like crown. *B*, Enamel pearl and pits. *C*, Incisor tooth with a double crown. *D*, Abnormal division of root. *E*, Distorted root. *F*, Branched root. *G*, Fused roots. *H*, Hyperdontia, with a supernumerary incisor tooth in the anterior region of the palate (arrow). *I*, Hyperdontia, with 13 deciduous teeth in the upper jaw instead of the normal 10.

erupted teeth are only small, loose, enamel caps covering a thin sheet of dentin. The cause of premature eruption is unknown, but endocrine factors may be involved because this condition sometimes occurs in infants with congenital adrenal hyperplasia.

Fused Teeth (Fig. 20–15C and G). Occasionally a tooth bud divides or two buds partially fuse to form fused, or joined, teeth. This condition is commonly observed in the mandibular incisors of the primary dentition. "Twinning" of teeth may result from *germination,* or division of the tooth germ. In some cases the permanent tooth does not form; this suggests that the deciduous and permanent tooth primordia may have fused.

Dentigerous Cyst (Tooth-Bearing Cyst). In rare cases, a cyst develops in a mandible, maxilla, or maxillary sinus that contains an unerupted tooth. The cyst develops owing to cystic degeneration of the enamel reticulum of the enamel organ of an unerupted tooth. Most of these cysts are deeply situated in the jaw and are associated with misplaced or malformed secondary (permanent) teeth that have failed to erupt.

Amelogenesis Imperfecta. The enamel is soft and friable because of hypocalcification, and the teeth are yellow to brown in color. This autosomal dominant trait affects about one in every 20,000 children. For more details of this condition, see Witkop (1965).

Dentinogenesis Imperfecta (Fig. 20–16). This condition is relatively common in white children. The teeth are brown to gray-blue with an opalescent sheen. The enamel tends to wear down rapidly, exposing the dentin. This malformation is inherited as an autosomal dominant trait (Thompson and Thompson, 1980).

Discolored Teeth. Foreign substances incorporated into the developing enamel will cause discoloration of the teeth. The *hemolysis* (liberation of hemoglobin) associated with *erythroblastosis fetalis* (see Chapter 7) may produce blue to black discoloration of the primary teeth.

All tetracyclines are extensively incorporated into the enamel of teeth; these drugs may produce ugly, brownish-yellow discoloration (mottling) and even hypoplasia of the enamel because they

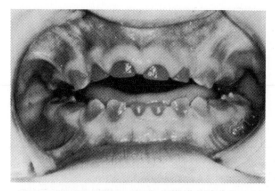

Figure 20–16 Photograph of the teeth of a child with dentinogenesis imperfecta. (From Thompson, J. S., and Thompson, M. W.: *Genetics in Medicine.* 3rd ed. Philadelphia, W. B. Saunders Co., 1980.)

interfere with the metabolic processes of the ameloblasts. The primary teeth are affected if the tetracyclines are given from 18 weeks (prenatal) to 10 months (postnatal), and the permanent teeth are affected from 18 weeks (prenatal) to 16 years. *Tetracyclines should not be administered to pregnant women or to children, if they can be avoided, because these drugs adversely affect tooth development* (Seeman et al., 1980).

SUMMARY

The skin and its appendages develop from ectoderm and mesoderm. The epidermis and its derivatives (hairs, nails, and glands) are derived from surface ectoderm. The melanocytes are derived from neural crest cells that migrate into the epidermis. The dermis develops from mesenchyme. Hairs develop from downgrowths of the epidermis into the dermis. By about 20 weeks, the fetus is completely covered with fine, downy hairs called *lanugo*. These hairs are shed before birth and are replaced by coarser hairs called *vellus*. The sebaceous glands develop as outgrowths from the side of hair follicles. The sweat glands are developed from epidermal downgrowths. Mammary glands also develop from downgrowths of the epidermis.

Congenital abnormalities of the skin are mainly disorders of keratinization (ichthyosis) and pigmentation (albinism). Abnormal blood vessel development results in various types of *angiomas*. Nails may be absent or malformed. Hair may be absent or excessive. Absence of mammary glands is rare, but supernumerary breasts (polymastia) or nipples (polythelia) are relatively common.

The teeth develop from ectoderm and mesoderm. The enamel is produced by cells derived from oral ectoderm; all other dental tissues develop from mesenchyme. The common congenital malformations of teeth are defective formation of enamel and dentin, abnormalities in shape, and variations in number and position.

All tetracyclines are extensively incorporated into the enamel of teeth and produce ugly brownish-yellow discoloration and hypoplasia of the enamel.

SUGGESTIONS FOR ADDITIONAL READING

Moynahan, E. J.: The developmental biology of the skin; *in* J. A. Davis and J. Dobbing (Eds.): *Scientific Foundations of Paediatrics.* Philadelphia, W. B. Saunders Co., 1974, pp. 526–543. A treatise on the histogenesis of the epidermis, the hair follicles, the mammary glands, the sebaceous glands, and the sweat glands. Many electron micrographs of the developing skin are included.

Smith, D. W., and Gong, B. T.: Scalp-hair patterning: Its origin and significance relative to early brain and upper facial development. *Teratology* 9:17, 1974. These clinical teratologists illustrate how aberrant scalp-hair patterning may be used as an indicator of altered size or shape (or both) of the brain prior to the twelfth week. They also describe the normal development of hair follicles and hair patterning.

Sperber, G. H.: Development of the dentition (odontogenesis); *in* Sperber, G. H.: *Craniofacial Embryology.* 2nd ed. Bristol, John Wright & Sons Ltd. (distributed by Year Book Medical Publishers, Inc. Chicago), 1976, Chapter 18. This chapter gives details of tooth development that are of immediate relevance to dental students. Written by an eminent oral biologist, this developmental account clarifies the structural complexities of the teeth and their supporting tissues.

CLINICALLY ORIENTED PROBLEMS

1. A newborn infant had two erupted teeth. What are these teeth called? How common is this? In what position do they usually occur? *Are they supernumerary teeth?* What problems and/or danger might be associated with the presence of teeth at birth?

2. *The primary dentition of an infant had a brownish-yellow color* and some hypoplasia of the enamel. The mother recalled that she had been given antibiotics during the second trimester of her pregnancy. What is the probable cause of the infant's discolored teeth? Dysfunction of what cells would cause the *hypoplasia of enamel?* Would the secondary dentition be discolored?

3. An infant was born with a small, irregularly shaped, light-red blotch on the nape (posterior surface of the neck). *It was level with the surrounding skin and blanched when light pressure was applied to it.* Name this malformation. What do these observations probably indicate? Is this condition common? Are there other names for this malformation?

4. A newborn infant had a *midline tuft of hair in the lumbosacral region of the*

back. What does this probably indicate? Is this condition common? Is it clinically important?

5. A newborn infant had a *collodion type of covering* that fissured and exfoliated shortly after birth. Later, *lamellar ichthyosis* developed. Briefly describe this condition. Is it common? How is it inherited?

The answers to these questions are given at the back of the book.

REFERENCES

Alter, M.: Dermatoglyphic analysis as a diagnostic tool. *Medicine 46*:35, 1966.

Boyd, J. D.: The embryology and comparative anatomy of the melanocyte; *in* A. Rook (Ed.): *Progress in the Biological Sciences in Relation to Dermatology*. London, Cambridge University Press, 1960, pp. 3–14.

Butler, P. M., and Joysey, K. A. (Eds.): *Development, Function and Evolution of Teeth*. New York, Academic Press, 1978.

Butterworth, T.: Hair and nails; *in* A. Rubin (Ed.): *Handbook of Congenital Malformations*. Philadelphia, W. B. Saunders Co., 1967, pp. 258–295.

Goldenring, H., and Crelin, E. S.: Mother and daughter with bilateral congenital amastia. *Yale J. Biol. Med. 33*:466, 1961.

Haagensen, C. D.: Breasts; *in* A. Rubin (Ed.): *Handbook of Congenital Malformations*. Philadelphia, W. B. Saunders Co., 1967, pp. 15–18.

Haagensen, C. D.: *Diseases of the Breast*. 2nd ed. Philadelphia, W. B. Saunders Co., 1971.

Hale, A. R.: Morphogenesis of volar skin in the human fetus. *Am. J. Anat. 91*:147, 1952.

Hurmerinta, K., Thesleff, I., and Saxen, L.: Inhibition of tooth germ differentiation *in vitro* by diazo-oxonorleucine (DON). *J. Embryol. Exp. Morphol. 50*:99, 1979.

Jain, S. R., Sepaha, G. C., and Khandelwal, G. D.: A case of unilateral amastia. *J. Anat. Soc. India 10*:45, 1961.

Jolly, H.: *Diseases of Children*. 2nd ed. Oxford, Blackwell Scientific Publications, 1968.

Leeson, T. S., and Leeson, C. R.: *Histology*. 4th ed. Philadelphia, W. B. Saunders Co., 1981, pp. 308–317.

Miles, A. E. W.: Malformations of the teeth. *Proc. R. Soc. Med. 47*:817, 1954.

Moore, K. L.: *Clinically Oriented Anatomy*. Baltimore, Williams & Wilkins Company, 1980, pp. 999–1002.

Moynahan, E. J.: Abnormalities of the skin; *in* A. P. Norman (Ed.): *Congenital Abnormalities in Infancy*. 2nd ed. Oxford, Blackwell Scientific Publications, 1971, pp. 289–349.

Pardo-Castello, V.: *Diseases of Nails*. Springfield, Ill., Charles C Thomas, Publisher, 1960.

Pinkus, H.: Embryology of hair; *in* W. Montagne and R. A. Ellis (Eds.): *The Biology of Hair Growth*. New York, Academic Press, Inc., 1958, pp. 1–32.

Provenza, D.: *Oral Histology. Inheritance and Development*. Philadelphia, J. B. Lippincott Co., 1964.

Rawles, M. E.: Origin of melanophores and their role in the development of color patterns in vertebrates. *Physiol. Rev. 28*:383, 1948.

Ronchese, F.: Peculiar nail anomalies. *Arch. Dermatol. 63*:565, 1951.

Schwarz, V.: The development of the sweat glands and their function; *in* J. A. Davis and J. Dobbing (Eds.): *Scientific Foundations of Paediatrics*. Philadelphia, W. B. Saunders Co., 1974, pp. 544–546.

Seeman, P., Sellers, E. M., and Roschlau, H. E.: *Principles of Medical Pharmacology*. 3rd ed. Toronto, University of Toronto Press, 1980, p. 501.

Shafer, W. G., Hine, M. K., and Levy, B. M.: *A Textbook of Oral Pathology*. 2nd ed. Philadelphia, W. B. Saunders Co., 1963, pp. 2–75.

Sicher, H., and Bhaskar, S. N.: *Orban's Oral Histology and Embryology*. 7th ed. St. Louis, The C. V. Mosby Co., 1972.

Sperber, G. H.: Genetic mechanisms and anomalies in odontogenesis. *J. Can. Dent. Assoc. 33*:433, 1967.

Ten Cate, A. R., Mills, C., and Solomon, G.: The development of the periodontium. A transplantation and autoradiographic study. *Anat. Rec. 170*:365, 1971.

Thesleff, I.: Role of the basement membrane in odontoblast differentiation. *J. Biol. Buccale 6*:241, 1978.

Thesleff, I.: Extracellular matrix and tooth morphogenesis; *in* R. M. Pratt and R. L. Christensen (Eds.): *Current Research Trends in Prenatal Craniofacial Development*. Amsterdam, Elsevier/North Holland, 1980, pp. 329–338.

Thompson, J. S., and Thompson, M. W.: *Genetics in Medicine*. 3rd ed. Philadelphia, W. B. Saunders Co., 1980.

Trier, W. C.: Complete breast absence. *Plastic Reconstr. Surg. 36*:430, 1965.

Turner, E. P.: The growth and development of the teeth; *in* J. A. Davis and J. Dobbing (Eds.): *Scientific Foundations of Paediatrics*. Philadelphia, W. B. Saunders Co., 1974, pp. 420–434.

Uchida, I. A., and Summit, R. L.: Dermatoglyphics; *in* V. C. Vaughan, R. J. McKay, and R. E. Behrman (Eds.): *Nelson Textbook of Pediatrics*. 11th ed. Philadelphia, W. B. Saunders Co., 1979.

Vaughan, V. C., McKay, R. J., and Behrman, R. E. (Eds.): *Nelson Textbook of Pediatrics*. 11th ed. Philadelphia, W. B. Saunders Co., 1979.

Warkany, J.: *Congenital Malformations: Notes and Comments*. Chicago, Year Book Medical Publishers, Inc., 1971.

Wilflingseder, P.: Skin hemangiomas; aggressive approach; *in* A. J. C. Huffstadt (Ed.): *Congenital Malformations*. Amsterdam, Excerpta Medica, 1980, pp. 75–90.

Witkop, C. J.: Genetic disease of the oral cavity; *in* R. W. Tiecke (Ed.): *Oral Pathology*. New York, McGraw-Hill Book Co., 1965, 786–843.

Zimmerman, A. A., and Becker, S. W.: *Melanoblasts and Melanocytes in Fetal Negro Skin*. Urbana, University of Illinois Press, 1959.

ANSWERS TO CLINICALLY ORIENTED PROBLEMS

Chapter 1

1. The development of a human being begins with *fertilization*, a process by which a sperm from a male unites with an ovum from a female.
2. A new human organism is called a *zygote* at the beginning of its development.
3. The term *conceptus* is used when referring to an embryo or a fetus and its membranes, i.e., *the products of conception*. The term *abortus* refers to any product or all *products of an abortion,* e.g., the embryo (or part of it) and/or its fetal membranes and placenta (or parts of them).
4. *Growth of secondary sexual characteristics occurs,* reproductive functions begin, and sexual dimorphism becomes more evident. The ages of presumptive puberty are 12 years in girls and 14 years in boys.
5. Literally, the term *embryology means study of the embryo*, but clinically it means the study of the embryo and the fetus, i.e., the study of prenatal development. The term *ontogeny* is used to describe the series of successive stages of development that occur during the complete life of a person. *Teratology* is the branch of embryology concerned with abnormal development and the factors that produce it, e.g., drugs and viruses.

Chapter 2

1. Numerical changes in chromosomes arise chiefly through the process of *nondisjunction,* either in a mitotic division or in the first or second meiotic division. Nondisjunction is the *failure of two members of a chromosome pair to disjoin* during anaphase of cell division; as a result, both chromosomes pass to the same daughter cell. *Trisomy 21* (Down syndrome) is the most common chromosomal disorder resulting in congenital malformations. It occurs about once in every 800 births.
2. An extra set of chromosomes (*triploidy*) in the embryonic cells would usually result from fertilization of an ovum by two sperms (*dispermy*), but it could also result from failure of a secondary oocyte to discard its second polar body.
3. *Blockage of the uterine tubes* resulting from infection is one of the major causes of infertility in women. Because this condition prevents secondary oocytes from

coming into contact with sperms, fertilization cannot occur.
4. *Mosaicism* results when nondisjunction of a pair of chromosomes occurs during an early cleavage division of the zygote rather than during gametogenesis. As a result, the embryo in whom this nondisjunction occurs has two or more cell lines with different chromosome numbers. Such persons are called *mosaics*. About 1 per cent of Down syndrome patients are 46/47 mosaics. These patients have relatively mild stigmata of the syndrome and are less retarded than the typical trisomy 21 patients. These mosaics may be at a high risk of having children with Down syndrome if the mosaicism affects the germ cells (oocytes or spermatogonia).

Chapter 3

1. Yes, it is reasonable because the patient's uterus and ovaries are not in the x-ray beam. The only radiation the ovaries would receive would be a *negligible and scattered amount* from the irradiated thorax. Furthermore, this small amount of radiation would be highly unlikely to damage the products of conception if the patient happened to be pregnant.
2. DES appears to affect the endometrium by rendering it unsuitable for implantation, a process regulated by a delicate balance between estrogen and progesterone. The large doses of estrogen given to the patient upset this balance. Progesterone makes the endometrium grow thick and succulent so that the blastocyst may become embedded and be nourished adequately. DES pills are referred to as *"morning after pills"* by laypeople.
3. *Over 90 per cent of ectopic pregnancies are in the uterine tube,* and 60 per cent of them are in its ampulla, or infundibulum. The doctor would perform a *laparotomy* promptly, removing the uterine tube containing the conceptus.
4. *No!* Exposure of an embryo during the second week to the slight trauma that might be associated with abdominal surgery would not cause a congenital malformation of the brain. Furthermore, the anesthetics used during the operation could not induce a gross malformation of the brain. *Teratogens are **not** known to induce congenital malformations during the first two weeks of development* (see Fig. 8–14).

Chapter 4

1. Chromosomal abnormalities may have caused the spontaneous abortion. *The incidence of chromosomal abnormalities in early abortions is about 61.5 per cent* (Boué et al., 1975). Carr (1971) observed a pronounced increase in polyploidy (cells containing three or more times the haploid number of chromosomes) in embryos expelled during spontaneous abortions when conception occurred within about two months after discontinuance of oral contraception. *Polyploidy is known to be fatal to the developing embryo.* This information suggests that it might be wise to use some other type of contraception for one or two menstrual cycles before attempting pregnancy after discontinuing oral contraceptives. In the present case, the doctor probably told the patient that her abortion was a natural screening process, i.e., that it was probably the spontaneous expulsion of an embryo that could not have survived because of severe chromosomal abnormalities.

2. *The presence of embryonic and/or chorionic tissue in the endometrial remnants would be an absolute sign of pregnancy,* but this tissue would be very difficult to find at such an early stage of pregnancy. By five days after the expected menses, i.e., about five weeks after the last menstrual period, the embryo would be in the third week of its development.

3. The central nervous system (brain and spinal cord) begins to develop during the third week. *Anencephaly* (see Fig. 18–19), in which most of the brain and cranial vault are absent, may result from environmental teratogens acting during the third week of development. This severe malformation of the brain results from almost complete failure of the cranial part of the neural tube to develop normally, which is caused by nonclosure of the neuropores.

4. *Sacrococcygeal teratomas commonly arise from remnants of the primitive streak,* probably in the region of the primitive knot. Because cells from the primitive streak are pleuripotent, the tumors often contain various types of tissue derived from all three germ layers. There is a clear-cut difference in the incidence of these tumors with regard to sex; i.e., they are three to four times more frequent in girls than in boys.

Chapter 5

1. The doctor would likely tell the patient that her embryo was undergoing a *critical stage of its development* and that it would be safest for her baby if she were to stop smoking and to avoid taking any unprescribed medication throughout her pregnancy. He would also likely tell her that *heavy cigarette smoking is known to cause underweight babies* and that the incidence of prematurity increases with the number of cigarettes smoked.

2. *The embryonic period is a critical period of development because all the main tissues and organs are forming.* Thus, it is the time during which the embryo is most vulnerable to the injurious effects of environmental agents (e.g., drugs and viruses).

3. *One can not always predict how a drug will affect the human embryo* because human and animal embryos may differ in their response to a drug. For example, thalidomide is extremely teratogenic to human embryos but has very little effect on some experimental animals, e.g., rats and mice.

4. *Information about the starting date of a pregnancy may be unreliable because it depends upon the patient's remembering an event* (last menses) that occurred two or three months earlier. In addition, she may have had some implantation bleeding, or *breakthrough bleeding,* at the time of her first missed period and might have thought that it was a light menses. This misinterpretation would result in incorrectly considering a pregnancy of 12 weeks (10-week-old fetus) to be an 8-week gestation (6-week-old embryo).

Chapter 6

1. *Doctors cannot rely completely on information about the time of the last menstrual period reported by the patient,* especially in cases in which determination of gestational age is extremely important. Patients' memories may be poor, and their reports may be based on bleeding that sometimes occurs at the time of the first missed period. One can determine with reasonable accuracy the *estimated date of delivery (EDD)* using diagnostic ultrasound to estimate the size of the fetal head and/or of the entire fetus and/or of the conceptus.*

2. The patient would likely undergo *amniocentesis* for study of the fetus's chromosomes. The most common chromosomal disorder detected in fetuses of women

---

*Sanders, R. C., and James, A. E.: *The Principles and Practice of Ultrasonography in Obstetrics and Gynecology.* New York, Appleton-Century-Crofts, 1980.

over 40 years of age is trisomy 21 (see Fig. 8–7). If the chromosomes of the patient's fetus were normal, but if congenital abnormalities of the brain or of the limbs were still suspected, *fetoscopy* might be performed. This method allows one to look for morphological abnormalities while scanning the entire fetus.

The sex of the fetus could be determined by examining the sex chromatin patterns in the nuclei of cells obtained by amniocentesis. *One can often determine sex using ultrasonography.* In the hands of persons with technical experience, this method can be used to identify sex (particularly male) with a certainty that approaches 100 per cent after about 30 weeks of gestation.

3. There is a danger when uncontrolled drugs (*over-the-counter drugs*, e.g., aspirin, cough medicine) are consumed excessively or indiscriminately by pregnant women. In a retrospective study, Nelson and Fofar (1971)* reported that significantly more infants with congenital malformations were born to mothers who took aspirins, antacids, barbiturates, and cough medicine than to mothers who did not consume these drugs.

    Withdrawal seizures have been reported in infants born to mothers who are heavy drinkers, and the *fetal alcohol syndrome* is present in some of these infants (see Chapter 8). The doctor would likely tell the patient not to take any drugs that he does not prescribe. He might tell her that those drugs that are most detrimental to her fetus are under legal control and that he dispenses them with great care and caution.

4. *Many factors (fetal, maternal, and environmental) may reduce the rate of fetal growth.* Examples of such factors are intrauterine infections, multiple pregnancies, and chromosomal abnormalities (see Chapter 8). Also, *cigarette smoking*, narcotic addiction, and consumption of large amounts of alcohol are well-established causes of *intrauterine growth retardation*. A mother interested in the growth and general well-being of her fetus will eat a good-quality diet and not use narcotics, smoke heavily, or drink alcohol.

Chapter 7

1. A simple method, called *Nägele's rule*, is to count back three months from the first day of the last menses (LMP), and then add one year and seven days. The biparietal diameter of the fetal head could be measured by *ultrasonography* because this measurement correlates well with fetal age.

2. *Polyhydramnios* (hydramnios) is the accumulation of an excessive amount of amniotic fluid (in excess of 2000 ml). When it occurs rapidly over the course of a few days, there is an associated high risk of severe fetal abnormalities, especially of the central nervous system (anencephaly and spina bifida cystica). Fetuses with gross brain defects do not drink the usual amounts of amniotic fluid, hence the amount of this liquid increases. *Atresia of the esophagus* is almost always accompanied by polyhydramnios because the fetus cannot swallow and absorb amniotic fluid. *Twinning* is also a predisposing cause of polyhydramnios.

3. *There is a well-known tendency for twins to "run in families."* It appears unlikely that there is a genetic factor in monozygotic twinning, but *a disposition to dizygotic twinning is genetically determined.* The frequency of dizygotic twinning rises sharply with maternal age up to age 35 and then declines, but the frequency of monozygotic twinning is affected very little by the age of the mother. Determination of twin zygosity can usually be made on the basis of the type of placenta and fetal membranes present. Later, one can determine zygosity by looking for genetically determined similarities and differences in a twin pair. *A single difference in a genetic marker proves twins to be dizygotic.* For more details, see Thompson and Thompson (1980).*

4. *A single umbilical artery occurs in about one of every 200 umbilical cords.* This anomaly is accompanied by a 15 to 20 per cent incidence of cardiovascular abnormalities.**

Chapter 8

1. *It is generally accepted that only 2 to 3 per cent of congenital malformations are caused by drugs and chemicals.* It is difficult for clinicians to assign specific defects to specific drugs because (1) the drug may be administered as therapy for

---

*Nelson, M. M., and Fofar, J. D.: Associations between drugs administered during pregnancy and congenital abnormalities. *Br. Med. J. 1*:523, 1971.

*Thompson, J. S., and Thompson, M. W.: *Genetics in Medicine*. 3rd ed. Philadelphia, W. B. Saunders Co., 1980.

**Page, E. W., Villee, C. A., and Villee, D. B.: *Human Reproduction. Essentials of Perinatal Medicine*. 3rd ed. Philadelphia, W. B. Saunders Co., 1981.

an illness that itself causes the malformation; (2) the fetal malformation may cause maternal symptoms that are treated with a drug; (3) the drug may prevent the spontaneous abortion of an already malformed fetus; and (4) the drug may be used commonly with another drug that causes the malformation.

2. It is well known that women over the age of 35 are more likely to have a child with Down syndrome or some other chromosomal disorder than are younger women. Nevertheless, *many women over the age of 35 have normal children.* The doctor caring for a pregnant 40-year-old woman would in all probability recommend *amniocentesis* to determine if the infant had trisomy 21 or some other chromosomal disorder.

3. *Penicillin has been widely used during pregnancy for over 20 years without any implication of teratogenicity.* Aspirin and other salicylates are ingested by most pregnant women, and when they are consumed as directed, the teratogenic risk associated with maternal ingestion of these substances is very low. Chronic consumption of large doses of aspirin during early pregnancy may be harmful. *Commonly used antinausea medications (e.g., meclizine and Bendectin) can be safely used during pregnancy,* but caution must always be exercised with new drugs. Remember that thalidomide, an "over-the-counter drug," was at one time commonly used as an antinauseant and sedative during early pregnancy.

4. The doctor would likely tell the mother that there was no danger that her child would develop *cataracts* as the result of *German measles*. He would undoubtedly explain that cataracts often develop in fetuses whose mothers contract the disease early in pregnancy. He might say that *it is not necessarily bad for a girl to contract German measles before her childbearing years* because this attack would probably confer permanent immunity to the disease. However, he would likely urge her to tell the girl to avoid exposure to German measles should she later become pregnant because of the common permanent defects caused by spread of the infection to the fetus.

Chapter 9

1. A diagnosis of *congenital diaphragmatic hernia* is most likely. This defect results from *failure of the left pericardioperitoneal canal to close completely* during the sixth week of development. As a result, herniation of abdominal organs into the thorax occurs. This compresses the lungs, especially the left one, resulting in *respiratory distress*. The diagnosis can usually be established by an x-ray examination of the chest. Characteristically, there are air- and/or fluid-filled loops of intestine in the left hemithorax of a newborn infant afflicted with this condition.

2. In the rare congenital malformation known as *retrosternal hernia,* the intestine may herniate into the pericardial sac, or, conversely, the heart may be displaced into the superior part of the peritoneal cavity. A hernia through the *sternocostal hiatus* (foramen of Morgagni) causes this condition.

3. *Posterolateral defect of the diaphragm,* usually on the left, occurs about once in every 2000 births. A newborn infant in whom a diagnosis of this defect is suspected would immediately be positioned with the head and thorax higher than the abdomen and feet to facilitate the downward displacement of the abdominal organs that would likely be in the thorax.

4. *Epigastric hernias* occur in the midline in the epigastric region. They are uncommon and resemble umbilical hernias. The defect through which herniation occurs results from failure of the lateral body folds to fuse completely during the fourth week (see Fig. 5–1).

Chapter 10

1. The mucoid material was probably discharged from an *external branchial sinus* (Fig. 10–10A), a remnant of the second branchial groove and/or *cervical sinus*. Normally, this groove and sinus disappear as the neck of the embryo forms. As evident in this case, the branchial sinus extends superiorly into the subcutaneous tissue.

2. *The position of the inferior parathyroid glands is variable.* They develop in close association with the thymus gland and are carried caudally with it during its descent through the neck. If the thymus fails to descend to its usual position in the *superior mediastinum*, one or both inferior parathyroid glands may be located near the bifurcation of the common carotid artery. If an inferior parathyroid gland does not separate from the thymus and adhere to the thyroid gland, it may be carried into the superior mediastinum with the thymus.

3. The patient very likely has a *thyroglossal cyst* that arose from a small remnant of the embryonic thyroglossal duct. When complete degeneration of this duct does not occur, a cyst may form anywhere along the midline of the neck between the foramen cecum of the tongue and the jugular notch.

4. *About 60 to 80 per cent of persons having a cleft lip with or without cleft palate are males.* When both parents are normal and have had one child with a cleft lip, the chances that the next infant will have the same malformation is about 4 per cent.

5. There is substantial evidence that *anticonvulsant drugs* (e.g., phenytoin, or diphenylhydantoin), when given to epileptic women during pregnancy, increase by two or three fold the incidence of cleft lip and cleft palate when compared with the general population.* Cleft lip with cleft palate is caused by many factors, some genetic and others environmental; i.e., this condition has a *multifactorial etiology*. In most cases, the environmental factor involved is not identifiable.

Chapter 11

1. Inability to pass a catheter through the esophagus into the stomach would indicate the presence of *esophageal atresia*. Because this malformation is commonly associated with *tracheoesophageal fistula* and respiratory distress, the pediatrician would probably suspect tracheoesophageal fistula. An x-ray examination, performed after injecting a small amount of radiopaque material via a catheter into the esophagus, would demonstrate the esophageal atresia and some types of tracheoesophageal fistula. Of course, the radiopaque material would be aspirated and the catheter left in place to keep saliva and other matter from entering the lungs. If certain types of tracheoesophageal fistula were present, there would also be air in the stomach that passed there from a connection between the lower esophagus and the trachea (Fig. 11-5).

2. *An infant in respiratory distress tries to overcome the ventilatory problem by increasing the rate and depth of respiration.* Intercostal, subcostal, or sternal retractions and nasal flaring are prominent signs of respiratory distress. *Hyaline membrane disease* is a leading cause of

the *respiratory distress syndrome* and death in liveborn premature infants. A deficiency of pulmonary surfactant is associated with *RDS*.

3. *The most common type of tracheoesophageal fistula connects the trachea with the inferior part of the esophagus* (Fig. 11-5A). This malformation is associated with atresia of the esophagus superior to the level of the fistula. Tracheoesophageal fistula results from *incomplete division of the foregut* by the tracheoesophageal septum into the esophagus and trachea.

4. In most types of tracheoesophageal fistula, air passes from the trachea through the fistula into the lower esophagus and stomach. *Pneumonitis* from aspiration of oral and nasal secretions into the lungs is a serious complication of this malformation. *Giving the baby water or food by mouth is contraindicated in such cases.*

Chapter 12

1. Complete absence of a lumen (*duodenal atresia*) usually involves the second (descending) and third (horizontal) parts of the duodenum. The obstruction usually results from *incomplete vacuolization of the lumen of the duodenum* during the eighth week. The obstruction causes distention of the stomach and proximal duodenum because the fetus swallows amniotic fluid, and the newborn infant swallows air, mucus, and milk.
Duodenal atresia is common in Down syndrome, as are other severe congenital malformations, e.g., anular pancreas, *cardiovascular abnormalities*, malrotation of the bowel (midgut loop), and anorectal malformations. *Polyhydramnios occurs because the duodenal atresia prevents normal absorption of amniotic fluid* from the fetal intestine. The fetus swallows amniotic fluid before birth, but, owing to blockage of the duodenum, this fluid cannot pass along the bowel and be absorbed into the fetal circulation and transferred across the placental membrane into the mother's circulation.

2. *Normally, the yolk stalk undergoes complete involution* before the tenth week of fetal development, at which time the intestines return to the abdomen. In 2 to 4 per cent of people, a remnant of this stalk persists as a *Meckel's diverticulum* of the ileum, but only a small number of these ever become symptomatic. In the present case, the entire yolk stalk persisted so that the Meckel's diverticulum was connected to the anterior abdominal wall and

---

*Golbus, M. S.: Teratology for the obstetrician: Current status. *Obstet. Gynecol.* 53:269, 1980.

the umbilicus by a *sinus tract* (see Fig. 12–19). This malformation is very rare, and its external opening may be confused with a *granuloma* (inflammatory lesion) of the stump of the umbilical cord.

3. The fistula was very likely connected to the blind end of the rectum. The condition is known as *imperforate anus with rectovaginal fistula*. This malformation results from failure of the urorectal septum to form a complete separation between the anterior and posterior portions of the *urogenital sinus*. Because the inferior one third of the vagina forms from the anterior part of the urogenital sinus, it joins the rectum, which forms from the posterior part of the urogenital sinus (see Fig. 12–23F).

4. This malformation is called *omphalocele* (exomphalos). A small omphalocele, like the present one, is sometimes called an *umbilical cord hernia,* but it should not be confused with an umbilical hernia that occurs after birth and is covered by skin. The thin membrane covering the mass in the present case would be composed of peritoneum and amnion. The mass, or hernia, would be composed of small intestinal loops (see Fig. 12–15).
Omphalocele results when the intestinal loops fail to return to the abdominal cavity from the umbilical cord during the tenth week of fetal life. In the present case, because the hernia is relatively small, the intestine may have entered the abdominal cavity and then herniated again when the rectus muscles did not approach each other and close the circular defect in the anterior abdominal wall.

5. The ileum was probably obstructed, and the condition is called ileal atresia. *Congenital atresia of the small bowel involves the ileum most frequently*; the next most frequently affected organ is the duodenum, and the jejunum is involved least often. A little *meconium* is formed distal to the obstructed area (*atretic segment*) from the fetal epithelium and mucus in the intestinal lumen. At operation, the atretic ileum would probably appear as a narrowed segment connecting the proximal and distal segments of small bowel. Atresia of the ileum could result from failure of recanalization of the lumen, but more *likely the atresia resulted from a prenatal interruption of the blood supply to the ileum*. Sometimes, a loop of small bowel becomes twisted, interrupting its blood supply and causing death (*necrosis*) of the affected segment. The damaged section of bowel usually becomes a fibrous cord connecting the proximal and distal segments of bowel. This type of atresia can be produced experimentally in animals by ligating the arteries to a segment of bowel during the fetal period.

Chapter 13

1. *Double renal pelves and ureters result from the formation of two metanephric diverticula,* or buds, on one side of the embryo. Subsequently, the primordia of these structures fuse together. Usually, both these ureters open into the urinary bladder, but, occasionally, the extra ureter opens into the urogenital tract inferior to the bladder. This occurs when the extra ureter is not incorporated into the base of the bladder with the other ureter. Instead, the extra ureter is carried caudally with the mesonephric duct and opens with it into the caudal part of the urogenital sinus. Because this part of the urogenital sinus gives rise to the urethra and the epithelium of the vagina, the *ectopic (abnormally placed) ureteric orifice* may be located in either of these structures.
A ureteric orifice that opens inferior to the bladder results in urinary incontinence because there is no urinary bladder or urethral sphincter between it and the exterior. Normally, the oblique passage of the ureter through the wall of the bladder allows the contraction of the bladder musculature to act like a *sphincter for the ureter*, controlling the flow of urine from it.

2. *Supernumerary renal arteries are relatively common*. About 25 per cent of kidneys receive two or more branches directly from the aorta, but more than two is exceptional. Supernumerary arteries enter either through the renal sinus or at the poles of the kidney, usually the inferior pole. Accessory renal arteries, more common on the left side (Fig. 13–12), represent *persistent fetal renal arteries*, which grow out in sequence from the aorta as the kidneys "ascend" from the pelvis to the abdomen. Usually, the inferior vessels degenerate as new ones develop (Fig. 13–8). *Supernumerary arteries are about twice as common as supernumerary veins,* and they usually arise at the level of the kidney.
The presence of a supernumerary artery is of clinical importance in other circumstances because it may cross the *ureteropelvic junction* (Fig. 13–12B) and

hinder the outflow of urine, leading to some degree of dilation of the calyces and pelvis on the same side (*hydronephrosis*). Hydronephrotic kidneys frequently become infected (*pyelonephritis*); infection may lead to destruction of the kidneys.

3. A rudimentary uterine horn is a very uncommon malformation. *Rudimentary horn pregnancies are very rare,* but they are clinically important because it is difficult to distinguish between this type of pregnancy and a tubal pregnancy. In the present case, the uterine malformation resulted from retarded growth of the right paramesonephric duct and incomplete fusion of this duct with its partner during development of the uterus. Although *most malformations resulting from incomplete fusion of the paramesonephric ducts do not cause clinical problems,* a rudimentary horn that does not communicate with the main part of the uterus may cause pain during the menstrual period because of distention of the horn by blood. Because most rudimentary uterine horns are thicker than uterine tubes, *rupture of a rudimentary horn pregnancy* is likely to occur much later than that of a tubal pregnancy.

4. *Glandular hypospadias* is the term applied to a penile malformation in which the urethral orifice is on the ventral surface near the glans penis. The ventral curving of the penis is called *chordee*. Glandular hypospadias results from failure of the *urogenital folds* on the ventral surface of the developing penis to fuse completely and to establish communication with the terminal part of the spongy urethra within the glans penis. *Hypospadias is associated with an inadequate production of androgens* by the fetal testes, but it is thought to have a *multifactorial etiology* (genetic and environmental factors) because close relatives of patients with hypospadias are more likely to have the malformation than persons in the general population. Hypospadias, a common malformation of the urogenital tract, occurs in about one of every 300 male infants.

5. *The embryological basis of indirect inguinal hernia is persistence of the processus vaginalis,* a fetal diverticulum, or outpouching, of peritoneum. This finger-like pouch evaginates the anterior abdominal wall and forms the inguinal canal. *A persistent processus vaginalis predisposes to indirect inguinal hernia* by creating a weakness in the anterior abdominal wall and a hernial sac into which abdominal contents may herniate if the intra-abdominal pressure becomes very high (as occurs during straining). The hernial sac would be covered by internal spermatic fascia, the cremaster muscle, and cremasteric fascia.

Chapter 14

1. *Ventricular septal defect is the most common cardiac malformation.* Keith et al. (1978) estimate that it occurs in about 20 per cent of children with congenital heart disease.

2. *Patent ductus arteriosus is the most common cardiac malformation associated with maternal rubella infection during early pregnancy.* When the ductus arteriosus remains patent, aortic blood is shunted into the pulmonary artery. In extreme cases, one half to two thirds of the left ventricular output may be shunted through the patent ductus arteriosus. This extra work for the heart results in *cardiac enlargement.*

3. The *tetrad of cardiac abnormalities* present in the malformation called tetralogy of Fallot is as follows: pulmonary stenosis, ventricular septal defect (VSD), overriding aorta, and right ventricular hypertrophy. *Angiocardiography* could be used to reveal the malpositioned aorta (straddling the ventricular septal defect) and the degree of pulmonary stenosis. *Cyanosis* occurs but may not be present at birth.

4. *Cardiac catheterization* would probably be used to confirm the diagnosis of transposition of the great vessels. If this cardiac abnormality were present, a bolus (mass) of contrast material injected into the right ventricle would enter the aorta, whereas contrast material injected into the left ventricle would enter the pulmonary circulation. The infant was able to survive after birth because the ductus arteriosus remains open in these patients, allowing some intermixing of blood between the two circulations. In other cases, there is also an ASD or VSD that permits intermixing of blood. *Complete transposition of the great vessels is incompatible with life if there are no associated septal defects or a patent ductus arteriosus.*

5. This would probably be a *secundum type ASD* located in the region of the *fossa ovalis*, because this is the most common type of clinically significant ASD. Large defects, as in the present case, often extend inferiorly toward the inferior vena

cava. *The pulmonary artery and its major branches were dilated as the result of the increased blood flow through the lungs* and the increased pressure within the pulmonary circulation. In these cases, a considerable shunt of oxygenated blood flows from the left atrium to the right atrium. This blood, along with the normal venous return to the right atrium, enters the right ventricle and is pumped to the lungs. Large ASDs may be tolerated for a long time, as in the present case, but progressive dilation of the right ventricle often leads to *heart failure*.

Chapter 15

1. The most common congenital malformation of the vertebral column is *spina bifida occulta* (see Fig. 18–12A). This defect of the vertebral arch of the first sacral and/or the last lumbar vertebra is present in about 10 per cent of people. The defect also occurs in cervical and thoracic vertebrae. *The spinal cord and nerves are usually normal and neurological symptoms are absent*. Spina bifida occulta does not cause back problems.

2. A rib on the seventh cervical vertebra is of clinical importance because it may compress the subclavian artery and/or the brachial plexus, producing symptoms of artery and nerve compression. *Cervical ribs produce no symptoms in most cases*. They result from development of the costal processes of the seventh cervical vertebra into ribs.

3. A *hemivertebra* can produce an abnormal lateral curvature of the vertebral column (*scoliosis*). A hemivertebra, composed of one half of a body, a pedicle, and a lamina, results when the mesenchymal cells from the sclerotomes on one side fail to form the primordium of half of a vertebra. As a result, there are more growth centers on the one side of the vertebral column; this imbalance causes the vertebral column to bend laterally (Fig. 15–9B).

4. *Craniosynostosis indicates premature closure of one or more of the cranial sutures*. This developmental abnormality results in malformations of the skull. *Scaphocephaly,* or a long, narrow skull, results from premature closure of the sagittal suture. This type of craniosynostosis accounts for about 50 per cent of the cases of premature closure of the cranial sutures.

5. *The features of the Klippel-Feil syndrome are short neck, low hair line, and restricted neck movements*. In most cases, the number of cervical vertebral bodies is less than normal. There is a lack of segmentation of several of the elements of the cervical region of the vertebral column.

Chapter 16

1. *Absence of the sternocostal portion of the left pectoralis major* muscle is surely the cause of the surface abnormalities observed. The costal heads of the pectoralis major and the pectoralis minor are usually present. Despite its numerous and important actions, absence of all or part of the pectoralis major usually causes no disability, but the deformity caused by the absence of the anterior axillary fold is striking, as is the low nipple (Fig. 16–3). The action of other muscles associated with the shoulder joint compensates for the absence of this part of the muscle.

2. *About 13 per cent of people lack a palmaris longus muscle* on one or both sides. Its absence causes no disability. This slender muscle is subject to much variation.

3. It would be the left sternocleidomastoid muscle that was prominent when tensed. The left one is the normal muscle and it does not pull the child's head to the right side. It is the short, contracted, right sternocleidomastoid that tethers the right mastoid process to the right clavicle and sternum. Hence, continued growth of the left side of the neck results in tilting and rotation of the head. This condition, called *congenital torticollis* (wryneck), probably resulted from injury to the muscle during birth (Fig. 16–4). Tearing of some muscle fibers probably occurred, resulting in bleeding into the muscle. Over several weeks, *necrosis* of some fibers occurred, and the blood was replaced by fibrous tissue. This resulted in shortening of the muscle and in pulling of the child's head to the side.

4. Absence of the development of striated musculature in the midline of the anterior abdominal wall is associated with *exstrophy of the urinary bladder* (see Fig. 13–17). This severe malformation is caused by (1) incomplete midline closure of the inferior part of the anterior abdominal wall, and (2) failure of mesenchymal cells to migrate from the somatic mesoderm between the surface ectoderm and the urogenital sinus during the fourth week of development (see Fig. 13–18). The absence of mesenchymal cells in the midline results in failure of striated muscles to develop.

Chapter 17

1. *The number of female infants affected with dislocation of the hip is approximately eight times that of male infants.* The hip joint in these infants is only rarely completely dislocated at birth or within the first postnatal month, but *the acetabulum is usually abnormal at birth.* Dislocation of the hip may not become obvious for several months after birth.
2. Severe malformations of the limbs (*amelia* and *meromelia*), similar to those produced by thalidomide, are very rare and are mainly of a hereditary nature. The *thalidomide malformation syndrome* consists of absence of limbs (*amelia*); gross deformities of the limbs (*meromelia*), e.g., hands and feet attached to the trunk by a small, irregularly shaped bone; intestinal atresia; and *cardiac defects.*
3. The most common type of clubfoot is *talipes equinovarus*, occurring in about one of every 1000 newborn infants. In this malformation, the soles of the feet are turned inward at the ankle (inverted) and the feet are adducted and plantarflexed. The feet are fixed in the tiptoe position, resembling the foot of a horse (L. *equinus*, horse).
4. *Syndactyly* (fusion of digits) is the most common type of limb malformation. It varies from cutaneous webbing to *synostosis* (union of bones). *Syndactyly is more frequent in the foot than in the hand.* This deformity results when the mesenchyme between the developing digits fails to break down. As a consequence, separation of the digits does not occur.

Chapter 18

1. *Ultrasound scanning* of the fetus may show absence of the calvaria (*acrania*) in an anencephalic fetus as early as 14 to 18 weeks' gestation. *Fetuses with anencephaly do not drink the usual amounts of amniotic fluid*, presumably because of impairment of the neuromuscular mechanism that controls swallowing. Inasmuch as fetal urine is excreted into the amniotic fluid at the usual rate, the amount of amniotic fluid increases. Normally, the fetus swallows amniotic fluid, which is absorbed by its intestines and mostly passed to the placenta for elimination. Anencephaly can be easily and safely detected by a plain radiograph, but radiographs of the fetus are not usually done during the first six months. It could also be detected by *fetoscopy* (see Chapter 8) or by *amniocentesis* (see Chapter 6). An elevated level of *alpha fetoprotein* in the amniotic fluid indicates an open *neural tube defect* such as spina bifida with myeloschisis or anencephaly.
2. *A neurological deficit occurs because the spinal cord and/or nerve roots are often incorporated into the wall of the sac.* This impairs the development of nerves to structures distal to the lesion. Paralysis of the lower limbs often occurs, and there may be *incontinence of urine and feces* owing to paralysis of the sphincters of the anus and the bladder.
3. The condition is called *internal hydrocephalus*. The block would most likely be in the *cerebral aqueduct* of the midbrain. Obstruction at this site (stenosis or atresia) interferes with or prevents passage of ventricular fluid from the lateral and third ventricles to the fourth ventricle. Hydrocephalus is sometimes recognized before birth, but most cases are diagnosed in the first few weeks or months after birth. *Hydrocephalus can be recognized if radiographs are taken of the mothers' abdomen during the last trimester.* Surgical treatment of hydrocephalus usually consists of shunting the excess ventricular fluid via a plastic tube to another part of the body (e.g., into the blood stream or into the peritoneal cavity), where it will subsequently be excreted by the kidneys.
4. *Hydrocephalus is not synonymous with a large head* because a large brain (*macroencephalon* or *megalocephalon*), a subdural hygroma, or a hematoma can cause enlargement of the head. Hydrocephalus may or may not enlarge the head. In a condition known as *hydrocephalus ex vacuo*, the ventricles are large owing to brain destruction, but the head is not enlarged. *Microencephaly* (small brain) is usually associated with *microcephaly* (small calvaria). Growth of the skull is largely dependent upon growth of the brain; hence, *arrest of brain development can cause microcephaly*. During the fetal period, cytomegalovirus, *Toxoplasma gondii*, herpes simplex virus, and high level radiation are environmental agents that are known to induce microencephaly and microcephaly. Measles and mumps occurring after birth can also cause cerebral atrophy and mental retardation.
5. *Agenesis of the corpus callosum*, partial or complete, is frequently associated with *seizures*. Some patients are asymptomatic and lead normal lives. As in the present case, a large third ventricle may be asso-

ciated with agenesis of the corpus callosum. A *large third ventricle* exists because it rises above the roofs of the lateral ventricles when the corpus callosum is absent. The lateral ventricles are usually moderately enlarged.

Chapter 19

1. The mother had almost certainly contracted *German measles* during early pregnancy because her infant had the characteristic *triad of malformations* resulting from infection of the embryo by the *rubella virus*. Cataract is common after such infections in the first six weeks of pregnancy, when the lens vesicle is forming and separating from the surface ectoderm. *Cataract is believed to result from direct invasion of the developing lens by the virus*. The most common cardiovascular lesion in infants whose mother had rubella in early pregnancy is *patent ductus arteriosus*. Although a history of a rash during the first trimester of pregnancy is helpful for diagnosing the *congenital rubella syndrome*, embryopathy (embryonic disease) can occur after *subclinical maternal infection* (i.e., without a rash).

2. *Congenital ptosis*, or drooping of the upper eyelids, is usually caused by abnormal development or failure of development of the *levator palpebrae superioris muscles*. Congenital ptosis is usually transmitted by autosomal dominant inheritance, but injury to the *superior branch of the oculomotor nerve* (CN III), which supplies the levator palpebrae superioris muscle, could also cause drooping of an upper eyelid.

3. The protozoon involved was most likely *Toxoplasma gondii, an intracellular parasite*. The congenital abnormalities result from invasion of the fetal blood stream by the toxoplasma parasites, which disrupt development of the central nervous system, including the eye, which develops from an outgrowth of the brain. The doctor would certainly tell the woman about *toxoplasma cysts* in meat and advise the woman to cook her meat well during her next pregnancy. He would also tell her about the *toxoplasma oocysts in cat feces* and the importance of carefully washing her hands after handling a cat or its litter box.

4. Very likely, the infant has *trisomy 18* because the characteristic phenotype is present (see Chapter 8). Low-set, malformed ears associated with mental retardation, prominent occiput, and failure to thrive are suggestive of the trisomy 18 syndrome. This numerical chromosomal abnormality results from *nondisjunction of the number 18 chromosome pair* during gametogenesis. Postnatal survival of these infants is poor; the mean survival time is only two months.

5. *Detachment of the retina* is a separation of the two embryonic retinal layers: the pigment epithelium derived from the outer layer of the optic cup, and the nervous portion of the retina derived from the inner layer of the optic cup. The *intraretinal space* (Fig. 19–3), representing the cavity of the optic vesicle, normally disappears as the retina forms (Fig. 19–5).

Proximal parts of the hyaloid artery persist as the *central artery of the retina,* but distal parts of this vessel normally degenerate. Persistence of the posterior end of the artery is more common than persistence of the anterior end. Treatment is usually not necessary.

Chapter 20

1. *Natal teeth* occur in about one of every 2000 newborn infants. Usually, there are two teeth in the position of the mandibular incisors. Natal teeth may be supernumerary ones, but they are often *prematurely erupted primary teeth*. If it is established radiologically that they are supernumerary teeth, they would probably be removed so that they would not interfere with the subsequent eruption of the normal primary teeth.

   Natal teeth may cause maternal discomfort owing to abrasion or biting of the nipple during nursing. They may also injure the infant's tongue, which, because the mandible is relatively small at birth, lies between the alveolar processes.

2. The discoloration of the infant's teeth's was likely caused by the administration of *tetracycline antibiotics* to the mother during her pregnancy. Tetracyclines become incorporated into the developing enamel of the teeth and cause discoloration. *Dysfunction of the ameloblasts*, as a result of tetracycline therapy, would cause *hypoplasia of the enamel* (e.g., pitting). It is very likely that the secondary dentition would also be affected because *enamel formation begins in the permanent teeth before birth* (at about 20 weeks in the medial incisors).

3. This is an *angiomatous malformation of the skin*, often called a flat *capillary angioma (hemangioma)*. It is formed by an overgrowth of small blood vessels, consisting mostly of capillaries, but there are

also some arterioles and venules in it. The blotch is red because oxygen is not taken from the blood passing through it. This type of angioma is quite common, and the mother should be assured that *this malformation is of no significance and requires no treatment*. It will more or less fade in a few years. Formerly, this type of angioma was called a *nevus flammeus* (flame-like birthmark), but these names are sometimes applied to other types of angiomas, and, to avoid confusion, it is better not to use them. *Nevus is not a good term* because it is derived from a Latin word meaning a mole or birthmark, which may or may not be an angioma.

4. A midline tuft of hair in the lumbosacral region usually indicates the presence of *spina bifida occulta* (Fig. 18–12A). This is the most common developmental abnormality of vertebrae and is present in L5 and/or S1 in about 10 per cent of people. *Spina bifida occulta is usually of no clinical significance,* but a small percentage of infants with this malformation also have a developmental defect of the underlying spinal cord and nerve roots (Vaughan et al., 1979).*

5. The superficial layers of the epidermis of infants with *lamellar ichthyosis* consist of fish-like, grayish-brown scales (Fig. 20–2) that are adherent in the center and raised at the edges. Fortunately, the condition is very rare. It is inherited as an *autosomal recessive trait.*

---

*Vaughan, V. C., McKay, R. J., and Behrman, R. E. (Eds.): *Nelson Textbook of Pediatrics.* 11th ed. Philadelphia, W. B. Saunders Co., 1979.

INDEX

Note: Page numbers in *italics* refer to illustrations; page numbers followed by (t) refer to tables.